Dictionary of Science

P. Hartmann-Petersen
J. N. Pigford

Edward Arnold
A division of Hodder & Stoughton
LONDON MELBOURNE AUCKLAND

© 1984 P. Hartmann-Petersen

First published in Great Britain 1984
Reprinted 1985, 1989

ISBN 0 7131 8168 0

All rights reserved. No part of this publication may be reproduced or transmitted in any form or by any means, electronically or mechanically, including photocopying, recording or any information storage or retrieval system, without either prior permission in writing from the publisher or a licence permitting restricted copying. In the United Kingdom such licences are issued by the Copyright Licensing Agency: 33-34 Alfred Place, London WC1E 7DP.

Acknowledgements
The publishers would like to thank the following for permission to include copyright material:

Addison-Wesley Publishing Company for a figure from Sears, Zemansky and Young: *University Physics;* Cambridge University Press for an extract from Miller and Powell: *The Cambridge Elementary Mathematical Tables 2nd Edition;* Gyldendal for extracts from Rasmussen: *Chemical and Physical Tables;* Granada Publishing Limited for figures from Nelkon: *Principles of Physics* and Nelkon: *CSE Physics;* Heinemann Educational Books for figures from Abbott: *Ordinary Level Physics;* Hulton Educational Publications Ltd for some line drawings from their catalogue: Longman Group Limited for a figure from Atkinson: *Certificate Chemistry 3rd Edition* and tables from Kaye and Laby: *Tables of Physical and Chemical Constants;* Don Mackean for figures from his *Introduction to Biology 5th Edition;* Macmillan, London and Basingstoke for material from Brewer and Burrow: *Life: Form and Function* and Thomas Nelson and Sons Ltd for a figure from Roberts: *Biology: a Functional Approach.*

> To our children and all students who may gain from this book.

Typeset in Times by Thomson Litho Ltd.
Printed and bound in Great Britain for Edward Arnold, the educational, academic and medical publishing division of Hodder and Stoughton Limited, 41 Bedford Square, London WC1B 3DQ by Athenaeum Press Ltd, Newcastle upon Tyne.

Preface

This illustrated dictionary has been specially prepared for students (foreign language students studying Science in English) studying Biology, Chemistry, General Science, Physical Science and Physics. It is the result of a careful analysis of the needs of such students and should be an invaluable aid in their day-to-day learning and in their preparation for examinations.

As well as containing definitions and explanations of the terms used in the above courses, with illustrations where appropriate, the dictionary contains explanations of terms used in Integrated and Combined Science together with many terms of an agricultural, astronomical, geological and medical nature.

As it is extremely difficult to limit the contents of such a dictionary to 'O' level standard, many headwords are included that are above this level and that may be of interest to 'A' level students. A few headwords that are not strictly scientific are also included because of their common usage in everyday scientific terminology. In books of this type, it is common practice to use italics or asterisks to indicate that a word used in an entry is found as a headword elsewhere in the book. This system has not been used here, as almost all scientific words used in the definitions and explanations of headwords are found as headwords elsewhere in the dictionary. Where cross-references are given for additional information, they appear at the end of an entry. SI units are used throughout. IUPAC names are used for chemicals with a few exceptions; however, alternative names which are still in common use are given as headwords with a cross-reference to the IUPAC name. The Appendix contains tables providing a wide range of information together with other items of scientific interest for both student and teacher.

It has been our aim to provide explanations that are easily understood, whilst maintaining a high academic standard.

We would like to thank the following for their criticism and suggestions and for reading many sections of the manuscript:

Mr. G. C. Clark, Ph.D., B. Sc.
Mr. F. Holmberg Ph.D., M.Sc., B.Sc.
Mr. D. G. Hutcins M.Sc., B.Sc.
Mr. G. E. T. Makunga B.Sc. (Senior Education Officer for Science, Ministry of Education, Botswana)
Mr. A. I. O'Rourke B.A.
Mr. H. Sheridan B.A.

P. Hartmann-Petersen
J. N. Pigford
London, 1984

Note to the reader

Nomenclature
Chemicals are normally referred to by their IUPAC names, with the main entry for the chemical under that name. Alternative names are listed in brackets after the headword, and are also given as headwords with a cross reference to the main entry.

Exceptions are made where the fully systematic name is considerably more complex than is required for a dictionary at this level; in the case of plastics, e.g. Perspex (poly(methyl 2-methylpropenoate)); and in the case of chemicals which are much more frequently used in biology, e.g. glycerol (propane-1,2,3-triol). In these cases the most commonly used name is the one used here.

Decimals
There are two international systems for denoting decimals, both of which are widely used. For purposes of completeness, this book employs both. Throughout the main text, decimal figures have been given using the decimal comma, followed by the same figure in brackets using the decimal point. For reasons of space, the decimal comma alone is used in the Appendix.

A

abdomen 1. In vertebrates, a region of the body containing the digestive organs, e.g. stomach and intestines; in mammals it is separated from the thorax by the diaphragm. **2.** In arthropods, the posterior part of the body.

Abegg's rule of eight See Periodic Table of the Elements.

aberration 1. The displacement of a heavenly body's true position due to the motion of the observer with the Earth. **2.** A certain defect in lenses or mirrors in which a true image is not formed. There are several types of aberration, e.g. (1) Chromatic aberration (chromatism), in which an image with coloured fringes is formed because the refractive index of glass is different for light of different colours (wavelengths). This causes the light to disperse into a coloured band. (2) Spherical aberration of lenses, in which rays of light refracted at the periphery of a lens cross the principal axis nearer to the lens than rays refracted near the centre. Spherical aberration is also observed in concave mirrors.

abiotic Non-living.

abomasum (reed) The fourth or true stomach in ruminants.

A-bomb Atomic bomb. See fission.

abortion The premature birth of a mammalian embryo or foetus. In humans, if the developing embryo or foetus is expelled before the 28th week of pregnancy, the premature birth is considered an abortion. After this period the birth is not considered an abortion. From a medical point of view, the term miscarriage means the same as abortion. An abortion may be either spontaneous or induced.

abrasive A substance used for rubbing or grinding down surfaces, e.g. carborundum, corundum, diamond powder.

abscisic acid A plant hormone which acts as a growth inhibitor. It is present in fruits, seeds, buds, leaves, etc. Sometimes abscisic acid is called abscisin or dormin.

abscisin See abscisic acid.

abscission The natural loss of parts of plants, e.g. the shedding of leaves and fruits.

abscission zone A region at the base of a leaf or other part of a plant. It consists of the abscission layer of loose, dry cells which become separated before the fall of leaves or fruit, and the protective layer of cork tissue.

absolute Not relative; independent.

absolute alcohol Ethanol (ethyl alcohol) containing not less than 99% pure ethanol by mass. See also alcohol.

absolute humidity The amount of water vapour present in the air (atmosphere), measured in kilograms or grams of water per cubic metre (m^3) of air. See also relative humidity.

absolute refractive index See refractive index.

absolute temperature See thermodynamic temperature.

absolute zero The lowest theoretically possible value of thermodynamic temperature, equal to 0 kelvin (K) or $-273,16°C$ ($-273.16°C$).

absorbance See transmittance.

absorption 1. The process in which a liquid or a gas is taken up and retained by a solid or a liquid, forming a uniform solution. Absorption in solids is sometimes called sorption. **2.** In spectroscopy, the process in which a substance receives and retains certain wavelengths of radiant energy. **3.** In atomic physics, the process in which some elements like cadmium and boron pick up ('capture') neutrons produced in fission processes. **4.** In biology, the passage of material through living cells or vessels.

absorption of radiation See radiant energy.

absorption spectrum See spectrum.

absorptivity See transmittance.

abyssal Inhabiting deep water, i.e. below approximately 1000 metres.

a.c. See alternating current.

Acarina (Acari) A large and varied order of arachnids, including ticks and mites; some are important parasites, e.g. cattle ticks.

acceleration Rate of change of velocity with time expressed in metres per second per second (m/s^2 or $m\,s^{-2}$). If the velocity is increasing, acceleration is usually considered as positive. When the velocity is decreasing, acceleration is considered as negative and is commonly called a deceleration or retardation.

acceleration of free fall (acceleration due to gravity) Symbol: g. The standard value of g is $9,80665\,m\,s^{-2}$ ($9.80665\,m\,s^{-2}$). This value varies slightly with latitude, from $9,78049\,m\,s^{-2}$ ($9.78049\,m\,s^{-2}$) at 0° to $9,83221\,m\,s^{-2}$ ($9.83221\,m\,s^{-2}$) at latitude 90°.

accelerator 1. A positive catalyst, i.e. a catalyst which increases (accelerates) the rate of a chemical reaction. **2.** A machine in which the kinetic energy of charged particles such as electrons and protons is increased by accelerating them in electric fields using a high potential difference. See also cyclotron.

acclimatise To become adapted to a new environment by slow changes in physiology.

accommodation The ability of the eye to produce clear images of objects at different distances by altering the focal length of the eye lens. This is brought about by the action of the ciliary muscles and the elasticity of the lens.

accumulator A device for storing electricity, consisting of one or more secondary cells. *See also* cell.

acetabulum The cavity on each side of the pelvic girdle into which fits the head of the femur, forming the hip joint in vertebrates.

acetaldehyde *See* ethanal.

acetate *See* ethanoate.

acetate rayon *See* rayon.

acetic acid *See* ethanoic acid.

acetone *See* propanone.

acetyl chloride (ethanoyl chloride) *See* acyl chloride.

acetylcholine (ACh) A tissue hormone found in the majority of synapses. It is a neurotransmitter in the parasympathetic nervous system and is believed to play a part in the transmission of nerve-impulses across a synapse. Acetylcholine is inactivated by the enzyme cholinesterase, which is found in all nervous tissue. *See also* choline.

acetylene *See* ethyne.

acetylsalicylic acid *See* 2-hydroxybenzoic acid *and* aspirin.

ACh *See* acetylcholine.

achena A dry, one-seeded, indehiscent fruit formed from a single carpel.

Achilles tendon (Achilles' heel) The tendon of the heel.

achromatic Having no colour. (1) White, black and grey are achromatic colours. (2) An achromatic lens is free from chromatic aberration.

acid A substance existing as molecules or ions which can donate hydrogen ions, H^+ (protons). Less accurate definitions include the following: (1) reacts with some metals to evolve hydrogen; (2) reacts with a base to form a salt and often water (neutralisation); (3) has a pH less than 7; (4) turns blue litmus red. *See also* acid–base theories.

acid–base theories 1. *Broensted–Lowry theory:* An acid is a substance which can donate one or more protons (hydrogen ions), H^+, to a base. A base is a substance which can accept one or more protons from an acid. The relationship between an acid and a base is therefore:
$$acid \rightleftharpoons H^+ + base$$
In aqueous solutions, in which there are no free hydrogen ions but H_3O^+ (oxonium ions) instead, the expression is:
$$acid + H_2O \rightleftharpoons H_3O^+ + base$$
i.e. H_2O is acting as a base and H_3O^+ as an acid. The change can then be generalised to:
$$acid(1) + base(2) \rightleftharpoons acid(2) + base(1)$$
in which acid(1) conjugates with base(1) and acid(2) conjugates with base(2). *Example:* with hydrochloric acid, HCl, one gets:
$$HCl + H_2O \rightleftharpoons H_3O^+ + Cl^-$$
2. *Lewis theory:* An acid is a substance which can accept a pair of electrons to form a coordinate bond. A base is a substance which can donate a pair of electrons to form a coordinate bond. E.g.

$H^+_{(acid)} + :\ddot{O}\!\!-\!\!H^-_{(base)} \rightleftharpoons H:\ddot{O}\!\!-\!\!H$ i.e.
$H^+ + OH^- \rightleftharpoons H_2O$ or

$$\underset{(acid)}{\overset{Cl}{\underset{|}{Cl\!-\!\underset{|}{B}}}} + :NH_{3(base)} \rightarrow Cl_3B:NH_3$$

In the Lewis theory of acids and bases, it is seen that substances containing no hydrogen can act as acids, i.e. the theory embraces reactions in which protons are not involved. The Broensted–Lowry theory is the most generally accepted of the two.

acid chloride *See* acyl chloride.

acid halide *See* acyl chloride.

acidic Having the properties of an acid.

acidic hydrogen One or more hydrogen atoms in molecules of acids which can be liberated in aqueous solution to form hydrogen ions, H^+ (protons). *Example:* in ethanoic acid (acetic acid), CH_3COOH, only the hydrogen atom present in the carboxyl group, —COOH, is an acidic hydrogen atom.

acidic oxide *See* oxide.

acidify To make a solution acidic.

acid radical In an acid, the group attached to the acidic hydrogen atom or atoms. *Example:* in ethanoic acid (acetic acid), CH_3COOH, the ethanoate group (acetate group), CH_3COO^-, is the acid radical.

acid salt *See* salt.

aclinic line *See* magnetic equator.

acoelomate Having no coelom.

acoustics The study of sound.

acquired character A character that develops during the life of an individual as it responds to the environment. It is not passed on to the next generation.

acrophobia An exaggerated, abnormal fear of high places.

ACTH Adrenocorticotrophic hormone. *See* corticotrophin.

actin *See* myosin.

actinides *See* actinoids.

actinoids (actinides, actinons) The series of fourteen elements following actinium, including the transuranic elements. The actinoids are all radioactive and have closely related chemical properties because the outer electron structure is almost the same for them all. From americium (atomic number 95), the characteristic oxidation number is +3. Experiments carried out to determine the electronic configurations of the actinoids have proved inconclusive.

actinons *See* actinoids.

action *See* Newton's laws of motion.

activated carbon A highly porous form of carbon (usually charcoal) with an enormous surface area, enabling it to adsorb large quantities of gases or dissolved or suspended substances. Activated carbon is commonly used in gas masks.

activated charcoal *See* activated carbon.

activation energy Symbol: E. The least amount of energy an atom, molecule, etc. must acquire before it is able to react chemically: i.e. the minimum amount of energy necessary to start a chemical reaction. A positive catalyst decreases the activation energy of a chemical reaction, thus providing a new pathway for it.

active mass *See* mass action (law of).

active transport A process involving the movement of materials into cells by means other than diffusion and osmosis. Energy is expended by the cell in this process, which often takes place against concentration gradients.

activity coefficient *See* mass action (law of).

activity series of metals (reactivity series of metals) Metallic elements arranged in order of their decreasing chemical reactivity with water or dilute acids. Hydrogen is included in this series. Metals placed above hydrogen liberate it from water and from certain dilute acids, whereas metals placed below hydrogen do not. Also, a metal placed above another in the series may displace this other metal from its compounds. *Example:* zinc, Zn, will replace copper, Cu, in copper(II) sulphate, $CuSO_4$, to form zinc sulphate, $ZnSO_4$, and free copper:
$Zn + CuSO_4 \rightarrow ZnSO_4 + Cu$
The activity series of metals should not be confused with the electrochemical series of metals. *Example:* in the activity series sodium is placed above calcium, but in the electrochemical series calcium is placed above sodium.

actomyosin *See* myosin.

acyl chloride (acid chloride) One of a class of organic compounds called acyl halides (acid halides) which can be prepared by the reaction between a carboxylic acid and phosphorus trichloride, PCl_3, or phosphorus pentachloride, PCl_5. Ethanoyl chloride (acetyl chloride), CH_3COCl, is an acyl chloride derived from ethanoic acid (acetic acid), CH_3COOH:
$3CH_3COOH + PCl_3 \rightarrow 3CH_3COCl + H_3PO_3$
From the reaction it is seen that the hydroxyl group in the acid has been replaced by a chlorine atom. Other acyl halides contain fluorine, bromine and iodine instead of chlorine. They are all used in the organic synthesis of other compounds.

acyl halide *See* acyl chloride.

Adam's apple The projection of thyroid cartilage of the larynx, especially prominent in men.

adaptation The process by which an organism becomes adjusted to its environment. *See also* specialisation.

Addison's disease A disease in which there is a lack of certain hormones produced by the cortex of the adrenal gland. Symptoms include weakness, loss of weight, vomiting, hypotension and a dark brown pigmentation of the skin. The function of the kidneys is impaired, causing an accumulation of urea in the blood. The disease is treated by giving the patient cortisone and derivatives of this hormone.

addition dimerisation *See* dimer.

addition polymerisation *See* polymerisation.

addition reaction A chemical reaction in which an unsaturated compound takes up atoms or groups of atoms. *Example:* when ethene, $CH_2\!\!=\!\!CH_2$, reacts with hydrogen, H_2, ethane, $CH_3\!\!-\!\!CH_3$, is formed:
$CH_2\!\!=\!\!CH_2 + H_2 \rightarrow CH_3\!\!-\!\!CH_3$
The double bond in ethene is converted into a single bond in ethane, i.e. the unsaturated ethene is converted into the saturated ethane. *Compare* elimination reaction; *See also* substitution reaction.

adenine A nitrogenous, cyclic organic base (purine base) which is part of the genetic code in DNA, where it pairs with thymine. It is also a part of RNA, NAD, AMP, ADP and ATP.

adenine nucleoside *See* adenosine.

adenoids In some mammals, gland-like structures situated where the nasal passage enters the throat. Lymph circulates through the adenoids, helping to remove bacteria from the blood. Together with the tonsils, the adenoids help in guarding the body against micro-organisms entering through the mouth or nose.

adenosine (adenine nucleoside) A nucleoside with adenine as its base.

adenosine diphosphate (ADP) A nucleotide associated with energy transfer in living organisms also involving adenosine triphosphate, ATP. ADP is a complex molecule consisting of adenine, a carbohydrate part (ribose) and two phosphate groups. Energy from respiration or from sunlight in photosynthesis is used to build up ATP from ADP and phosphate. This energy transfer takes place in the mitochondria of the cell. *See also* Krebs' cycle.

adenosine monophosphate (AMP) A nucleotide consisting of adenine, a carbohydrate part (ribose) and one phosphate group. It is formed by hydrolysis of ADP, by which energy is released together with one phosphate group.

adenosine triphosphate (ATP) A nucleotide consisting of adenine, a carbohydrate part (ribose) and three phosphate groups. It is the energy carrier of the living cell and is formed from ADP. Hydrolysis of ATP releases energy, at the same time yielding ADP and phosphate.

ADH Antidiuretic hormone. *See* antidiuretic.

adhesion The interaction between surfaces of different materials in contact, which causes them to cling together. *Compare* cohesion.

adhesive A substance used for sticking surfaces together, e.g. glue and cement.

adiabatic Occurring without heat loss or heat gain to a system.

adipose tissue In animals, connective tissue whose cells contain large quantities of fat. *See also* areolar tissue.

ADP *See* adenosine diphosphate.

adrenal gland (suprarenal gland) An organ of hormone secretion in vertebrates, situated just above the kidney. It is a ductless gland which secretes adrenalin(e) and noradrenalin(e) into the bloodstream. In mammals, the adrenal gland consists of two main parts, the medulla and the cortex. Adrenalin(e) and noradrenalin(e) are produced by the medulla, whereas the cortex produces hormones such as sex hormones and cortisone.

adrenalin(e) A hormone secreted by the adrenal gland. It causes excitement and stimulation, affecting circulation and muscular action. In medicine, adrenalin(e) is used as a heart stimulant and to constrict blood vessels. *See also* fear *and* glycogenolysis.

adrenocorticotrophic hormone (ACTH) *See* corticotrophin.

adsorption The attachment of molecules of gases or liquids to the surface of another substance (usually a solid). Adsorption occurs on substances like silica gel and activated carbon.

adventitious Describes tissues and organs which occur in an unusual place. *See* adventitious roots.

adventitious roots Roots which are not developed from the radicle of the seed, but produced on some other part of the plant. They may grow directly from a stem, e.g. in tulip and onion. Some plants may even produce adventitious roots from leaves.

aerate Expose to the mechanical or chemical action of air.

aerial 1. Inhabiting the air; e.g. roots growing above ground. **2.** (antenna) The part of a radio system which transmits radio waves or receives them.

aerobe An organism which requires free oxygen (either gaseous or dissolved) in order to live. *Compare* anaerobe.

aerobic Describes organisms which require free oxygen (gaseous or dissolved) in order to live. *Compare* anaerobic; *see also* aerobic respiration.

aerobic respiration Respiration taking place in the presence of free oxygen (gaseous or dissolved in water). *Compare* anaerobic respiration.

aerodynamics The study of gases in motion. The branch of science concerned with the motion and control of solid objects (aircraft, rockets, missiles) in air.

aerosol 1. A suspension of extremely small particles of a liquid or solid in air or other gases. **2.** A pressurised can with a spray mechanism for causing the suspension of particles.

aestivation 1. A period of dormancy in animals during summer or a dry season. **2.** The arrangement of the parts in a flower-bud.

afferent Conducting towards, e.g. nerves conducting impulses to nervous centres. *Compare* efferent.

afferent neurone *See* sensory neurone.

affinity The tendency of two substances to combine. Chemical attraction.

afterbirth The placenta and foetal membranes discharged after the offspring's birth.

Ag Chemical symbol for silver.

agar A material obtained from certain red algae, a form of seaweed. It is a mixture of polysaccharides, commonly used as a base for bacterial, fungal and tissue cultures.

Agaricus *See* mushroom.

agate A very hard, naturally occurring form of silicon(IV) oxide, SiO_2. Agate consists of thin varying layers of many colours, the outer layers being oldest.

ageing process The gradual degenerative process of living cells, i.e. of tissue and tissue function. The causes of this process are still uncertain and many theories have been put forward in order to explain it. Current theories suggest that the ageing process in higher animals involves free radicals. These form naturally within the body and are very reactive, sometimes combining with each other. If they react with protein molecules or nucleic acids, the result will be a malfunction of these molecules. This eventually results in the death of the cell, then of tissue and finally of the entire organism. Some scientists think that vitamin E slows down the ageing process by keeping the free radical concentration low.

agglutination The process in which bacteria or red blood cells clump together. *See also* blood group.

agglutinin *See* blood group.

agglutinogen *See* blood group.

agoraphobia An exaggerated, abnormal fear of open places. *Compare* claustrophobia.

agranulocyte A lymphoid leucocyte with non-granular cytoplasm. About 30% of all leucocytes are agranulocytes, of which there are two types: lymphocytes (25%) and monocytes (about 5%).

agriculture The science or practice of cultivating the soil and keeping animals.

AIDS (Acquired Immune Deficiency Syndrome) A very serious disease which is probably caused by a virus. The disease was first diagnosed in the USA in 1980 and since then it has been diagnosed in other countries and 36 American states. Physicians fear that AIDS will come to affect more and more people in the future. Symptoms include weakness, loss of appetite, loss of weight, fever and swelling of the lymph nodes. The patient becomes more liable to catch infectious diseases and cancer. The number of certain lymphocytes in the blood of the patient is clearly below average and those which are present are not working properly. About 40% of the patients die within a year or two. The origin of the disease is not known, but it is assumed that it comes from West Africa or the southern part of Europe, as a certain serious skin disease found in these places is a complication to AIDS. Treatment with interferon has so far had some effect.

air The invisible mixture of gases surrounding the Earth. At sea level, the composition (in per cent by volume) is nitrogen, 78,1 (78.1); oxygen, 20,9 (20.9); argon, 0,94 (0.94); carbon dioxide, 0,03 (0.03). There are also minute amounts of other gases, such as helium, neon, krypton, xenon and radon, as well as varying amounts of water vapour. *See* Appendix.

air bladder *See* swim bladder.

air pore *See* stoma.

air pressure *See* pressure.

air pump A device for transferring air or other gases from one place to another. In some air pumps pressures of the order of 10^{-3} mm of mercury can be achieved. *See also* vacuum pump.

air sac In birds, a thin-walled, air-filled extension of the lungs, often extending into the bones. In some insects, a thin-walled widening of tracheae. In mammals the bronchioles terminate in air sacs which have thin elastic walls. *See* alveolus.

Al Chemical symbol for aluminium.

albedo The proportion of solar light which is reflected from the atmosphere and surface of a planet back into space.

albinism The absence of pigmentation in animals or plants which are normally pigmented.

albino Any animal or plant with a deficiency of pigment. In animals, the absence of colouring pigment (melanin) is seen in skin, hair and feathers, which are white, and eyes, which are usually pink. An albino is unusually sensitive to light.

albumen White of egg, of birds and some reptiles; the nutritive material surrounding the yolk. *See also* albumin.

albumin Any of a group of water-soluble proteins found in egg-white, blood serum and milk.

alburnum *See* sapwood.

Alcad-accumulator® Trade name for a nickel-cadmium alkaline cell (accumulator).

alchemy The medieval forerunner of modern chemistry. Alchemists sought the philosopher's stone, a substance that could change base metals into gold, and a liquid that could prolong life indefinitely (elixir of life).

alcohol The general name for a group of organic substances containing carbon, hydrogen and oxygen. An alcohol can be thought of as being derived from a hydrocarbon in which one or more hydrogen atoms have been replaced by a hydroxyl group, OH. Alcohols are either aliphatic or cyclic molecules. Examples of alcohols are methanol, CH_3OH; ethanol, C_2H_5OH; and phenyl methanol (benzyl alcohol), $C_6H_5CH_2OH$ (cyclic). Alcohols are divided into primary alcohols, containing the functional group —CH_2OH; secondary alcohols, containing the functional group

—CHOH;
|

and tertiary alcohols, containing the functional group

—COH
|

The term 'alcohol' is often used for ethanol.

alcoholate *See* metal alkoxide.

alcohol thermometer A thermometer containing alcohol (usually ethanol). The instrument is useful at low temperatures, as ethanol has a lower freezing-point ($-117°C$) than mercury ($-39°C$).

aldehyde The general name for an organic substance containing the carbonyl group,

—C=O
|

in which one of the bonds of the carbon atom is attached to a hydrogen atom and the other one to a carbon atom (methanal is an exception). Aldehydes are either aliphatic or cyclic molecules, e.g. methanal (formaldehyde), H—CHO, and benzaldehyde, C_6H_5—CHO.

The group —CHO in aldehydes is called the aldehyde group. *Compare* ketone.

aldohexose *See* sugar.

aldopentose *See* sugar.

aldose A carbohydrate containing an aldehyde group, —CHO, e.g. glucose, $C_6H_{12}O_6$ or $CH_2OH(CHOH)_4CHO$. *See also* sugar.

aldrin $C_{12}H_6Cl_6$ A cyclic compound consisting of two fused six-membered rings. It is a pale yellow crystalline solid which is insoluble in water, but soluble in most organic solvents. Aldrin is used as an insecticide.

aleuroplast *See* leucoplast.

alfalfa *See* lucerne.

alga (pl. algae) A primitive, non-vascular photosynthetic plant, found growing aquatically and in damp situations, e.g. on tree trunks, soil and damp walls. Algae have neither stems, roots nor leaves. The most common varieties are the green, brown and red algae, all of which contain chlorophyll as well as other pigments. *See also* Cyanophyta.

alimentary canal (gut) The tube, from mouth to anus, concerned with ingestion, digestion and absorption of food. *See* Fig. 49.

alimentary system *See* digestive system.

aliphatic Describes organic compounds with carbon atoms arranged in straight or branched chains. *Compare* cyclic.

alive (live) An electrical conductor which is not at earth potential.

alkali A basic hydroxide which is soluble in water. Examples include sodium hydroxide, NaOH; potassium hydroxide, KOH; calcium hydroxide, $Ca(OH)_2$; and ammonium hydroxide, NH_4OH. Ammonium hydroxide is a solution of ammonia, NH_3, in water. The solution contains very few hydroxide ions, OH^-, as NH_4OH is not fully ionised: it is therefore a weak alkali. *See also* base.

alkali metals The metals (elements) found in Group IA of the Periodic Table of the Elements. These metals are lithium, sodium, potassium, rubidium, caesium and francium. They are very electropositive, soft, less dense than water and have low melting-points.

alkaline Having the properties of an alkali.

alkaline accumulator *See* Ni–Fe cell *and* nickel–cadmium alkaline cell (accumulator).

alkaline cell *See* Ni–Fe cell *and* nickel–cadmium alkaline cell (accumulator).

alkaline earth metals The metals (elements) found in Group IIA of the Periodic Table of the Elements. These metals are beryllium, magnesium, calcium, strontium, barium and radium. They are rather electropositive and harder and denser than the alkali metals.

alkaloid Any of a class of nitrogenous, organic bases found in certain plants. An alkaloid has a strong physiological effect; examples of alkaloids are caffeine, cocaine, nicotine and quinine.

alkane The general name for a group of hydrocarbons which are saturated, i.e. have only single bonds between the carbon atoms. The general formula of the alkanes is C_nH_{2n+2}, where $n \geq 1$. Examples of alkanes are methane, CH_4, ethane, C_2H_6, and propane, C_3H_8.

alkanoic acid *See* fatty acid.

alkene The general name for a group of unsaturated hydrocarbons which have one double bond between carbon atoms in each molecule. The general formula of the alkenes is C_nH_{2n}, where $n \geq 2$. Examples of alkenes are ethene, C_2H_4, and propene, C_3H_6.

alkine *See* alkyne.

alkoxide *See* metal alkoxide.

alkyl An aliphatic hydrocarbon group with the general formula C_nH_{2n+1}—, i.e. containing one less hydrogen atom than the corresponding alkane. *Example:* CH_3—, an alkyl group called methyl.

alkyl halide An organic compound in which a halogen atom is attached to an alkyl group. CH_3Cl, methyl chloride, and C_2H_5I, ethyl iodide. Alkyl halides are of great importance in organic synthesis because of the variety of compounds which can be made from them. *See also* amine *and* Williamson synthesis.

alkyne (alkine) The general name for a group of unsaturated hydrocarbons which have one triple bond between carbon atoms in each molecule. The general formula of the alkynes is C_nH_{2n-2}, where $n \geq 2$. Examples of alkynes are ethyne (acetylene), C_2H_2, and propyne, C_3H_4.

allantois An embryonic organ consisting of a membranous sac. In placental mammals, the allantois grows around the tail of the embryo. It supplies blood to the placenta and acts as an organ of nutrition, respiration and excretion.

allele (allelomorph) One of a set of alternative forms of the same gene. The alleles of a gene occupy the same relative position (locus) on homologous chromosomes, are able to mutate one to another and control the same characteristic, but do not necessarily produce the same effect.

allelomorph *See* allele.

allergen A substance, usually a protein, which causes an allergy. Examples of allergens are hair and pollen.

allergy An unusual reaction to a particular substance (allergen) which may be a food, pollen, an insect bite, a metal, a medicine, hair, house dust, etc. Hay fever is a common form of allergy. Symptoms of allergy can include a running nose, breathing difficulty, a rash and oedema. *See also* asthma.

allo- Prefix meaning other.

allogamy Cross-fertilisation. *Compare* autogamy.

allomerism A similarity in the crystalline structure of molecules of different chemical composition.

allomorph *See* allomorphism.

allomorphism A variability in the crystalline structure of certain molecules. Different crystalline forms of the same substance are called allomorphs.

allotrope *See* allotropy.

allotropy The existence of several forms of an element in the same state, but with different physical rather than chemical properties, e.g. oxygen, O_2, and ozone (trioxygen), O_3. The different forms are called allotropes. *See also* polymorphism.

alloy A mixture of two or more metals, or of a metal and a non-metal. The properties of an alloy are different from its components' properties.

alluvial Describes deposits of finely divided material, such as earth and sand, left by flood.

Alnico® An alloy used to make permanent magnets. The alloy contains the following metals in varying proportions: aluminium, Al; nickel, Ni; cobalt, Co; iron, Fe; and copper, Cu.

alpha (α) The first letter of the Greek alphabet.

Alpha Centauri A very bright triple star system, often called Rigil Kent. It can be seen with the naked eye, appearing to be a single star. It is 4,26 light-years (4.26 light-years) away from Earth. *See also* Proxima Centauri.

alpha decay A spontaneous radioactive disintegration in which a parent nucleus of an element decays into an alpha particle, $_2^4$He, and a daughter nucleus. This daughter nucleus will have two neutrons and two protons fewer than the parent; it will have a mass number four atomic mass units less and an atomic number two less.

alpha-naphthol test *See* Molisch's test.

alpha particle The nucleus of a helium atom, $_2^4$He. Alpha particles are emitted from the nuclei of certain radioactive elements. Each particle has a double positive charge. *See also* alpha decay.

alpha ray A stream of fast-moving alpha particles, with a relatively low penetrating power. Alpha rays produce ionisation in gases through which they pass.

alternate Describes leaves, branches, etc., which occur at different levels successively on opposite sides of a stem.

alternating current (a.c.) An electric current which varies in strength and periodically reverses its direction. *See also* frequency.

alternation of generations In the life cycles of certain organisms, the occurrence of two or more generations in which a form of sexual reproduction alternates with a form of asexual reproduction. This occurs in coelenterates and some arthropods, but is more clearly seen in plants such as ferns and mosses. In the life cycles of these plants a haploid phase alternates with a diploid phase, i.e. a haploid gametophyte alternates with a diploid sporophyte.

alternator A device for producing an alternating current. *See also* generator.

altimeter An instrument for measuring height above sea-level. It is usually an aneroid barometer, calibrated to read directly in metres or feet.

altitude Height above sea-level or horizon.

alum A traditional name used for several double salts. It is most commonly used for aluminium potassium sulphate-12-water (potash alum), $AlK(SO_4)_2 \cdot 12H_2O$, which is used in dyeing, for the production of mordants and pigments and in water purification as a coagulant. In ammonium alum (aluminium ammonium sulphate-12-water) the potassium is replaced by the ammonium group (ion), NH_4^+. Originally the name 'alum' indicated the presence of the trivalent aluminium ion, Al^{3+}. However, it is now also used for other double salts containing trivalent ions. An example is chrome alum (chromium potassium sulphate-12-water), $CrK(SO_4)_2 \cdot 12H_2O$, which contains the trivalent chromium ion, Cr^{3+}.

alumina A naturally occurring form of aluminium oxide, Al_2O_3. Alumina is also called corundum and in an impure form, emery. Both are used as abrasives. *See also* ruby *and* sapphire.

aluminium An element with the symbol Al; atomic number 13; relative atomic mass 26,98 (26.98); state, solid. It is a very electropositive metal, mainly extracted from bauxite. Aluminium is used in making light and strong alloys, food containers, foil for wrapping, cooking utensils, overhead electric cables, paint, etc. In air, a very thin layer of aluminium oxide, Al_2O_3, is formed on the surface of aluminium. This oxide layer protects the aluminium from further atmospheric corrosion and renders it less reactive chemically.

aluminium oxide Al_2O_3. *See* aluminium; *see also* alumina *and* oxide.

aluminium sulphate $Al_2(SO_4)_3$ (anhydrous); $Al_2(SO_4)_3 \cdot 18H_2O$ (hydrated) An aluminium salt with two major uses: in textile dyeing and

alveolus

as a flocculating agent in water and sewage purification.

alveolus (pl. alveoli) **1.** A terminal air sac, in lungs, where gaseous exchange takes place between air and blood. It is a cup-shaped cavity surrounded by a dense network of capillaries. An alveolus has a thin wall covered with a film of moisture in which air dissolves. The air then diffuses through the epithelium, the capillary wall and the plasma, and into red blood cells. Here oxygen combines with haemoglobin. Carbon dioxide in the blood diffuses into the alveoli and is eventually exhaled. *See* Fig. 1. **2.** A cavity in a gland. **3.** A cavity in the jaw-bone forming a tooth socket.

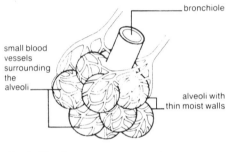

Fig. 1 Alveoli

AM Amplitude modulation. *See* modulation.

amalgam A liquid or solid alloy of mercury with one or more metals or non-metals. Important uses of amalgams are in the repair of dental caries and in the recovery of gold and silver from their ores.

amber (succinite) A yellow, translucent fossil resin, derived from an extinct species of pine. Amber is often found on sea-shores and is used for ornamental purposes. Often amber contains trapped insects or pieces of plants.

Ames' test A biological method used in the testing of chemicals, mainly for their carcinogenic effect on human beings. The test is carried out using salmonella bacteria and is based on the theory that if a certain chemical is able to induce a hereditary change in salmonella bacteria, it is probably able to do the same in human beings, i.e. it is likely to be carcinogenic.

amethyst A form of quartz, SiO_2, which is pale to dark violet or purple in colour. The colour is caused by impurities and may disappear on heating. Amethyst is a semi-precious stone.

amide The general name for a class of organic compounds containing the functional group
—C=O
 |
 NH$_2$
the amide group. Amides can be prepared by reacting an acyl halide with ammonia, NH$_3$, e.g.
$CH_3COCl + 2NH_3 \rightarrow CH_3CONH_2 + NH_4Cl$
Here ethanoyl chloride (acetyl chloride), CH_3COCl, is converted into ethanamide (acetamide), CH_3CONH_2. Amides are white solids with a neutral reaction when dissolved in water. They are used in the organic synthesis of other compounds. *See also* urea.

amide group *See* amide.

amine The general name for a class of organic compounds containing the functional groups —NH$_2$, —NH or —N—
 |
Amines may be considered to be derived from ammonia, NH$_3$, in which one or more hydrogen atoms are replaced by an organic group such as an alkyl or aryl. Examples are CH_3NH_2, methylamine, and $C_6H_5NH_2$, phenylamine (aniline). Amines can be divided into primary, secondary and tertiary amines depending on the number of alkyl or aryl groups in the molecule: CH$_3$—NH$_2$ (methylamine) is a primary amine,
CH$_3$—NH
 |
 CH$_3$
(dimethylamine) is a secondary amine and
CH$_3$—N—CH$_3$
 |
 CH$_3$
(trimethylamine) is a tertiary amine. The alkyl or aryl groups may be similar (as in these examples) or different, and amines containing both alkyl and aryl groups exist. Amines have an alkaline reaction when dissolved in water. They can be prepared by reacting an alkyl halide with ammonia, NH$_3$, and a strong alkali:
$C_2H_5I + NH_3 + OH^- \rightarrow C_2H_5NH_2 + I^- + H_2O$
Here ethyl iodide, C_2H_5I, gives ethylamine, $C_2H_5NH_2$. Amines are used in the organic synthesis of other compounds. *See also* phenylamine.

aminoacetic acid *See* glycine.

amino acid An organic acid containing the carboxyl group —COOH and the amino group —NH$_2$. All peptides and proteins contain units of amino acids. About twenty different amino acids occur in nature. Ten of these are called essential amino acids as they cannot be synthesised in the human body for conversion into proteins; therefore they must be present in the diet. The chief amino acids from which natural proteins are made all contain an amino group in the α-position: i.e. the amino group is attached to the carbon atom adjacent to the carboxyl group. An example is
CH$_3$—CH$_2$—CH—COOH,
             |
             NH$_2$
α-aminobutanoic acid. The simplest of all

amino acids is glycine, H$_2$N—CH$_2$—COOH. *See also* zwitterion.

aminobenzene *See* phenylamine.

4-aminobenzenesulphonamide *See* sulphanilamide.

aminoethanoic acid *See* glycine.

amino group The group —NH$_2$ (primary amino group), as found in peptides, proteins and several other organic compounds. *See also* amine.

ammeter An instrument for measuring electric current, having a low internal resistance. Two common types are the moving coil ammeter and the moving iron ammeter. *See also* moving coil instrument *and* moving iron instrument.

ammonia NH$_3$ A pungent-smelling gas, very soluble in water, giving a weak alkaline solution; i.e. in the chemical equilibrium
$$NH_3 + H_2O \rightleftharpoons NH_4^+ + OH^-$$
the reverse reaction is favoured. Ammonia is prepared industrially by the Haber–Bosch process. It is used as a fertiliser and in the production of other compounds such as nitric acid.

ammoniacal liquor A solution of ammonia in water, produced during the manufacture of coal-gas. When treated with sulphuric acid, H$_2$SO$_4$, it produces ammonium sulphate, (NH$_4$)$_2$SO$_4$, an important fertiliser.

ammonia-soda process *See* Solvay process.

ammonite A member of an important group of fossils used for dating rocks of the Mesozoic age. Ammonites have flat, spiral shells.

ammonium alum *See* alum.

ammonium chloride NH$_4$Cl (sal ammoniac) An ammonium salt. When heated it sublimes. It is used in dry cells and as a flux in soldering.

ammonium cyanate *See* Wöhler synthesis.

ammonium dichromate(VI) (NH$_4$)$_2$Cr$_2$O$_7$ An ammonium salt. When heated, it decomposes into chromium(III) oxide, Cr$_2$O$_3$, nitrogen, N$_2$, and water vapour:
$$(NH_4)_2Cr_2O_7 \rightarrow Cr_2O_3 + N_2 + 4H_2O$$

ammonium hydroxide NH$_4$OH A solution of ammonia in water. It contains no molecules of NH$_4$OH; instead it contains ammonium ions, NH$_4^+$, hydroxide ions, OH$^-$, unionised ammonia, NH$_3$, and water. *See also* ammonia.

ammonium ion NH$_4^+$ A monovalent group of atoms which has a positive electrical charge. The ammonium ion forms compounds which are similar to the salts of monovalent metals.

ammonium nitrate NH$_4$NO$_3$ An ammonium salt used as a fertiliser and in making explosives. When heated it decomposes into dinitrogen oxide (laughing gas), N$_2$O, and water vapour:
$$NH_4NO_3 \rightarrow N_2O + 2H_2O$$

ammonium nitrite NH$_4$NO$_2$ An ammonium salt. When heated it decomposes into nitrogen, N$_2$, and water vapour:
$$NH_4NO_2 \rightarrow N_2 + 2H_2O$$

ammonium sulphate (NH$_4$)$_2$SO$_4$ An ammonium salt, mainly used as a nitrogenous fertiliser, obtained as a by-product of coal-gas manufacture. It is now manufactured from an aqueous ammonia solution, NH$_3$, saturated with carbon dioxide, CO$_2$, and mixed with powdered calcium sulphate, CaSO$_4$:
$$CaSO_4 + 2NH_3 + CO_2 + H_2O \rightarrow (NH_4)_2SO_4 + CaCO_3$$
The calcium carbonate, CaCO$_3$, is filtered off and the ammonium sulphate crystallised. *See also* ammoniacal liquor.

amnion A foetal membrane of mammals, birds and reptiles. In birds and mammals it is the innermost of three membranes and encloses the amniotic cavity. This is filled with amniotic fluid which serves to cushion the embryo.

amniote One of a group of vertebrates, comprising mammals, birds and reptiles, whose embryos possess an amnion. *Compare* anamniote.

amniotic fluid *See* amnion.

Amoeba A single-celled aquatic protozoan, which is constantly changing shape by projecting temporary 'feet' or pseudopodia. It is just visible to the naked eye as an opaque white speck. *See also* phagocytosis.

amoeboid Resembling an *Amoeba*. Moving by pseudopodia. *See also* pseudopodium.

amorphous Without definite form or shape; lacking a crystal structure. Amorphous substances have no fixed melting-points, e.g. glass.

amorphous sulphur A non-crystalline form of sulphur which is often white and consists of very small particles which are difficult to filter. With water, amorphous sulphur may form a colloidal solution called milk of sulphur. Amorphous sulphur may be prepared by acidifying an aqueous solution of sodium thiosulphate, Na$_2$S$_2$O$_3$:
$$Na_2S_2O_3 + 2H^+ \rightarrow S + SO_2 + 2Na^+ + H_2O$$

AMP *See* adenosine monophosphate.

ampere Symbol: A. The SI unit of electric current. The ampere is the current which, if flowing in two straight parallel conductors of infinite length, placed one metre apart in a vacuum, will produce a force of 2×10^{-7} newtons per metre length on each conductor.

ampere-hour The quantity of electricity flowing in a conductor when a current of one ampere flows through it for one hour.
1 ampere-hour = 3600 coulombs

Amphibia A class of cold-blooded, tetrapod vertebrates adapted for life on land and in water. Examples are frogs, toads and salamanders. Characteristics include pentadactyl limbs, a soft, moist skin without scales and external fertilisation. They are egg-laying, and undergo metamorphosis from the larval stage to the adult stage. The larvae are aquatic, breathing with gills, while the adults live on land.

amphimixis The normal method of sexual reproduction by fusion of gametes. *Compare* apomixis.

Amphioxus (Branchiostoma) A genus of primitive, marine, invertebrate animals belonging to the cephalochordates, the lancelets. The species *Amphioxus lanceolatus* is about 5 cm long and is pointed at both ends. It has a dorsal fin along the whole length of its body and a caudal and ventral fin. It is a ciliary feeder, spending most of its life burrowed in sand with only its head end exposed. Amphioxus is the only chordate whose notochord is retained throughout life.

ampholyte (amphoteric electrolyte) A substance which is capable of exhibiting both acidic and basic properties. *See also* amphoteric *and* oxide.

amphoteric Having both acidic and basic properties. *Example:* water can behave as an acid:
$H_2O \rightleftharpoons H^+ + OH^-$
or as a base:
$H_2O + H^+ \rightleftharpoons H_3O^+$

amphoteric electrolyte *See* ampholyte.

amphoteric oxide *See* oxide.

amplexus In frogs and toads, the embrace of the female by the male, during which ova are squeezed from the female's body into the surrounding water prior to fertilisation.

amplifier A device for increasing the power of an electrical input. *See also* transistor.

amplitude The maximum extent of vibration or oscillation from the position of equilibrium. (1) The amplitude of an alternating current is the peak value of the current. (2) The amplitude of a pendulum is half the length of the swing.

amplitude modulation (AM) *See* modulation.

ampulla (pl. ampullae) A swelling at the end of each semicircular canal in the ear, containing a crista.

amputation The removal of a part of the body, e.g. a limb. Amputation is carried out when the part is seriously damaged or diseased.

a.m.u. *See* atomic mass unit.

amylase (diastase) A general name for any enzyme capable of breaking down polysaccharides into smaller carbohydrate units. Amylases are widely distributed in plants and animals. An example of an amylase is ptyalin, found in saliva.

amylopectin *See* starch.
amyloplast *See* leucoplast.
amylose *See* starch.

anabolism The building up of complex molecules from simpler ones, involving the taking up and storing of energy (gaseous or dissolved in water). *Compare* catabolism; *see also* metabolism.

anabolite A substance taking part in anabolism.

anaemia A deficiency of red cells in the blood, which may arise from a large loss of blood, a deficiency of iron in the diet, failure of the marrow to produce erythrocytes (red blood cells) or excessive destruction of the red blood cells.

anaerobe An organism living only in the absence of free oxygen (gaseous or dissolved in water). *Compare* aerobe.

anaerobic Describes organisms living in the absence of free oxygen (gaseous or dissolved in water). *Compare* aerobic; *see also* anaerobic respiration.

anaerobic respiration Respiration taking place in the absence of free oxygen (gaseous or dissolved in water). *Compare* aerobic respiration.

anaesthesia Insensibility to pain.

anaesthetic A substance used in relieving pain, either partial or complete. General anaesthetics affect the whole body, usually with loss of consciousness, whereas local anaesthetics only affect a limited part of the body.

anal Pertaining to or situated near the anus.

anal fin In fish, a median fin controlling rolling and yawing movements.

analgesia The absence or relief of pain. *See also* anaesthesia.

analgesic A pain-killing drug. *See also* anaesthetic.

analogous Describes parts of animals or plants which are similar in function or appearance but differ in structure and development, i.e. have a different origin. *Example:* the legs of a fly have the same function as those of a bird, i.e. they enable both animals to walk, but they have no common structural features. *Compare* homologous.

analysis (pl. analyses) Determination of the properties of matter. *Compare* synthesis; *see also* qualitative analysis *and* quantitative analysis.

anamniote One of a group of vertebrates, comprising amphibians and fish, whose embryos do not possess an amnion. *Compare* amniote.

anaphase The stage of mitosis or meiosis, following metaphase, when chromosomes separate and move towards opposite poles of the spindle.

anatomy The study of the structure of plants and animals, as determined by dissection.

androecium Collective name for the male reproductive organs (stamens) of a plant.

androgen One of a group of male sex hormones. Androgens are responsible for developing and maintaining many secondary sexual characteristics. *See also* testosterone.

anemometer An instrument for measuring the velocity of a fluid, particularly wind velocity.

aneroid Without liquid.

aneroid barometer An instrument used to measure air pressure. It consists of a partially evacuated metal drum which varies in width as the air pressure changes. The movement of the drum walls is transmitted through gears to a pointer which moves over a scale indicating the air pressure.

aneurin *See* thiamine.

angina pectoris A condition in which the supply of blood (oxygen) to the heart becomes inadequate. This causes acute chest pains which sometimes spread to the arms (particularly the left one) and to the neck and jaw. Angina pectoris may be a result of hardening and thickening of the coronary blood vessels (arteriosclerosis). It is treated with nitroglycerine tablets placed under the tongue, from where they are quickly absorbed into the blood stream. Nitroglycerine dilates the coronary blood vessels, allowing more blood to pass through. Sometimes angina pectoris is treated surgically. Many things may trigger an attack, such as physical exertion, the effort involved in eating and digesting a heavy meal, or anything that is strenuous. Attacks are often brief, lasting only a couple of minutes.

angiosperm A member of a major division of the plant kingdom. Angiosperms are flowering plants, distinguished from gymnosperms by having their ovules carried within a closed cavity, the ovary. *See also* Spermatophyta.

angle of declination (angle of variation) The angle between the geographic meridian and the magnetic meridian at a given place on the Earth. *See* Fig. 2.

angle of deviation The angle between the incident ray and the refracted ray when a ray of light passes from one medium to another. *See* Fig. 3.

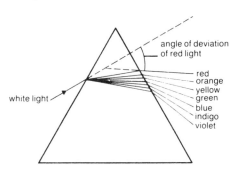

Fig. 3 Angle of deviation

angle of dip (angle of inclination) The angle made with the horizontal by a freely suspended magnetic needle at a place on the Earth's surface.

angle of incidence The angle between the incident ray of light, striking a reflecting or refracting surface, and the normal to the surface at the point of incidence. *See* Figs. 4 and 5.

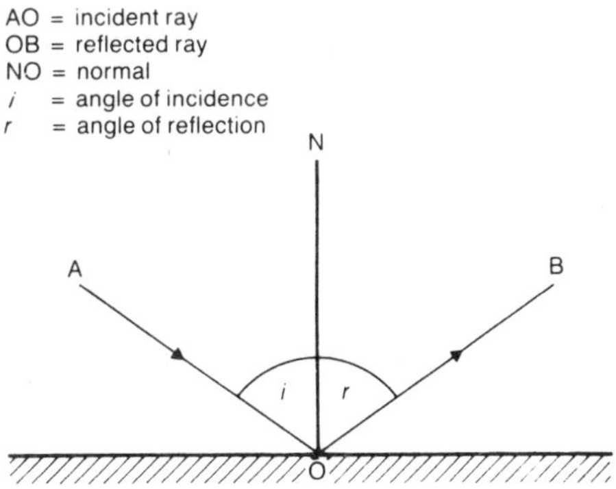

AO = incident ray
OB = reflected ray
NO = normal
i = angle of incidence
r = angle of reflection

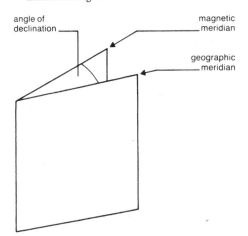

Fig. 2 Angle of declination

Fig. 4 Angle of incidence and angle of reflection

angle of inclination *See* angle of dip.

angle of reflection The angle between the reflected light ray from a surface and the normal to the surface at the point of reflection. *See* Fig. 4.

angle of refraction The angle between the refracted light ray and the normal to the surface at the point of refraction. *See* Fig. 5.

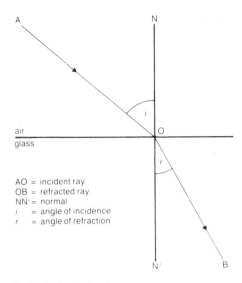

AO = incident ray
OB = refracted ray
NN' = normal
i = angle of incidence
r = angle of refraction

Fig. 5 Angle of refraction

angle of variation *See* angle of declination.

angstrom (Ångström or angstrom unit; Å, ÅU, AU) A unit of length equal to 10^{-8} cm (10^{-10} m).

angular magnitude The angle which an object subtends at the eye. This angle governs the apparent size of the object, as it determines the size of the image formed on the retina. *Example:* telegraph-poles appear to be shorter the further away they are, although they are the same height. *See* Fig. 6.

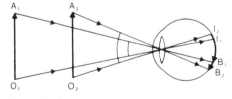

Fig. 6 Angular magnitude

angular velocity Symbol: ω; unit: radians per second (rad s^{-1}). The rate of motion of a body through an angle about an axis.

anhydride The substance obtained when the elements of water (hydrogen and oxygen) are removed from a compound. Examples of anhydrides are: sulphur trioxide, SO_3, the anhydride of sulphuric acid, H_2SO_4; and calcium oxide, CaO, the anhydride of calcium hydroxide, $Ca(OH)_2$. Usually an anhydride takes up water easily, and some of them are therefore good drying agents.

anhydrite A naturally occurring anhydrous form of calcium sulphate, $CaSO_4$.

anhydrous Without water. The term is often applied to salts which have no water of crystallisation. *Compare* hydrated.

aniline *See* phenylamine.

animal cell *See* cell.

animal charcoal (bone black, bone char) A substance obtained by charring animal material, especially bones.

animal pole 1. The region of an egg cell containing the nucleus and clear, active cytoplasm. 2. The side of a blastula where micromeres collect. *Compare* vegetal pole.

animal starch *See* glycogen.

anion An atom or group of atoms carrying a negative electrical charge. In the presence of an electric field, e.g. in electrolysis, an anion moves towards the positive electrode, the anode. *Compare* cation.

anneal To harden metals and glass by heating followed by slow cooling. The process also relieves strains in the material.

Annelida (Annulata) A phylum of ringed or segmented worms. Examples include earthworms and leeches. *See also* chaeta.

annual plant A plant which completes its life-cycle from seed germination to seed production and subsequent death in one year, e.g. maize.

annual ring The growth of secondary xylem (wood) in the stem of a woody plant in a temperate climate during one year. Annual rings appear as a series of concentric lines (rings) in a cross-section of a stem. One light ring and one darker ring are produced each year. From the number of annual rings the approximate age of the plant may be determined.

Annulata *See* Annelida.

anode The positive electrode of an electrolytic cell or discharge tube. *Compare* cathode.

anodise To produce a layer of oxide on the surface of metals by making the metal the anode in an electrolytic bath. The process hardens the surface of the metal, makes it more resistant to corrosion and enables it to absorb dyes.

anomalous expansion of water The irregular expansion of water. In the temperature range 0–4°C, water decreases in volume with increasing temperature, reaching its maximum density at 4°C. Above 4°C water expands when heated. *See* Appendix.

ANS *See* autonomic nervous system.

ant *See* Hymenoptera.

antabuse An organic drug containing several sulphur atoms which is used in the treatment of chronic alcoholism. It interferes with the normal metabolism of alcohol, resulting in the accumulation of ethanal, CH_3CHO, in the blood. This causes vomiting, nausea and other unpleasant effects. An antabuse treatment requires a positive and co-operative attitude in the patient.

antacid A substance which neutralises excess stomach acid, e.g. milk of magnesia, bicarbonate of soda.

antagonism Active opposition between structures, organisms or substances. *Examples:* the relationship between a pair of antagonistic muscles, such that one contracts when the other relaxes; the effect of an antitoxin on a toxin. *Compare* synergism.

antenna (pl. antennae) **1.** A long, jointed sense organ on the heads of insects, crustaceans and myriapods, bearing receptors for touch, smell, taste, vibration, etc. In some crustaceans, the antennae are used for swimming. **2.** *See* aerial.

anterior Situated at or near the head (front) end of an animal. In human anatomy, the anterior side is equivalent to the ventral side of other mammals.

anther The part of a stamen producing pollen. *See* Fig. 75.

antheridium (pl. antheridia) The male sex organ of bryophytes, pteridophytes, algae and fungi. *See* Fig. 191.

anthracite The hardest and oldest form of coal, containing a high percentage of carbon. It is a smokeless fuel.

anthrax A disease of sheep and cattle caused by the bacterium *Bacillus anthracis*. The disease can be transmitted to humans by contact with the hides or meat of infected animals. Symptoms include skin ulcers, fever and swelling of the lymph nodes. Anthrax is treated with antibiotics.

anthropoid Similar in appearance to a human being.

anthropology The study of the societies, customs, structure and evolution of mankind.

anti- Prefix meaning against.

antibiotic A drug obtained from living cells (e.g. moulds) which stops the growth of micro-organisms (bacteriostatic) or kills micro-organisms (bactericidal). Antibiotics may be antibacterial or antifungal or both. The first antibiotic to be discovered was penicillin. *See also* sulphonamide.

antibody (antitoxin) A protein produced by vertebrates as a result of the presence of an antigen. When an animal has produced antibodies, it is said to be sensitised to the antigen. A period of a few weeks must elapse before full sensitivity is developed, but once established, it usually lasts for a long time. An antibody combines chemically with an antigen (antibody–antigen reaction). Antibodies are highly specific, only combining with antigens of a particular kind. *Example:* the antibody against mumps virus antigen will destroy only this virus and not any other. The combination between the antigen and the antibody takes place at certain small areas on the surface of the molecules. Antibodies form a part of the body's natural immunity system by destroying pathogens, e.g. by clumping them together, thus making them easier for phagocytes to ingest, or by neutralising the toxins released from them. Antibodies are produced in different parts of the body, e.g. in the lymph nodes and in the spleen. They exercise their effect in the blood. *See also* gamma globulin *and* immunisation.

antichlor A substance, such as sodium thiosulphate, $Na_2S_2O_3$, used to remove chlorine from materials after chlorine bleaching.

anticoagulant A drug used to prevent blood from clotting, e.g. heparin.

anticodon *See* codon.

antidiuretic Reducing the volume of urine. Antidiuretic hormone (ADH) is produced by the hypothalamus and released by the posterior lobe of the pituitary gland. It stimulates water reabsorption by kidney tubules thus diminishing the volume of urine. *See also* vasopressin.

antidote A substance which counteracts the effects of a poison (toxin).

antifreeze A substance, usually glycol (ethane-1,2-diol), which is added to water to lower its freezing-point. Antifreeze is commonly used in motor-car radiators.

antigen A substance which stimulates the production of antibodies. Parasites and most of their toxic products are antigens.

antigibberellin One of a group of organic compounds which cause plants to grow with short, thick stems; when applied to grass they retard its growth.

antihistamine A drug which counteracts most of the effects produced by histamine. Antihistamines are commonly used to relieve allergies.

anti-knock *See* inhibitor.

antineutron *See* antiparticle.

antinode A point of maximum displacement in a standing wave. Antinodes are situated a distance of a quarter of the wavelength of the wave motion from their adjacent nodes. *See also* node.

antiparticle In experiments within nuclear physics, it has been revealed that some (if not all) elementary particles have antiparticles. *Examples:* the positron (discovered in 1932) and the electron are each other's antiparticles. The antiproton (discovered in 1955) has the same mass and spin as the proton, but its charge is opposite, i.e. −1. The antineutron (discovered in 1956) is identical to the neutron except that its magnetic moment is of opposite sign. On collision, antiparticles annihilate each other, yielding a variety of decomposition products.

antiperspirant A substance which inhibits perspiration, for instance by constricting the pores of the skin. An antiperspirant is often a salt of aluminium or zinc.

antiproton See antiparticle.

antipyretic A substance which lowers the body temperature, e.g. aspirin.

antiseptic 1. Preventing infection. 2. A drug destroying harmful micro-organisms.

antiserum A serum with high antibody content.

antitoxin A type of antibody which neutralises a toxin by combining with it.

Anura An order of amphibians including frogs and toads.

anus Posterior opening of the alimentary canal through which undigested food is expelled. See Fig. 49.

anvil See incus.

aorta The large, long artery which arises from the left ventricle of the heart and through which oxygenated blood passes to all parts of the body via arteries and their branches. The aorta is divided into an ascending portion, an arch and a descending portion. See Fig. 27.

apatite A mineral which may be represented by the formula $CaF_2 \cdot 3Ca_3(PO_4)_2$. However, in some forms of apatite there may be chlorine or hydroxide groups instead of fluorine. Apatite is used in the manufacture of fertilisers and as a source of phosphorus.

ape See Primates.

aperture 1. An opening. 2. The size of the opening admitting light to an optical instrument. 3. The diameter of the reflecting or refracting surface in spherical mirrors or lenses.

apex Tip. In biology, the term is applied to the heart, lungs, roots, stems, wings, etc.

aphelion The point in a planet's orbit when it is furthest from the Sun. Compare perihelion.

aphid A plant-louse of great importance to agriculture. It damages crops by sucking the cell-sap and by transmitting certain virus diseases. The commonest aphid is the greenfly.

apiary A place where bees are kept.

apical meristem See meristem.

apogee The point in the orbit of a moon, a planet or an artificial satellite when it is furthest from the Earth. Compare perigee.

apomixis A form of reproduction in plants in which seeds are formed without the fusion of gametes. Compare amphimixis.

apparent depth The depth of a liquid viewed from above appears to be less than the real depth. This is caused by the refraction of light. The relation between real depth, apparent depth and the refractive index of the liquid is given by the equation

$$\frac{\text{real depth}}{\text{apparent depth}} = n$$

where n is the refractive index of the liquid. See Fig. 7.

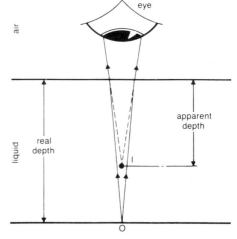

Fig. 7 Real and apparent depth

appendage A part or organ attached to the main body or trunk of a plant or animal, e.g. a branch or limb.

appendicitis An inflammation of the appendix, caused by the blocking of the passage from the apex of the caecum to the appendix. Why this blocking occurs is not known.

appendicular skeleton The limbs and limb girdles of a vertebrate skeleton.

appendix A worm-like, blind tube (i.e. closed at one end) extending from the apex of the caecum. In humans it is about 8–10 cm long and its position is variable. See Fig. 49.

apposition In plant cells, the thickening of cell walls caused by the deposition of successive layers of cellulose. Compare intussusception.

aprotic solvent A solvent which is neither an acid nor a base, i.e. a solvent which can neither donate nor accept a proton. An example of an aprotic solvent is benzene, C_6H_6. An acid dissolving in an aprotic solvent does not produce any ions.

aqua- (aqu-) Prefix meaning water.

aqua fortis Concentrated nitric acid, HNO_3.

aqua regia A mixture of one part concentrated nitric acid, HNO_3, and three parts concentrated hydrochloric acid, HCl. Aqua regia is capable of dissolving such metals as gold and platinum.

aquarium (pl. aquaria) A tank or container, with transparent sides, for keeping and displaying aquatic animals and plants.

aquatic Living in water.

aqueduct A channel containing a fluid, e.g. the cochlea of the ear.

aqueous Containing water.

aqueous humour A thin, watery fluid filling the space between the cornea and the lens of the eye. It is of a similar composition to blood serum. See Fig. 66.

aqueous solution A solution in which the solvent is water.

Ar Chemical symbol for argon.

arable Land which is ploughed or suitable for ploughing in preparation for crop production.

Arachnida A class of carnivorous animals comprising spiders, scorpions, ticks and mites. Characteristics include no antennae, four pairs of walking legs (except in some of the mites), a body divided into two parts, and simple eyes.

arachnoid A thin membrane enveloping the brain and spinal cord, situated between the dura mater and the pia mater. See also meninges.

arboreal Living in or pertaining to trees.

arc See electric arc.

archaeology The study of ancient things, especially of prehistoric times.

archegonium (pl. archegonia) The female sex organ of bryophytes, pteridophytes and most gymnosperms. See Fig. 191.

Archimedes' principle A principle stating that when a body is totally or partially immersed in a liquid, its apparent loss in weight is equal to the weight of the liquid displaced. See also flotation (law of).

area Symbol: A. A measure of surface. Units of area are: square centimetres (cm^2); square metres (m^2); and square kilometres (km^2).

arene The general name for aromatic hydrocarbons, e.g. benzene, C_6H_6. The term includes derivatives of benzene such as methylbenzene, $C_6H_5CH_3$, or aromatic hydrocarbons containing two or more rings.

areola Circular pigmented area surrounding the nipple of the breast and the pupil of the eye.

areolar tissue Connective tissue found throughout the body of vertebrates. It consists of several types of cell and protein fibres embedded in a matrix. The cells include mast cells which secrete the matrix, flattened cells called fibroblasts which secrete the protein fibres, and fat cells which store fat and macrophages. There are two types of protein fibres, and these give the tissue its strength. The protein fibres consist of bundles of white collagen fibres interwoven with a network of yellow elastic fibres made from the protein elastin. When great numbers of fat cells fill the spaces between the fibres and fibre bundles, the tissue becomes known as adipose tissue.

argentite See silver.

argentum The Latin name for silver.

arginase See arginine.

arginine An essential amino acid which appears to be a constituent of all proteins and which, in certain organisms, is an essential requirement for growth and the synthesis of urea. It is formed in the liver and kidney and is hydrolysed by the enzyme arginase to ornithine and urea.

argon An element with the symbol Ar; atomic number 18; relative atomic mass 39,95 (39.95); state, gas. It is a very inert gas which makes up 0,94% (0.94%) of the atmosphere. Argon belongs to the noble gases in the Periodic Table of the Elements. The reason why argon is found in a much larger amount in the atmosphere than any of the other noble gases is that a naturally occurring radioactive isotope of potassium, $^{40}_{19}K$, catches an electron from its inner shell (K-capture) and produces the isotope $^{40}_{18}Ar$:
$$^{40}_{19}K + e^- \rightarrow {}^{40}_{18}Ar$$
Argon is used for filling electric lamps, in fluorescent tubes and, like helium, to provide an inert atmosphere for welding operations.

armature 1. The part of an electric motor in which the driving current flows or the part of a generator in which the electric current is induced. These parts are usually rotating coils. 2. Any moving part of an electrical apparatus in which a voltage is induced by a magnetic field, or which closes a magnetic circuit. 3. A piece of magnetic material held across the poles of an electromagnet to transmit force to support a load. 4. (keeper) A piece of magnetic material placed across the poles of a permanent magnet to complete the magnetic circuit when the magnet is out of use, thus preventing loss of magnetism.

aromatic compound One of a group of organic compounds derived from benzene, C_6H_6, which are either carbocyclic or heterocyclic. The term 'aromatic' was originally given to compounds of this nature because some of them have a pleasant smell.

arsphenamine See Salvarsan.

arterial Pertaining to an artery.

arteriole A very small artery.

arteriosclerosis The hardening and thickening of artery walls, causing loss of elasticity. The process occurs with advancing age. See also angina pectoris.

artery A thick-walled, elastic blood vessel which conveys blood from the heart. Blood in arteries is oxygenated, apart from that carried by the pulmonary artery. *Compare* vein.

arthritis 1. (neuropathic) A degenerative joint disease which is caused by the loss of the sensory nerve supply. 2. (rheumatoid) A chronic inflammatory disease of connective tissues, predominantly affecting joints, but also involving other sites. 3. (tuberculous) A blood-borne tuberculous infection involving bone or joint or both. The primary origin of the infection is often the lungs, less commonly the intestines.

Arthropoda A phylum of animals with segmented bodies, hard exoskeletons and jointed legs. Examples include crabs, spiders, millipedes, centipedes and insects.

articulate 1. Jointed. 2. To form a joint.

artificial insemination The artificial injection of semen into the uterus.

artificial kidney *See* kidney machine.

artificial parthenogenesis *See* parthenogenesis.

artificial respiration A manual or mechanical method of forcing air into the lungs to restore breathing. *See also* kiss of life.

Artiodactyla An order of herbivorous, placental mammals commonly called the even-toed ungulates. Members include the pig, the hippopotamus and the ruminants (cattle, deer, goat, sheep, camel, etc.) They have either two toes (cattle) or four toes (pigs). *Compare* Perissodactyla.

aryl An aromatic hydrocarbon group which contains one hydrogen atom fewer than the corresponding arene; e.g. C_6H_5- is an aryl group called phenyl.

aryl halide An organic compound in which a halogen atom is attached to an aryl group, e.g. C_6H_5Cl is phenyl chloride. (In the IUPAC system these compounds are named as arene compounds, e.g. C_6H_5Cl is chlorobenzene.) Aryl halides are used in the organic synthesis of other compounds.

asbestos A fibrous mineral, mainly calcium magnesium silicate. It was formerly of great importance as a heat-insulating material and in fire-proof fabrics. Because of its carcinogenic properties its use has been restricted.

asbestosis A lung disease arising from inhalation of the fibres of asbestos.

ascorbic acid *See* vitamin.

asdic Abbreviation for Allied Submarine Detection Investigation Committee. *See* echo-sounder.

-ase Suffix denoting an enzyme.

aseptic Free from bacteria.

asexual reproduction Reproduction which does not involve fusion of gametes. Confined to certain plants and simpler animals. *See also* vegetative reproduction.

ash The non-volatile, inorganic residue left after the combustion of some substances.

asphalt A viscous mixture of hydrocarbons (bitumens) obtained as a residue of the distillation of petroleum and from natural sources. Mainly used for surfacing roads.

asphyxia (suffocation) A condition brought about by depriving an organism of oxygen.

aspirator An apparatus for drawing a gas through a liquid. When the water runs out of the bottle, air is sucked in through the bent tube. *See* Fig. 8.

Fig. 8 Aspirator

aspirin The common name for 2-ethanoyloxy-benzoic acid (acetylsalicylic acid) or salts of this acid. Aspirin relieves mild pain, lowers the body temperature, relieves inflammation, stimulates respiration (breathing) and causes a fall of blood sugar. *See also* antipyretic.

assimilation The absorption and conversion of nutrients into complex constituents of the body.

astatine An element with the symbol At; atomic number 85; relative atomic mass 210; state, solid. Astatine is radioactive and belongs to the halogens. It occurs only in extremely minute quantities in nature in some ores of uranium. The estimated amount of astatine in the Earth's crust is 30 g. The longest-lived isotope of astatine is $^{210}_{85}$At, which has a half-life of 8,3 hours (8.3 hours). Astatine is more metallic than iodine; like iodine, it probably accumulates in the thyroid gland.

aster A star-shaped system of fibres radiating from the centriole, seen in the cells of simpler plants and most animals during cell division. The function of an aster is obscure.

asteroid A minor planet or planetoid. They are very small, dim objects of the Solar System mainly found between the orbits of the planets Mars and Jupiter. Ceres is the largest asteroid. Vesta, the brightest, can occasionally be seen by the naked eye.

asthma Breathlessness due to bronchial constriction and an obstructive secretion. There are two major causes of asthma. The first is an infection of the tubes leading to the lungs or of the lungs themselves; the second is an allergic reaction, the stimuli of which include pollen, house dust, drugs, bacteria and animal hair. Inhaling certain chemicals may suppress the symptoms of asthma in many patients.

astigmatism An aberration of lenses and mirrors preventing light rays from being brought to a common focus. In the eye it is usually caused by a defect in the cornea, only rarely by a defect in the lens.

astringent A substance which causes contraction of body tissues thus reducing the discharge of fluids such as mucus and blood.

astrology The study of the position of heavenly bodies in the belief that they influence human destiny. Astrology is no longer considered a science.

astronaut A person who travels in a spacecraft.

astronomical telescope (Kepler telescope) *See* telescope.

astronomical unit Symbol: a.u. The basic unit of distance within the Solar System, being the average distance of the Earth from the Sun. The vaule of 1 a.u. is 149 597 870 km.

astronomy The study of heavenly bodies, e.g. stars and planets. This includes their motions, relative positions and nature. Astronomy is the oldest of all sciences.

asymmetric carbon atom *See* optical activity.

-ate Suffix denoting a salt of the corresponding -ic acid; e.g. sodium sulphate, Na_2SO_4, is a salt of sulphuric acid, H_2SO_4. However, salts of hydrochloric acid, HCl, are called chlorides. *See also* -ite.

athlete's foot An infection of the skin between the toes, caused by a ringworm type of fungus. In severe cases the disease may spread to the whole foot or to other parts of the body.

atlas In humans and other tetrapods, the first of the seven cervical vertebrae, appearing as a ring of bone. Together with the second cervical vertebra, the axis, it is modified to enable the head to move.

atmosphere 1. A region of gas surrounding a planet or other heavenly body. The Earth's atmosphere consists of a layer of air which is a few hundred kilometres thick. The atmosphere is actually divided into several layers which have different physical properties. Beginning with the layer next to the Earth, these layers are the troposphere, the stratosphere, the mesosphere, the ionosphere and the exosphere. *See also* air. **2.** A unit of pressure. The atmosphere exerts a pressure at the surface of the Earth and, at sea level and at 0°C, this pressure is 101 325 newtons per square metre (N m^{-2}) or 101 325 pascals (Pa) or 101,325 (101.325) kilopascals (kPa). This pressure is also called one atmosphere. However, atmospheric pressure usually fluctuates about this value from day to day. An older unit of pressure is millimetres of mercury. Atmospheric pressure at sea level and at 0°C is 760 mm of mercury (mmHg).

atmospheric pressure *See* atmosphere.

atmospherics A term used to describe the radio interference produced by electrical discharges in the atmosphere, e.g. lightning flashes. *See also* static.

atom The smallest particle of an element which can be said to exhibit the properties of that element. *See also* atomic structure.

atomic bomb (A-bomb) *See* fission.

atomic clock *See* clock.

atomic energy *See* nuclear energy, fission *and* fusion.

atomic heat *See* Dulong and Petit's law.

atomicity The atomicity of an element is the number of atoms in one molecule of it. *Examples:* the atomicity of helium, He, is one; of oxygen, O_2, two; of ozone, O_3, three; of phosphorus, P_4, four; and of sulphur, S_8, eight. The term also applies to chemical compounds, e.g. the atomicity of ammonia, NH_3, is four.

atomic mass unit Symbol: a.m.u. is $\frac{1}{12}$ the mass of one atom of the most abundant variety of carbon atoms. It is approximately $1,66 \times 10^{-27}$ kg (1.66×10^{-27} kg). The true atomic mass of carbon, when the masses of its isotopes are averaged, is 12,011 a.m.u. (12.011 a.m.u.).

atomic number (proton number) Symbol: Z. The number of protons in the nucleus of an atom. When the atom is neutral, i.e. not electrically charged, the atomic number equals the number of electrons in its shells.

atomic orbital *See* orbital.

atomic structure An atom consists of a nucleus surrounded by one or more negatively charged particles called electrons. The nucleus contains positively charged particles called protons; in all atoms except hydrogen (but including heavy hydrogen) it also contains neutral particles called neutrons. The electrons circling around the nucleus form a diffuse region containing a negative charge equal and opposite to the charge on the nucleus. This means that the atom remains neutral. *See also* Bohr atom.

atomic theory A hypothesis stating that all matter is made up of small particles called atoms.

atomic time *See* clock.

atomic weight *See* relative atomic mass.

ATP *See* adenosine triphosphate.

ATPase A phosphatase enzyme which catalyses the breakdown of adenosine triphosphate (ATP) by hydrolysis to adenosine diphosphate (ADP). This reaction liberates large amounts of energy.

atrium (pl. atria) (auricle) One of the anterior chambers of the heart receiving blood from the veins. *See also* heart.

atrophy The wasting away of a cell, tissue or organ due to under-nourishment or disease.

atropine An alkaloid which can be extracted from the plant *Atropa belladonna*, also called 'deadly nightshade'. It also occurs in a number of other plants belonging to the family Solanaceae, and it has been synthesised. Atropine acts on the central nervous system. Applied to the eye, it causes dilatation of the pupil and paralysis of accommodation. Atropine is used to aid in the examination of the retina and also in the treatment of inflammation in the cornea. As atropine inhibits the secretion of the bronchial glands, it is sometimes used in the treatment of asthma. However, it also reduces or eliminates the secretion of tears, sweat, saliva and digestive juices.

atto- Symbol: a. A prefix meaning 10^{-18}.

Au Chemical symbol for gold.

a.u. *See* astronomical unit.

audiofrequency Any frequency to which the human ear responds, normally ranging from 20 to 20 000 hertz (Hz).

audiometer An instrument used to test hearing.

auditory Pertaining to hearing.

auditory nerve The nerve conveying impulses of sound from the inner ear to the brain. It is the eighth cranial nerve. *See* Fig. 55.

auricle The outer visible part of the ear, often called the pinna. *See also* atrium.

aurum The Latin name for gold.

autocatalysis The catalysis of a chemical process by one of its own products; e.g. ethanedioate ions (oxalate ions), $C_2O_4^{2-}$, and manganate(VII) ions (permanganate ions), MnO_4^-, are, when they react with one another, autocatalysed by manganese(II) ions (manganous ions), Mn^{2+}:
$$5C_2O_4^{2-} + 2MnO_4^- + 16H^+ \rightarrow$$
$$2Mn^{2+} + 10CO_2 + 8H_2O$$

autoclave A laboratory version of the domestic pressure-cooker. It is a thick-walled container with a tightly fitting lid, in which substances may be heated under pressure. It is commonly used to sterilise items of apparatus at a temperature above 100°C, and is therefore equipped with a manometer and a thermometer.

autogamy Self-fertilisation. *Compare* allogamy.

autolysis The self-destruction of tissues by the action of their own enzymes.

autonomic nervous system (ANS) This is a system of motor (efferent) nerve fibres supplying all smooth muscles of the body, the cardiac muscle of the heart and the glands of the body. It is divided into the sympathetic and parasympathetic nervous systems.

autopsy *See* post-mortem.

autosome A chromosome which is not a sex chromosome.

autotomy An automatic or voluntary self-amputation of a part of an animal's body. *Example:* the lizard can shed its tail when seized by a predator.

autotransplantation The transplantation of an organ or tissue from one part of an organism to another part of the same organism.

autotroph(e) An organism which uses inorganic material as its main source of carbon. *Compare* heterotroph(e).

auxanometer An instrument used to measure plant growth.

auxin One of a group of hormones promoting plant growth. Very low concentrations of auxin are sufficient to show an appreciable effect. *See also* cytokinin, gibberellin *and* indoleacetic acid.

avalanche A process such as that which occurs in a Geiger–Müller counter, when a large number of ions are produced from a single ionising collision.

Aves A class of vertebrates, the birds.

aviary A large cage where birds are kept.

avitaminosis A disease or condition caused by vitamin deficiency.

Avogadro constant (Avogadro number) Symbol: L or N_A. The number of particles (e.g. atoms, molecules, ions, electrons) in one mole of any substance. It is approximately equal to $6,023 \times 10^{23}$ (6.023×10^{23}). *See also* mole.

Avogadro's hypothesis (Avogadro's law) A law stating that equal volumes of all gases at the same temperature and pressure contain the same number of molecules.

axerophthol *See* vitamin.

axial skeleton The part of the skeleton consisting of the skull, the backbone (vertebral column) and the ribs.
axil The angle between a leaf and the stem to which it is attached.
axillary 1. Growing in or from the axil. 2. Pertaining to the armpit.
axillary bud (lateral bud) A bud formed in the axil of a leaf. *See* Fig. 86.
axiom A proposition considered to be true or self-evident.
axis (pl. axes) 1. In humans and other amniotes, the second of the seven cervical vertebrae. The head turns at the atlas-axis joint. *See also* atlas *and* odontoid process. 2. The imaginary line about which a body is considered to rotate.
axle A rod upon or with which a wheel revolves; a rod or bar connecting a pair of wheels in a vehicle.
axon A long appendage of a nerve-cell which conducts nerve-impulses away from the cell body. *See* Fig. 150.
azeotropic mixture (constant boiling mixture) A mixture of two or more liquids in such a proportion that it boils without any change in composition, i.e. the composition of the liquid is the same as the composition of the vapour. An azeotropic mixture is also called a constant boiling mixture since it distils at a given constant temperature, at a given pressure. An example of an azeotropic mixture is ethanol (b.p. 78,5°C; 78.5°C) and water (b.p. 100,0°C; 100.0°C). This mixture has a boiling-point of 78,2°C (78.2°C) and a composition of 95,6% (95.6%) ethanol and 4,4% (4.4%) water. An azeotropic mixture may be broken if it is distilled with a liquid different from its constituents, or by a chemical reaction.
azo compound An organic compound containing the group —N=N—, called the azo group; e.g. C_6H_5—N=N—C_6H_4OH is 4-hydroxyphenylazobenzene or *p*-hydroxyphenylazobenzene. The azo group connects aromatic rings. Many azo compounds are dyes; in fact they make up in number about half the dyes produced. Azo dyes are made by treating a strongly acidic solution of an aromatic amine with nitrous acid, HNO_2:
$C_6H_5NH_2 + HNO_2 + HCl \rightarrow C_6H_5N_2Cl + 2H_2O$
The product $C_6H_5N_2Cl$ is called benzenediazonium chloride and is a diazonium salt. The reaction is a diazotisation. Diazonium salts react with phenols or aromatic amines to give azo compounds (azo dyes):
$C_6H_5N_2Cl + C_6H_5OH + NaOH \rightarrow$
C_6H_5—N=N—$C_6H_4OH + NaCl + H_2O$
This reaction is called a coupling. It is brought about in the cold.
Azotobacter A genus of bacteria which can fix atmospheric nitrogen, N_2, to form nitrogenous organic matter. *See also* nitrogen fixation.

B

B Chemical symbol for boron.
Ba Chemical symbol for barium.
bacillus (pl. bacilli) A rod-shaped, spore-producing bacterium.
backbone The vertebral column.
back cross The mating (cross) of a hybrid with one of its parents.
back e.m.f. An electromotive force opposing the flow of current in an inductive circuit. (1) In a simple cell, polarisation sets up a back e.m.f. In other cells polarisation is either reduced or eliminated. (2) In an electric motor, a back e.m.f. is induced in the armature coils by the rotation of the armature in the field magnets.
background The signal registered by a radiation counter (e.g. a Geiger–Müller tube) when it is not near a radioactive source. This is mainly due to natural radioactivity in the soil and to cosmic rays.
bacteria *See* bacterium.
bactericidal Causing death of bacteria.
bactericidin A substance that kills bacteria without the breaking down or dissolution of cells.
bacteriochlorophyll *See* bacterium.
bacteriology The study of bacteria.
bacteriolysis The destruction of bacteria by antibodies.
bacteriophage (phage) A bacterial virus capable of penetrating the cell, multiplying and killing the bacterium.
bacteriostatic Inhibiting the growth of bacteria without killing them.
bacteriotropin A substance, e.g. opsonin, which is found in blood serum and which makes bacteria more readily ingestible by phagocytes.
bacterium (pl. bacteria) A unicellular procaryotic micro-organism without chlorophyll. A few bacteria have bacteriochlorophyll which has a slightly different chemical composition to chlorophyll. Bacteria have formerly been classified as plants, but are distinct from both plants and animals. In a bacterium there is a plasma membrane surrounded by a rigid cell wall containing proteins, fats and polysaccharides. Around the cell wall there is, at some time during the life cycle of a bacterium, a slime layer which is secreted through the cell wall. The cell wall may have one or more flagella for locomotion. However, sometimes there are no flagella; the bacterium is then non-motile. It is notable that there is no nuclear membrane in a bacterium and therefore no well-defined nucleus. Nuclear

material consists of DNA which floats in the cytoplasm. Bacteria respire either aerobically or anaerobically. Usually they reproduce asexually by binary fission under favourable conditions. This reproduction may result in millions of new bacteria in a few hours. Under unfavourable growth conditions, bacteria may form endospores. Bacteria have no mitochondria. There are five recognisable shapes of bacteria: (1) cocci (spherical), (2) rods (straight rods), (3) vibrios (curved rods; comma-shaped), (4) spirals (coiled) and (5) filaments. *See also* Protista, spirochaetes *and* murein.

Bakelite® A type of plastic which may be prepared from phenol and methanal (formaldehyde) mixed with a filler such as wood shavings and a pigment. Other types of Bakelite are also known. The high electrical resistance of Bakelite makes it useful for insulation in such things as electric plugs, switches and tools. Bakelite was first patented in 1909 by Leo Hendrik Baekeland, after whom the material is named.

baking-powder Any mixture used in baking as a substitute for yeast. When heated or wetted it produces carbon dioxide, CO_2. The commonest form of baking-powder is a mixture of sodium hydrogencarbonate, $NaHCO_3$, and tartaric acid, $C_4H_6O_6$.

baking-soda Sodium hydrogen carbonate, $NaHCO_3$.

balance An instrument for measuring mass. The classic balance consists of a lever (beam) with a central pivot. At each end of the lever hangs a pan. The object whose mass is to be found is placed in one pan and standard masses are placed in the other until the beam is horizontal.

balanced diet A diet sufficient in both quantity and quality. It should contain a sufficient number of joules of energy and a correct proportion of proteins, carbohydrates and fats, as well as vitamins, mineral salts, water and roughage.

balance-wheel The part of a clock or watch which acts as a time control.

ball and socket joint A joint allowing movement in several planes, e.g. shoulder joints and hip joints.

ballistics The study of projectiles, especially when moving in air.

Balmer series A series of distinct lines in the hydrogen spectrum. The series lies approximately between the wavelengths 365 nm and 655 nm. Other series in the hydrogen spectrum are the Brackett series and the Paschen series, both lying in the infra-red, and the Lyman series, lying in the ultraviolet. *See also* line spectrum.

band spectrum An emission or absorption spectrum, characteristic of molecules, which is composed of a large number of closely spaced lines forming bands. Each distinct band results from the movement of electrons between energy levels. Single lines which make up a band result from the different molecular vibrational energies. When a sample producing a band spectrum is heated strongly so that it decomposes, the band spectrum tends to become a line spectrum.

bar Symbol: b. A unit of pressure equal to 10^5 pascals, Pa (newtons per square metre, $N\,m^{-2}$). The millibar, mb (1 bar = 1000 millibars) is commonly used by meteorologists. 1013,25 mb (1013.25 mb) = 1 atmosphere
= 760 mm of mercury

barb One of many hair-like projections on each side of the rachis of a feather. The barbs together make up the vane of a feather. *See also* feather.

barbiturate A derivative of barbituric acid, an organic acid which is a heterocyclic compound containing two nitrogen atoms and one hydroxyl group. Barbiturates are used as drugs, acting as sedatives in small doses. In larger doses they induce sleep, and in still larger doses they are able to produce deep anaesthesia.

barbule One of many hair-like filaments situated at each side of a barb. The barbules at one side bear hooks, those of the other bear slots into which the hooks fit. This makes the vane of the feather firm. *See also* feather.

Barfoed's solution A solution containing copper(II) ethanoate (copper(II) acetate), $(CH_3COO)_2Cu$, and ethanoic acid (acetic acid), CH_3COOH. It is used as a specific test for monosaccharides. A red precipitate of copper(I) oxide, Cu_2O, is observed when a mixture of the solution and a monosaccharide solution is heated. Barfoed's solution does not give any positive reaction with reducing disaccharides because these are less reducing than monosaccharides. *See also* Benedict's solution *and* Fehling's solution.

barium An element with the symbol Ba; atomic number 56; relative atomic mass 137,34 (137.34); state, solid. It is a very electropositive metal, belonging to the alkaline earth metals in the Periodic Table of the Elements.

barium chloride $BaCl_2 \cdot 2H_2O$ A solid, poisonous, soluble salt. In the laboratory an aqueous solution of $BaCl_2$ is used to test for sulphate, SO_4^{2-}. The sample to be tested is dissolved in water and a little dilute hydrochloric acid and some barium chloride solution are added to it. If SO_4^{2-} is present, a white

precipitate of barium sulphate, $BaSO_4$, is produced:
$$SO_4^{2-} + BaCl_2 \rightarrow BaSO_4 + 2Cl^-$$
The hydrochloric acid present prevents the formation of other white precipitates such as barium carbonate, $BaCO_3$, or barium phosphate, $Ba_3(PO_4)_2$, if carbonate ions, CO_3^{2-}, or phosphate ions, PO_4^{3-}, should also be present in the sample.

barium hydroxide $Ba(OH)_2 \cdot 8H_2O$ A colourless, poisonous solid which is considerably more soluble in water than calcium hydroxide, $Ca(OH)_2$. Barium hydroxide forms a strong alkali when dissolved in water; this gives a white precipitate of barium carbonate, $BaCO_3$, with carbon dioxide, CO_2:
$$Ba(OH)_2 + CO_2 \rightarrow BaCO_3 + H_2O$$

'barium meal' This is a liquid (suspension) containing barium sulphate, $BaSO_4$, which is given to patients who are about to have their alimentary canal X-rayed. Since X-rays cannot pass through barium sulphate molecules, the alimentary canal shows up clearly on the developed X-ray film. *See also* barium sulphate.

barium peroxide BaO_2 A greyish-white powder which can be prepared by heating barium oxide, BaO, carefully in oxygen:
$$2BaO + O_2 \rightarrow 2BaO_2$$
It is used in the laboratory preparation of hydrogen peroxide, H_2O_2, by treating it with dilute sulphuric acid, H_2SO_4:
$$BaO_2 + H_2SO_4 \rightarrow H_2O_2 + BaSO_4$$

barium sulphate $BaSO_4$ An insoluble white powder. It is a barium salt which is commonly used as a pigment and as the basis of a 'barium meal' in X-ray photography.

bark A tough layer of thick protective tissue made up of dead cells. It is formed by cork cambium and present on the outside of woody plant stems and roots, e.g. on tree trunks.

Barlow's wheel An apparatus for demonstrating the conversion of electrical energy into mechanical energy. It consists of a star-shaped copper disc free to rotate in a vertical plane between the poles of a horseshoe magnet. The tips of the disc are arranged so as to dip into a pool of mercury. The disc (wheel) rotates if a voltage is applied between the mercury and the axle of the disc.

barograph An instrument giving a continuous record of pressure of air.

barometer An instrument for measuring atmospheric pressure. The two main types of barometers are the mercury barometer and the aneroid barometer.

barometric height *See* mercury barometer.

basal Pertaining to the base.

basalt A dark, igneous rock.

base 1. A substance existing as molecules or ions which can receive (take up) hydrogen ions (protons), H^+. Less accurate definitions include the following: (1) contains oxide (O^{2-}) or hydroxide (OH^-) ions; (2) reacts with an acid to form a salt; (3) has a pH above 7; (4) turns red litmus blue. *See also* acid–base theories. **2.** *See* transistor.

base metals Metals such as lead, zinc, copper and iron which become oxidised when heated in air and which are therefore distinct from precious metals such as gold and silver.

base pair A base pair consists of a purine base and a pyrimidine base which pair in the genetic code of DNA and RNA. In both DNA and RNA cytosine and guanine pair; in DNA adenine pairs with thymine and in RNA it pairs with uracil. *See also* deoxyribonucleic acid.

basic Having the properties of a base.

basicity of an acid The number of hydrogen ions which can be formed from one molecule of the acid. *Examples:* hydrochloric acid, HCl, is monobasic (has a basicity of one); sulphuric acid, H_2SO_4, is dibasic; and phosphoric acid, H_3PO_4, is tribasic.

basic oxide *See* oxide.

basic salt *See* salt.

basophil *See* granulocyte.

bast The inner fibrous bark of certain trees.

bat A member of the order Chiroptera; a furry, flying mammal whose forelimbs are modified for flight.

Batesian mimicry *See* mimicry.

battery A group of two or more electric cells connected either in series or in parallel.

bauxite A natural deposit (a mineral) of hydrated aluminium oxide; a mixture of $Al_2O_3 \cdot H_2O$ and $Al_2O_3 \cdot 3H_2O$. It is the chief commercial source of aluminium.

BCG vaccine (Bacille Calmette–Guérin; Calmette vaccine) A vaccine named after Calmette and Guérin who developed it in France in 1909 and first used it in 1921. The vaccine is used against tuberculosis. It is usually given to children if they show no immunity to the disease. *See also* Mantoux test *and* patch test.

beak The horny projecting jaws of birds. *See also* mandible.

beat A regular fluctuation in sound intensity, heard when two tones of almost the same frequency are sounded simultaneously. The frequency of the beats is equal to the difference in frequency of the two tones.

Beaufort scale A numerical scale of wind speed ranging from 0 (calm) to 12 (hurricane).

Beckmann thermometer A very sensitive and accurate thermometer used in measuring temperature differences. It consists of a large bulb and a very thin, long capillary tube, in both of which mercury moves. Usually it is graduated in hundredths of a degree Celsius, and thousandths can be estimated by using a magnifying glass. The range of its temperature is very small, seldom more than 8°C. At the top of its capillary tube it has an extra, smaller bulb which may be U-shaped and which is used to vary the temperatures registered by the scale; i.e. the amount of mercury in the larger bulb can be adjusted by transferring more or less mercury to the U-shaped bulb.

bedbug A small, oval flattish insect which usually inhabits bedding. It pierces the skin of mammals and birds using its powerful needle-like mouth-parts, and sucks their blood for food.

bee *See* honey-bee.

Beer's law *See* transmittance.

beeswax *See* wax.

beetle An insect of the order Coleoptera.

beet-sugar A carbohydrate obtained from sugar-beet. It is a disaccharide with the chemical name saccharose or sucrose and the chemical formula $C_{12}H_{22}O_{11}$. *See also* cane-sugar.

bel *See* decibel.

Benedict's solution A blue solution containing copper(II) sulphate, sodium carbonate and sodium citrate, used in a test for reducing sugars. When an aqueous solution of a reducing sugar and Benedict's solution are heated to around boiling-point, the colour of the mixture changes from blue to green, yellow or orange; and finally a brick-red precipitate of copper(I) oxide, Cu_2O, is formed. *See also* Fehling's solution *and* Barfoed's solution.

benign Describes a disorder that is not likely to get worse. A tumour is said to be benign (non-cancerous) if it is considered harmless and unlikely to recur after removal. *Compare* malignant.

benthic Describes organisms (benthon) living on or near the bottom of an ocean or lake. *Compare* pelagic.

benthon The plant and animal life found at the bottom of an ocean or lake.

benthos The bottom of an ocean or lake.

benzaldehyde *See* benzenecarbaldehyde.

benzene C_6H_6 A colourless, organic liquid found in coal-tar. It is a member of a group of aromatic hydrocarbons which play an important role in organic chemistry. In a molecule of benzene the six hydrogen atoms are attached to the six carbon atoms, the latter being at the corners of a hexagon so that all twelve atoms lie in one plane. This arrangement is called a benzene ring. Fig. 9(a) and (b) show two different symbols for the benzene molecule. Since the three double bonds in benzene are delocalised the symbol in Fig. 9(b) is often used. However, the carbon and hydrogen atoms are not usually shown in either formula. Benzene is used as a solvent for many organic substances and in the manufacture of several organic compounds. *See also* resonance.

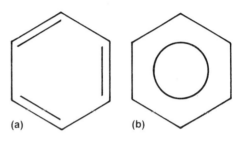

Fig. 9 Benzene

benzenecarbaldehyde C_6H_5CHO (benzaldehyde) An aromatic aldehyde which is a yellowish, oily liquid with a pleasant smell of almonds. Kernels of bitter almonds contain benzenecarbaldehyde. It may be prepared by oxidising methylbenzene (toluene), $C_6H_5CH_3$:
$C_6H_5CH_3 + O_2 \rightarrow C_6H_5CHO + H_2O$
Further oxidation of benzenecarbaldehyde produces benzoic acid, C_6H_5COOH. Benzenecarbaldehyde is used in flavours, perfumery and in the manufacture of dyes.

benzene-1,2-dicarboxylate (phthalate) A salt of benzene-1,2-dicarboxylic acid (phthalic acid), $C_6H_4(COOH)_2$.

benzene-1,2-dicarboxylic acid $C_6H_4(COOH)_2$ (phthalic acid) A white, soluble, solid, aromatic acid which may be prepared by oxidation of 1,2-dimethylbenzene (xylene), $C_6H_4(CH_3)_2$:
$C_6H_4(CH_3)_2 + 3O_2 \rightarrow C_6H_4(COOH)_2 + 2H_2O$
There are three isomers which have the formula $C_6H_4(COOH)_2$. They are distinguished from one another according to where the carboxyl groups are placed in the benzene molecule. Benzene-1,3-dicarboxylic acid is also called isophthalic acid and benzene-1,4-dicarboxylic acid is also called terephthalic acid. Salts of benzene-1,2-dicarboxylic acid are called benzene-1,2-dicarboxylates (phthalates). Benzene-1,2-dicarboxylic acid can be dehydrated to form benzene-1,2-dicarboxylic anhydride (phthalic anhydride), $C_6H_4(CO)_2O$, which has the structure

This is an important substance in the plastic industry and in the manufacture of dyes. With phenol, benzene-1,2-dicarboxylic anhydride produces phenolphthalein.

benzene-1,2-dicarboxylic anhydride (phthalic anhydride) *See* benzene-1,2-dicarboxylic acid.

benzene hexachloride *See* lindane.

benzene ring *See* benzene.

benzenesulphonic acid *See* sulphonation.

benzene-1,2,3-triol $C_6H_3(OH)_3$ (pyrogallol, pyrogallic acid, 1,2,3-trihydroxybenzene) A white, crystalline, aromatic solid (a phenol) which is soluble in water. It is a strong reducing agent, used in photography and, in alkaline solution, as an absorber of oxygen.

benzyl alcohol *See* phenylmethanol.

bequerel Symbol: Bq. The SI unit for the activity of a radioactive source. One bequerel is defined as one radioactive distintegration per second. *See also* curie.

Bergius' process A liquefaction process in which coal is turned into oil. This is done by making a paste of coal and oil which is then heated to about 450°C with hydrogen and a suitable catalyst under a pressure of about 250 atmospheres. The carbon present in the coal now reacts with the hydrogen to give a mixture of different hydrocarbons, mainly present in the liquid state.

beriberi A deficiency disease caused by lack of vitamin B_1. It produces degeneration to nerve-endings and paralysis. In rice grains vitamin B_1 occurs mainly in the husk, so people who eat polished rice (white rice) as a major part of their diet are likely to develop beriberi. The disease is treated by giving the patient a balanced diet or extra vitamin B_1.

Bernouilli effect An effect seen in a pipe with a constriction in it; the pressure exerted by a fluid flowing through the constriction is lower than the pressure it exerts in the wider parts of the pipe where the fluid moves slowly, i.e. the faster the flow, the lower the pressure. This effect also produces the pressure that lifts an aircraft. *See* Fig. 10.

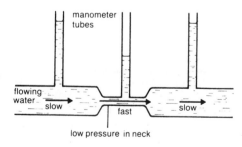

Fig. 10 Bernouilli effect

berry A fleshy fruit containing one or more seeds. The pericarp consists of the inner, soft membranous portion (the endocarp), the middle fleshy region (the mesocarp) and the outer skin (the epicarp). Examples include the banana and the tomato. *Compare* drupe.

Bessemer process A process by which pig-iron from the blast furnace is converted into steel. Molten pig-iron is poured into an egg-shaped vessel called a converter and a strong blast of air is blown through the molten pig-iron, Impurities such as sulphur and carbon are oxidised to gaseous products. Other impurities such as manganese and silicon react with the lining of the converter and form a slag. The lining is made up of oxides of calcium and magnesium. When the iron is considered pure, measured amounts of carbon, manganese, etc. are added to produce the required type of steel. *See also* Kaldo process, L–D process and Open-Hearth process.

beta (β) The second letter of the Greek alphabet.

beta decay A radioactive disintegration in which a neutron, $_0^1n$, of an unstable nucleus of an element changes into a proton, $_1^1H$, with the emission of an electron, $_{-1}^0e$ (the beta particle). The atomic number increases by one, while the mass number remains unchanged.

beta particle An electron, $_{-1}^0e$, emitted from certain radioactive elements. *See also* beta decay.

beta ray A stream of high-energy electrons. The more energetic electrons can penetrate a layer of several millimetres of aluminium. In gases through which they pass they produce ionisation.

bhang *See* cannabis.

BHC *See* lindane.

bi- Prefix meaning two, formerly used in chemical nomenclature.

bicarbonate An old name for salts containing the hydrogencarbonate group, HCO_3^-.

bicarbonate indicator A solution containing sodium hydrogencarbonate (sodium bicarbonate), thymol blue, cresol red, ethanol and water. It is a pH indicator commonly used to detect the presence or absence of carbon dioxide, depending for its action on the acidity of a solution of carbon dioxide. Note: It is not specific for carbon dioxide.

bicarbonate of soda Sodium hydrogencarbonate, $NaHCO_3$. It is used as an antacid.

biceps Any muscle with two heads or origins. There is a biceps muscle in the upper arm in front of the humerus and another in the upper leg behind the femur.

biconcave *See* lens.

biconvex *See* lens.

bicuspid A tooth with two points, e.g. a premolar.

bicuspid valve (mitral valve) The heart valve separating the left atrium from the left ventricle. The bicuspid valve has two flaps.

biennial plant A plant which completes its life cycle every two years. The food is kept in storage organs at the end of the first year and flowers are produced during the second year. An example is the carrot.

bifocal Having two foci. Describes spectacles in which each lens is made up of two parts, the upper one for distant vision the lower one for near vision.

'big bang' theory Most galaxies are moving rapidly away from each other (and from the Earth) as if a gigantic explosion has taken place. At one time the galaxies must have been much closer to each other and may even have been one big mass. This forms the basis of the 'big bang' theory, which estimates that an explosion initiated about 15 thousand million years ago created the Universe. The movement of the galaxies is therefore considered to be the repercussion of the 'big bang'. The cause of the 'big bang' is not known.

bilaterally symmetrical Capable of being cut along the midline, in one plane only, to produce two halves which are (almost) mirror images of one another. Most freely moving animals are bilaterally symmetrical. The term also applies to plants. *See also* radially symmetrical.

bile (gall) A greenish liquid manufactured by the liver and stored in the gall-bladder before passing into the duodenum. It consists of bile pigments and bile salts. The bile breaks up fats and oils into an emulsion upon which enzymes can act and it also helps to neutralise acids coming from the stomach.

bile duct A duct through which bile from the gall-bladder is transported to the duodenum, a part of the small intestine. *See* Fig. 49.

bile pigment A waste product formed from the breakdown of haemoglobin by the liver cells. The main bile pigments are bilirubin (red) and biliverdin (green), the latter being an oxidation product of bilirubin. They are found in bile and, in small amounts, in blood and urine. After they have been in contact with bacteria in the alimentary canal they give colour to the faeces.

bile salt A sodium salt formed from an organic acid. Bile salts are important in digestion since they emulsify fats and oils. *See also* bile.

bilharzia Parasitic flatworms causing bilharziasis.

bilharziasis (schistosomiasis) A disease caused by parasitic flatworms called bilharzia. It is spread through water contaminated with human sewage. If the disease is not treated a chronic state develops: fibrous tissue and scarring form in the liver, bladder and intestine, eventually causing death.

bilirubin *See* bile pigment.

biliverdin *See* bile pigment.

bimetallic strip A strip of metal consisting of two different metals or alloys, or one of each, e.g. iron and brass. The metals are riveted or welded together. When heated, the bimetallic strip bends since the metals it is made up of expand by different amounts. Bimetallic strips are commonly used as a temperature control. *See* thermostat.

binary compound A chemical compound formed from two elements: e.g. ammonia, NH_3, or sodium chloride, NaCl.

binary fission A type of asexual reproduction in which a cell divides into two approximately equal parts.

binary system Two stars travelling around each other under their mutual gravitational attraction. In some binary systems both stars are visible to the naked eye, one usually brighter than the other; in others they are so close that only analysis of their combined spectrum reveals that two stars are present. Binary systems are much commoner than is generally realised. Mizar, the middle star in the handle of the Plough, is a binary system; this is easily visible with a low-powered telescope.

binding energy *See* mass defect.

binocular Involving the use of two eyes.

binoculars *See* prismatic binoculars.

binomial system of nomenclature A widely used and internationally accepted system of nomenclature for the scientific naming of animals and plants. The system was first suggested by Linnaeus, who gave each species a double Latin name. The first word is the generic name indicating the genus, the second word indicates the specific name (trivial name). The generic name begins with a capital letter whereas the specific name begins with a small letter. Both names are conventionally written in italics. *Examples:* the tiger is *Panthera tigris* and the lion, another species of the same genus, is *Panthera leo*. Sometimes the name of the person who first named and described the animal or plant is indicated: e.g. the wolf, *Canis lupus*, may be called *Canis lupus* Linnaeus or just *Canis lupus* Linn. A trinomial system may be used in cases where sub-species have been defined.

bio- Prefix meaning life.

biochemistry The chemistry of living things.

biodegradable Capable of being broken down by micro-organisms.

biogas A combustible gas, e.g. methane, CH_4, produced by the controlled decomposition of organic matter.

biological oxygen demand (BOD) A measure of organic pollution in water. It is the amount of oxygen used up in a sample containing a known amount of dissolved oxygen, which is kept at 20°C, usually for five days. It is expressed in milligrams of oxygen per litre of water. The oxygen is consumed by micro-organisms feeding on organic matter in the sample. This can be written:
micro-organisms + organic material + $O_2 \rightarrow$ CO_2 + H_2O + demolished organic material + more micro-organisms
In the laboratory, the oxygen consumption is measured using a mercury manometer connected to a bottle that contains the sample. The sample is stirred and the carbon dioxide, CO_2, produced is absorbed in liquid potassium hydroxide, KOH, thus resulting in a pressure drop. This can be read on a scale calibrated in milligrams of oxygen per litre of water. A low BOD signifies low pollution.

biology The science of life.

bioluminescence The emission of light radiation by living organisms such as the firefly. Light is produced as a result of chemical reactions in the organism. *See also* luciferin.

biomass The total mass of organisms in a given area.

biome A climatic region which supports a particular type of ecological community. Biomes may be terrestrial or aquatic. Examples of terrestial biomes are tropical rain forest, desert and grassland. Aquatic biomes are either freshwater or marine. Terrestrial biomes are identified by the predominant floral types present, whereas aquatic biomes are identified by physical criteria such as distance from shore, depth and salinity.

biophysics The application of physics to biology.

biopsy Examination of a small piece of living tissue cut from the body in order to diagnose a disease.

biosphere The part of the Earth and its atmosphere inhabited by living things.

biosynthesis The formation of organic compounds by living things.

biosystem *See* ecosystem.

biotic Living.

biotic factors Environmental effects of living organisms on an ecosystem. *Example:* grazing animals determine the length of the grass in a certain area of land.

biotin A vitamin of the B-complex group. It occurs chiefly in egg-yolk, liver, kidney and yeast. A deficiency of biotin causes skin inflammation in many animals; however, no deficiency has been recorded in a human being.

bird A warm-blooded vertebrate belonging to the class Aves. Characteristics include a feather-covered body, a horny beak containing no teeth, and fore-limbs modified into wings or wing-like structures. Eggs are laid in a nest or a hollow in the ground and incubated by one or both parents until hatching occurs.

birefringence *See* double refraction.

birth canal The passage formed by the cervix and vagina through which a foetus passes during the process of birth.

birth rate *See* crude birth rate.

bisexual Having both male and female reproductive organs. *See* hermaphrodite.

bitumen Any one of numerous mixtures of hydrocarbons usually obtained from petroleum distillation. *See also* asphalt.

bituminous Containing bitumen or tar.

bituminous coal *See* coal.

biuret An insoluble, crystalline substance which is formed together with ammonia when urea is heated.

biuret reaction A chemical reaction used as a test for proteins and urea. A purple colour is seen when a solution of dilute sodium hydroxide, NaOH, and one or two drops of a dilute solution of copper(II) sulphate, $CuSO_4$, are added to a sample containing a protein or urea.

bivalent 1. A pair of homologous chromosomes lying side by side at meiosis. **2.** *See* divalent.

bivalve An animal with a shell consisting of two parts (valves) hinged together, e.g. the oyster. *Compare* univalve.

black body A body which is a perfect absorber of radiations of all frequencies falling on it. A practical approximation of a black body is a cavity consisting of a hollow sphere with a roughened and blackened inside wall and a small hole in the surface through which radiation enters and leaves. A black body is therefore also a perfect emitter of radiation where the emission only depends on its temperature. If the cavity is at 2000 K the black body emits visible radiation whose colour is characteristic of that temperature. Below about 800 K the black body emits mainly infrared radiation. The energy, M, radiated over all wavelengths (frequencies) from a black body per unit area per second is proportional to the fourth power of the thermodynamic temperature, T:
$M = \sigma T^4$
This is called Stefan–Boltzmann's law. σ (sigma) is the Stefan–Boltzmann constant, with value $5.67 \times 10^{-8} \text{W m}^{-2} \text{K}^{-4}$ ($5.67 \times 10^{-8} \text{W m}^{-2} \text{K}^{-4}$).

Black Death *See* bubonic plague.

black hole In space, an enormous field of force which emits no radiation at all because it has a fantastically strong gravitational field which absorbs all matter coming near it. Neutron stars may end their 'lives' as black holes. It is believed that the centres of many galaxies are in fact black holes.

black lead *See* graphite.

blackwater fever An acute tropical disease related to a form of malaria unsuccessfully treated with quinine. Symptoms include high fever, severe headache, vomiting, jaundice and the passage of dark brown urine containing blood pigments from disintegrated red blood cells. In severe cases, the kidneys fail to function. The disease may be treated with steroid drugs and blood transfusion. Any accompanying malaria is treated with an antimalarial drug other than quinine. A patient with severe kidney damage may be put on a kidney machine.

bladder (urinary) A hollow, pear-shaped organ which has muscular walls and lies in the pelvis. The urine formed in the kidneys passes down two long tubes, the ureters, to the bladder where it is stored until it is expelled from the body. From the bladder the urine passes through the urethra to the outside. In humans, the average volume of the bladder is 500 cm³ (ml).

blade 1. (lamina) The flat part of a leaf. 2. A flat, triangular bone, e.g. the shoulderblade.

blast furnace A tower approximately 30 m high and 6 m wide, used in the extraction of iron from its ores. The furnace is charged from the top with a mixture of iron ore, e.g. iron(III) oxide, Fe_2O_3, coke, C, and limestone, $CaCO_3$. Pre-heated air is blown into the base of the furnace through tubes called tuyères, causing the coke to burn. Several chemical reactions take place, resulting in the reduction of the ore. The limestone decomposes to form calcium oxide, CaO, which combines with impurities in the ore such as silica, SiO_2, forming a molten slag of calcium silicate, $CaSiO_3$, which floats on the surface of the molten iron at the base of the furnace. The chemical reactions are:

$C + O_2 \rightarrow CO_2$ (at the base)
$C + CO_2 \rightarrow 2CO$
$Fe_2O_3 + 3CO \rightleftharpoons 2Fe + 3CO_2$
$CaCO_3 \rightarrow CaO + CO_2$
$CaO + SiO_2 \rightarrow CaSiO_3$

See Fig. 11.

blastocoel (blastocoele) *See* blastula.

blastomere One of the cells formed by cleavage of an animal egg. Blastomeres are of two types: small micromeres and larger megameres.

blastula A hollow ball of cells surrounding a fluid-filled cavity (the blastocoel). The blastula is formed by cleavage of the zygote.

bleaching A process whereby colour is removed from coloured substances by a chemical which is either an oxidising agent, e.g. chlorine, Cl_2, or a reducing agent, e.g. sulphur dioxide, SO_2.

bleaching powder (chloride of lime) A white powder consisting of a mixture of hydrated calcium chloride, $CaCl_2 \cdot 6H_2O$, calcium hydroxide, $Ca(OH)_2$, and calcium chlorate(I) (calcium hypochlorite), $Ca(OCl)_2$. With dilute acid, bleaching powder liberates chlorine, Cl_2, which is the bleaching agent. Bleaching powder may also be used as a disinfectant.

blende A naturally occurring sulphide of a metal, e.g. zinc blende, ZnS.

blight A plant disease caused by fungi or insects. The term can also refer to the fungi and insects themselves.

blind spot (optic disc) The place at the back of the inner eye where the optic nerve pierces the retina. At this place there are no nerve-endings (rods and cones), so the blind spot is insensitive to light. *See* Fig. 66.

block and tackle An important pulley system consisting of a fixed block and a movable tackle. *See* Fig. 12.

blood The circulating fluid in animals, carrying oxygen, waste products, nourishment, hormones, etc.

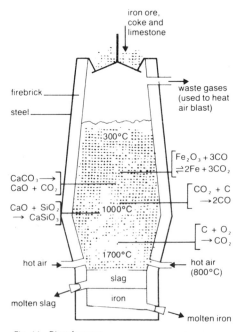

Fig. 11 Blast furnace

blood pressure

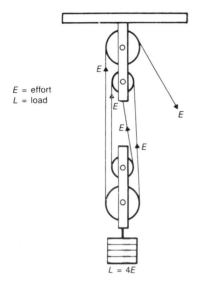

E = effort
L = load

$L = 4E$

Fig. 12 Block and tackle

blood bank A place where blood, taken from a blood donor, is preserved and stored until needed.

blood cell *See* blood corpuscle.

blood clotting The process in which blood is changed from the liquid to the solid state. This complex process takes place when a blood vessel is injured. It is caused by soluble fibrinogen in the blood plasma changing into insoluble fibrin, brought about by an enzyme called thrombin. However, many other substances, such as calcium ions and blood platelets, play a role in blood clotting. Sometimes clotting takes place under abnormal conditions in uninjured vessels. This is very dangerous, as the clot formed may cause disturbances in the proper circulation of blood. *See* thrombosis *and* thrombokinase.

blood corpuscle (blood cell) One of the cells making up the blood. There are red blood cells, called erythrocytes, and white blood cells, called leucocytes. The red cells give colour to the blood. They carry oxygen from the lungs to the body tissue and carbon dioxide from the tissue to the lungs. There are about 5 000 000 red cells per cubic millimetre (mm^3) of blood. The red cells are biconcave discs without nuclei; they are formed in the red bone marrow, especially at the upper end of the femur. There are between 7000 and 10 000 white cells per mm^3 of blood, the number varying according to the physiological state of the body. The main function of the white cells is to ingest and destroy bacteria and dead cells. White cells are irregular in shape and have nuclei. There are several different types of white blood cells: some of them are formed in the lymph system, others in the red bone marrow. *See also* haemoglobin *and* granulocyte.

blood donor *See* blood transfusion.

blood group There are four main blood groups, referred to as A, B, AB and O. Blood from some groups is incompatible with blood from other groups; if they are mixed, the red blood cells stick together in clumps (agglutination). This happens because antigens (called agglutinogens) in the red blood cells react with antibodies (called agglutinins) in the plasma. It is therefore important in transfusions to give blood which is compatible with the patient's blood group. *See also* Rhesus factor.

blood plasma A watery liquid which makes up about 55% of the blood's total volume. It consists of about 90% water and 10% dissolved substances, including proteins, amino acids, fats, cholesterol, glucose, antibodies, urea, hormones, mineral salts and traces of many other substances. Blood corpuscles (cells) and blood platelets are suspended in the blood plasma. Carbon dioxide, CO_2, produced during respiration, diffuses into the plasma and passes into the red blood cells. By means of an enzyme, carbonic anhydrase, carbon dioxide combines with water in the red cells to form carbonic acid, H_2CO_3. This dissociates into hydrogencarbonate (bicarbonate) ions, HCO_3^-, and hydrogen ions, H^+, most of which are found in the plasma. In this way some of the carbon dioxide is carried in the red blood cells and some in the plasma. Blood plasma may be obtained when blood is centrifuged or allowed to stand in a container. The red blood cells settle to the bottom leaving the plasma on top of them.

blood platelet (thrombocyte) A small coin-shaped particle, found in blood, which is active in blood clotting by liberating the enzyme thrombokinase. There are about 250 000 platelets per cubic millimetre (mm^3) of blood. Blood platelets are formed in the red bone marrow.

blood-poisoning (septicaemia) A condition in which micro-organisms multiply in the blood in sufficient number to cause illness.

blood pressure The pressure of the blood in the main arteries. It depends on the strength of the pumping action of the heart and the elasticity of the blood vessels. Usually blood pressure is measured in millimetres of mercury using a sphygmomanometer. There are two types of blood pressure that are measured: the systolic and the diastolic pressures, the former

blood serum

resulting from the contraction of the heart (this is the higher of the two) and the latter resulting from the relaxation of the heart. *See also* hypertension *and* hypotension.

blood serum The clear, yellowish liquid which remains when fibrinogen is removed from blood plasma. Blood serum contains no clotting factors.

blood smear A microscope slide with a thin layer of blood on it, used in the microscopic examination of blood. It is made by smearing a drop of blood along a slide with the edge of a cover slip or another slide.

blood sugar Glucose present in the blood. *See also* blood plasma.

blood test A laboratory test carried out to find the composition of blood, e.g. to check for the presence of drugs, alcohol or disease.

blood transfusion The transfer of blood from one person, called the donor, into the circulatory system of another, called the recipient or patient. Blood taken from a donor is kept in a blood bank until needed. To prevent clotting an anti-coagulant such as sodium citrate is added. The citrate ions form a complex compound with calcium ions present in the blood, and without calcium ions the blood does not clot. Transfusions of blood save many lives by restoring blood lost in operations, accidents, etc. *See also* blood group *and* Rhesus factor.

blood vessel A tube through which blood passes. There are three types of vessels: arteries, veins and capillaries.

blood volume In humans the blood volume is approximately 4–5 litres.

blotting-paper A special type of absorbent paper containing many fine pores which act as tiny capillary tubes.

blubber Whale or seal fat.

blue-green algae *See* Cyanophyta.

bob Any weight at the end of a pendulum.

BOD *See* biological oxygen demand.

body temperature The normal body temperature of a human being remains almost constant in the range of 36–37°C. Usually this temperature is measured by placing a clinical thermometer in the mouth or rectum. The rectal temperature is about 0,5°C (0.5°C) higher than the mouth temperature. In a healthy person, heat production is rapidly balanced by heat loss. Almost all the heat produced by the body comes from the oxidation of food. *See also* metabolism.

Bohr atom A theoretical model for the atomic structure of a hydrogen atom. The theory states that the electron travels in orbits around the nucleus (proton). The electrostatic force of attraction between the electron and the proton provides the centripetal force that retains the electron in its orbits. If the atom gains energy, the electron moves to an orbit further out from the nucleus. If the atom gains sufficient energy, the electron is removed altogether and a hydrogen ion (proton), H^+, is formed. The electron can only occupy certain specified orbits around the nucleus. If the electron moves to orbits nearer the nucleus, energy is given out; i.e. when the electron changes energy levels towards the nucleus, energy is radiated. The lowest energy state of a hydrogen atom (or any other atom) is called the ground state. Atoms with more energy than they possess in the ground state are said to be in an excited state. The terms ground state and excited state also apply to ions and molecules. This theory for the atomic structure of a hydrogen atom also applies to other atoms.

boil A pus-filled area on the skin caused by bacterial infection.

boiling The state of a liquid at which its saturated vapour pressure is equal to the external pressure.

boiling-point (b.p.) The temperature of a liquid at which its saturated vapour pressure equals the external atmospheric pressure. Boiling-points are always quoted for the standard atmospheric pressure, 101 325 Pa. At reduced external pressure the boiling-point of a liquid decreases. Conversely, an increased external pressure increases the boiling-point. *See also* elevation of boiling-point.

bole The trunk of a tree.

boll A spherical seed capsule, as in the cotton plant.

bolometer An electrical instrument used for measuring heat radiation. Its action depends upon the change of resistance of platinum strips with temperature.

bolus A rounded lump of chewed food.

bomb calorimeter A calorimeter in which a thick-walled steel cylinder is placed in the inner vessel and immersed in a certain known mass of water. A known mass of the sample, e.g. a fuel, is placed inside the steel container which is then filled with compressed oxygen. By igniting the sample electrically using a thin iron wire which is in contact with the sample, the temperature rise of the water can be measured on a thermometer (often a Beckmann thermometer) and the calorific value of the sample calculated. The heat due to the combustion of the iron wire is subtracted from the final result.

bond A linkage by which one atom is attached to another in a molecule. *See also* covalent bond *and* ionic bond.

bond angle The angle between linked atoms in a molecule. *Example:* the atoms in a water molecule, H_2O, are not in a straight line, and in an ammonia molecule, NH_3, the atoms are in a pyramidal arrangement. The bond angle is taken as the angle between imaginary lines through the nuclei of the atoms. See Fig. 13.

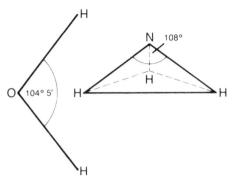

Fig. 13 Bond angle

bond energy The minimum amount of energy required to separate two atoms in a molecule, i.e. to break the bond between the atoms.

bond length The distance from the centre of one nucleus of an atom to the centre of the nucleus of another, joined by a chemical bond.

bone Specialised connective tissue making up the skeleton of vertebrates. It consists of cells distributed in a matrix of collagen fibres, which give the bone tensile strength, and calcium salts, especially phosphate and carbonate, which make the bone hard. Usually bones are first formed in cartilage which is then later replaced by bone. See Fig. 14; see also cancellous bone, compact bone and ossification.

bone ash The ash obtained from heating bones in air. The main constituent of bone ash is calcium phosphate.

bone black See animal charcoal.

bone char See animal charcoal.

bone marrow Soft, gelatinous connective tissue found in the marrow cavities and cancellous tissue in long bones. It consists of a network of blood vessels, fibres, blood-producing cells, fats and red and white blood corpuscles in various stages of development. However, the composition of the bone marrow differs in different bones. It is divided into red and yellow marrow. The red marrow is actively engaged in the production of blood corpuscles, the yellow marrow contains fat cells and has the potential of producing blood corpuscles.

bone-meal Crushed or powdered bone, used especially as a fertiliser.

bony labyrinth (of mammalian ear) The bony labyrinth consists of the vestibule, the semicircular canals and the cochlea.

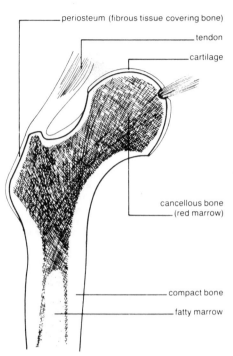

Fig. 14 Bone

boo See cannabis.

borate A salt of boric acid, H_3BO_3.

borate test A few millilitres of concentrated sulphuric acid, H_2SO_4, is added to a small amount of the sample containing a borate, e.g. the tetraborate ion, $B_4O_7^{2-}$. Some ethanol, C_2H_5OH, is put into an evaporating dish and the above mixture is added whilst stirring. When the dish is warmed over a Bunsen burner, the vapour from the dish can be lit. A green colour in the flame indicates the presence of a borate in the sample. The colour is caused by the volatile compound triethyl borate, $B(OC_2H_5)_3$:
$Na_2B_4O_7 + H_2SO_4 + 5H_2O \rightarrow$
$\qquad\qquad\qquad\qquad 4H_3BO_3 + Na_2SO_4$
and
$H_3BO_3 + 3C_2H_5OH \rightarrow B(OC_2H_5)_3 + 3H_2O$

borax $Na_2B_4O_7 \cdot 10H_2O$ (disodium tetraborate-10-water; sodium tetraborate) A sodium salt which occurs naturally, precipitated by the evaporation of water from certain saline lakes. It is mainly used in the manufacture of glass and as a mild antiseptic. It is a source of boron.

borazon See boron nitride.

Bordeaux mixture A liquid mixture containing copper(II) sulphate, $CuSO_4$, calcium oxide, CaO, and water. It is used as a fungicide and insecticide for plant diseases.

bordered pit *See* pit.

boric acid H_3BO_3 (orthoboric acid) A white, solid acid. It is a very weak acid used in aqueous solution as a mild antiseptic. Salts of boric acid are called borates.

boron An element with the symbol B; atomic number 5; relative atomic mass 10,81(10.81); state, solid. It is a non-metal, but still has some metallic properties. It is extracted from borax. One of its main uses is as a neutron absorber for controlling atomic reactions in nuclear reactors.

boron nitride BN A slippery, white, solid semiconductor with a layer structure almost similar to that of graphite. However, the hexagonal rings contain alternate boron and nitrogen atoms and the substance has no metallic lustre. Boron nitride may be prepared by the reaction of boron with ammonia, NH_3, at white heat:
$$2B + 2NH_3 \rightarrow 2BN + 3H_2$$
Under high temperature and pressure, boron nitride can be converted into a cubic form with a diamond-like structure. In this form it is called borazon and is extremely hard (it will scratch diamond) and very resistant to oxidation.

Bosch process A process for the industrial manufacture of hydrogen, H_2. Steam and water gas (a mixture of carbon monoxide, CO, and hydrogen) are mixed and passed over a heated catalyst. The carbon monoxide reacts chemically with the steam giving hydrogen and carbon dioxide, CO_2:
$$H_2O + CO + H_2 \rightarrow 2H_2 + CO_2$$
The carbon dioxide is finally removed by dissolving it in water under pressure.

botany The study of plants.

botulism A form of food poisoning which is often fatal. It is caused by the anaerobic bacterium *Clostridium botulinum*, which produces a powerful neurotoxin.

Bourdon gauge An instrument used to measure pressure above atmospheric pressure. It consists of a flattened metal tube closed at one end and bent into a circular arch. When air is blown into the tube, the pressure in it increases. The tube tends to straighten and the movement of the closed end of the tube is transmitted, through a lever and a gear, to a pointer which moves over a scale indicating the pressure. *See* Fig. 15.

bowel A common name for the intestines, especially the large intestine.

Bowman's capsule A cup-shaped structure surrounding a glomerulus in a nephron of a vetebrate kidney. It collects the filtered fluid that passes through the capillary walls of the glomerulus. *See* Fig. 149.

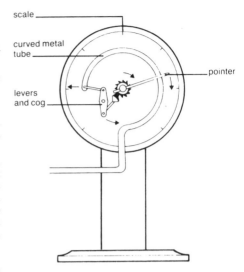

Fig. 15 Bourdon gauge

Boyle's law A law stating that the volume of a given mass of gas is inversely proportional to its pressure, provided the temperature remains constant. This can be expressed as
$$V \propto 1/p \quad \text{or} \quad pV = \text{constant}$$
where p is the pressure and V is the volume.

b.p. *See* boiling-point.

Br Chemical symbol for bromine.

brachial Pertaining to or like an arm.

Brackett series *See* Balmer series.

bract A leaf-like structure, from the axil of which arises a flower or inflorescence.

Braille A system of reading and writing used by blind people. The letters and words consist of raised dots which are read by passing the fingertips over them.

brain (encephalon) The anterior part of the central nervous system, contained within the cranial cavity. It is the upper extension of the spinal cord and is divided into three areas: hindbrain, midbrain and forebrain. *See* Fig. 16; *see also* grey matter.

brain-stem The expansion of the upper part of the spinal cord into a thicker, more complex stalk. It consists of the medulla oblongata, the pons and the midbrain.

Bramah's press *See* hydraulic press.

Branchiostoma *See* Amphioxus.

brass An alloy in which the principal constituents are copper and zinc.

brazing *See* solder.

brazing solder *See* solder.

breastbone *See* sternum.

breathing (external respiration) In many vertebrates, a series of movements which leads

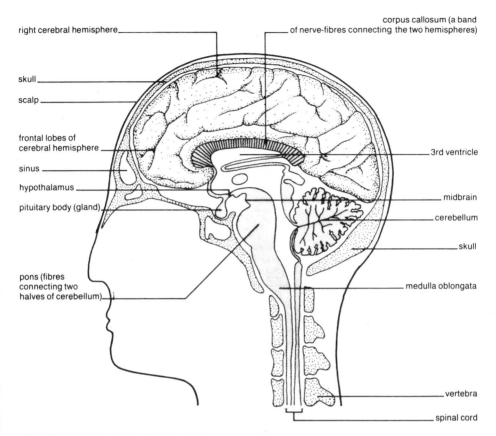

Fig. 16 Section through head to show brain

to the intake of air to the lungs (inspiration) and the removal of stale air from them (expiration).
Bright's disease *See* nephritis.
brine A strong solution of sodium chloride, NaCl, in water.
brittle Describes material which is easily broken.
bromination *See* halogenation.
bromine An element with the symbol Br; atomic number 35; relative atomic mass 79,91 (79.91); state, liquid. It is a rather electronegative non-metal belonging to the halogens. It is extracted from bromide salts by passing chlorine through an aqueous solution of a bromide:
$2Br^- + Cl_2 \rightarrow Br_2 + 2Cl^-$
An important use of bromine is in making silver bromide, AgBr, for photographic emulsions.
bronchiole A branch of a bronchus terminating in alveoli. *See* Fig. 118.
bronchitis An inflammation of the bronchi, which may also involve the bronchioles.
bronchus (pl. bronchi) One of the main branches of the trachea (windpipe) leading to the lungs. *See* Fig. 118.
bronze An alloy in which the principal constituents are copper and tin.
Brownian movement The continuous, random movement of tiny solid particles in liquids or gases. This is caused by the impact of moving liquid or gas molecules pushing at the solid particles from all sides. *See also* smoke cell.
brown ring test A chemical test for nitrates, NO_3^-. The sample under test is dissolved in water in a test tube and a solution of iron(II) sulphate, $FeSO_4$, is added. After mixing, a slow continuous stream of concentrated sulphuric acid, H_2SO_4, is poured down the side. The acid forms a separate layer underneath the aqueous layer, and a brown ring is seen at the junction of the two if nitrate

is present in the sample. The brown ring disappears if the test tube is shaken.

brucite *See* magnesium hydroxide.

brush A conductor serving to provide electrical contact in a generator or electric motor. Brushes are made of specially prepared carbon.

bryology The science of mosses and liverworts.

Bryophyta A major division of the plant kingdom, comprising mosses, liverworts and hornworts.

bubble cap *See* fractionating tower.

bubble chamber An apparatus for observing ionising particles. It consists of a container filled with superheated liquid, usually liquid hydrogen, through which ionising particles pass. Gas bubbles are formed along the path of the moving particles, thus making their tracks visible. *See also* cloud chamber.

bubonic plague (Black Death) An acute, infectious, epidemic disease caused by bacteria and transmitted to human beings by fleas from infected rats such as the black and brown rats, or wild rodents. Symptoms include fever, vomiting and painful swelling of the lymph nodes, especially those in the groin and armpit. The swollen lymph nodes are called buboes. The disease has been called Black Death because of the dark spots of diseased tissue developing from bleeding under the skin. If the lungs are affected, the disease is called pneumonic plague. A vaccine is available and the disease can be treated successfully with antibiotics. To avoid an outbreak of plague, both the rats and the fleas should be killed.

buccal Pertaining to or in the mouth.

Büchner funnel A porcelain or plastic funnel which has a flat, circular base perforated with many small holes, on top of which a filter paper is placed. It is used for fast filtration by suction. *See* Fig. 17; *see also* filter flask.

bud An undeveloped shoot which may bear a flower.

budding 1. A method of artificial propagation in which a bud from one plant is inserted under the bark of another. **2.** A form of asexual reproduction in unicellular plants, e.g. yeast, and in primitive animals, e.g. *Hydra*. In yeast, a new cell is formed as an outgrowth or bud of a parent cell. This bud may detach itself or may remain to form a chain of cells by further budding. In *Hydra*, the bud is derived from a small group of cells on the parent which grow out from the parent and eventually separate from it. *See also* gemmation.

buffer capacity The ability of a buffer solution to resist any tendency to change its pH. This ability decreases by dilution; to cause a change of pH of, for example, 0,1 (0.1), smaller

Fig. 17 Büchner funnel

amounts of acid or alkali are necessary after dilution than before.

buffer solution A solution capable of retaining an almost constant pH value when moderate amounts of acid or alkali are added or the solution is diluted. Buffers are present in body fluids, e.g. in blood. They are important in practically all biological processes as well as in many other chemical processes. A typical buffer solution is one made up of equimolar amounts of sodium ethanoate (sodium acetate), CH_3COONa, and ethanoic acid (acetic acid), CH_3COOH. The former is highly ionised, the latter only partially. If an acid (i.e. hydrogen ions, H^+) is added to such a buffer solution, the hydrogen ions combine with ethanoate ions, CH_3COO^-, to form undissociated ethanoic acid:

$H^+ + CH_3COO^- \rightarrow CH_3COOH$

This removes H^+ from the solution. If an alkali (i.e. hydroxide ions, OH^-) is added to the same buffer solution, more CH_3COOH dissociates to form hydrogen ions which then combine with the hydroxide ions to form water:

$CH_3COOH \rightarrow CH_3COO^- + H^+$ and
$H^+ + OH^- \rightarrow H_2O$

Again, the hydroxide ions are removed from the solution.

bugs The common name for the Hemiptera.

bulb 1. A spherical structure. **2.** A specialised organ of vegetative reproduction consisting of a short underground stem bearing fleshy scale leaves. The whole structure encloses next year's bud. *See* Fig. 18. **3.** The reservoir for the liquid in a thermometer. **4.** The glass container of an electric lamp.

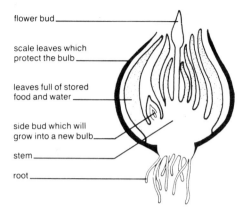

Fig. 18 Onion bulb

Bunsen burner A common laboratory burner which burns a mixture of gas and air. It consists of a metal tube with an adjustable air-valve. *See* Fig. 19.

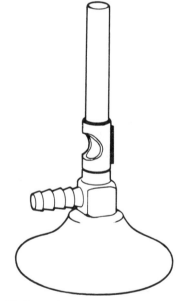

Fig. 19 Bunsen burner

buoyancy The upward force exerted on a body immersed in a fluid. This force is equal to the weight of the fluid displaced. *See also* Archimedes' principle.

burette A long, graduated glass tube fitted with a tap. It is used for the accurate measurement of the volume of liquid running from it. *See* Fig. 20; *see also* titration.

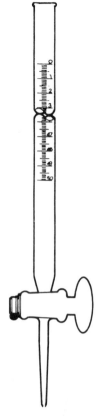

Fig. 20 Burette

burning *See* combustion.
butadiene *See* buta-1,3-diene.
buta-1,3-diene (butadiene) A diene: i.e. an unsaturated, aliphatic hydrocarbon containing two conjugated double bonds. Buta-1,3-diene has the formula $CH_2\!\!=\!\!CH\!\!-\!\!CH\!\!=\!\!CH_2$. It is a colourless, insoluble gaseous compound which polymerises easily. It may be prepared by catalytic dehydrogenation of butanes or butenes. It is used in the manufacture of synthetic rubber. *See also* methylbuta-1,3-diene.

butane C_4H_{10} A saturated hydrocarbon belonging to the alkanes. Butane is a flammable gas used as a fuel. M.p. $-138{,}4°C$ ($-138.4°C$), b.p. $-0{,}5°C$ ($-0.5°C$).

butterfly A member of the order Lepidoptera. The vast majority of butterflies have the following characteristics: they fly by day, have brightly coloured wings, have thickened or club-shaped antennae, have slender bodies and rest with their wings held together above the body. Butterflies have a specialised proboscis for sucking nectar, which is coiled under the head when not in use.

buttress roots Roots, often adventitious, growing from a stem or trunk before entering the soil. They give additional support to the plant.

by-product A substance obtained in addition to the main substance during a physical or chemical reaction. *Example:* ammonia, coke and coal-tar are important by-products in the manufacture of coal-gas.

C

C Chemical symbol for carbon.
Ca Chemical symbol for calcium.
cadmium cell *See* Weston cell.
caecum The first part of the large intestine. It has a blind lower end connecting it with the appendix. In many herbivorous animals both the caecum and the appendix are large and contain bacteria capable of digesting cellulose. *See* Fig. 49.
Caesarean section The surgical delivery of a child by cutting the abdomen walls. This operation is usually performed when the mother's birth canal is too narrow.
caesium clock *See* clock.
caffeine (theine) A colourless, crystalline substance (an alkaloid) found in several plants such as coffee, tea and cacao. It has a purine structure. Caffeine stimulates all parts of the central nervous system, lessens fatigue and increases the muscular power to do physical work.
Cainozoic The geological age (era) of recent life, estimated to have commenced approximately 65 million years ago. *See* Appendix.
calamine *See* calamine solution *and* zinc.
calamine solution A soothing liquid containing zinc carbonate, $ZnCO_3$. It is applied to the skin to relieve sunburn and itching caused by insect bites.
calcareous Containing or made entirely of calcium carbonate, $CaCO_3$.
calcicole Describes a plant which grows best in chalk soils, i.e. which is confined to alkaline conditions. *Compare* calcifuge.
calcification The hardening of tissue due to deposits of calcium salts. This is a normal process taking place when bones are formed and kept in repair. Late in life, calcification may affect the articular cartilage of joints, causing much pain.

calcifuge Describes a plant which grows best in acid soils. *Compare* calcicole.

calcination A process in which an inorganic substance, usually a carbonate, is strongly heated to produce an oxide and carbon dioxide, CO_2: e.g.
$CaCO_3 \rightarrow CaO + CO_2$

calcite (Iceland spar) A mineral consisting of crystals of calcium carbonate, $CaCO_3$. Certain clear calcite crystals have the property of double refraction. When such a crystal is placed over a black dot on a paper, two images are seen. If the crystal is rotated, one image remains stationary and the other rotates with the crystal.

calcium An element with the symbol Ca; atomic number 20; relative atomic mass 40,08 (40.08); state, solid. It is a very electropositive metal, belonging to the alkaline earth metals. It is extracted by electrolysis of fused calcium chloride, $CaCl_2$. Calcium has no important use.

calcium carbide *See* calcium dicarbide.

calcium carbonate $CaCO_3$ A calcium salt which occurs naturally in several forms, e.g. marble, limestone, chalk and calcite. Marble, limestone and chalk are microcrystalline forms of calcium carbonate, composed of the shell remains of dead marine animals. Calcite forms large crystals. Calcium carbonate is practically insoluble in pure water; when heated, it decomposes into calcium oxide, CaO, and carbon dioxide, CO_2:
$CaCO_3 \rightarrow CaO + CO_2$
Like all carbonates, it dissolves in acids to give a salt, carbon dioxide and water. Calcium carbonate is a raw material in the Solvay process for the manufacture of soda (sodium carbonate), Na_2CO_3. It is also used in the manufacture of cement, glass, quicklime (calcium oxide) and slaked lime (calcium hydroxide).

calcium chlorate(I) $Ca(OCl)_2$ (calcium hypochlorite) A calcium salt which is used as a constituent in bleaching powder. It is formed when chlorine is passed through a cold suspension of calcium hydroxide:
$2Ca(OH)_2 + 2Cl_2 \rightarrow Ca(OCl)_2 + CaCl_2 + 2H_2O$

calcium chloride $CaCl_2$ A soluble and rather deliquescent calcium salt used as a drying agent. It can take up six molecules of water of crystallisation to form $CaCl_2 \cdot 6H_2O$. Calcium chloride can be prepared by reacting calcium carbonate with hydrochloric acid:
$CaCO_3 + 2HCl \rightarrow CaCl_2 + CO_2 + H_2O$

calcium dicarbide CaC_2 (carbide, calcium carbide) A calcium compound of considerable

industrial importance. It is made by a reaction between coke and calcium oxide, CaO, in an electric furnace at a temperature of about 2000°C. With water, calcium dicarbide liberates ethyne (acetylene), C_2H_2:
$CaC_2 + 2H_2O \rightarrow C_2H_2 + Ca(OH)_2$
Ethyne is an important raw material in the organic chemical industry.

calcium hydrogencarbonate $Ca(HCO_3)_2$ A soluble calcium salt which cannot be isolated in the pure state as its aqueous solution decomposes on heating into calcium carbonate, $CaCO_3$, carbon dioxide, CO_2, and water:
$Ca(HCO_3)_2 \rightarrow CaCO_3 + CO_2 + H_2O$
It is one of the causes of temporary hardness in water: when rain-water, containing dissolved carbon dioxide from the atmosphere, attacks calcium carbonate, which is found in large quantities in most soil, the insoluble calcium carbonate is converted into soluble calcium hydrogencarbonate:
$CaCO_3 + CO_2 + H_2O \rightarrow Ca(HCO_3)_2$

calcium hydroxide $Ca(OH)_2$ (slaked lime) A crystalline, soluble, white powder. It can be obtained by adding water to calcium oxide:
$CaO + H_2O \rightarrow Ca(OH)_2$
This process is called slaking and is highly exothermic. Calcium hydroxide in aqueous solution forms a strong alkali called lime water, commonly used in a laboratory test for carbon dioxide. Calcium hydroxide has many uses: e.g. in the Solvay process for the manufacture of soda (sodium carbonate), Na_2CO_3; in treating an 'acid soil'; and in mortar.

calcium hypochlorite *See* calcium chlorate (I).

calcium nitrate $Ca(NO_3)_2$ A soluble, deliquescent calcium salt obtained by reacting calcium oxide, calcium carbonate or calcium hydroxide with nitric acid. When heated, it decomposes into calcium oxide, CaO, nitrogen dioxide, NO_2, and oxygen, O_2:
$2Ca(NO_3)_2 \rightarrow 2CaO + 4NO_2 + O_2$
It is widely used as a nitrogenous fertiliser and also in the manufacture of matches and explosives.

calcium oxide CaO (quicklime, lime, calx) A solid, white powder which is formed when calcium carbonate is strongly heated:
$CaCO_3 \rightarrow CaO + CO_2$
With water it gives calcium hydroxide (slaked lime):
$CaO + H_2O \rightarrow Ca(OH)_2$
This process is called slaking. Calcium oxide is used in the laboratory for drying ammonia, NH_3. Calcium oxide is also used in building and agriculture and in the chemical industry.

calcium phosphate(V) $Ca_3(PO_4)_2$ One of several phosphates of calcium occurring in animal bones and rocks. It is an insoluble calcium salt. When treated with sulphuric acid, H_2SO_4, it is converted into a soluble compound called calcium dihydrogenphosphate(V) (calcium hydrogen orthophosphate), $Ca(H_2PO_4)_2$:
$Ca_3(PO_4)_2 + 2H_2SO_4 \rightarrow Ca(H_2PO_4)_2 + 2CaSO_4$
The mixture of $Ca(H_2PO_4)_2$ and $CaSO_4$ is called superphosphate and is an important fertiliser. *See also* bone ash.

calcium sulphate $CaSO_4$ A slightly soluble calcium salt, occurring naturally as anhydrite, $CaSO_4$, and as gypsum, $CaSO_4 \cdot 2H_2O$. Anhydrite is sometimes used in the manufacture of sulphuric acid, H_2SO_4; gypsum, heated to around 120°C, is used in the manufacture of plaster of Paris, $CaSO_4 \cdot \frac{1}{2}H_2O$:
$CaSO_4 \cdot 2H_2O \rightarrow CaSO_4 \cdot \frac{1}{2}H_2O + 1\frac{1}{2}H_2O$

calibration The graduation of an instrument or apparatus so that it can take readings in definite units; e.g. a galvanometer may be calibrated to read in amperes.

caliche *See* Chile saltpetre.

callipers An instrument used for measuring distances on solid objects where an ordinary rule cannot be used directly. Callipers may consist of a pair of steel jaws hinged together. After the instrument has been used, the distance between the jaws can be measured on an ordinary scale. Slide or vernier callipers are another form of callipers. They consist of a fixed jaw at one end of a steel scale, and a sliding jaw. The object to be measured is placed between the jaws. On the sliding jaw there is a short scale of ten divisions called a vernier. Using this scale, together with the scale on the fixed jaw, it is possible to obtain accurately the second decimal place in a measurement. *See* Fig. 21.

callus 1. A hardened area of the skin, usually appearing on the palms of the hands or soles of the feet. 2. Bony tissue formed at the site of a bone fracture. 3. Tissue formed over a cut or damaged plant surface to protect the injured surface.

Calmette vaccine *See* BCG vaccine.

calomel *See* mercury chlorides.

calomel electrode A reference electrode consisting of mercury, Hg, in contact with solid dimercury(I) chloride, Hg_2Cl_2 (calomel), and a saturated solution of potassium chloride, KCl, and dimercury(I) chloride. A platinum wire is used as a connector. The potential of a calomel electrode varies only slightly with temperature and is independent of the hydrogen ion concentration (pH) of the solution in which it is immersed. For this reason it is widely used in electrical pH measurements, in which the other electrode is a glass electrode.

calorie

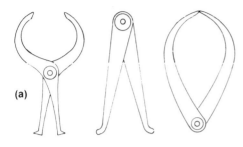

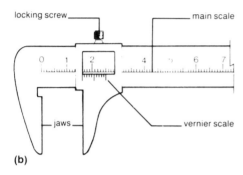

Fig. 21 (a) Simple callipers
(b) Vernier callipers

calorie An old unit of heat, defined as the quantity of heat required to raise the temperature of 1 gram of water by 1°C (from 14,5°C (14.5°C) to 15,5°C (15.5°C)). *See also* joule.

calorific value (energy value) The quantity of heat obtained per unit mass when a fuel is burnt or when a food is consumed; e.g. the calorific value of wholemeal bread is 10^7 joules per kg.

calorimeter A metal vessel used in heat measurements. It is made of a good conductor such as copper or brass so that it reaches the same temperature as its contents within a short time and it is designed so that negligible heat exchanges take place between the contents and the surroundings. In order to reduce heat loss by radiation the vessel is polished; to reduce heat losses by conduction and convection the vessel is placed in a larger container and surrounded by a layer of lagging. Finally an insulating lid is used to cover the calorimeter. *See also* bomb calorimeter.

calorimetry The measurement of heat as used in determining specific heat capacities, calorific values, heat of solution, etc.

calx The metal oxide formed when a metallic sulphide (ore or mineral) is roasted. *See also* calcium oxide.

calyx The collective name for the sepals of a flower.

cambium A layer of actively dividing cells (a meristem) situated between the xylem and phloem. Cambium is responsible for secondary thickening in stems and roots. *See also* bark.

Cambrian A geological period of approximately 70 million years, occurring at the beginning of the Palaeozoic. It ended approximately 500 million years ago.

camera A device for producing photographs. It consists of a light-proof box, usually with a lens at one end and a light-sensitive film at the other. Situated behind the lens is an aperture, the size of which can be controlled by a diaphragm. This allows light to enter the camera. The time for which light is allowed to pass through the aperture is controlled by a shutter. Focusing is usually obtained by moving the lens nearer to or further away from the film. The earliest camera was the pinhole camera. This camera has neither lens, diaphragm nor shutter; it simply makes use of the fact that light travels in straight lines. *See* Fig. 22 (a) *and* (b).

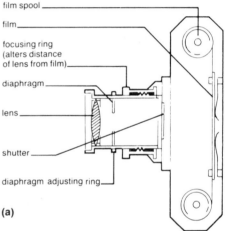

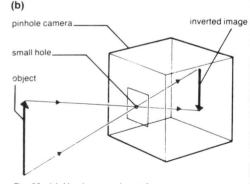

Fig. 22 (a) Hand camera in section
(b) Image formed by a pinhole camera

Canada balsam A clear, yellowish substance obtained from fir trees and soluble in dimethylbenzene (xylene). It is used for the preparation of microscope slides and as an adhesive for optical instruments, as its refractive index is similar to that of glass.

cancellous bone The spongy tissue in the epiphyses of long bones and in the centre of most others. It is situated beneath compact bone. The spaces in this tissue are filled with bone marrow. See Fig. 14.

cancer A disease caused by the uncontrolled growth of tissue leading to the production of malignant cells.

candela Symbol: cd. The SI unit of luminous intensity.

cane-sugar A carbohydrate made from sugar-cane. It is a disaccharide with the chemical name saccharose or sucrose and the chemical formula $C_{12}H_{22}O_{11}$. See also beet-sugar.

canine teeth (cuspids) Strong, pointed teeth situated between the incisors and the molars. They are the longest of the teeth; there are four in the human. They are designed for holding and tearing. A canine tooth has one root.

cannabis A drug obtained from the herbaceous plant called *Cannabis sativa* (Indian hemp). It is hallucinogenic and accumulates in the body. Cannabis is known under many names in various parts of the world. Examples include marijuana, grass, pot, weed, tea, boo, Mary Jane, bhang, ganja, kif, hashish (hash) and dagga.

canopy The cover formed by tree tops in a dense forest.

cap See diaphragm.

capacitance Symbol: C. The ability of a system of electrical conductors and insulators to store electric charge as its potential rises. The SI unit of capacitance is the farad (F). The capacitance of a conductor is defined as the ratio of its charge to its potential:

$$\text{capacitance} = \frac{\text{charge}}{\text{potential}}, \quad C = \frac{Q}{V}$$

capacitor (condenser) A system of electrical conductors and insulators capable of storing electric charge. A common capacitor is the parallel-plate capacitor, consisting of two parallel metal plates separated by air or some other insulator. Another capacitor is the variable capacitor which is commonly used to tune radio sets. See also dielectric.

capillarity See capillary action.

capillary 1. Having a hair-like bore. 2. A capillary tube is a tube with a narrow bore and thick walls, usually made of glass; e.g. the stem of a thermometer is made from a glass capillary tube. 3. An extremely small (narrow) blood vessel which has very thin walls and a uniform diameter. Capillaries form an extensive network throughout all living cells in the body and connect arteries and veins. Through the walls of the capillaries exchange of materials takes place between the blood and the tissue fluids. There are also capillaries in the body which convey lymph and bile.

capillary action (capillarity) The movement of liquids through capillary tubes as a result of surface tension.

capillary tube See capillary.

capillary water Water lying in the pore-spaces between soil particles. It may completely fill these spaces or be held on the surfaces of the particles. Water ascends into the pore-spaces by capillary action and it is this water which is most easily available to plants.

capitulum 1. A rounded, articular surface situated at the end of a bone. *Examples:* the capitulum of the humerus articulates with the head of the radius; the capitulum of a rib articulates with a vertebra. 2. The flattened or rounded flower-head of Compositae, consisting of numerous florets. See Fig. 179.

capsule A dry, dehiscent fruit; in liverworts and mosses, a structure containing spores; in some kinds of bacteria, the slimy coating surrounding the cell walls.

caramel A brownish, complex substance obtained when solid sugar is heated.

carapace The bony or chitinous shield covering the whole or part of the bodies of certain animals; e.g. the shell of crabs and tortoises. See also Chelonia.

carat 1. A unit of mass expressing the mass of diamonds and other precious stones. One carat is equal to 0,200 g (0.200 g). 2. A measure of the purity of gold, based on the number of parts of gold by mass in 24 parts of a gold alloy. *Examples:* 24 carat gold is pure gold; 18 carat gold contains 18/24 parts of gold and so is 75% gold. The remaining 25% may be other metals such as silver or copper.

carbamide See urea.

carbide 1. A substance composed of carbon and a more electropositive (less electronegative) element. Carbon compounds with oxygen, sulphur, phosphorus, nitrogen and the halogens are not considered to be carbides, and by convention neither are those with hydrogen. Examples of carbides are calcium dicarbide, CaC_2; silicon carbide (carborundum), SiC; aluminium carbide, Al_4C_3; and boron carbide, B_4C. A carbide may be produced by heating its constituents in an electric furnace. Some carbides are extremely hard and therefore used as abrasives and in cutting tools. See silicon carbide. 2. See calcium dicarbide.

carbocyclic compound An organic, cyclic compound which contains only carbon atoms as part of the ring structure; e.g. benzene, C_6H_6, and cyclohexane, C_6H_{12}. *See also* homocyclic compounds *and* heterocyclic compounds.

carbohydrase An enzyme which hydrolyses higher carbohydrates to simpler carbohydrates; e.g. amylase and lactase.

carbohydrate (saccharide) One of a large group of organic compounds containing carbon, hydrogen and oxygen. Many carbohydrates have the general formula $C_x(H_2O)_y$. They are usually divided into monosaccharides, e.g. glucose; disaccharides, e.g. saccharose; and polysaccharides, e.g. starch and cellulose. Carbohydrates are important in the metabolism of living organisms. Cellulose forms supporting tissue which makes up the structure of plants.

carbolic acid *See* phenol.

carbon An element with the symbol C; atomic number 6; relative atomic mass 12,01 (12.01); state, solid. Carbon is present in all organic substances. As an element it occurs in several allotropic forms such as diamond and graphite. Diamond is non-metallic, whereas graphite has some metallic properties: it has metallic lustre and is a good conductor of heat and electricity. Coal, like diamond and graphite, is a naturally occurring form of carbon. It consists mainly of carbon, together with compounds containing hydrogen, oxygen, nitrogen and sulphur; i.e. coal, unlike diamond and graphite, is an impure form of carbon. In the past, all forms of carbon except diamond and graphite were referred to as amorphous. However, this is misleading as they have been shown to be microcrystalline.

carbon assimilation The process of photosynthesis in green plants, resulting in the uptake of carbon dioxide from the air and its conversion into organic substances.

carbonate A salt of carbonic acid, H_2CO_3.

carbonation The process in which carbon dioxide, CO_2, is dissolved in a liquid under pressure. When the pressure is released, the carbon dioxide is liberated from the liquid. This process is used in the manufacture of fizzy drinks, e.g. lemonade and soda-water.

carbon black (lampblack, soot) Extremely small particles of pure carbon made from incomplete combustion of a hydrocarbon such as methane, CH_4. Carbon black is used in the manufacture of certain kinds of rubber and plastic, shoe polish and printer's ink.

carbon cycle The balance in nature between atmospheric carbon dioxide, CO_2, used up during photosynthesis and the carbon dioxide returned to the atmosphere as a result of respiration and decay of green plants, animals and micro-organisms. The extensive use of fuels, mainly of plant origin, also liberates carbon dioxide, as do volcanoes. *See* Fig. 23.

carbon dating *See* radio-carbon dating.

carbon dioxide CO_2 A colourless, dense gas present in the atmosphere (air) to the extent of 0,03% (0.03%) by volume. Carbon dioxide is slightly soluble in water. Some of the dissolved

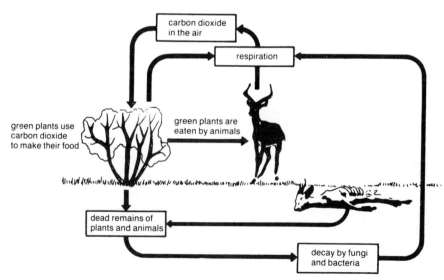

Fig. 23 Carbon cycle

carbon dioxide produces the carbonic acid, H_2CO_3, a weak acid:
$CO_2 + H_2O \rightleftharpoons H_2CO_3$
Carbon dioxide can be prepared by adding a not too weak acid to a carbonate, e.g. hydrochloric acid to calcium carbonate:
$2HCl + CaCO_3 \rightarrow CaCl_2 + CO_2 + H_2O$
Carbon dioxide is also formed by oxidation of carbon and some carbon compounds. At $-78.5°C$ ($-78.5°C$) carbon dioxide solidifies, forming 'dry ice' which sublimes and is used as a refrigerant. Carbon dioxide does not support combustion. When passed through lime water, $Ca(OH)_2$, a white insoluble substance, calcium carbonate, $CaCO_3$, is formed:
$Ca(OH)_2 + CO_2 \rightarrow CaCO_3 + H_2O$
This is a test serving to distinguish carbon dioxide from any other gas. Carbon dioxide is used in fire extinguishers, in the manufacture of fizzy drinks and in welding. See also carbon cycle.

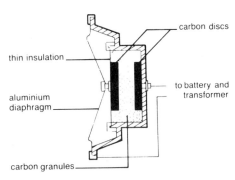

Fig. 24 Carbon microphone

carbon disulphide CS_2 A colourless, flammable liquid with an unpleasant smell. It is made by heating carbon with sulphur and is used as a solvent for substances such as rubber and sulphur.

carbonic acid H_2CO_3 A very weak acid formed from carbon dioxide and water. It cannot be isolated in the pure state as it decomposes on heating into carbon dioxide and water:
$H_2CO_3 \rightarrow CO_2 + H_2O$
Two series of salts can be obtained from carbonic acid: carbonates and hydrogencarbonates (bicarbonates).

carbonic anhydrase See blood plasma.

Carboniferous An important period in the Palaeozoic era because of its association with coal-forming vegetation.

carbon microphone A microphone used in the telephone transmitter. The instrument makes use of the fact that the electric resistance of carbon granules varies with the pressure put upon them. It consists of two carbon discs separated by carbon granules. One of the discs is fixed, the other is attached to a metal diaphragm and both discs are connected to a battery. Sound vibrations falling on the diaphragm subject the carbon granules to variable compressions, varying the resistance between the carbon discs and causing corresponding changes of current. The changes of current are transmitted along lines to the receiver, where electrical energy is converted back to sound energy. See Fig. 24 and Fig. 225.

carbon monoxide CO A colourless, poisonous gas with practically no smell. It burns in air with a blue flame forming carbon dioxide, CO_2. When inhaled, it combines with haemoglobin in the red blood cells forming carboxyhaemoglobin which is chemically stable. The tissue cells are then starved of oxygen and death may result. Carbon monoxide is formed by partial combustion of carbon and carbon fuels. It occurs in the exhaust gas of motor vehicles. Carbon monoxide is used as a fuel and in the synthesis of certain organic compounds.

carbon steel Steel whose properties depend mainly on the amount of carbon it contains. Carbon steels contain small amounts of manganese, phosphorus and sulphur, but no chromium, nickel or vanadium.

carbon tetrachloride See tetrachloromethane.

carbonyl A complex substance containing the carbonyl group attached to a metal, e.g. tetracarbonylnickel(0) (nickel carbonyl), $[Ni(CO)_4]$. Carbonyls are covalent compounds which are soluble in organic solvents. They are prepared by passing carbon monoxide, CO, over the heated metal.

carbonyl chloride $COCl_2$ (phosgene) A colourless, extremely poisonous gas, m.p. $-104°C$, b.p. $8.3°C$ ($8.3°C$). It may be prepared from carbon monoxide, CO, and chlorine, Cl_2, in the presence of light:
$CO + Cl_2 \rightarrow COCl_2$
Carbonyl chloride is used in organic synthesis. It has also been used as a weapon in war.

carbonyl group The functional group
$$-\overset{|}{C}=O,$$
found in aldehydes, ketones and carbonyls.

carborundum See silicon carbide.

carboxyhaemoglobin A stable substance consisting of carbon monoxide, CO, and haemoglobin. It is formed when carbon monoxide is inhaled. Carboxyhaemoglobin cannot take up oxygen and therefore takes no part in any respiratory function. Smokers have a raised level of carboxyhaemoglobin in their blood. This increases the permeability of the

carboxyl group

blood vessels, leading to a higher rate of deposition of fats in the artery walls and so increasing the risk of a coronary heart attack.

carboxyl group The group —COOH, as found in carboxylic acids.

carboxylic acid An organic acid containing one or more carboxyl groups, —COOH. Carboxylic acids are either aliphatic, e.g. ethanoic acid (acetic acid), CH_3COOH and ethanedioic acid (oxalic acid), $(COOH)_2$; or cyclic, e.g. benzoic acid, C_6H_5COOH. In more complex carboxylic acids there is at least one more other functional group apart from the carboxyl group. Like most organic acids, carboxylic acids are weak. *See also* amino acid.

carburettor A device attached to an internal combustion engine. Its function is to mix petrol vapour with air entering the engine.

carcinogen A chemical able to induce cancer.

carcinoma A malignant tumour arising in epithelial tissue such as the membrane lining the lungs and stomach. *See also* sarcoma.

cardiac Pertaining to the heart or the anterior part of the stomach.

cardiac muscle (heart muscle) A special kind of striated muscle found in the walls of the hearts of vertebrates. It undergoes involuntary rhythmical contractions.

cardiac orifice The opening at the upper end of the stomach between the oesophagus and the stomach. The sphincter around this orifice is less well-developed than the pyloric sphincter. *See also* pyloric orifice.

cardiac sphincter A circular muscle situated around the cardiac orifice. It is able to open and close the cardiac orifice thus controlling the entry of food to the stomach. *See also* pyloric sphincter.

caries Tooth decay brought about by colonies of bacteria in the mouth. The bacteria combine with food residue forming a sticky layer on the teeth called plaque. In this layer of plaque the bacteria produce enzymes that break down polysaccharides to lactic acid which will dissolve the enamel (the protective covering of the teeth) and later the dentine. Eventually this results in the development of a cavity. Brushing the teeth after meals is essential in preventing dental caries. A toothpaste containing fluorides is beneficial. *See also* fluoridation.

carina A ridge-shaped structure, especially the keel of a bird's breastbone.

carnallite A mineral consisting of hydrated potassium magnesium chloride, $KMgCl_3 \cdot 6H_2O$. It occurs as a saline residue in the Stassfurt deposits and is used as a fertiliser. Carnallite is a source of magnesium.

carnassial teeth In carnivores, molar or premolar teeth modified for cutting flesh.

carnauba wax *See* wax.

Carnivora An order of placental mammals which have well developed incisor and canine teeth and which eat meat and fish. Examples include dogs, cats, bears and seals.

carnivore A flesh-eating animal.

carnivorous Flesh-eating. Used of animals belonging to the Carnivora, and also of certain plants which trap and feed on insects or other small animals; e.g. Venus's fly-trap.

Carnot cycle A hypothesis stating that for an ideal heat engine in which heat is converted into work, the maximum efficiency does not depend on the nature of the substance employed, but entirely on the operating temperatures T_1 and T_2 between which the heat engine works; i.e. the engine absorbs a quantity of heat at some temperature T_2 and rejects the waste heat at a lower temperature T_1. The second law of thermodynamics is based on this statement.

carnotite *See* vanadium.

carotene (carotin) The general name for a series of unsaturated hydrocarbons with the formula $C_{40}H_{56}$ occurring as an orange, yellow or red pigment in animals and plants. In green plants carotene probably plays a small part in photosynthesis. In some plants it appears to be essential for phototropism. In animals carotene plays an important part in the synthesis of vitamin A and for this reason is sometimes known as pro-vitamin A. Xanthophylls are oxygenated derivatives of carotenes. Carotenes and xanthophylls are called carotenoids.

carotenoids *See* carotene.

carotid artery One of two main arteries in the neck, carrying blood from the heart to the head.

carotin *See* carotene.

carpal A wrist bone. In a human being there are eight such bones in the wrist. *See* Fig. 162.

carpel The female reproductive organ in flowering plants. It consists of an ovary containing one or more ovules which become seeds after fertilisation, a style and a stigma to receive pollen. *See* Fig. 75.

carpus 1. A medical name for the wrist. 2. The region of the fore-limb of vertebrates containing the carpal bones.

carrier 1. A person or animal that transmits a disease without suffering from it, e.g. a typhoid carrier. 2. (eluent) In chromatography, the solvent (mobile phase) passing through the column, paper, thin-layer, etc. In gas chromatography, the gas is the carrier.

carrier wave An electromagnetic wave of constant amplitude or frequency which is emitted (sent out) from a radio transmitter. *See also* modulation.

Cartesian diver A small float containing air which can be made to rise or fall in a liquid when the surface of the liquid is subjected to varying pressure. It is used to demonstrate the transmission of pressure by liquids.

cartilage Flexible, white connective tissue lining the joints and found in other parts of the body. There are three types of cartilage: hyaline cartilage, commonly known as gristle; fibrous cartilage, which makes up the discs between the vertebrae of the spinal column; and elastic cartilage, which forms the framework of the external ear.

caryopsis An achene whose testa and ovary wall are fused. They are found in grasses, e.g. maize and wheat.

casein A mixture of proteins found in milk. It is a pale, odourless, tasteless solid which can be obtained from milk by adding an acid or an enzyme called rennet (rennin). Casein is used in cheese-making, and in making plastics and adhesives.

cassiterite See tin.

cast iron (pig-iron) Impure iron obtained from the roasting of iron ores in a blast furnace. Its name is derived from the fact that, when molten, it can be poured into sand moulds to make castings.

Castner–Kellner cell (process) (mercury cathode cell) An electrolytic cell used primarily for the manufacture of sodium or potassium hydroxide, but also producing chlorine and hydrogen. A strong solution of sodium chloride, NaCl, (brine) flows through a cell. The cathode is mercury flowing slowly through the cell in the same direction as the brine. The anode is made of a number of graphite blocks. When an electric current is passed through the cell, sodium ions are discharged at the cathode and form an alloy (amalgam) with the mercury which then flows out of the cell; hydrogen ions cannot discharge easily on a mercury surface. Chloride ions are discharged at the anode as chlorine gas. When the amalgam is treated with water, the sodium reacts to form a solution of sodium hydroxide, NaOH, and the mercury is returned to the cell. The processes are:
(at the cathode) $2Na^+ + 2e \rightarrow 2Na$
(at the anode) $2Cl^- - 2e \rightarrow Cl_2$
(with the mercury)
$Hg + Na \rightarrow Hg/Na$ (amalgam)
(amalgam with water)
$2Hg/Na + 2H_2O \rightarrow 2NaOH + 2Hg + H_2$
Potassium hydroxide, KOH, is prepared in the same way using a concentrated solution of potassium chloride, KCl, in place of the brine. See also diaphragm cell.

castor oil A colourless, rather thick vegetable oil extracted from the seeds of the castor plant. It is a powerful laxative but should not be taken regularly as it irritates the intestine. It is also used in making plastics and lubricants.

castrate To remove the sex glands of a male or female animal.

catabolism (katabolism) The breaking down of organic substances by living organisms thus liberating energy. Compare anabolism; see also metabolism.

catabolite A product formed during catabolism.

catalase An enzyme capable of catalysing the breakdown of hydrogen peroxide, H_2O_2, into oxygen and water.

catalysis The acceleration or retardation of the rate of a chemical reaction when a catalyst is present.

catalyst A substance which accelerates or retards the rate of a chemical reaction. The catalyst participates in the reaction but is chemically unchanged at the end of it, although it may change physically. However, a catalyst may be destroyed either as a result of the process it catalyses or by combining with one or more of the products. For reversible reactions, a catalyst alters both the forward and the backward reaction to the same extent: i.e. the catalyst alters the rate at which equilibrium is obtained. Catalysts may be divided into positive and negative catalysts. The former accelerate the rates of chemical reactions, the latter retard them. Negative catalysts are often called inhibitors. However, the term 'inhibitor' is also used of a substance which lowers the efficiency of a positive catalyst. See also activation energy and promoter.

cataphoresis See electrophoresis.

cataract A condition which causes clouding of the eye lens so that less light reaches the retina. This condition may be caused by diabetes or by over-exposure of the eyes to heat.

caterpillar The larva of certain insects, e.g. butterflies and moths. A caterpillar has a soft segmented body which has three pairs of true legs on the thorax and short, unjointed prolegs on the abdomen. The two prolegs situated on the final segment of the abdomen are called claspers.

cathode The negative electrode of an electrolytic cell or discharge tube. Compare anode.

cathode ray A stream of fast-moving electrons emitted from the cathode of a discharge tube or vacuum tube. The cathodes of modern discharge tubes are made of a mixture of barium and strontium oxides.

cathode-ray deflection tube See cathode-ray tube, CRT and Perrin tube.

cathode-ray oscilloscope (CRO) An instrument which enables electrical signals to

41

be displayed on a fluorescent screen for visual examination.

cathode-ray tube (CRT) An electronic tube shaped rather like a funnel. It enables electrical signals to be displayed on a fluorescent screen and consists of an electron gun directing a beam of electrons onto the screen. The beam can be deflected up and down and from side to side by magnetic or electrostatic fields. It is used in television sets and in radar.

cation An atom or group of atoms carrying a positive electrical charge. In the presence of an electric field, e.g. during electrolysis, a cation moves towards the negative electrode, the cathode. *Compare* anion.

caudal Pertaining to the tail, e.g. caudal fin.

cauliflory The production of flowers on the trunk of a tree or on older leafless branches. This is seen on cocoa trees.

caustic 1. A curve or surface caused by the intersection of reflected or refracted light rays when parallel light rays strike a spherical mirror or pass through a convex lens. Such a curve may be seen on the surface of a liquid in a cup. *See* Fig. 25. **2.** Corrosive to organic substances. The term does not apply to acids.

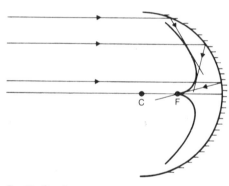

Fig. 25 Caustic

caustic curve *See* caustic.
caustic soda The common name for sodium hydroxide, NaOH.
cavity A hole, hollow or empty space in a solid body.
celestial Pertaining to the sky or heaven.
celestine *See* strontium.
cell 1. A protoplasmic unit consisting of a nucleus surrounded by cytoplasm and enclosed by a plasma membrane. Important differences between plant and animal cells are: (1) plant cells have a cell wall, whereas no cell wall is present in an animal cell; (2) plant cells are usually larger than animal cells; (3) many plant cells contain chloroplasts, whereas chloroplasts are not present in animal cells; (4) plant cells contain a large vacuole, whereas animal cells contain many small vacuoles or none at all. The cell is the basic unit of life. *See* Fig. 26 (a), (b) *and* (c).

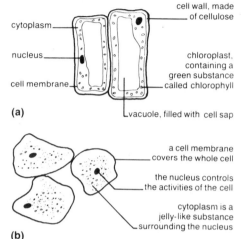

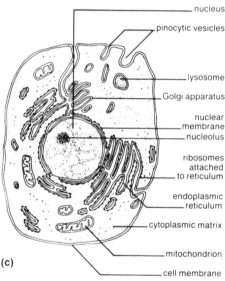

Fig. 26 (a) Plant cells (b) Animal cells (from the cheek) (c) Animal cell in detail

2. A vessel containing an electrolyte into which are dipped two electrodes for the purpose of producing electricity or for performing electrolysis. A cell is a unit of a battery. There are three main types of cell: (1) Primary cell. This cell produces electricity as a result of chemical action (usually irreversible) in the cell. Examples include the voltaic cell and the Daniell cell. *See also* fuel cell. (2) Secondary cell. This cell must be 'charged' by passing an electric current through it in the reverse direction to its discharge. The chemical actions in this cell are reversible. The lead-acid cell (accumulator) is a well-known example of a secondary cell. (3) Electrolytic cell. This cell does not produce electricity, but its electrodes are connected to a source of direct current. It is used to decompose an electrolyte. *See* electrolysis.

cell body Part of a neurone. A neurone is made up of a cell body and nerve fibres. Cell bodies have many different shapes. A cell body consists of a central nucleus lying in a cytoplasmic mass from which many filaments or dendrites branch. *See* Fig. 150.

cell division The splitting of a cell into two parts. Normally there are three consecutive stages in a cell division. These are division of the nucleus, cleavage of the cytoplasm and cell separation, in that order. *See also* meiosis *and* mitosis.

cell membrane (plasma membrane) An extremely thin membrane, of submicroscopic dimensions, enclosing the mass of protoplasm in a cell. It allows certain substances to pass into and out of the cell. The cell membrane is made up mainly of fat and protein.

cellobiose A reducing disaccharide, $C_{12}H_{22}O_{11}$, which is formed from cellulose by catalytic hydrolyses. Cellobiose gives two molecules of glucose on further hydrolyses. *See also* cellulase.

cellophane A product formed from cellulose. It is a clear, sparkling film which has almost the same composition as cellulose. It is used in wrapping.

cell organ A part of a cell which has a special function.

cell sap A watery solution found in vacuoles of plant cells. It is of very varied composition, containing substances such as organic acids, alkaloids, glycogen and pigments.

cell theory The idea that all plants and animals are made up of box-like structures called cells.

cellular 1. Consisting of cells. **2.** Pertaining to cells.

cellulase An enzyme present in seeds, bacteria, moulds and certain invertebrates. Cellulase hydrolyses cellulose to cellobiose.

celluloid A highly flammable substance obtained when cellulose dinitrate is treated with ethanol and camphor. It is used in the manufacture of photographic film.

cellulose A carbohydrate which is a long chain polysaccharide (a polymer) and a principal constituent of the cell walls in higher plants. It is manufactured from wood and used in making paper, cellulose ethanoate (acetate), cellophane, rayon and explosives. Cellulose is the most abundant organic compound in nature. It is insoluble in water, but dissolves in a solution of tetraamminecopper(II) hydroxide (copper(II)-ammonium hydroxide, Schweizer's solution). Schweizer's solution is made by dissolving copper(II) hydroxide, $Cu(OH)_2$, in concentrated ammonia solution, NH_3. *See also* cellobiose *and* cellulase.

cellulose acetate *See* cellulose ethanoate.

cellulose ethanoate (cellulose acetate) A solid, flammable substance (an ester) formed when cellulose reacts with ethanoic acid (acetic acid) or ethanoic anhydride, using sulphuric acid as a catalyst. It is used in the manufacture of lacquers, magnetic tape, photographic film, rayon, etc. It is insoluble in water, but dissolves in propanone (acetone) and in certain other organic solvents.

cellulose nitrate *See* cellulose trinitrate.

cellulose trinitrate (cellulose nitrate, guncotton) A highly flammable material made by treating cellulose with a mixture of concentrated nitric acid and concentrated sulphuric acid. It is used as an explosive and in lacquers. Sometimes cellulose trinitrate is inaccurately called nitro-cellulose; however, it is not a nitro-compound but an ester.

cell wall A rigid structure bounding a plant cell, i.e. surrounding the cell membrane. It is formed by the protoplasm and is made mainly of cellulose which gives the cell mechanical support. During the development of the cell, a first cell wall (the primary wall) is laid down, followed by more cellulose which constitutes the secondary wall. At intervals, the cell wall is perforated by tiny pores (pits) through which neighbouring cells are interconnected by fine cytoplasmic threads (plasmodesmata). The cell wall is freely permeable to aqueous solutions and gases in either direction. *See also* lignin *and* suberin.

Celsius scale A scale of temperature using a lower fixed point called the ice-point and an upper fixed point called the steam-point. The difference between these two points is one hundred equal degrees. The ice-point is the temperature of pure melting ice and is said to be zero degrees Celsius, 0°C. The steam-point is the temperature of steam from boiling water at a pressure of 101 325 Pa and is said to be one hundred degrees Celsius, 100°C. Formerly the Celsius scale was called the centigrade scale.

cement 1. Any substance which is used as a bonding material. **2.** A mixture of calcium silicate and calcium aluminate, used for building. It is made by heating a mixture of limestone, $CaCO_3$ and clay in a kiln. Clay contains aluminium silicates. After heating, the cement is ground into a fine powder. When building cement is mixed with water and allowed to stand, it sets forming a hard, rigid solid. The chemical processes involved in the setting of cement are not fully known, but they seem to involve a slow hydration of silicates and aluminates.   3. A hard layer covering the root of a tooth. It resembles bone, but contains no blood vessels.

centi- Symbol: c. A prefix meaning one hundredth, 10^{-2} (0,01; 0.01); e.g. one centimetre (cm) is equal to 10^{-2} metres.

centigrade scale See Celsius scale.

centimetre Symbol: cm. A unit of length equal to 10^{-2} metres (0,01 m; 0.01 m).

centipede An arthropod belonging to the class Chilopoda. It has a segmented body, each segment bearing only one pair of jointed legs. Usually there are fifteen to twenty pairs of jointed legs. The first pair of legs is modified as fangs and these are used when the centipede attacks insects, slugs, earthworms and sometimes other centipedes. Its bite is poisonous, but seldom fatal to human beings. The head bears a pair of antennae, three pairs of jaws and groups of simple eyes. Centipedes are usually nocturnal and some are luminescent at night.

central atom See ligand.

central force A force on a moving body which is always directed towards a fixed point or centre, e.g. the gravitational attraction of the Sun acting on a planet.

central nervous system (CNS) A nervous system in vertebrates consisting of the brain and the spinal cord, enclosed within the skull and the backbone. It is built up of a mass of nervous tissue and co-ordinates the activity of an animal.

centre of curvature See lens and mirror.

centre of gravity The fixed point in an object through which its total weight appears to act, irrespective of the position of the object.

centre of mass The fixed point in an object at which the mass of the object can be considered to act. For small objects, the centre of mass coincides with the centre of gravity.

centrifugal force A body's outward reaction to the inward centripetal force acting on it when it is moving in a circular path. This can be expressed as

$$F = m\frac{v^2}{r}$$

where F is the centrifugal force and m is the mass of the body moving with velocity v in a circle of radius r.

centrifuge An apparatus used for rotating an object at a high velocity. It is commonly used in separating suspended particles from a liquid by increasing the rate of sedimentation. Some high-speed centrifuges (ultracentrifuges) can be operated at speeds as high as 150 000–200 000 revolutions per minute.

centriole In a resting animal cell or simple form of plant cell, a single, minute rod-shaped body, situated near the nucleus. During meiosis and mitosis the centriole divides into two parts. These move to opposite poles of the cell and a spindle of protein fibres forms between them.

centripetal force A force acting radially inwards on a body rotating in a circle round a central point. It is equal in magnitude but opposite in direction to the centrifugal force (the reaction to this force). It keeps the body moving in a circular path.

centromere The region in a chromosome where its two parallel strands (chromatids) are joined. Centromeres are the points of attachment of the chromosomes to the spindle.

centrosome A region containing a centriole, situated near the nucleus in the cytoplasm of an animal cell or simpler plant cell.

centrum The massive part of a vertebra where the main strength lies. See Fig. 242.

Cephalochorda(ta) A subphylum of Chordata, commonly called the lancelets and including Amphioxus. Cephalochordates are primitive, marine invertebrates whose characteristics include no brain, a notochord persisting throughout life and no paired fins.

Cephalopoda A class of molluscs which are carnivores. Members of this class have large eyes and heads surrounded by mobile tentacles armed with suckers and sometimes hooks. They have a highly centralised nervous system. Examples include the octopus, cuttlefish and squid.

cephalothorax The name given to the organ formed through fusion of the head and thorax in some invertebrates, e.g. spiders.

ceramic A hard, inorganic material which has a high melting-point. Examples include pottery, porcelain and enamels. Most ceramics are made from clay. When the product has been shaped, it is fired in a kiln where it becomes hard, durable, stable and resistant to most chemicals. It also has a high electrical resistivity.

cereal A term applied to plants which are members of the family Gramineae and which are cultivated for their seeds, used as food. Examples of such plants include wheat, maize, oats and rice. Cereal crops are the world's most important source of food. See also gluten.

cerebellum The second-largest part of the brain. It is a segment of the hindbrain, connected to the midbrain and also to the underlying nerve-tissue called the brain-stem by bundles of fibres. The cerebellum is located towards the back of the head below the cerebrum. It is the centre of co-ordination of muscular movements, including the important factor of balance. *See* Fig. 16.

cerebral Pertaining to the brain.

cerebral cortex A thin layer of grey matter covering the entire surface of the cerebral hemispheres. It controls all conscious actions.

cerebral hemispheres The biggest and most complicated part of the central nervous system, filling the major part of the skull. Each of the two cerebral hemispheres is largely responsible for the sensation and movement of the opposite half of the body. *See* Fig. 16; *see also* cerebrum.

cerebrospinal fluid An alkaline, clear, colourless tissue fluid which circulates slowly through the subarachnoid space, the central canal of the spinal cord and the ventricles of the brain. It serves as a hydraulic shock absorber for the brain and spinal cord. *See also* lumbar puncture.

cerebrum The largest part of the brain. It is made up of two lobes called cerebral hemispheres.

cerumen A wax-like secretion from the ceruminous glands of the external ear.

ceruminous gland A tubular gland, thought to be a modified sweat gland. The ceruminous glands are situated beneath the skin lining the external auditory meatus and their function is to secrete ear wax.

cerussite Naturally occurring lead(II) carbonate, $PbCO_3$. It is commonly found in oxidised portions of ore bodies containing galena (lead(II) sulphide), PbS.

cervical Pertaining to the cervix or neck.

cervical vertebra One of the vertebrae forming the neck. In humans, there are seven. *See* Fig. 242.

cervix The neck or mouth of an organ, e.g. the posterior part of the uterus leading to the vagina.

Cestoda A class of parasitic, segmented flatworms without a mouth or gut. They are commonly called tapeworms.

Cetacea An order of placental mammals which includes whales, dolphins and porpoises. Their bodies are highly adapted for swimming.

c.g.s. system A system of units based on the centimetre, the gram and the second. The system has been superseded by SI units.

chaeta (pl. chaetae) A bristle characteristic of Annelida. Chaetae project from the skin and occur in groups. Their function is to assist in locomotion by gripping the earth.

Chagas' disease A disease caused by a protozoon, transmitted to humans by the bite of certain bugs which defaecate as they suck the blood of the victim. The bite causes itching and during scratching the infected faeces is rubbed into the wounds. The protozoon infects muscles, including the heart muscle. The disease responds badly to drugs and is often fatal. Chagas' disease occurs in South America.

chain isomerism *See* structural isomerism.

chain reaction A nuclear reaction in which energy is continuously released as a result of neutrons, emitted by the fission of an atomic nucleus of an element, each splitting another nucleus and causing the emission of more neutrons.

chalaza 1. In angiosperms, a part of the nucellus situated opposite the micropyle. **2.** One of the two twisted threads of albumen (egg-white) holding the yolk in the centre of an egg.

chalcocite *See* copper glance.

chalcopyrite *See* copper pyrites.

chalcosine *See* copper glance.

chalk A natural, porous form of calcium carbonate, $CaCO_3$. Usually chalk is a white substance, but it may also have other colours such as red, due to staining by iron compounds. Chalk is formed from the shells of small, dead marine organisms. *See also* calcium carbonate.

change of state A physical process in which a substance changes from one state to another. *Example:* if heated, ice will change from the solid state to the liquid state (water). If heated further, it changes to the gaseous state (steam). These processes are reversed by cooling. The following processes are involved in the change of state of matter: melting, evaporation, boiling, condensation, freezing, crystallisation and sublimation.

char *See* charcoal.

charcoal (char) A black, porous form of carbon made up of extremely small particles, giving it a large surface area. It is made by heating wood, bone or sugar in the absence of air (oxygen). It is used in gas masks as several poisonous gases adsorb to its surface. *See also* activated carbon.

charge Symbol: Q; unit: coulomb (C). The property of certain atoms or groups of atoms which causes them to exert forces on each other. An electron possesses the smallest unit of negative charge; the proton possesses an equal amount of positive charge. Like charges repel and unlike charges attract each other.

Charles's law (Gay-Lussac's law of heat) The volume of a fixed mass of a gas at constant pressure increases by 1/273 of its volume at 0°C per kelvin rise in temperature. The law can also be expressed as follows: at constant pressure the volume of a given mass of gas is directly proportional to its kelvin temperature: $V_1/T_1 = V_2/T_2$, where V_1 and V_2 are the volumes of the gas at kelvin temperatures T_1 and T_2 respectively.

chela (pl. chelae) A pincer borne on an appendage, e.g. the claw of a lobster.

chelate A metal-organic complex in which certain metal ions are bound to non-metallic atoms, such as carbon, oxygen or nitrogen, in a chelating agent, forming a cyclic structure. The most widely used chelating agent is edta, which is capable of forming six coordinate bonds with a metal ion. This produces very stable complexes.

chelating agent *See* chelate.

Chelonia An order of reptiles comprising tortoises, turtles and terrapins. The bodies of these animals are completely covered by a firm shell from which only head, limbs and tail emerge. The upper portion of the shell is called the carapace; the lower portion is called the plastron.

chemical affinity *See* affinity.

chemical change A change in the chemical composition of one or more substances resulting from a chemical reaction. A chemical change always produces one or more new substances, is generally not easily reversible and is usually accompanied by a noticeable heat change. The respective masses of the new substances formed (the products) are different from those of the reacting substances (the reactants). However, the sum of the masses of the products is equal to the sum of the masses of the reactants. *Compare* physical change.

chemical combination (laws of) Three laws describing how elements combine to form chemical compounds. These laws are: (1) *The Law of Definite Proportions* (the Law of Constant Composition, Proust's Law): a law stating that all pure samples of the same chemical compound contain the same elements combined in the same proportions by mass. (2) *The Law of Multiple Proportions*: a law stating that if two elements A and B combine to form more than one chemical compound, then the various masses of A which separately combine with a fixed mass of B, are in a simple ratio. (3) *The Law of Combining Masses* (the Law of Reciprocal Proportions): a law stating that elements combine in the ratio of their combining masses (equivalent masses), or in some simple multiple or sub-multiple of that ratio.

chemical compound A substance composed of different elements combined together chemically. *Examples:* water, H_2O, is composed of the elements hydrogen and oxygen; common salt (sodium chloride), NaCl, is composed of the elements sodium and chlorine.

chemical energy The stored energy in atoms or molecules which is released during a chemical reaction.

chemical equation An equation representing a chemical reaction. *Example:* when sodium, Na, reacts with water, H_2O, sodium hydroxide, NaOH, and hydrogen, H_2, are formed. This can be written in a chemical equation as:
$2Na + 2H_2O \rightarrow 2NaOH + H_2$
The substances to the left of the arrow are called the reactants and those to the right of the arrow are called the products.

chemical equilibrium A state of a reversible reaction at which both the forward and backward reactions proceed at the same rate. When a reversible reaction is in equilibrium, there is no apparent change with time in the amounts of reactants and products. *Example:* if iron, Fe, is brought to react with steam, H_2O, by heating them in a closed vessel, iron(II) di-iron(III) oxide (tri-iron tetroxide), Fe_3O_4 and hydrogen, H_2 are formed:
$3Fe + 4H_2O \rightleftharpoons Fe_3O_4 + 4H_2$
However, as soon as the products, Fe_3O_4 and H_2, are formed, they start to react with one another forming iron and steam. After some time the vessel will contain iron, steam, iron(II) di-iron(III) oxide and hydrogen in definite amounts. As long as the temperature is kept constant, there will be no further change.

chemical equivalent *See* equivalent mass.

chemical formula *See* formula.

chemical process *See* chemical reaction.

chemical reaction A process in which one or more substances react chemically to form new substances. In a chemical reaction bonds are broken and formed, resulting in a chemical change in the reacting substances. *Example:* when magnesium, Mg, burns in air, it combines with oxygen, O_2, to form magnesium oxide, MgO:
$2Mg + O_2 \rightarrow 2MgO$
See also chemical equation.

chemiluminescence *See* luminescence.

chemistry The study of the elements and their chemical compounds. This involves the study of how atoms (elements) and various chemical compounds react with each other. Chemistry is more concerned with the behaviour of the outer electrons of atoms than with their nuclei. Chemistry is often divided into branches such as inorganic chemistry, organic chemistry, physical chemistry and biochemistry.

chemoautotroph(e) An organism which uses inorganic material, e.g. carbon dioxide, as its main source of carbon and derives energy from the oxidation of certain inorganic compounds. *Compare* chemoheterotroph(e); *see also* photoautotroph(e).

chemoheterotroph(e) An organism which uses organic material as its main source both of carbon and of energy. *Compare* chemoautotroph(e); *see also* photoheterotroph(e).

chemonasty A plant movement in response to chemical stimuli; e.g. the bending of the long tentacles on sundew leaves towards the middle of the leaf in response to the presence of organic nitrogenous compounds.

chemosynthesis The process by which an organism derives its energy from an exothermic oxidation initiated by itself. Such an organism may live on inorganic materials alone, using carbon dioxide, CO_2, or some form of carbonate as its sole source of carbon.

chemotaxis The reaction of cells and organisms to chemical stimuli, e.g. the attraction of leucocytes to substances diffusing from bacteria. The curvature of plants or plant organs in response to chemical stimuli is often referred to as chemotropism.

chemotherapy The use of chemicals to prevent or treat a disease. *Examples:* penicillin is used in the treatment of many types of infections; drugs are used in the treatment of certain types of cancer. *See also* Salvarsan.

chemotropism *See* chemotaxis.

chiasma (pl. chiasmata) A point of juncture on homologous chromosomes at which crossing-over occurs and genetic material is exchanged.

chicken-pox (varicella) A mild, contagious, infectious disease, usually occurring in childhood. It is caused by a virus. Symptoms include fever and a skin rash that becomes blistered.

Chile saltpetre (caliche) Sodium nitrate, $NaNO_3$, usually mixed with clay and small amounts of sodium iodate(V), $NaIO_3$. It occurs naturally in large deposits in Chile; it is used as a fertiliser and in making nitric acid, HNO_3.

Chilopoda A class of arthropods comprising centipedes.

China clay *See* clay.

chip A very small piece of semiconductor incorporating an electronic component, e.g. a transistor, or an integrated circuit.

Chiroptera An order of placental mammals including bats and flying foxes, characterised by membranes attached to the limbs and sometimes to the tail.

chitin A carbohydrate (polysaccharide) with a structure similar to that of cellulose. However, chitin also contains nitrogen. Chitin is found in association with proteins and other substances. It is present in the exoskeleton of arthropods, in shells of crustaceans and in some plants, especially fungi. *See also* glucosamine.

Chlamydia A genus of intracellular, Gram-negative, non-motile bacteria with a spherical shape. The species *Chlamydia trachomatis* is the most frequent cause of sexually transferred infections in the industrialised countries. It is also the cause of the eye disease trachoma.

chloral *See* trichloroethanal.

chloral hydrate *See* trichloroethanal.

chlorate(I) (hypochlorite) A salt of chloric(I) acid, HClO.

chlorate(V) A salt of chloric(V) acid, $HClO_3$.

chlorate(VII) (perchlorate) A salt of chloric(VII) acid, $HClO_4$.

Chlorella A genus of unicellular algae.

chlorenchyma Plant tissue containing chlorophyll.

chloric(I) acid HClO (hypochlorous acid) A weak acid which is a strong oxidising agent. Its salts are called chlorate(I)s (hypochlorites). They are widely used as bleaching agents and in water purification.

chloric(V) acid $HClO_3$ A strong acid whose salts are called chlorate(V)s. $HClO_3$ is a colourless liquid which can be made by a reaction between barium chlorate(V), $Ba(ClO_3)_2$, and dilute sulphuric acid, H_2SO_4:
$Ba(ClO_3)_2 + H_2SO_4 \rightarrow 2HClO_3 + BaSO_4$
It is a strong oxidising agent which causes an explosion with organic material.

chloric(VII) acid $HClO_4$ (perchloric acid) A very strong acid whose salts are called chlorate(VII)s. $HClO_4$ is a fuming, colourless liquid which can be made by a reaction between potassium chlorate(VII), $KClO_4$, and concentrated sulphuric acid, H_2SO_4:
$2KClO_4 + H_2SO_4 \rightarrow 2HClO_4 + K_2SO_4$
It is a strong oxidising agent which causes an explosion with organic material. $HClO_4$ is used in electroplating and in explosives.

chloride A salt of hydrochloric acid, HCl.

chloride of lime *See* bleaching powder.

chlorination 1. Treatment with chlorine or a chlorinating agent, e.g. calcium chlorate(I) (calcium hypochlorite), $Ca(OCl)_2$, as in the disinfection of water. **2.** The introduction of one or more chlorine atoms into an organic molecule. *See also* halogenation.

chlorine An element with the symbol Cl; atomic number 17; relative atomic mass 35,45 (35.45); state, gas. It is a very electronegative non-metal, belonging to the halogens. It is produced commercially by the electrolysis of sodium chloride solution (brine). Chlorine is a greenish-yellow gas with an irritating smell and is very poisonous if inhaled. It is used as a bleaching agent, in water purification and in the manufacture of many chemicals. *See also* Castner–Kellner cell (process).

chloroethene *See* PVC.

chloroform *See* trichloromethane.

Chlorophyceae A class of algae known as green algae, characterised by containing chlorophyll, storing starch and possessing a cellulose cell wall.

chlorophyll Any of several complex, organic compounds all containing magnesium. It is the green pigment in plants, responsible for absorbing light energy during photosynthesis.

chloroplast A plastid containing chlorophyll. Chloroplasts occur in the cells of higher plants and algae, except blue-green algae. See also chromoplast, leucoplast *and* granum.

chloroprene See Neoprene.

chlorosis 1. A disease of green plants characterised by the absence of chlorophyll and caused by lack of light, by iron or magnesium deficiency, or by genetic factors. **2.** (green sickness) A rare anaemic disease in humans.

choke An inductor consisting of a low resistance coil with an air or laminated iron core. When an alternating current flows through the coil, an induced e.m.f. is set up which opposes that of the supply. This apparent resistance is known as inductive reactance. Chokes are used in circuits of radio and television sets and to smooth the outputs of rectifying circuits.

cholecystokinin See gall-bladder.

cholera An acute, epidemic disease caused by a bacterium called *Vibrio cholerae*, found in the intestinal contents of cholera patients. The disease is spread through the faeces of infected people and is usually caught by drinking polluted water or eating contaminated food. Symptoms of cholera are sudden and severe vomiting and diarrhoea, causing dehydration which may prove fatal. Cholera is treated using antibiotics and by replacing the water and salts lost from the tissue.

cholesterol $C_{27}H_{45}OH$ An organic, white, waxy compound (a sterol) found in almost all animal tissue, but particularly in the brain, nerves and adrenal gland (suprarenal gland). It is also found in yolk and in some plants such as red algae. In humans, cholesterol serves physiologically as a vehicle for the transport of fatty acids, it takes part in the production of hormones and it has some importance in immunological processes. If present in excessive amounts in the blood, it is believed to be associated with arterial diseases such as coronary thrombosis. See also lanolin.

choline $HO-CH_2-CH_2-N(CH_3)_3OH$ A rather strong, organic base which is of importance in preventing the accumulation of fat in the liver. Often choline is included among the B vitamins. Choline also plays an important role in carbohydrate and protein metabolism and is necessary for growth. A derivative of choline is acetylcholine. Choline is found in lecithins.

cholinesterase See acetylcholine.

chondroblast A cell secreting cartilage. During ossification chondroblasts precipitate calcium salts, causing calcification of the cartilage.

chordae tendineae Delicate, cord-like structures connecting the papillary muscles with the valves of the heart. They prevent blood from leaking back into the atria when the ventricles contract.

Chordata A phylum of animals comprising all five vertebrate classes plus three invertebrate groups. They possess a notochord and gill slits at some stage in their life cycle.

chorion The outermost foetal membrane enclosing the amnion in mammals, birds and reptiles.

choroid A pigmented, vascular layer in vertebrate eyes situated between the retina and the sclera (sclerotic coat). The anterior extension of the choroid makes up the ciliary body and the iris. See Fig. 66.

chromate(VI) A salt of chromic(VI) acid, H_2CrO_4.

chromatic aberration (chromatism) See aberration.

chromatid One half of a chromosome, observable during mitosis or meiosis. The two strands of chromatids make up a chromosome. They are held together near their midpoint, in a small region called the centromere. During anaphase, the chromatids separate and each passes to a pole.

chromatin Nucleoprotein which, if stained, appears as tangled threads in definite units called chromosomes.

chromatism Chromatic aberration. See aberration.

chromatography A method of separating and identifying certain substances. Several chromatographic techniques exist, e.g. paper chromatography, thin-layer chromatography and gas chromatography. Two phases, a mobile and a stationary, are involved. The mobile phase is either an inert solvent or a gas containing (carrying) the sample to be analysed. The stationary phase may be a solid or a liquid adsorbed to a solid. The method depends on differences in affinity (solubility) of the constituents of the sample for the mobile and stationary phases. Chromatography is also used in quantitative analysis. See also R_F-value.

chrome alum See alum.

chromic(VI) acid H_2CrO_4 This acid has never been isolated, but it is likely to exist in an aqueous solution of chromium(VI) oxide, CrO_3. Salts of chromic(VI) acid are called chromate(VI)s.

chromite A mineral with the formula $FeCr_2O_4$, i.e. containing iron, chromium and oxygen. It is the main ore of chromium, often found together with magnesium and aluminium.

chromium An element with the symbol Cr; atomic number 24; relative atomic mass 52,00 (52.00); state, solid. It is a hard white transition metal, very resistant to chemical attack at room temperature. It is extracted from chromite by reducing it with carbon or aluminium. Chromium is used as an alloying agent in the manufacture of stainless steel, tool steel and in chromium plating. All chemical compounds of chromium are coloured.

chromium plating The process of depositing a thin layer of chromium on another metal in order to make this other metal more resistant to chemical attack. This is brought about by electrolysis using a bath (an electrolyte) of chromic(VI) acid and an anode of pure chromium. The metal to be plated is made the cathode.

chromium steel A very hard, tough alloy containing chromium and several other metals such as nickel, vanadium and tungsten (wolfram). It is used in making tools.

chromophore A group of atoms (usually in an organic compound) which is responsible for the distinctive colour of a certain substance, e.g. in a dye.

chromoplast A plastid which contains no chlorophyll but contains other pigments, usually orange, yellow or red. Chromoplasts are responsible for the colour of different plant parts, e.g. for the red-orange colour of tomatoes and carrots. *See also* chloroplast *and* leucoplast.

chromosome A relatively long, thin, threadlike body found in the nucleus of animal and plant cells. Bacteria and viruses contain structures with a similar function. Chromosomes can be observed during cell division. They contain DNA, RNA and protein, the genetic material of the cell. Groups of these substances form the genes situated in a line along the chromosomes; a gene occupies a short length of a chromosome. Each species of plants and animals has a constant number of chromosomes, usually found in pairs. In Man there are 23 pairs in every cell, in a mouse 20 pairs and in rye 7 pairs. The members of each pair of chromosomes are called homologous chromosomes; the total number of chromosomes in each cell is called the diploid number.

chromosome mutation *See* mutation.

chromosphere The lower atmosphere of the Sun, which becomes visible as a rose-coloured ring during a total eclipse. The dark absorption lines in the solar spectrum originate in the chromosphere owing to some relatively cool gases absorbing wavelengths of light which normally would be emitted. These lines are called Fraunhofer lines.

chronometer An accurate clock used in navigation to determine longitude. The chronometer is regulated by a balance-wheel and a hair spring. *See also* clock.

chrysalis A pupa, especially of a butterfly or moth.

chyle Milky lymph found in the capillaries (lacteals) of the small intestine when it contains fat. *See also* lacteal.

chyme The name given to the contents of the stomach when they pass into the small intestine.

cilia *See* cilium.

ciliary body A circular band, continuous with the retina of the vertebrate eye. It contains the ciliary muscle and secretes aqueous humour. *See* Fig. 66.

ciliary muscle A muscle situated in the ciliary body of the vertebrate eye. Its contraction or relaxation causes the suspensory ligament to change the shape of the eye lens. *See* Fig. 66.

cilium (pl. cilia) A small, threadlike organ projecting from a cell and occurring in large numbers when present. Cilia move in an orderly rhythm in a constant direction. They have several functions, e.g. locomotion; causing a liquid to flow past a cell; or removing solid particles, as in human air passages, where cilia sweep away dust particles and bacteria by an upward movement. Cilia are much shorter than flagella but have the same internal structure.

cinnabar A scarlet to brownish mineral containing mercury(II) sulphide, HgS. It is the most common mercury mineral and is therefore an important ore of this metal.

circadian rhythm *See* diurnal rhythm.

circuit A number of conductors or semiconductors connected together through which an electric current can flow. *See also* magnetic circuit.

circuit breaker Usually an automatic device which opens (breaks) an electric circuit when there is a large surge of current. *See also* trip switch.

circulatory system In vertebrates, the system comprising the heart, arteries, veins and capillaries, through which blood flows. *See* Fig. 27.

circumcision *See* prepuce.

circumnutation *See* nutation.

cirrhosis A chronic disease of the liver characterised by the destruction of liver cells and their replacement by fatty or fibrous tissue.

***cis*-isomer** *See* geometrical isomerism.

***cis*-octadec-9-enoic acid** *See* oleic acid.

cisterna

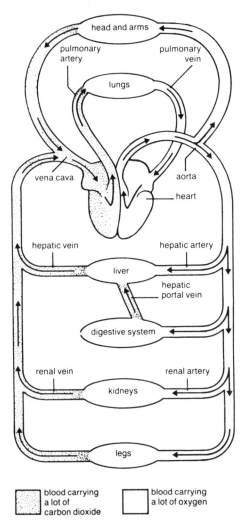

Fig. 27 Circulatory system

cisterna (pl. cisternae) A flattened sac-like vesicle of the Golgi apparatus and the endoplasmic reticulum.

cistron A length of DNA made up of the smallest number of nucleotides which have to be present on the same chromosome to permit the synthesis of a specific polypeptide chain.

citrate A salt of citric acid, $C_6H_8O_7$.

citric acid $C_6H_8O_7$ (2-hydroxypropane-1,2,3-tricarboxylic acid) An organic acid containing three carboxyl groups, —COOH, and one hydroxyl group, —OH. It is a white, crystalline solid found in solution in the juice of lemons and some other fruits. Citric acid plays an important role in metabolism. It is extracted from citrus fruits or made by fermentation of molasses, and it is used as a flavouring in soft drinks and in the manufacture of effervescent powders. See also Krebs cycle and polybasic.

citric acid cycle See Krebs cycle.

clasper See caterpillar.

class A group of living organisms into which a phylum or division is divided. A class is subdivided into orders.

classification The arrangement of living organisms into a series of related groups according to their similarities and differences.

clathrate A substance in which a 'host' crystal lattice contains cavities like cages where a 'guest' atom or molecule is trapped. The two substances do not form a chemical compound and the only attraction between them may be hydrogen bonds or Van der Waals' forces; i.e. a clathrate can be considered to be a certain type of solid mixture. Clathrates form when the host compound crystallises in the presence of guest atoms or molecules of a certain suitable size. An example of a clathrate is $[C_6H_4(OH)_2]_3Ar$, where $C_6H_4(OH)_2$ is benzene-1,4-diol (quinol), the host, and Ar is argon, the guest. The guest atom can be liberated by melting or dissolving the crystalline product. Clathrates are useful in certain separation processes. See also zeolite and occlusion.

claustrophobia An exaggerated, abnormal fear of confined places. Compare agoraphobia.

clavicle (collar-bone) The anterior, long curved bone of the shoulder girdle of many vertebrates. In humans it is situated just above the first rib and connected to the shoulder-blade and the sternum (breastbone). See Fig. 207.

clay A type of soil consisting of very small particles, formed by weathering of certain rocks. It is mainly composed of hydrated aluminium silicates with small amounts of impurities such as the oxides of iron, calcium and magnesium. Very pure clay is white and is called kaolin or China clay. Clay is used in making ceramics, bricks, cement, paper, pencil leads, etc. Clay soil is hard when dry, smooth when moderately moist, heavy, sticky and plastic when wet. It is not easily permeable to water.

cleavage 1. Repeated cell division of the zygote immediately after fertilisation, resulting in a mass of cells. In plants this process is often called segmentation. **2.** The splitting of a crystal into two parts, each showing clean, smooth surfaces and lying parallel to the cleavage plane. This is due to a relative weakness of bonds along certain directions in the crystal.

climacteric (menopause) The period in a woman's life during which her menstruation becomes irregular and eventually ceases; i.e. egg cells are no longer produced in the ovaries and therefore pregnancy can no longer occur. The climacteric usually begins when a woman is between 45 and 55 years of age. It may last from less than one year to up to three years. At any age, surgical removal of the ovaries will result in an artificial climacteric.

climate The temperatures, humidity levels, wind speeds etc. prevailing in a particular geographical region.

clinical thermometer A mercury thermometer specially designed for measuring body temperatures. There is a constriction in the capillary above the bulb. Mercury can expand past this constriction, but will not pass through it after the thermometer has been used unless the thermometer is shaken. The sudden cooling and contraction of the mercury in the bulb causes the thread to break at the constriction, which enables the instrument to indicate the maximum temperature reached. A clinical thermometer is usually placed with its bulb under the tongue or in the anus. *See* Fig. 28.

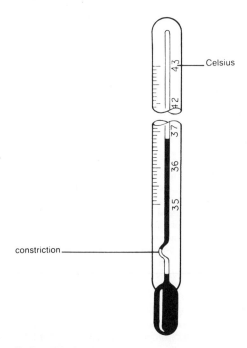

Fig. 28 Clinical thermometer

clinostat (klinostat) An instrument used for experimental investigation of geotropism and phototropism in seedlings and cut shoots. It consists of a cork disc attached to the spindle of a clockwork or an electric motor. The motor is adjustable to any angle from vertical to horizontal.

clitellum The swollen, saddle-like region of the skin of some Annelida. It contains glands secreting mucus to form a temporary sheath around copulating worms and also a cocoon in which fertilisation and development of eggs take place.

clitoris An organ in female mammals, homologous to the penis, situated at the upper part of the vulva above the urethral opening. It consists of two, short, fused, cylindrical masses of tissue. The free end of the clitoris is a small, rounded object (the glans) containing many sensory nerve-endings and consisting of erectile tissue.

cloaca A chamber in vertebrates (except most mammals) into which the intestinal, urinary and genital canals open.

clock An instrument for measuring time.

(1) *Early clocks* The sundial was the earliest device for measuring the passage of time. As the Sun moved across the sky, the shadow of the sundial's stick or gnomon was cast on a scale marked off in hours. Another device was the water clock which indicates the passage of time by the gradual flow of water from a container with a small hole in its base. Similar to the water clock is the hourglass. This clock contains sand used for measuring hours of time as it gradually runs from one container to another.

(2) *Mechanical clocks* These obtain the rotational force to move their hands from a falling weight or a spring. The rotation may be regulated by a pendulum or by the coiling and uncoiling of a fine spiral spring known as a hairspring, mounted in a balance-wheel.

(3) *Electric clocks* These may be driven by a battery-powered motor or a mains-powered motor. Battery-powered clocks are usually regulated by balance-wheels, whereas mains-powered clocks are precisely controlled by the frequency of the mains supply.

(4) *Crystal clocks* Crystal clocks are controlled by the oscillations of a quartz crystal which is made to vibrate by the application of a small voltage to it. The oscillations of the crystal produce voltages which can be used to drive a motor geared to the hands, or to power liquid crystal or light-emitting diode displays. Such clocks can have an accuracy of 0,001 (0.001) seconds a day. *See* piezo-electric effect.

(5) *Atomic clocks* These measure atomic time and are even more accurate than crystal clocks. They can have an accuracy of 1 second in 30 000 years. Atomic clocks are regulated by the vibrations of atoms. An example of this type of clock is the caesium clock which is used in the SI definition of the second.

clone A group of organisms produced asexually from a single parent and having identical genetical constitutions, apart from genetic variations due to mutation.

cloning A procedure designed to give rise to recognisable distinct clones.

Clostridium A genus of rod-shaped bacteria. Most of the species are Gram-positive, motile and anaerobic. They include *Clostridium botulinum*, which causes botulism, and *Clostridium tetani*, which causes tetanus.

clotting See blood clotting.

cloud A mass of condensed water vapour floating in the air. Clouds are formed when air is cooled and its water vapour condenses into tiny droplets of water or ice. Low-level clouds are known as mist or fog.

cloud chamber (Wilson cloud chamber) An apparatus for observing ionising particles by the tracks they produce passing through air saturated with water vapour. The chamber may be fitted with a piston whose outward movement produces an adiabatic expansion of the saturated vapour, cooling it and causing supersaturation. Particles such as electrons, protons and α-particles, are capable of ionising the air molecules as they pass through the chamber. The ions left in the path of the particles serve as nuclei for the condensation of the supersaturated water vapour, producing a visible track. See also bubble chamber.

CNS See central nervous system.

coagulation The process in which small particles in a solution collect together into larger particles (aggregates), eventually reaching such a size that they become visible under a microscope. When the particles form a precipitate or a gel the solution is said to coagulate.

coagulation of a protein When a protein is heated or treated with an acid or alkali, it becomes 'denatured'; i.e it coagulates and becomes insoluble. In most cases a denatured protein cannot be reconverted to the original protein. The denaturation is due to a change in the structure of the protein.

coagulin A substance capable of coagulating proteins.

coagulum A coagulated mass.

coal A fossil fuel of plant origin. Coal is found in certain types of sedimentary rocks and consists mainly of carbon, but also includes compounds containing hydrogen, oxygen, nitrogen and sulphur. In its initial formation, coal passes through several stages: peat, lignite, bituminous coal and anthracite. Principal coal fields date from the Carboniferous period, when the climate was warm and damp. Huge ferns and trees grew in swampy areas; when these plants died, they were covered by stagnant water, where lack of oxygen prevented their complete decay. Over a period of many millions of years the sea covered and uncovered large areas of land. In the course of this, layers of sedimentary rocks were laid down. The top layers of plant material and sedimentary rocks subjected the lower material to enormous pressure and the different stages of coal were formed.

coal-gas A fuel gas obtained by the destructive distillation of coal. Coal-gas contains hydrogen, H_2 (50%), methane, CH_4 (35%) and carbon monoxide, CO (8%). The rest is made up of nitrogen, N_2, carbon dioxide, CO_2, oxygen, O_2, and hydrocarbons other than methane.

coal-tar A black, thick, very complex mixture obtained by the destructive distillation of coal. Fractional distillation of coal-tar yields a large number of other valuable products such as benzene, C_6H_6, methylbenzene (toluene), $C_6H_5CH_3$, phenol, C_6H_5OH, and naphthalene, $C_{10}H_8$.

coarse chemical See fine chemical.

cobalamin(e) Vitamin B_{12}; abundant in liver and kidney, but absent in vegetables. This vitamin is very important in red blood cell formation, growth and reproduction in Man. It is used in the treatment of pernicious anaemia. Cobalamin(e) contains cobalt.

cobalt An element with the symbol Co; atomic number 27; relative atomic mass 58,93 (58.93); state, solid. Cobalt is a transition element (metal) occurring naturally in the minerals smaltite, $(Co, Ni)As_3$, and cobaltite (cobalt glance), CoAsS, which occur together with nickel ores. Smaltite belongs to a series of cobalt-nickel-arsenic minerals also containing some iron. The extraction of cobalt from the above-mentioned ores is rather complex. However, cobalt(II) dicobalt(III) oxide (tricobalt tetroxide), Co_3O_4, is eventually formed and then reduced by reacting it with carbon or aluminium at a certain high temperature:

$3Co_3O_4 + 8Al \rightarrow 9Co + 4Al_2O_3$

Cobalt is mainly used in alloys for the manufacture of permanent magnets, cutting tools and heating elements. An isotope of cobalt, cobalt-60, $^{60}_{27}Co$, is radioactive, having a half-life of 5,258 years (5.258 years). It is used in the treatment of cancer as it emits gamma rays. This isotope does not occur in nature, but is artificially produced by bombarding nuclei of cobalt with neutrons. See also Alnico.

cobalt(II) chloride $CoCl_2 \cdot 6H_2O$ (cobaltous chloride) A red cobalt salt of hydrochloric acid. The anhydrous salt is blue, but it turns red when it absorbs water. This property is used in testing for the presence of water. *See also* silica gel.

cobalt glance *See* cobalt.

cobaltite *See* cobalt.

cocaine A white, crystalline alkaloid which is found in the coca plant. Cocaine is a very potent local anaesthetic and is therefore of great value in minor surgical operations on the eye, nose, ear etc.

cocci *See* coccus.

coccus (pl. cocci) A rounded or roughly spherical bacterium.

coccygeal Pertaining to the coccyx.

coccygeal vertebrae The vertebrae forming the coccyx.

coccyx In humans, the last four, fused vertebrae of the spinal column. *See* Fig. 242.

cochlea A spirally coiled tube, like a snail shell, situated in the inner ear. It is an organ concerned with hearing, different parts of it responding to different frequencies. Nerve-endings in the cochlea unite to form the cochlear nerve which meets the vestibular nerve from the semicircular canals to form the auditory nerve. *See* Fig. 55.

cochlear nerve *See* cochlea.

cockroach *See* Dictyoptera.

cocoa A powder remaining after fatty substances have been extracted from crushed cacao seeds (cacao beans).

cocoon A protective covering produced by many larvae before they become pupae. Cocoons are also formed by many animals to protect their eggs. In the cocoon, fertilisation and development of the eggs may occur. In the commercial production of silk, the cocoons from mulberry silk moths are gathered and the silk is reeled. The larvae feed on mulberry leaves; when they are ready to pupate they spin thick, strong, silken cocoons. The thread of one cocoon may be three to four hundred metres long.

code *See* genetic code.

codeine An alkaloid which is a white, crystalline solid. It has a similar effect to that of morphine, and is found together with morphine in opium.

codon In messenger RNA (mRNA), a unit which consists of a particular sequence of three adjacent nucleotides and which codes for a given amino acid. One or several distinct codons exist for each amino acid. A similar triplet of nucleotides in transfer RNA (tRNA) is complementary to that of the codon and is called the anticodon. The anticodon 'recognises' the complementary sequence on the mRNA coding for an amino acid. The specific association between codon and anticodon ensures the accuracy of the protein synthesis. *See also* nonsense codon.

coefficient of friction *See* friction.

coefficient of viscosity *See* viscosity.

Coelenterata A phylum of aquatic (mostly marine) animals characterised by the absence of a definite head, respiratory, circulatory and excretory organs. Their nervous system is not centralised. They are also diploblastic. Examples include jellyfish, sea anemones, *Hydra* and corals. *See also* medusa.

coelom In many triploblastic animals, the main body cavity enclosed by the mesoderm and lined with epithelium. The coelom contains a fluid which, in the earthworm, gives the coelom a skeletal function because of its incompressibility. In many animals, it plays an important part in excretion.

coelomate A tripoblastic animal possessing a coelom at some stage in its life cycle.

coenzyme (cofactor) A small non-protein molecule which serves a specific function in several enzyme-catalysed reactions. A coenzyme activates an enzyme and may be firmly bound to or loosely associated with it. *See also* prosthetic group.

cofactor *See* coenzyme.

coffee A powder obtained from the seeds of the coffee plant. The seeds are roasted and ground before use.

cohesion The interaction between surfaces of the same material in contact, which causes them to cling together. *Compare* adhesion.

coil *See* solenoid.

coitus The act of sexual intercourse.

coke A grey, porous solid obtained from destructive distillation of coal. It is used as a fuel and in the extraction of metals from their ores, e.g. in blast furnaces. Coke is also used for making gaseous fuels.

cold-blooded (poikilothermic) Describes animals whose body temperature varies according to the surrounding temperature. All animals except birds and mammals are cold-blooded. *Compare* warm-blooded.

Coleoptera A very large order of insects commonly called beetles. They are characterised by modified fore-wings which act as horny covers to the hind part of the body. The membranous hind-wings may be small or absent. *See also* elytron.

coleoptile The sheath surrounding and protecting the plumule of some monocotyledons, especially grasses.

colic 1. Pertaining to the colon. 2. Severe abdominal pain.

collagen Fibrous protein forming white fibres in the connective tissue of vertebrate bones. It yields gelatin(e) on boiling. *See also* bone *and* scleroprotein.

collar-bone *See* clavicle.

collecting tubule A small tube which conveys urine from the cortex to the pelvis of the kidney. The collecting tubules are concerned with reabsorption of water. *See* Fig. 149.

collector *See* transistor.

collenchyma Supporting tissue in growing plant structures, especially in stems and leaves. Collenchyma cells may contain chloroplasts.

collimator In a spectrometer, the combination of a slit and the first lens. The collimator produces parallel light rays from the light source.

colloid A substance which is made up of very small particles intermediate between those of suspensions and those of true solutions. The size of colloidal particles is 1–100 nanometres; they may be single particles or several particles clinging together, all exhibiting Brownian movements. Milk and fruit jellies are types of colloids. *See also* sol.

colloidal particle *See* colloid.

colon Part of the large intestine in vertebrates, situated between the caecum and the rectum. It absorbs water and mineral salts from the faeces. *See* Fig. 49.

colony A collection of animals or plants living together, e.g. bees and ants. The term can also be used of bacteria.

colorimetry A method of analysis in which the concentration of the coloured sample is determined by the amount of coloured light it absorbs or transmits. In practice this may be done by measuring the amount of light at a certain wavelength which is absorbed by a solution of the sample and then comparing this value to the amount of light absorbed by a series of solutions whose concentrations are known. Colorimetric methods of analysis are often very simple to perform, using a photometer and a cuvette for the sample. The method is especially useful in finding the concentrations of small amounts of metals which form intensively coloured compounds. *See also* transmittance.

colostrum In mammals, the milk produced during the first few days after the birth of offspring. It is rich in proteins and contains antibodies.

colour The visual sensation arising from a particular wavelength of light striking the retina of the eye. White light can be separated into the colours of the spectrum, each having its own wavelength. These colours are red, orange, yellow, green, blue, indigo and violet.

colour blindness In humans, the inability to distinguish certain colours. This may be caused by missing or defective cones in the retina. The most common type is the inability to distinguish red from green. This defect is often inherited and sex-linked. It is more common in men than in women.

colour mixing 1. *Lights* The mixing of coloured lights is called colour mixing by addition. Mixing the three primary colours, red, green and blue, will produce a white patch on a white screen. Mixing any two primary colours results in the production of a secondary colour: red and blue produce magenta, blue and green produce cyan (peacock blue), red and green produce yellow. *See* Fig. 29 (a); *See also* complementary colours.

2. *Pigments* The mixing of pigments is called colour mixing by subtraction. The three basic colours are the three secondary colours resulting from colour mixing by addition. Mixing magenta, cyan and yellow produces black. Mixing any two of these basic colours results in the production of one of the primary colours: magenta and yellow produce red, magenta and cyan produce blue, yellow and cyan produce green. *See* Fig. 29 (b).

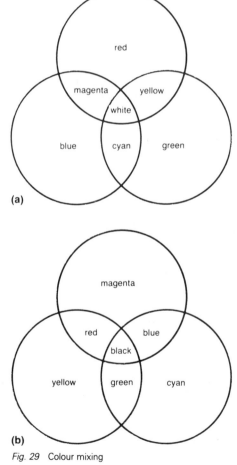

Fig. 29 Colour mixing

colours of the spectrum *See* colour.

columbium An obsolete name for the element niobium.

columella 1. The central pillar, composed of sterile tissue, in the sporangia of mosses and liverworts. **2.** A dome-shaped structure separating the sporangium from the sporangiophore in many fungi, e.g. *Mucor*.

columella auria In amphibians, birds and reptiles, an ear ossicle connecting the ear drum to the inner ear.

coma 1. A state of deep unconsciousness during which the brain ceases to operate normally and the body does not respond to physical stimuli. This can be brought about by disease, alcoholism, poison or injury. **2.** One of the three parts of a comet, consisting of a cloud of gas and dust.

combination A process in which two or more substances combine to form a single substance. *Example:* hydrogen combines with oxygen to form water:

$2H_2 + O_2 \rightarrow 2H_2O$

The term also applies to the combination of chemical compounds.

combining mass *See* equivalent mass.

combining power *See* valency.

combustion A process in which substances combine producing heat, light and sometimes flame. In this process at least one of the reactants is a gas. *Examples:* carbon burns in air to give carbon dioxide:

$C + O_2 \rightarrow CO_2$

Hydrogen combines with chlorine to give hydrogen chloride:

$H_2 + Cl_2 \rightarrow 2HCl$

The term also applies to combustion of chemical compounds.

comet A minor member of the Solar System, moving around the Sun in an elliptical or eccentric orbit. A comet consists of three parts: the nucleus (head), the coma and the tail. Most of the mass of a comet is situated in the nucleus, which is relatively small and brighter than the rest of the comet. The coma consists of a cloud of gas and dust surrounding the nucleus. The tail can be seen only when the comet is close to the Sun. It may consist of ionised molecules or unionised dust particles, and it always points away from the Sun. *See also* solar wind.

commensal *See* commensalism.

commensalism An association between two organisms in which one organism benefits and the other does not, nor is it harmed. *Example:* saprophytic bacteria may live in an animal's alimentary canal without benefiting or harming the animal. The benefiting organism is called the commensal. *See also* mutualism *and* parasitism.

common cold An infectious respiratory disease especially affecting the nose, throat and bronchi. It is caused by a number of viruses. Symptoms include headache, cough, occasionally fever, a blocked or running nose and reduced sensations of taste and smell.

common salt *See* sodium chloride.

community A group of interdependent organisms inhabiting a particular environment.

commutator A device which is used to reverse the direction of the current flowing in the armature coil of a d.c. motor and also to convert a.c. to d.c. in a generator. The simplest form of commutator consists of a split copper ring whose two halves are insulated from each other. Each half is connected to the wires of the armature coil and rotates with the coil. Connection to external circuits is made by carbon brushes held in contact with the outer surface of the commutator.

compact bone A dense layer of bone situated beneath the periosteum on the outside of the shafts and epiphyses. *See* Fig. 14.

companion cells Small cells with prominent nuclei and dense cytoplasm lying alongside sieve tubes in the phloem of flowering plants. It is thought that they are essential to the transport function of the sieve tubes. *See* Fig. 205.

compass An instrument used to indicate directions. The magnetic compass consists of a small magnet, freely pivoted at its centre, which moves in a horizontal plane over a scale marked with North, South, East, West and intermediate points (the points of the compass). Lines of force in the Earth's magnetic field cause the magnet to line up along the magnetic meridian which joins the magnetic North and South Poles of the Earth, with the north pole of the compass pointing to the magnetic North Pole. *Note:* The magnetic North Pole and the geographic North Pole do not coincide. *See also* angle of declination.

compensated pendulum A clock pendulum constructed in such a manner that its period is independent of temperature changes. One type of compensated pendulum is called Graham's pendulum. This consists of a cylindrical glass vessel of mercury supported by an iron rod. A rise in air temperature causes the iron rod to expand downwards. However, this downward expansion is compensated by the upward expansion of mercury. Conversely, a drop in air temperature causes the iron rod and mercury to contract. *See* Fig. 30.

compensation point In green plants, the point at which the rate of photosynthesis equals the rate of respiration, i.e. there is neither evolution nor absorption of either oxygen or carbon dioxide.

complemental air

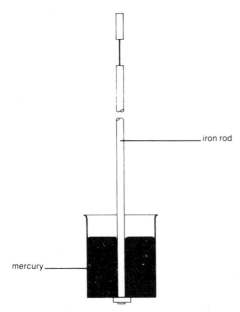

Fig. 30 Compensated pendulum (Graham's pendulum)

complemental air The additional intake (volume) of air, above tidal air volume, to fill the lungs completely.

complementarity The principle stating that a system can be explained in terms of particles or waves. *Example:* the behaviour of the electron can be explained either in terms of particles or in terms of wave motion. The particle model of the electron is said to be complementary to the wave model.

complementary colours Two colours producing white light when mixed by addition: e.g. blue and yellow, green and magenta, red and cyan. *See* Fig. 29 (a).

complete metamorphosis *See* metamorphosis.

complex A special type of chemical compound in which atoms or groups of atoms (ligands) form coordinate bonds with a metal atom or ion. *Examples:* a copper(II) ion, Cu^{2+}, forms a complex ion, $[Cu(NH_3)_4]^{2+}$, with ammonia, NH_3:
$Cu^{2+} + 4NH_3 \rightarrow [Cu(NH_3)_4]^{2+}$
An iron(II) ion, Fe^{2+}, forms the complex ion $[Fe(CN)_6]^{4-}$ with cyanide ions, CN^-:
$Fe^{2+} + 6CN^- \rightarrow [Fe(CN)_6]^{4-}$
An iron(III) ion, Fe^{3+}, forms the complex ion $[Fe(CN)_6]^{3-}$ with cyanide ions:
$Fe^{3+} + 6CN^- \rightarrow [Fe(CN)_6]^{3-}$
Tetracarbonylnickel(0) (nickel carbonyl), $[Ni(CO)_4]$, is an example of a neutral complex.

It is formed when metallic nickel, Ni, reacts with carbon monoxide, CO, at about 60°C:
$Ni + 4CO \rightarrow [Ni(CO)_4]$
The formation of complexes is especially seen among the transition metals. *See also* cyanoferrates, lone pair *and* coordination number.

complex ion *see* complex.

complex salt *See* salt.

Compositae The largest family of dicotyledons. A composite flower has a dense inflorescence (capitulum) made up of numerous florets. Examples include the daisy, the thistle and the sunflower.

composite flowers *See* Compositae.

compost Decayed organic material, usually of plant origin, applied to soil to increase its humus content.

compound *See* chemical compound.

compound eye (ommateum) The compound eye of an insect consists of thousands of hexagonal units called ommatidia. Each ommatidium contains a lens and light-sensitive cells. *See* Fig. 31; *compare* simple eye.

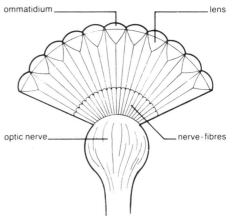

Fig. 31 Compound eye in section

compound leaf A leaf whose blade is divided into leaflets. *Compare* simple leaf.

compound microscope *See* microscope.

compression Part of a sound wave (longitudinal wave) in which molecules of a gas, liquid or solid are crowded together, making the pressure at that place a little above normal at that instant. *Compare* rarefaction. *See* Fig. 238.

computer An electronic machine which is able to perform calculations very quickly. A computer usually consists of an input unit, a processor, a main storage unit (memory) and an output unit. Computers are given instructions in the form of programs which are entered into the computer via the input unit. Programs can be written in any one of a number of specially designed programming languages, and are entered either indirectly as a card, paper tape, magnetic tape or disc, or directly via a keyboard. Once entered, the program is 'assembled' by the computer, and translated into machine code which is based on the binary arithmetic system. The computer then performs the calculations according to the instructions given in the program, and outputs the results via an output device, which may be a high-speed printer, a visual display unit (VDU), a typewriter, etc. The first computers used vacuum tubes as the basic electronic components. Now computers contain advanced and complicated integrated circuits as the basic semi-conductor part which make them faster, smaller and more reliable.

concave Curved inwards, like the inside of a sphere. The term is usually applied to lenses and mirrors. A concave lens is thinner at its centre and is a diverging lens, whereas a concave mirror is a converging mirror. *See also* lens *and* mirror.

concentrated Describes a solution containing a high proportion of solute.

concentration A term used to describe how much solute a given solution contains at a given temperature. There are several units of concentration, e.g. grams of solute per litre of solution, molarity and per cent.

concentric Describes circles which have the same centre and lie in the same plane.

concrete A mixture of stone or gravel, sand, cement and water used as a building material. To increase the tensile strength of concrete, steel rods can be embedded in it. The concrete is then called reinforced concrete.

condensate The liquid product of condensation.

condensation A process in which a liquid is formed from its vapour, usually by cooling.

condensation dimerisation *See* dimer.

condensation polymerisation *See* polymerisation.

condenser 1. An apparatus for converting a vapour into a liquid during distillation. A common type of condenser is the Liebig condenser. This consists of a glass tube surrounded by an outer jacket through which cooling water flows. The vapour passing through the tube is cooled and condenses. *See* Fig. 32. **2.** A mirror or lens combination in optical instruments which concentrates light on to an object, strongly illuminating it. In the projector, the condenser consists of two plano-convex lenses which concentrate rays of light from the projection lamp onto a slide which is the object. **3.** *See* capacitor.

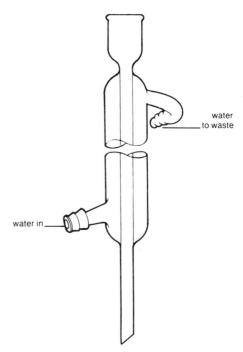

Fig. 32 Liebig condenser

conditioned reflex A reflex which has become modified by experience. This was first demonstrated in 1903 by the Russian physiologist Ivan Petrovich Pavlov, who used dogs in his experiments. He showed that, if a bell was rung each time a dog received food, the natural reflex of salivating at the sight and smell of food was conditioned to act by the artificial stimulus of ringing a bell, without any food near. Conditioned reflexes tend to become weaker with time if not practised. Other examples of conditioned reflexes are walking, swimming and writing. *Compare* unconditioned reflex.

condom A sheath of thin rubber or plastic to be pulled on to the erect penis before sexual intercourse. Ejaculated semen is retained in the condom, thus not entering the female reproductive system. The condom also provides some protection against infection by a venereal disease.

conductance Symbol: G. The ability of a conductor to conduct an electric current. The conductance of a conductor is the reciprocal of its resistance (for a direct current). For a circuit in which an alternating current flows, the conductance is its resistance divided by the square of its impedance. The SI unit of conductance is the siemens (S). The former unit, the reciprocal ohm or mho, is now obsolete.

conduction 1. The transfer of heat energy through a substance from a region of high temperature to a region of low temperature by molecular motion (vibration) between adjacent molecules, but without movement of the substance itself. In metals, which are good conductors of both heat and electricity, the mechanism is different because of the presence of free electrons. These electrons gain kinetic energy when heated and move towards cooler regions, losing some of this energy in collisions with cooler molecules. *See also* Ingenhousz's apparatus. **2.** The transfer of electrical energy through metallic conductors by the movement of free electrons migrating from atom to atom. In electrolytes, positive and negative ions act as charge carriers. In gases, both positive ions and electrons act as charge carriers.

conduction band The energy band in a crystalline solid in which electrons are free to move under the influence of an electric field.

conductivity, electrical A measure of the ability of a substance to conduct electricity. It is the reciprocal of the resistivity and is measured in siemens per metre.

conductivity, thermal *See* thermal conductivity.

conductor A material which is able to conduct heat and electricity. *See also* conduction.

condyle A rounded structure which fits into a socket. It may be joined directly to the shaft of a long bone. *Examples:* condyles are situated at the lower end of the femur, allowing a proper articulation with the tibia. There is also a condyle at each side of the lower jaw, articulating with the upper jaw.

cone 1. (strobilus) A reproductive structure bearing spores or seeds and consisting of tightly packed sporophylls grouped on a central axis, e.g. the cones of a pine tree. Usually distinct male and female cones are present. **2.** Specialised, light-sensitive cells in the retina of the eye of most vertebrates. Cones are concerned in colour vision and vision in bright light. They are conical in shape and are connected to the brain by nerve-fibres in the optic nerve. *Compare* rod.

configuration The three-dimensional arrangement in space of the atoms or groups of atoms in a molecule with regard to distances and angles. The configuration of a molecule is best illustrated using three-dimensional molecular models. *See also* conformation *and* electron configuration.

conformation A particular shape assumed by a molecule. Because of the free rotation about single bonds and a certain flexibility of bond angles, molecules which have the same structure and configuration may have different shapes in space. Since the rotation involves only small energy changes, the transition from one conformation to another may easily take place. However, certain conformations may be preferred to others. *Examples:* in 1,2-dichloroethane, $CH_2Cl—CH_2Cl$, the most likely conformation is the one in which the distance between the two chlorine atoms is farthest. In cyclohexane, C_6H_{12}, the preferred conformation is a 'chair' structure. When the cyclohexane ring has this structure it is strainless with bond angles of about 109°.

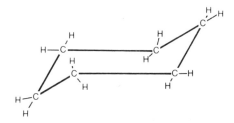

congenital A condition present at birth. *Example:* an impaired blood circulation due to a defect in the heart at birth is a congenital disease.

congestion Accumulation of blood in blood vessels of tissue. *Example:* a tight bandage around a finger constricts the blood vessels in the finger, producing congestion.

conical flask (Erlenmeyer flask) A cone-shaped flask made of glass. It is extensively used in the chemical laboratory. *See* Fig. 33.

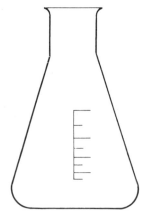

Fig. 33 Conical flask

conifer (softwood) A tree bearing cones.

Coniferales (Coniferae) The largest order of gymnosperms. Its members include *Pinus* (pine) and *Taxus* (yew). They are trees and shrubs which are mostly evergreen and are shaped like pyramids. Their leaves are simple and they bear separate male and female cones.

conjugated compound *see* conjugated system.

conjugated double bond *See* conjugated system.

conjugated protein *See* protein.

conjugated system A system of alternate single and double bonds between carbon atoms in an organic compound. *Examples:* buta-1,3-diene (butadiene)
$CH_2=CH-CH=CH_2$,
and methylbuta-1,3-diene (isoprene),
$CH_2=C-CH=CH_2$,
 $\quad\;\;|$
 $\;\;\;CH_3$
are examples of compounds containing a conjugated system; i.e. they are conjugated compounds.

conjugation 1. (zygosis) The fusion of two gametes to form a zygote. 2. The transfer of genetic material as exhibited by *Paramecium*. This is a form of sexual reproduction. 3. The one-way transfer of genetic material in *Spirogyra* when two filaments lie side by side. This is also a form of sexual reproduction. This type of conjugation takes place in certain bacteria.

conjunctiva A thin, mucous membrane lining the inner surface of the eyelid and covering the cornea of the vertebrate eye. *See* Fig. 66.

conjunctivitis (pink-eye) An inflammation of the conjunctiva causing the eyelids to swell and stick together, usually producing a pus-containing discharge.

connective tissue Fibrous vertebrate tissue supporting all organs in the body and connecting other tissues. Examples include bone, cartilage, areolar tissue and adipose tissue.

conservation (laws of) (1) *Law of conservation of energy* A law stating that, in any system, energy can neither be created nor destroyed.
(2) *Law of conservation of matter (mass)* A law stating that, in any system, matter (mass) can neither be created nor destroyed. *See also* Einstein's equation.
(3) *Law of conservation of momentum* A law stating that the total momentum of colliding bodies before impact is equal to their total momentum after impact, provided that no external forces are acting.

constantan (eureka) An alloy of copper (60%) and nickel (40%). Its electrical resistance does not vary with temperature. It is used in electrical equipment.

constant boiling mixture *See* azeotropic mixture.

constant pressure gas thermometer *See* gas thermometer.

constant volume gas thermometer *See* gas thermometer.

constellation A group of stars forming a recognisable pattern in the sky, e.g. Orion and the Plough.

constriction A narrowing of a tube or vessel, e.g. the constriction of the capillary in a clinical thermometer.

consumer A living organism, normally an animal, feeding directly or indirectly on plant material. Consumers are divided into two groups, primary and secondary. Primary consumers are herbivores, secondary consumers are carnivores or omnivores.

contact process The process used in the industrial manufacture of sulphuric acid, H_2SO_4. First iron pyrites, FeS_2, is roasted to make sulphur dioxide, SO_2:
$4FeS_2 + 11O_2 \rightarrow 2Fe_2O_3 + 8SO_2$
The sulphur dioxide is then catalytically oxidised with air to form sulphur(VI) oxide, SO_3:
$2SO_2 + O_2 \rightarrow 2SO_3$
The sulphur(VI) oxide is absorbed in concentrated sulphuric acid, forming oleum (fuming sulphuric acid), $H_2S_2O_7$:
$SO_3 + H_2SO_4 \rightarrow H_2S_2O_7$
When oleum reacts with water, sulphuric acid is formed:
$H_2S_2O_7 + H_2O \rightarrow 2H_2SO_4$

contagious Describes an infectious disease which is transmitted by direct contact.

continuous cropping Cultivation of the same piece of land year after year, often leading to soil exhaustion, erosion and low productivity.

continuous spectrum A spectrum which is composed of a continuous region of emitted or absorbed wavelengths. The spectra of hot solids have continuous emission in the visible or infra-red regions. Continuous spectra occur as a result of electrons moving between different energy levels. *See also* line spectrum.

contour feather One of the types of feathers which cover a bird's body. The contour feathers give it shape and form the wings and tail. The larger contour feathers are called the quill feathers or flight feathers. The smaller contour feathers are called coverts; they cover the bases of the flight feathers. *See* Fig. 67.

contraceptive A device for preventing pregnancy, e.g. the pill, the diaphragm or cap, and the condom.

contractile Capable of contraction.

contractile root A root, found on corms and bulbs, which shortens with age. These roots pull the plant deeper into the soil.

contractile vacuole In Protozoa, particularly freshwater species, a vacuole which expands and contracts at regular intervals. The vacuole fills with water from the cytoplasm through the vacuole membrane and expels its contents to the exterior of the cell during contraction. Excess water entering the protozoon by osmosis is expelled from it by the action of the contractile vacuole. *See also* diastole.

contraction A movement which leads to either a shortening of length or a lessening of diameter. Contraction is seen in muscles, the pupil of the eye, etc.

control experiment (control) When an experiment is set up to test the effect of a particular set of conditions, a control experiment is often set up which is identical except that the important conditions are replaced by normal conditions. This is necessary to show that the conditions being tested are responsible for the results which are observed. *Example:* in an experiment to test the effects of growing plants in darkness, identical plants are grown in normal light as a control.

convection The transfer of heat energy in fluids by the motion of the fluid molecules. Molecules in contact with the heat source gain heat energy, move faster and further apart, and rise. Colder, more densely packed molecules move to take their place, thus setting up convection currents.

conventional current An electric current considered to flow from positive to negative, i.e. in the opposite direction to electron flow. This convention is useful in describing certain electromagnetic phenomena. The convention dates from a time when the electronic nature of current flow was not understood.

converging lens *See* lens.

converter 1. A machine which converts a.c. to d.c., or, less frequently, d.c. to a.c. **2.** The large egg-shaped vessel used in the Bessemer process.

convex Curved outwards, like the outside of a sphere. The term is usually applied to lenses and mirrors. A convex lens is thicker at its centre and is a converging lens, whereas a convex mirror is a diverging mirror. *See also* lens *and* mirror.

convoluted tubule Part of the uriniferous tubule in the kidney concerned with the reabsorption of glucose, water and amino acids.

coolant Any fluid used in reducing the temperature of a system.

coordinate bond *See* complex *and* lone pair.

coordination number 1. The number of bonds from a metal atom or ion to its ligands in a complex. *Example:* in the complex ion $[Cu(NH_3)_4]^{2+}$ the coordination number of copper is four. **2.** In an ionic crystal, the number of ions of opposite charge which surround another ion. *Examples:* in sodium chloride, NaCl, the cations have the coordination number 6; in calcium fluoride, CaF_2, they have the coordination number 8.

copper An element with the symbol Cu; atomic number 29; relative atomic mass 63,55 (63.55); state, solid. A reddish-brown transition metal occurring as the free metal and as copper pyrites, $CuFeS_2$, copper glance, Cu_2S, and cuprite, Cu_2O. It is often extracted from copper pyrites by roasting the ore:
$2CuFeS_2 + 4O_2 \rightarrow Cu_2S + 3SO_2 + 2FeO$
The copper(I) sulphide, Cu_2S, is then reduced to copper by heating it in a certain amount of air:
$Cu_2S + O_2 \rightarrow 2Cu + SO_2$
The extraction processes are more complex than described here. The crude copper is refined electrolytically. After silver, copper is the best conductor of heat and electricity. For this reason it is used to make electric wires, cooking utensils, alloys, etc. In air, copper is only superficially oxidised, producing a green coating of basic copper(II) carbonate or copper(II) sulphate. *See also* verdigris.

copper(II) carbonate $CuCO_3$ (cupric carbonate) A copper salt which is unknown in the pure state since it always occurs as a green, insoluble, basic carbonate with the formula $CuCO_3 \cdot Cu(OH)_2$. It is soluble in dilute acids and in ammonia solution. When heated, it decomposes into copper(II) oxide, CuO, carbon dioxide, CO_2, and water vapour:
$CuCO_3 \cdot Cu(OH)_2 \rightarrow 2CuO + CO_2 + H_2O$
See also malachite.

copper chlorides Copper(I) chloride (cuprous chloride), CuCl, and copper(II) chloride (cupric chloride), $CuCl_2$.
Copper(I) chloride, CuCl, is a white, insoluble solid copper salt formed by boiling a mixture of copper turnings, copper(II) chloride solution and concentrated hydrochloric acid, HCl, and pouring the dark solution formed into water:
$Cu + CuCl_2 \rightarrow 2CuCl$
Copper(II) chloride, $CuCl_2$, is an anhydrous, soluble brown solid copper salt giving a brownish aqueous solution when concentrated. When diluted, the solution changes its colour to green and then blue. $CuCl_2$ is formed when copper(II) oxide, CuO, is treated with hydrochloric acid, HCl:
$CuO + 2HCl \rightarrow CuCl_2 + H_2O$
There is also a green, soluble copper(II) chloride, $CuCl_2 \cdot 2H_2O$.

copper glance Cu_2S (cuprous sulphide, copper(I) sulphide, chalcocite, chalcosine) A black, insoluble copper salt often associated with other copper minerals.

copper(II) hydroxide $Cu(OH)_2$ (cupric hydroxide) A blue-green, insoluble, gelatinous base which can be made by mixing an aqueous solution of a copper(II) salt with sodium hydroxide solution. When heated, it decomposes into copper(II) oxide, CuO, and water vapour:
$Cu(OH)_2 \rightarrow CuO + H_2O$

copper(II) nitrate $Cu(NO_3)_2$ (cupric nitrate) A blue, deliquescent, soluble copper salt. When heated, it decomposes into copper(II) oxide, CuO, nitrogen dioxide, NO_2, and oxygen, O_2:
$2Cu(NO_3)_2 \rightarrow 2CuO + 4NO_2 + O_2$
Usually copper(II) nitrate has three molecules of water of crystallisation: $Cu(NO_3)_2 \cdot 3H_2O$.

copper oxides Copper(I) oxide (cuprous oxide), Cu_2O, and copper(II) oxide (cupric oxide), CuO.
Copper(I) oxide, Cu_2O, is an insoluble, red solid which can be obtained by reduction of an alkaline solution of copper(II) sulphate, $CuSO_4$.
Copper(II) oxide, CuO, is an insoluble, black solid which can be obtained, for example, by heating copper(II) nitrate, $Cu(NO_3)_2$:
$2Cu(NO_3)_2 \rightarrow 2CuO + 4NO_2 + O_2$

copper pyrites $CuFeS_2$ (chalcopyrite) A rather abundant copper ore which is an insoluble yellow mineral.

copper(II) sulphate $CuSO_4 \cdot 5H_2O$ (cupric sulphate) A blue, soluble copper salt which can be obtained in many ways, e.g. by treating metallic copper with hot, concentrated sulphuric acid, H_2SO_4, and then crystallising the solution:
$Cu + 2H_2SO_4 \rightarrow CuSO_4 + SO_2 + 2H_2O$
or by treating copper(II) oxide, CuO with dilute sulphuric acid:
$CuO + H_2SO_4 \rightarrow CuSO_4 + H_2O$
Copper(II) sulphate is used in Bordeaux mixture and as a wood preservative. The anhydrous form of copper(II) sulphate is greyish-white. When it reacts with water, the blue pentahydrate, $CuSO_4 \cdot 5H_2O$ is formed. This property is used as a test for the presence of water.

coprecipitation The carrying down of normally soluble impurities during precipitation of an insoluble compound. Coprecipitation occurs to some extent in any precipitation, but is especially marked with colloidal precipitates. The effect of coprecipitation may be minimised by thorough washing of the precipitate. *See also* occlusion.

copulation Sexual union in which spermatozoa are transferred from a male to a female animal and which may result in fertilisation of ova.

coracoid A paired, ventral bone in vertebrates. In birds, the coracoid acts as a strut between the breastbone and the spine. The action of the coracoid causes the wing bone to be pulled down rather than the breastbone pulled up when the flight muscles contract. In mammals, the coracoid is reduced to a small, thick, curved process (coracoid process) on the anterior surface of the scapula.

coracoid process *See* coracoid.

coral A mass of white or pink calcareous material, perforated by tiny holes. This material is of animal origin, secreted by certain polyps belonging to the phylum Coelenterata. The term is also used as a name for the polyps. Corals produce skeletons formed from calcite, $CaCO_3$. They usually live in colonies in warm, clear water at a depth of seldom more than ten metres, where they form a great thickness of rock.

cord *See* spinal cord, umbilical cord *and* vocal cord.

cordate Shaped like a heart. *See* Fig. 111.

core 1. A piece of soft iron around which a coil of insulated wire is wound. **2.** The central part of a nuclear reactor. **3.** That part of an apple, pear, etc., which contains the seeds.

cork (phellem) A waterproof, protective layer of dead, air-filled cells situated on the outside of the stems of woody plants. These cells are produced by cork cambium (phellogen).

cork cambium (phellogen) A layer of cells giving rise to cork and phelloderm. *See also* bark.

corkscrew rule *See* Maxwell's corkscrew rule.

corm A specialised organ of vegetative reproduction consisting of a short, swollen underground stem containing stored food. *See* Fig. 34.

cornea A transparent continuation of the sclera (sclerotic) of the vertebrate eye. Its outer surface is covered by the conjunctiva. The cornea is the first of the four refracting media in the eye and the greatest refraction takes place at the cornea. *See* Fig. 66.

corolla The collective name for the petals of a flower.

corona 1. The outer region of the Sun's atmosphere, visible to the naked eye during a total eclipse of the Sun. *See also* halo. **2.** A luminous, electrical discharge which is seen as a glow round the surface of a charged conductor at a sufficiently high potential to cause ionisation of the surrounding air.

coronary Shaped like a crown; forming a circle round a structure, e.g. coronary arteries encircling the heart.

coronary thrombosis (heart attack) A condition in which one of the arteries supplying the heart with blood becomes blocked by a

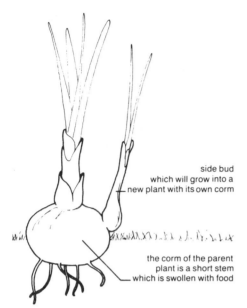

Fig. 34 Crocus corm

blood clot. Symptoms are severe chest pain, accompanied by pallor and sweating. *See also* cholesterol.

coronary vessels Arteries and veins of vertebrates carrying blood to and from the tissues of the heart muscle.

corpus (pl. corpora) A homogenous structure forming part of an organ.

corpus callosum In placental mammals, a band of nerve-fibres connecting the two cerebral hemispheres. *See* Fig. 16.

corpuscle A minute particle which is a cell floating in a fluid, e.g. a blood corpuscle.

corpus luteum A solid mass of tissue forming a temporary organ in the mammalian ovary. It forms when an ovum (egg) has been discharged from the ovary and functions as a temporary ductless gland to secrete the hormone progesterone, which initiates changes in the womb preparing it for pregnancy. If the ovum is not fertilised the corpus luteum degenerates. However, if fertilisation does take place it persists for some time. *See also* relaxin *and* luteinising hormone.

corrosion The gradual destruction of a metal surface by the action of chemicals, air or water. Most reactions leading to corrosion are solid–liquid or solid–gas reactions. However, solid–solid corrosion also occurs. *See* rusting of iron *and* electrolytic corrosion.

cortex (pl. cortices) The outer layer of a structure or organ, e.g. cortex of kidney, cortex of root.

corticotrophin (ACTH) A polypeptide hormone secreted by the pituitary gland and stimulating the adrenal cortex so that the adrenal glands release other hormones such as cortisone.

cortisone A hormone (a steroid) produced in the outer layer (cortex) of the adrenal glands. It is concerned with carbohydrate metabolism and the maintenance of connective tissue. Cortisone is used in the treatment of several diseases which are not caused by lack of natural cortisone, such as rheumatic fever, rheumatoid arthritis, some types of asthma, Addison's disease and skin ailments. However, it has a number of side-effects, including swelling of tissue, diabetes and high blood pressure.

Corti's organ An organ consisting of hair cells and supporting tissue, situated in the cochlear duct of the mammalian ear. It is concerned with the reception of stimuli and transmission of sound impulses.

corundum *See* alumina.

corymb A type of inflorescence in which the lower pedicels are longer than the upper, thus forming a flattened cluster of flowers at the same level. *See* Fig. 179.

cosmic rays Rays consisting of high energy particles, usually protons with smaller amounts of heavier nuclei. They travel through space at very high speeds. When they enter the Earth's atmosphere they collide mainly with nuclei of nitrogen and oxygen, giving rise to secondary particles such as neutrons which can reach ground level. The origin of cosmic rays is still uncertain, although a small proportion come from the Sun.

cosmogony A branch of cosmology concerned with the origin and evolution of the universe.

cosmology The study of the universe as a whole, including its origin, its evolution and theories of its future development.

cotton Almost pure cellulose formed in the cell walls of cotton fibres. These fibres are produced by the seed pods of the cotton plant. Cotton is mainly used in the manufacture of textiles.

cotyledon The seed-leaf or primary leaf of a plant embryo, playing an important part in seedling development. In some seeds, e.g. peas and beans, the cotyledon is a food storage organ; in other seeds, e.g. grasses, the cotyledon absorbs food from the endosperm and passes it on to the seedling.

coulomb Symbol: C. The SI unit of electric charge, defined as the charge transported by an electric current of one ampere in one second.

coulombmeter (voltameter) An apparatus used for measuring quantities of electric current by depositing a metal from an electrolyte during electrolysis. The metal is deposited on the cathode and the total amount of electricity which is passed through the electrolyte can be calculated from the increase in the mass of the cathode and the electrochemical equivalent of the metal. Formerly a silver coulombmeter was used in defining the unit of quantity of electricity. *See also* Hofmann voltameter.

Coulomb's law A law stating that the force of attraction or repulsion between two point charges is directly proportional to the product of the charges and inversely proportional to the square of the distance between them:

$$F = k \frac{Q_1 Q_2}{r^2}$$

where k is a proportionality constant, Q_1 and Q_2 are the charges and r is the distance between them. The magnitude of k depends on the units in which F, Q_1, Q_2 and r are expressed.

coumarin $C_9H_6O_2$ An aromatic, heterocyclic compound occurring in certain plants such as clover and grasses. It is synthetically prepared and used in perfumery and in flavours.

counter A device for detecting ionising particles or radiation, used in association with an electronic circuit for counting the electric pulses resulting from the impacts of individual particles. *See also* Geiger-Müller counter.

counterstain *See* staining.

couple Two equal and parallel forces acting together in opposite directions, thus producing a turning effect. The moment of a couple is the product of either force and the perpendicular distance between them.

coupling *See* azo compound.

covalent bond A single bond between two similar or different atoms, consisting of two electrons (one pair) with one electron contributed by each of the atoms. The two electrons are shared between the atoms. In a covalent double bond four electrons (two pairs) are shared between the atoms, and in a triple bond six electrons (three pairs) are shared. In the hydrogen molecule, H_2, the two atoms are held together by a covalent bond. Covalent compounds usually have low boiling-points and melting-points, are liquids or gases and consist of molecules. *Compare* ionic bond; *see also* octet.

covalent compound *See* covalent bond.

covalent crystal A crystal in which the atoms are held together by covalent bonds. A typical example is diamond.

coverslip A very thin piece of glass used to cover specimens prepared on a microscope slide.

covert *See* contour feather.

Cr Chemical symbol for chromium.

cracking In the petroleum industry, a method for converting large hydrocarbon molecules into small ones or to alkenes. This is brought about by heating and using catalysts. *Examples:* butane, C_4H_{10}, can be converted by cracking to ethane, C_2H_6, and ethene, C_2H_4:

$CH_3\text{—}CH_2\text{—}CH_2\text{—}CH_3 \rightarrow$
$\phantom{CH_3\text{—}CH_2\text{—}CH_2\text{—}}CH_3\text{—}CH_3 + CH_2\text{=}CH_2$

or it can be converted to but-1-ene, C_4H_8, and hydrogen, H_2:

$CH_3\text{—}CH_2\text{—}CH_2\text{—}CH_3 \rightarrow$
$\phantom{CH_3\text{—}CH_2\text{—}}CH_3\text{—}CH_2\text{—}CH\text{=}CH_2 + H_2$

depending on the catalyst and the temperature. Cracking is an important process because distillation of crude oil does not give a large enough yield of petroleum products to meet the demand, and it also gives far more of the higher-boiling hydrocarbons than required.

cranial Pertaining to the skull.

cranial nerves In vertebrates, paired nerves arising from the brain. Most of these nerves are concerned with the sense organs and muscles of the head: however, the vagus nerve connects the brain with other parts of the body. In mammals there are twelve pairs of cranial nerves, including the auditory nerve, olfactory nerve and optic nerve.

Craniata *See* Vertebrata.

craniology The study of the skull.

cranium The name for the skull or the part of the skull enclosing the brain.

cream of tartar $C_4H_5O_6K$ (potassium hydrogen tartrate) A white, crystalline solid used in baking powder and in medicine. It is a potassium salt.

creatine An amino acid found in the muscles of vertebrates. It is formed during muscle contraction when phosphocreatine (phosphagen) is hydrolysed with the liberation of phosphoric acid, H_3PO_4, and energy.

creatinine A waste nitrogenous substance formed from the dehydration of creatine; i.e. creatinine is the anhydride of creatine. It is found in muscles, blood and urine.

creep The gradual enlargement of a material when exposed to a certain (large) strain for a certain time. This effect is highly dependent on temperature. At temperatures above a certain level, all materials will creep if exposed to a strain; i.e. creep may occur at even low strains.

crenate Having a margin or edge with small, shallow, regular curves. *See* Fig. 111.

Cretaceous A geological period of approximately 72 million years occurring at the end of the Mesozoic. It ended approximately 64 million years ago.

cretin A person with a deformity caused by cretinism.

cretinism A deficiency, before puberty, of the hormone thyroxine secreted by the thyroid gland. This leads to a type of dwarfism with marked mental retardation. In adults, a lack of thyroxine is known as myxoedema. Symptoms are thickening of the skin, and mental and physical indolence. *See also* goitre.

cricket The common name for some members of the order Orthoptera.

crista (pl. cristae) **1.** In a mitochondrion, a fold in the inner membrane. Cristae divide up the interior of the mitochondrion. **2.** A raised ridge of sensory cells and supporting cells in the ampulla at the end of each semicircular canal. A tuft of hairs on the crista supports the cupula. This is a gelatinous mass of tissue that moves when movements of the head cause currents in the endolymph in the ampulla. The sensory cells of the crista are stimulated by movements of the cupula.

critical angle Symbol: c. The least angle of incidence of a ray of light passing from a dense to a less dense medium, e.g. from glass to air, at which grazing refraction occurs, i.e. the angle of refraction is equal to 90°. When this angle of incidence is exceeded, total internal reflection occurs. *See* Fig. 35.

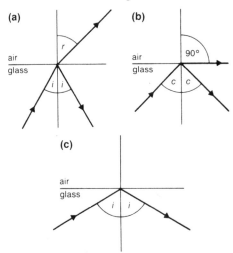

Fig. 35 Critical angle: what happens when light passes from glass to air
(a) Refraction and internal reflection $i < c$
(b) Critical internal reflection $i = c$
(c) Total internal reflection $i > c$

critical mass The minimum mass a radioactive isotope must have in order to maintain a spontaneous fission chain reaction. The value of the critical mass depends on the type and purity of the isotope. On conditions at which the production of neutrons by the fission process just equals loss by escape and capture by the isotope and impurities, the critical mass or size is said to have been reached. Under subcritical conditions, that is when too few neutrons are produced, the chain reaction will not occur. Production of too many neutrons may lead to a nuclear explosion.

critical pressure Symbol: P_c. The saturated vapour pressure of a liquid at its critical temperature.

critical temperature Symbol: T_c. The temperature above which a gas cannot be liquefied by an increase of pressure.

critical volume Symbol: V_c. The volume of a given mass of gas at its critical temperature and critical pressure.

CRO *See* cathode-ray oscilloscope.

Crookes' radiometer An instrument for detecting heat radiation. It consists of a partially evacuated glass bulb, containing at its centre a fine pivot supporting four lightweight metal arms, each carrying a vane of mica painted black on one surface. When molecules of air strike the warmer blackened vane surfaces and rebound, they carry away more momentum than those which rebound from the cooler unpainted vane surfaces. This results in the rotation of the metal arms in such a direction that the blackened vane surfaces turn away from the source of radiation. *See also* radiant energy.

Crookes' spinthariscope *See* spinthariscope.

crop 1. A sac-like enlargement of the gullet of birds where food is stored temporarily and softened by absorption of water. A similar structure is found in the alimentary canal in many invertebrates, where food is stored and digested. **2.** The produce of cultivated plants, e.g. cereals, fruits and legumes.

crop rotation A system of agriculture in which a field is planted with a succession of crops in such a way that the soil is not depleted of one particular group of minerals. Crop rotation also discourages pests, e.g. weeds, insects and fungi.

cross A product of cross-fertilisation.

cross-fertilisation (allogamy) The process in which a male and a female gamete fuse, the two gametes being derived from different individuals of the same species. *Compare* self-fertilisation.

crossing A term used for cross-fertilisation or cross-pollination.

crossing-over The mutual exchange of chromatid material between pairs of homologous chromosomes during meiosis. *See also* chiasma.

cross-pollination The transfer of pollen from the anther of one individual flower to the stigma of another on a different plant of the same species. *See* Fig. 36. *Compare* self-pollination.

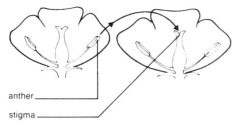

Fig. 36 Cross-pollination

crown 1. The part of a tooth that is covered by enamel. **2.** The leafy, upper part of a tree.
CRT *See* cathode-ray tube.
crucible A vessel made from porcelain or another heat-resistant material, used when a substance has to be heated to a high temperature. *See* Fig. 37.

Fig. 37 Crucible

crude birth rate The total number of live births in a year divided by the total population at the middle of that year, and multiplied by one thousand:

crude birth rate =

$\dfrac{\text{number of live births in a year} \times 1000}{\text{total population at the middle of that year}}$

crude death rate The total number of deaths in a year divided by the total population at the middle of that year, multiplied by one thousand:

crude death rate =

$\dfrac{\text{number of deaths in a year} \times 1000}{\text{total population at the middle of that year}}$

crude oil (petroleum) A thick, black liquid mixture of gaseous, liquid and solid hydrocarbons, obtained from deposits in the ground. It is refined by fractional distillation to give hydrocarbon mixtures of less complex composition such as petrol and paraffin.
crumbs In soil, collections of inorganic particles and humus which have a diameter of about 3 mm or less.
crust The outermost and thinnest layer of the Earth (lithosphere), which has a maximum thickness of about 40 kilometres.
Crustacea A large class of arthropods, mainly aquatic. In addition to the normal structure of arthropods, crustaceans have a carapace and two pairs of antennae in front of the mouth. Examples include crabs, lobsters, shrimps and woodlice.
cryolite Sodium aluminium fluoride, Na_3AlF_6. A white or colourless mineral formerly used as a flux in the electrolytic manufacture of aluminium. The previously vast deposits of cryolite found in Greenland are now almost exhausted and an artificial mixture of sodium, aluminium and calcium fluorides has now almost entirely replaced it as a flux.
cryometer A thermometer designed for measuring very low temperatures.
cryophyte An alga, bacterium, fungus or moss growing on snow or ice.
Cryptogamia An older group name used by botanists for plants which have no flowers and produce no seeds, e.g. bryophytes, pteridophytes and thallophytes.
crystal A substance which has an orderly inner arrangement of atoms, ions or molecules. The pattern of atoms, ions or molecules in a crystal is regular and repeating in all dimensions giving rise to the flat faces of a crystal; i.e. crystals have a regular geometrical shape (a lattice). *See* Fig. 210; *see also* covalent crystal, ionic crystal *and* molecular crystal.
crystal clock *See* clock.
crystalline Having a crystal structure. Crystalline substances have fixed melting-points.
crystalline lens *See* eye lens.
crystallisation The process of forming crystals from a supersaturated solution at a given temperature.
crystallography The study of the structure of crystals, including crystal forms and properties. *See also* X-ray crystallography.
Cu Chemical symbol for copper.
cubic centimetre A unit of volume, cm^3. One cubic centimetre (cm^3) is equal to one millimetre (ml).
cubic expansivity Symbol: γ (gamma). The cubic expansivity of a substance is approximately equal to three times its linear expansivity:
$\gamma = 3\alpha$
cud In ruminants, partially digested food regurgitated from the rumen to the mouth for further chewing.

cultivar A plant variety produced in agriculture or horticulture.

cultivation 1. The preparation and use of soil to produce crops. **2.** The improvement of plants.

cultivator An implement or machine used to cultivate soil or plants. The term is also applied to a person who cultivates.

culture The growth of micro-organisms or tissues, e.g. bacteria and fungi, in prepared media for scientific purposes.

culture medium A sterile preparation, e.g. nutrient agar, for the growth and cultivation of micro-organisms.

cupric Term formerly used in the names of copper compounds to indicate that the copper is divalent (bivalent); e.g. cupric oxide, CuO, now called copper(II) oxide. *Compare* cuprous.

cupric carbonate *See* copper(II) carbonate.

cupric chloride *See* copper chlorides.

cupric hydroxide *See* copper(II) hydroxide.

cupric nitrate *See* copper(II) nitrate.

cupric oxide *See* copper oxides.

cupric sulphate *See* copper(II) sulphate.

cuprite *See* copper.

cuprous Term formerly used in the names of copper compounds to indicate that the copper is monovalent (univalent); e.g. cuprous oxide, Cu_2O, now called copper(I) oxide. *Compare* cupric.

cuprous chloride *See* copper chlorides.

cuprous oxide *See* copper oxides.

cuprum The Latin name for copper.

cupula *See* crista.

curare A poison obtained from a number of different plant species and existing in many varieties. The precise chemical nature of most of these varieties is not completely known; however, one variety was isolated and its structure established in 1935. Since then it has been used in medicine, e.g. to produce muscle relaxation during surgery.

curie Symbol: Ci. A unit used to express the quantity of radioactivity associated with a given product. One curie is defined as being equal to 3.7×10^{10} (3.7×10^{10}) distintegrations per second. This is almost equal to the radioactivity of 1 gram of the radium isotope $^{226}_{88}Ra$. For any radioactive element, its activity in curies can be calculated from the expression:

$$\left(Ci = \frac{1.13 \times 10^{16} \times M}{A \times t} \right)$$

where M is the mass of the sample in kilograms, A is the mass number and t is the half-life of the element in seconds. The curie is not an SI unit. The SI unit of radioactivity is the bequerel (Bq).
1 Ci = 3.7×10^{10} Bq
(1 Ci = 3.7×10^{10} Bq)

Curie point *See* Curie temperature.

Curie temperature (Curie point) The temperature at which a ferromagnetic material becomes merely paramagnetic, i.e. loses its magnetism. This happens when the energy associated with thermal motion becomes large enough to make the alignment of the domains disappear. The Curie temperature is usually lower than the melting-point. For iron the Curie temperature is 770°C, for cobalt 1131°C and for nickel 358°C. *See also* domain theory.

current *See* electric current.

current balance An instrument for the accurate determination of the size of an electric current by measuring the force produced between conductors carrying the current. Laboratory ammeters are calibrated during manufacture and repair by connecting them in series with a standard current balance. *See also* ampere.

cusp A rounded projection or sharp point, e.g. a cusp of a tooth.

cuspids *See* canine teeth.

cutaneous Pertaining to, or of, the skin.

cuticle The thin outer skin covering animals and plants. It is composed of non-cellular material secreted by the epidermis. In higher plants, the cuticle forms a continuous layer of a waxy material, cutin, over the aerial parts with gaps for stomata and lenticels. Its chief function is the prevention of excessive water loss, but it also protects against mechanical injury. In Arthropoda it forms the exoskeleton and is composed of chitin and protein. In insects its outer surface is covered with a thin layer of wax preventing evaporation of water from the body.

cutin A wax-like substance forming the cuticle of plants. It is composed of a complex mixture of oxidation and condensation products of fatty acids.

cutting A piece of branch or stem, detached from the parent plant and used as a means of vegetative reproduction.

cyan *See* colour mixing.

cyanide A salt of hydrocyanic acid, HCN.

cyanoferrates Hexacyanoferrate(II) (ferrocyanide), $[Fe(CN)_6]^{4-}$, and hexacyanoferrate(III) (ferricyanide), $[Fe(CN)_6]^{3-}$. *Hexacyanoferrate(II)*, $[Fe(CN)_6]^{4-}$, is a yellow complex ion formed by adding cyanide ions to an aqueous solution containing iron(II) ions, Fe^{2+}. If hexacyanoferrate(III) ions are added to a solution of an iron(II) salt, a dark blue precipitate called Turnbull's blue is formed. An almost identical precipitate called Prussian blue is obtained by mixing hexacyanoferrate(II) ions and iron(III) ions.

Hexacyanoferrate(III), $[Fe(CN)_6]^{3-}$, is a reddish complex ion formed by adding cyanide ions, CN^- to an aqueous solution containing iron(III) ions, Fe^{3+}.

cyano-group See nitrile.

Cyanophyta A division comprising blue-green algae. These algae contain a blue pigment which gives them their colour. However, in some species other pigments may be present in varying proportions, producing colours such as red and yellow. Blue-green algae live mainly in fresh water and are most often planktonic. They store food in the form of glycogen and protein, have a procaryotic cell structure and have cellulose and pectin in their cell walls. They possess no flagella: however, a sliding motion may occur. Blue-green algae are unicellular or more often colonial, reproducing by binary fission. They have no mitochondria. Some species are able to fix nitrogen from the atmosphere.

cyanosis A purple or blue tint of the skin, especially the skin around the lips, ears and finger tips. It is caused by an insufficient supply of oxygen to the blood or by the blood giving up an unusually large proportion of its oxygen to the tissues. In both instances the capillary blood has a very high concentration of reduced haemoglobin which is darker in colour than oxygenated haemoglobin. Cyanosis may be caused by a heart or lung disorder.

cybernetics The science of control and communications in animals and machines. The science was founded by Norbert Wiener, an American mathematician.

cycle per second See hertz.

cyclic Having a ring structure. *Compare* aliphatic; *see also* benzene *and* fused ring.

cyclisation The process in which an aliphatic organic compound is converted into a cyclic compound; e.g. one part of the molecule reacts with another part and ring closure occurs.

cyclohexane C_6H_{12} A homocyclic, organic compound. It is a colourless, flammable liquid which may be prepared by hydrogenation of benzene, C_6H_6. A mixture of benzene vapour and hydrogen, H_2, is passed over a nickel catalyst at 150°C:
$C_6H_6 + 3H_2 \rightarrow C_6H_{12}$
Cyclohexane is used as a solvent and in the manufacture of nylon. *See also* conformation.

cyclosis The circulation of protoplasm in a cell.

cyclotron An electromagnetic accelerator producing an intensive beam of charged, high-energy particles, e.g. protons and electrons. The particles are accelerated in a vacuum in spiral orbits between two dee-shaped metal sections of a large electromagnet. Particles leaving an ion source in the centre of the machine are accelerated towards one of the dees; the field is reversed, causing them to accelerate towards the other dee, and so on. A deflector electrode draws the particles out of the machine by the time they have reached maximum acceleration. A cyclotron may be used to obtain particle energies required to disintegrate heavier atomic nuclei; e.g. to produce artificial isotopes, neutrons or X-rays.

cylinder 1. A geometrical shape which is a prism with a circular cross-section. **2.** In pumps and engines, a cylindrical vessel containing a piston.

cyme A type of inflorescence in which each main axis ends in a flower. A monochasial cyme is one in which each branch of the inflorescence forms a single lateral branch. A dichasial cyme is one in which each branch of the inflorescence forms two lateral branches. See Fig. 38.

cymose Describes an inflorescence whose growth is limited by the formation of a flower at the apex of each main axis, i.e. further growth takes place by the development of lateral branches. See Fig. 38; *compare* racemose.

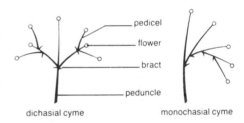

dichasial cyme   monochasial cyme

Fig. 38 Cymose inflorescences

cyst A lump-like structure consisting of a sac without an opening, containing a fluid. Cysts may occur in cavities, tissues and organs of the body. Cysts are usually considered to be benign. However, some cysts may become malignant (cancerous) and should therefore be removed when diagnosed.

cyto- Prefix meaning cell.

cytochemistry A branch of chemistry dealing with the study of chemical processes and chemical composition of cells.

cytochrome The general name for a group of cell pigments which are found in almost all living tissues and are involved in cellular respiration. Cytochromes are sometimes divided into three main types, *a*, *b* and *c*, of which the structure of *c* has been most extensively studied. In cells of animals, plants and fungi, the cytochromes are almost exclusively found in the mitochondria, whereas in bacteria they are located in the

cytokinin

plasma membrane. Cytochromes are conjugated proteins containing iron in the prosthetic group, and are mainly acting as coenzymes.

cytokinin (phytokinin) The general name for a group of hormones which promote cell division, growth and organ formation in plants. Cytokinins also stimulate bud formation and delay the process of senescence. Chemically, cytokinins are purines. One of the first cytokinins discovered was kinetin. Cytokinins require the presence of auxin for their function.

cytolemma Plasma membrane.

cytology The study of cells, including their structure, function and behaviour.

cytolysin A substance causing cytolysis.

cytolysis The dissolution or degeneration of a cell, particularly by the destruction of the cell membrane.

cytoplasm A slightly viscous, transparent fluid (often granular) found in all plant and animal cells. Together with the plasma membrane and the nucleus, the cytoplasm makes up the protoplasm of the cell.

cytosine A pyrimidine base found as part of nucleic acids in RNA and DNA. In DNA cytosine pairs with a purine base called guanine.

cytotoxin A cell poison such as a cytolysin.

D

D Chemical symbol for deuterium.
Dacron® *See* Terylene.
dagga *See* cannabis.
Dalton's atomic theory The first modern attempt to explain the formation of compounds from elements. In 1808 Dalton postulated the following: (1) all elements are composed of small particles, which Dalton named atoms; (2) all atoms of the same element are identical; (3) atoms can neither be created nor destroyed; (4) atoms combine in simple ratios to form 'compound atoms' (molecules). This theory could explain the laws of chemical combination and conservation of mass. However, it has since been proved that atoms can be created and destroyed in nuclear reactions.

Dalton's law of partial pressures A law stating that, at constant temperature, the total pressure of a mixture of gases is equal to the sum of the individual pressures (partial pressures) each gas in the mixture would exert if it were present alone and occupied the same total volume as the whole mixture.

dance of bees A form of communication among honey-bees consisting of certain movements they perform on returning to the hive. These series of movements inform the other bees in the hive of the location of a food source.

Daniell cell A primary cell consisting of a copper vessel (the positive electrode) containing a saturated solution of copper(II) sulphate and an amalgamated zinc rod (the negative electrode) immersed in a saturated solution of zinc sulphate or dilute sulphuric acid, contained in a porous pot standing in the copper(II) sulphate solution. This cell has a fairly constant e.m.f. of 1,1 volts (1.1 V).

Darwinism *See* Darwin's theory.

Darwin's theory (Darwinism) A theory of evolution stating that different species arise by the process of natural selection. Among the slight variations that occur in members of a species, the successful ones are 'selected': i.e. successful variants have a greater chance than others of surviving and so of passing on their characteristics to the next generation. In this way a species can change or a new species can arise.

dating *See* radio-carbon dating.

dative bond *See* lone pair.

daughter cell One of the two cells (or nuclei) that derive from the division of a single cell, usually by mitosis.

Davy lamp A miner's safety lamp which, in its earliest form, consisted of an oil burner inside a cylindrical wire gauze made of copper. Later models were fitted with a thick glass window below the gauze. The copper gauze conducts the heat of the flame away from any flammable gas (e.g. methane, CH_4) which may be outside the lamp, thus preventing an explosion. The flammable gas burns harmlessly inside the lamp. If this gas is methane, a bluish haze is seen around the flame. *See* Fig. 39.

day The time taken by the Earth to make one rotation on its own axis.

day-neutral plants *See* photoperiodism.

dB *See* decibel.

d.c. *See* direct current.

DDT Dichlorodiphenyltrichloroethane. A double-cyclic organic compound with the formula $(C_6H_4Cl)_2CHCCl_3$. It is a colourless solid substance used as a powerful insecticide. DDT is insoluble in water, but soluble in several organic solvents. Since it is not biodegradable and accumulates in the fatty tissue of many animals, its use has been somewhat restricted. *See also* trichloroethanal.

deficiency disease

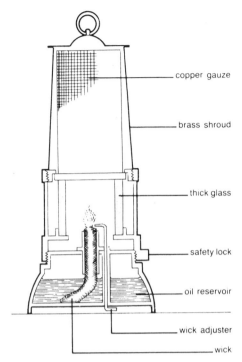

Fig. 39 Davy safety lamp

dead 1. Describes a conductor or electric circuit which is at earth potential. *See also* earth. **2.** A term used in acoustics to describe a room whose walls are covered with pyramid shapes made from or covered with sound-absorbing material such as felt, fibreglass or polystyrene. This reduces internal sound reflections to a bare minimum. The room is used for acoustics experiments. **3.** *See* death.

deaminating enzyme *See* deamination.

deamination The removal of the amino group, —NH_2, during the breakdown of an amino acid by means of deaminating enzymes. The process occurs chiefly in the liver. The main product of deamination is ammonia. However, a non-nitrogenous substance is also formed. The ammonia may be excreted as urea during the ornithine cycle.

death The state of an organism in which there is a complete and permanent ceasing of vital functions.

death rate *See* crude death rate.

deca- Symbol: da. A prefix meaning ten; e.g. one decametre (dam) is equal to ten metres.

decantation The process of pouring off a liquid from a precipitate or sediment so that the latter is left undisturbed.

decay 1. A spontaneous disintegration of radioactive nuclei of certain elements. A parent nucleus decays into a daughter nucleus by emitting alpha, beta or gamma rays. *See also* alpha decay, beta decay, half-life *and* radioactivity. **2.** The breakdown of organic matter from dead plants and animals. This is brought about by bacteria and fungi using dead material as a source of energy with the liberation of carbon dioxide, CO_2. *See also* saprophyte.

deceleration *See* acceleration.

deci- Symbol: d. A prefix meaning one-tenth, 10^{-1} (0,1; 0.1); e.g. one decilitre (dl) is equal to 10^{-1} litres.

decibel Symbol: dB. The decibel is a logarithmic unit, especially used in comparing sound intensities, but generally used in comparing two amounts of power. If two notes have sound intensities I_1 and I_2 respectively, then the difference in their intensity level is $10\log_{10} I_1/I_2$ decibels or $\log_{10} I_1/I_2$ bels. One decibel is equal to 0,1 bel (0.1 bel).

deciduous (hardwood) Describes plants (trees and shrubs) which shed their leaves at a particular season, e.g. autumn. *Compare* evergreen.

deciduous teeth Milk teeth, i.e. the first of the two sets of teeth of many mammals, shed before puberty. *See also* permanent teeth.

decimal system (denary system) The Hindu-Arabic number system using the base ten and digits 0–9.

declination *See* angle of declination.

decomposer An organism feeding on dead plant or animal material, thus breaking them down and, in this way, recycling organic and inorganic matter to the environment.

decomposition The breakdown of a chemical compound into simpler substances, which may be either elements or compounds. *Example:* ammonium nitrite, NH_4NO_2, decomposes when heated into nitrogen, N_2, and water (steam):
$$NH_4NO_2 \rightarrow N_2 + 2H_2O$$

decontamination The removal of dirt or infectious organisms from a certain area, material or person. The term is also used to describe the removal of radioactivity.

defaecation *See* defecation.

defecation The process in an animal whereby solid matter (faeces) is removed from the alimentary canal. The faeces are usually passed out through the anus, the cloaca or an anal pore.

deficiency disease A disease caused by inadequate intake of vitamins, minerals or other essential food factors. Examples include beriberi (lack of vitamin B_1), night-blindness (lack of vitamin A) and scurvy (lack of vitamin C).

definite proportions (law of) *See* chemical combination (laws of).

deflagrating spoon A device in which a substance (solid or liquid) may be burnt, usually in a gas jar. *See* Fig. 40.

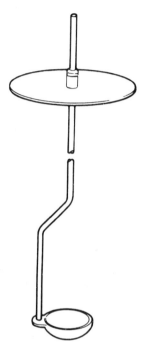

Fig. 40 Deflagrating spoon

deflection magnetometer *See* magnetometer.

deflection tube *See* cathode-ray tube, CRT *and* Perrin tube.

deflorate Describes the stage of a plant after flowering.

deformation An alteration in the size or the shape of an object.

degeneration A change (deterioration) in cells causing them to become less specialised and less active in function.

degree An interval in certain scales, e.g. temperature and hardness.

dehiscent Describes fruits which burst open to liberate seeds. *Compare* indehiscent.

dehydration The removal of water, usually chemically combined, from a substance. *Example:* if the elements hydrogen, H_2, and oxygen, O_2, are removed from concentrated sulphuric acid, H_2SO_4, in the ratio 2:1 (two parts of hydrogen to one of oxygen), then sulphur(VI) oxide, SO_3, remains:
$H_2SO_4 - H_2O \rightarrow SO_3$

dehydrogenase An enzyme which can remove hydrogen from a substance (substrate), thus oxidising it; i.e. it is an enzyme which catalyses oxidation. *See also* dehydrogenation.

dehydrogenation The chemical process in which hydrogen atoms are removed from a substance. This is a form of oxidation since each hydrogen atom contains one electron. *Example:* when two atoms of hydrogen are removed from one molecule of ethanol (ethyl alcohol), CH_3CH_2OH, ethanal (acetaldehyde), CH_3CHO, is formed:
$CH_3CH_2OH - H_2 \rightarrow CH_3CHO$

deionise To remove ions from an electrolyte. *See also* demineralisation.

deliquescence The process in which a solid substance absorbs water vapour from the air, thus passing into solution. Examples of deliquescent substances are sodium hydroxide, $NaOH$; potassium hydroxide, KOH; and calcium chloride (anhydrous), $CaCl_2$. *Compare* efflorescence.

delirium tremens An acute physical and mental disorder (commonly known as DTs) due to prolonged alcohol consumption. The symptoms include extreme excitement, fever, trembling, hallucinations (the feeling of having disgusting creatures crawling over one's body), and a rapid and irregular pulse. This disorder requires immediate and intensive hospital care.

delivery Childbirth.

delta (Δ) The fourth letter of the Greek alphabet.

delta G (ΔG) *See* entropy.

delta H (ΔH) *See* enthalpy *and* entropy.

delta S (ΔS) *See* entropy.

deltoid muscle A large, triangular muscle covering the shoulder joint. Its function is to lift the upper arm.

demagnetise To remove magnetism. Magnets can best be demagnetised by placing them inside a coil (solenoid), placed with its axis pointing East–West, through which an alternating current is flowing. After a few seconds the magnet is slowly withdrawn to a distance of a few metres from the coil. Another method is to heat the magnet until it is red-hot and then allow it to cool while lying in an East–West direction. Magnets can also be weakened by hammering them while they lie in an East–West direction. *See* Fig. 41.

demineralisation The removal of positive and negative ions from an aqueous solution with the aim of making pure water (soft water). This is brought about by passing the solution through certain substances (either organic or inorganic). These substances have a 'cage-like' molecular structure and are insoluble. They are able to accept positive ions (cations) in

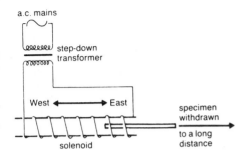

Fig. 41 Demagnetising a magnet using alternating current

exchange for hydrogen ions, H^+, and negative ions (anions) in exchange for hydroxide ions, OH^-, thus leaving pure water when H^+ and OH^- combine. This method of exchanging one ion with another of the same charge is called ion exchange. It is a reversible process; after the ion exchange material has been used for some time, the trapped positive and negative ions can be washed out of it by treating it with strong acid and strong alkali respectively. Ion exchange is also used in purification of chemicals and in separation. *See also* zeolite.

demodulation The process of separating or extracting audiofrequency waves from a modulated carrier wave, e.g. by a radio receiver. *Compare* modulation.

denary system *See* decimal system.

denaturant A substance which is added in small amounts to ethanol (ethyl alcohol), C_2H_5OH, to make it unfit for human consumption. This is done for taxation purposes. Very often the poisonous methanol (methyl alcohol), CH_3OH, is used as a denaturant. *See also* methylated spirit(s).

denaturation 1. The process of making ethanol (ethyl alcohol), C_2H_5OH, unsuitable for human consumption. *See* denaturant.
2. Usually an irreversible process in which a protein is exposed to drastic conditions such as heating, extremes of pH, X-rays or other ionising radiation. During this process the structure of the protein is broken down, mainly because of damaged hydrogen bonds, and the molecule loses its shape and physiological activity.

dendrite A nerve-fibre which is fairly broad when it leaves the cell body of a neurone. Dendrites usually have many branches and convey impulses towards the cell body. The term dendron is used as a synonym for dendrite. However, 'dendron' is sometimes used to refer to the broad part of the nerve-fibre leaving the cell body, while the branches are called dendrites. *See* Fig. 150.

dendron *See* dendrite.

dengue A usually tropical, viral disease of humans transmitted by Culex mosquitoes. The symptoms include fever and severe limb and back pains. Patients usually recover after about five days.

denitrification *See* denitrifying bacteria.

denitrifying bacteria Anaerobic bacteria which cause denitrification, i.e. the breakdown of nitrogen compounds (nitrates and nitrites) in soil to gaseous nitrogen. Denitrification especially occurs in poorly aerated soils. *Compare* nitrifying bacteria.

dens *See* odontoid process.

density Symbol: ρ (rho). The mass per unit volume of a substance. The SI unit of density is the kilogram per cubic metre, $kg\,m^{-3}$. In the c.g.s. system, the unit of density is the gram per cubic centimetre, $g\,cm^{-3}$. *See also* relative density *and* vapour density.

density bottle A small bottle used in the accurate determination of relative densities of liquids. The bottle has a ground glass stopper with a fine hole through it. When the bottle is filled with a liquid and the stopper inserted, the excess liquid rises up through the hole and overflows. So long as it is filled in this manner it will always contain the same volume of liquid at a certain temperature. *See* Fig. 42.

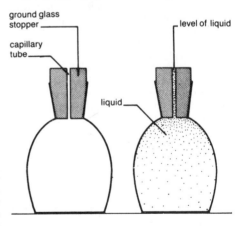

Fig. 42 Density bottle

dental Pertaining to teeth.

dental formula A formula expressing types and numbers of teeth in a mammal. It consists of a series of fractions, each fraction preceded by an abbreviation for the type of tooth and indicating the number of teeth of each type in one half of the upper jaw divided by the number of those in one half of the lower jaw.

dentate

The dental formula for the deciduous or milk teeth of Man is:

$i\frac{2}{2} \cdot c\frac{1}{1} \cdot p\frac{0}{0} \cdot m\frac{2}{2}$ Total 20.

where i = incisors, c = canines, p = premolars and m = molars. The dental formula for the permanent teeth of Man is;

$i\frac{2}{2} \cdot c\frac{1}{1} \cdot p\frac{2}{2} \cdot m\frac{3}{3}$ Total 32.

dentate With toothed margin or edge. *See* Fig. 111.

dentine The hard tissue, resembling bone, which forms the main part of a tooth. It surrounds the pulp cavity and in the area of the crown it is covered by a layer of enamel; in the area of the root it is covered by a layer of cement. *See also* odontoblast *and* odontoclast.

dentistry The study of teeth, including their care.

dentition The arrangement, number and kind of teeth in the jaws.

denture A set of artificial teeth (false teeth).

deoxy- Prefix meaning 'oxygen removed'. When oxygen is removed from a substance the substance is reduced; e.g. when copper(II) oxide, CuO, is heated in a stream of carbon monoxide, CO, the copper in the copper(II) oxide is reduced to free copper:
$CuO + CO \rightarrow Cu + CO_2$

deoxyribonucleic acid (DNA) The material of inheritance of almost all living things. It is found in the chromosomes of plants and animals and in bacteria and viruses. DNA is a polymer and consists of two strands (chains) of nucleic acid, each in the form of a helix, which usually are intertwined forming a so-called double helix. The two interwound, helical strands of polynucleotides are very long. Each nucleic acid consists of alternating units of carbohydrate and phosphate groups with a nitrogen base attached to each carbohydrate unit. The carbohydrate is deoxyribose and there are four different bases: adenine (A), guanine (G), thymine (T) and cytosine (C). Adenine and guanine are purine bases, whereas thymine and cytosine are pyrimidine bases. The four bases in a strand of nucleic acid pair with the four bases in another strand in the following way: adenine always pairs with thymine and guanine always pairs with cytosine. This means that the sequence of bases on one strand determines the sequence of bases on the other. Thus a sequence of bases such as A A C G A T will be matched by T T G C T A. The two strands (helices) are held together by hydrogen bonds between the nitrogen bases in each. Since there is a great variety of possible sequences for the four bases in a DNA molecule, there is a large number of individual compounds of DNA.

DNA has two main functions in a cell. First, when the cell reproduces itself by division, the strands split apart and new nucleotides (there are always free ones in the cell) float in to bond with the newly unlinked bases. Since each base on one strand can only pair with a certain base on the other, the two new double strands are identical to the original. Secondly, DNA controls the manufacture of RNA. The structure of DNA was discovered by Crick and Watson in 1953. *See* Fig. 43.

A = adenine

T = thymine

C = cytosine

G = guanine

X = phosphate

0 = deoxyribose

The dotted lines between the bases indicate hydrogen bonds

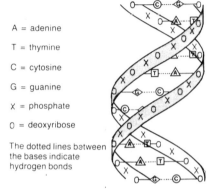

Fig. 43 Deoxyribonucleic acid

deoxyribonucleotide A nucleotide consisting of a purine or pyrimidine base attached to the carbohydrate deoxyribose which is attached to a phosphate group.

deoxyribose $C_5H_{10}O_4$ A carbohydrate found in DNA. It is a colourless, soluble crystalline solid.

depolarisation The prevention of polarisation in an electric cell. *Example:* in the Leclanché cell, the carbon anode is surrounded by a mixture of manganese dioxide, MnO_2, and powdered carbon. The manganese dioxide is the depolariser or depolarising agent and oxidises hydrogen, the cause of polarisation. *See also* polarisation.

depolariser *See* depolarisation.

depolarising agent *See* depolarisation.

depression of freezing-point A decrease in the freezing-point of a solvent by the addition of a solute. Measurement of the depression is used in relative molecular mass determinations. For pure substances, the depression of freezing-point is proportional to the number of molecules or ions which dissolve in the solvent, and the depression which is produced by the same molecular concentration of a substance is a constant for a particular solvent. *Compare* elevation of boiling-point; *see also* Appendix.

depth of field The zone in which the lens in a camera or other optical instrument will produce a distinct image of an object.

derivative A substance obtained from another substance. *Example:* phenol, C_6H_5OH, is a derivative of benzene, C_6H_6.

dermal Pertaining to, or derived from, the skin.

dermis In vertebrates, a layer of tough, elastic, fibrous connective tissue lying under the epidermis. It is thicker than the epidermis and contains blood vessels, lymphatic vessels, nerve-endings, sweat glands, hair follicles and sebaceous glands. *See* Fig. 208.

dermoskeleton *See* exoskeleton.

desalination The process in which salts are removed from sea-water to make it fit for drinking (potable) and for irrigation. Several methods are used, such as distillation and ion exchange.

desert A large, dry area of virtually uninhabitable, uncultivated land consisting almost entirely of sand and rock.

desiccant A substance used as a drying agent.

desiccation The removal of moisture (water) from a material by means of a hygroscopic substance such as silica gel or calcium oxide, CaO. Desiccation is often carried out in a closed vessel (a desiccator) under vacuum.

desiccator A glass or plastic vessel with a close-fitting lid. It is used for drying and keeping substances dry. The substance to be dried or kept dry is placed on a perforated metal or porcelain stand below which is placed a hygroscopic substance such as silica gel. Some desiccators are fitted with a tap for evacuation purposes. *See* Fig. 44.

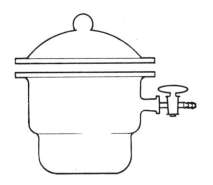

Fig. 44 Desiccator

destarching The process in which starch disappears from the leaves of a plant when it is put in darkness for a few days. The starch is changed into glucose which is carried to other parts of the plant.

destructive distillation A process in which a complex organic substance (liquid or solid) is heated (usually) in the absence of air leading to decomposition and the formation of volatile products. The destructive distillation of coal produces four major products: coal-gas, ammoniacal liquor, coal-tar and coke. These are formed in varying proportions depending on the temperatures used in the process. Wood may also be destructively distilled.

detector A device or instrument used to detect or measure the presence of a physical property, e.g. electromagnetic radiation.

detergent A substance which can be used as a cleansing agent. Soap is a detergent, but the term is usually applied to synthetic substitutes for soap. A detergent lowers the surface tension of water and emulsifies oils and fats. Synthetic (soapless) detergents have some advantages over soap: they are more soluble in water and their calcium and magnesium salts are soluble. The corresponding salts of soap are insoluble and form a precipitate (scum) in hard water. *See also* soap.

detonator An explosive device often used to initiate a main explosion by producing a high-pressure shock wave.

deuterium (heavy hydrogen) Symbol: D or ^2_1H. An isotope of hydrogen. The nucleus of the deuterium isotope contains one proton and one neutron, unlike hydrogen whose nucleus contains one proton only. Deuterium occurs naturally in very small amounts in water, where it is linked to oxygen forming heavy water (deuterium oxide, D_2O). The properties of deuterium are almost the same as those of hydrogen.

deuterium oxide *See* heavy water.

deuteron The name given to the nucleus of a deuterium atom (isotope).

Devarda's alloy An alloy of copper (50%), aluminium (45%) and zinc (5%) which is used in the laboratory as a reducing agent.

developing *See* photography.

deviation, angle of *See* angle of deviation.

Devonian A geological period of approximately 50 million years which occurred immediately after the Silurian, and which ended approximately 345 million years ago.

dew Drops of water which have been precipitated by cooling air saturated with water vapour. This formation of dew takes place when the air temperature slowly falls as at late evening or night.

Dewar flask (Thermos, vacuum flask) A flask (vessel) used to keep liquids either cold or hot, i.e. either below or above the temperature of the surrounding air. It consists of a double-walled glass container (*see* Fig. 45) with a vacuum between the walls A and B. This prevents loss or gain of heat by convection or conduction. The sides of the walls in the vacuum are silvered, preventing loss or gain of heat by radiation. C is a poor conductor of heat, used to support the glass container and to reduce heat loss or gain by conduction to and from the outside part of the flask, M. The flask is fitted with a stopper which is also a poor conductor of heat.

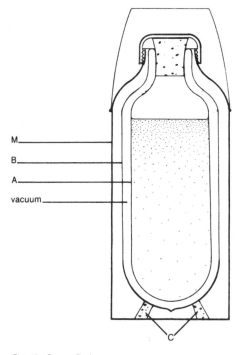

Fig. 45 Dewar flask

dew point The temperature at which water vapour present in the air is just sufficient to saturate it, i.e. the temperature at which dew is formed.

dextrin A general name for a group of small, soluble polysaccharides obtained by the partial hydrolysis of starch. Dextrins are used as adhesives.

dextrorotatory compound *See* optical isomerism.

dextrose *See* glucose.

di- Prefix meaning two.

diabetes mellitus This is the most common form of diabetes, which is a disease caused by lack of the hormone insulin. Insulin is produced by the pancreas and released into the blood stream following an increase in the concentration of blood sugar (glucose), thus enabling glucose to be stored and used in the tissues. In diabetes, insulin production is insufficient or lacking and glucose accumulates in the blood, eventually being passed unused in the urine. Symptoms include the passing of large quantities of urine, the feeling of thirst, weight loss and fatigue. Diabetes mellitus is treated by giving the patient a diet containing a carefully measured amount of carbohydrate, and by injecting insulin. If the disease is improperly treated severe complications may arise, eventually leading to early death.

diagnosis The identification of a disease by means of a patient's symptoms.

diakinesis *See* meiosis.

dialysis A method of separating colloid particles from smaller particles in a liquid mixture by diffusion of the latter through a semi-permeable membrane. This method may be used to separate salts from colloids. The term also applies to the purification of blood from a patient suffering kidney failure. *See* kidney machine.

dialysis tubing *See* Visking tubing.

diamagnetic materials *See* ferromagnetic materials.

diamond A naturally occurring allotrope of carbon, which is transparent and, when pure, colourless. The crystal structure of diamond is responsible for its properties, including high melting-point and hardness (10 on Mohs's scale). About 20% of the diamonds found are suitable for jewellery; the remainder are used in industry for tools or are crushed to a fine abrasive powder. The largest diamond ever found had a mass of 3106 carats (621,2 g; 621.2 g). It was found in South Africa in 1905 and was named the Cullinan. For some years, artificial diamonds have been prepared from graphite using great heat and pressure. Large amounts of industrial diamonds are now made synthetically. *See* Fig. 46.

diaphragm 1. In mammals, a thin, dome-shaped muscle covered by serous membrane which acts as a partition between the thorax and the abdomen. It has a number of openings for structures which pass between the thorax and the abdomen. It is attached to the lower ribs, the breastbone and the backbone. The diaphragm contracts and relaxes rhythmically, thus playing an important part in breathing. *See* Fig. 118.   2. A contraceptive device (also called the cap) which fits over the cervix, preventing the entrance of sperm into the womb.   3. A device for controlling the size of the aperture in optical instruments such as the camera. *See also* iris.

diaphragm cell An electrolytic cell used primarily in the manufacture of sodium and potassium hydroxide, but also producing chlorine and hydrogen. The anode is made of a

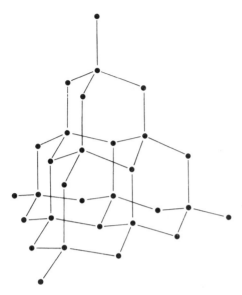

Fig. 46 Arrangement of carbon atoms in diamond

ring of graphite rods and the cathode is an iron gauze cylinder. The electrodes are separated by a diaphragm of porous asbestos to prevent the chlorine mixing with the hydroxide. A concentrated solution of sodium chloride, NaCl, (brine) is passed into the anode chamber and from here it flows through the diaphragm into the cathode chamber. At the anode, chloride ions, Cl$^-$, are discharged and at the cathode water molecules receive electrons. The processes are:
NaCl → Na$^+$ + Cl$^-$
at the anode: 2Cl$^-$ − 2e → Cl$_2$
at the cathode: 2H$_2$O + 2e → 2OH$^-$ + H$_2$
Potassium hydroxide, KOH, is prepared in the same way using a concentrated solution of potassium chloride, KCl, in place of the brine. Hydroxides of sodium and potassium are more cheaply produced in this way than in the Castner–Kellner process. However, the hydroxide is often contaminated with chloride ions which have passed through the asbestos diaphragm.

diaphysis (pl. diaphyses) *See* ossification.

diarrhoea A condition in which excessive and frequent discharge of fluid leaves the bowel. This may soon lead to a dangerous dehydration (loss of water). Diarrhoea may be caused by careless eating or drinking, by food poisoning or by infections. To restore the body's fluid balance, plenty of liquid, e.g. water, tea or thin soup, should be taken.

diastase *See* amylase.

diastereoisomer *See* racemic mixture.

diastole 1. In vertebrates, the phase of heartbeat in which the atria and ventricles are relaxed, allowing blood from the main veins to fill the heart. *Compare* systole. 2. In a contractile vacuole, the phase in which it refills with fluid (water). *Compare* systole.

diathermy A heat treatment of tissue by passing a high-frequency current through it, thus causing a controlled rise in temperature. This is brought about by placing two metal electrodes on the skin. Diathermy is used in the treatment of certain inflammations and muscular disorders. It can also be used to stop bleeding from small injured blood vessels during surgery.

diatom The common name for members of a class of unicellular algae abundant in plankton. Their cell walls are impregnated with silica and consist of two overlapping halves (like the two halves of a petri dish). The cell walls persist long after death and form deposits of diatomaceous earth (kieselguhr).

diatomaceous earth *See* diatom and kieselguhr.

diatomic Describes a molecule consisting of two like or unlike atoms: e.g. hydrogen, H$_2$, and hydrogen chloride, HCl.

diatonic scale A series of eight notes (musical sounds) commonly known as: doh-ray-me-fah-soh-lah-te-doh^1. One such scale ranges from middle C, 256 Hz, to upper C, 512 Hz.

diazonium salt *See* azo compound.

diazotisation *See* azo compound.

dibasic Describes an acid having two acidic hydrogen atoms which can be replaced by a metal or a metal-like group such as the ammonium group, NH$_4$, thus forming salts. Examples of dibasic acids are sulphuric acid, H$_2$SO$_4$, carbonic acid, H$_2$CO$_3$, and ethanedioic acid (oxalic acid), (COOH)$_2$. These acids give rise to normal salts and acid salts, e.g. sodium sulphate, Na$_2$SO$_4$, and sodium hydrogensulphate, NaHSO$_4$.

dicarboxylic acid A carboxylic acid (organic acid) containing two carboxyl groups. A dicarboxylic acid is either aliphatic or cyclic. *Examples:* ethanedioic acid (oxalic acid) (COOH)$_2$ is aliphatic; benzene-1,2-dicarboxylic acid (phthalic acid), C$_6$H$_4$(COOH)$_2$, is cyclic.

dichasial cyme *See* cyme.

dicotyledon A member of the larger class of angiosperms, Dicotyledoneae, having an embryo with two cotyledons. Other characteristics include: net venation of the leaves and vascular tissue of the stem usually arranged in the form of a ring of open bundles. Dicotyledons are either herbs, shrubs or trees. Examples include mangoes, pawpaws, oaks and buttercups. *Compare* monocotyledon.

Dictyoptera An order of insects whose members include cockroaches and mantises. Characteristics include biting mouth parts and incomplete metamorphosis.

dictyosome A unit of the Golgi apparatus.

dielectric An insulating material or medium placed in an electric field. The air or other insulating material between the plates of a capacitor is called the dielectric.

diencephalon In vertebrates, the posterior part of the forebrain, giving rise to the thalami. It connects the cerebral hemispheres with the midbrain.

diene An organic compound containing two double bonds between carbon atoms. Buta-1,3-diene (butadiene) and methylbuta-1,3-diene (isoprene) are examples of dienes.

diesel engine A type of internal combustion engine. The fuel used is a heavy oil which is ignited by the heat of compressed air. A sparking-plug is not required and the engine is sometimes referred to as a compression-ignition engine.

diesel fuel (diesel oil, gas oil) A petroleum distillate mainly containing higher aliphatic hydrocarbons (alkanes with 14–20 carbon atoms). It is used as the fuel in diesel engines.

diesel oil *See* diesel fuel.

diethyl ether *See* ether.

differentiation The modification taking place in a cell causing it to become specialised for a particular function. Differentiated cells group together in large numbers to form tissues. During differentiation there is a change in the structure of the cell.

diffraction The bending of waves, e.g. light and sound waves, when they pass the edge of an obstacle or through small openings. The extent of the bending depends on the wavelength of the waves.

diffraction grating A device used to demonstrate light diffraction. It consists of a piece of transparent film with many thousands of equally spaced, parallel, opaque lines marked on it. Between the lines are very narrow clear spaces (slits). *Example:* there may be 5000 lines per centimetre, so that the slits are $\frac{1}{5000}$ cm apart. When the grating is arranged as in Fig. 47, light from the lamp is diffracted and emerges from the clear spaces in several different directions, forming multiple images.

diffuse reflection Reflection from an irregular reflecting surface where the angle of incidence does not equal the angle of reflection. *See* Fig. 48; *compare* regular reflection.

diffusion The process by which fluids mix with one another (liquids with liquids and gases with gases) by random kinetic movement of their molecules, eventually forming uniform mixtures. Diffusion does occur in solids, but is a very slow process at normal temperature. Diffusion plays an important part in the passage of substances through the cell membranes of living things.

diffusion pressure deficit (DPD; suction pressure, water potential) The net capacity of plant cells to absorb water. This can be written as

$$DPD = (OP_1 - OP_2) - TP$$

where OP_1 represents osmotic pressure inside the cell, OP_2 osmotic pressure outside the cell and TP turgor pressure; $(OP_1 - OP_2)$ is the net osmotic pressure. If $OP_2 > OP_1$, then DPD will be negative, i.e. the cell will lose water. The cell is fully turgid when DPD is zero, i.e. when the net osmotic pressure is equal to TP.

digestion The breakdown of nutrients (foodstuffs) making them soluble and absorbable. This is brought about by chemical reactions involving various enzymes.

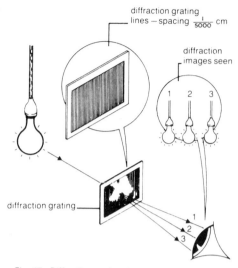

Fig. 47 Diffraction grating

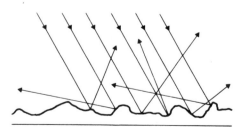

Fig. 48 Diffuse reflection

digestive juices Juices, containing enzymes, secreted in the alimentary canal to aid digestion. Examples include saliva, gastric juice, pancreatic juice and bile.

digestive system (alimentary system) The alimentary canal and associated glands, including the liver and the pancreas. *See* Fig. 49.

digit A finger or toe of a pentadactyl limb.

digital display A method of giving the reading of a measuring instrument, e.g. a clock or voltmeter. The reading appears as numbers on a screen, instead of being given by a pointer moving over a scale.

digitalin *See* digitalis.

digitalis Any of several poisonous compounds (e.g. digitalin) obtained from dried leaves of the foxglove (*Digitalis purpurea*). Digitalis is used in the treatment of certain heart and circulatory disorders. It slows down the rate at which the heart beats at the same time strengthening each heartbeat.

dihybrid The offspring of parents which are genetically different in two characteristics. *Compare* monohybrid.

dihydric Describes a molecule containing two hydroxyl groups, —OH. *Example:* ethane-1,2-diol (glycol), CH_2OH—CH_2OH, is a dihydric alcohol.

(−)-2,3-dihydroxybutanedioic acid *See* tartaric acid.

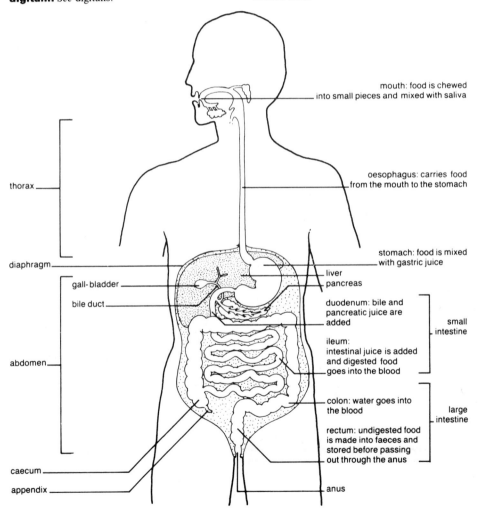

Fig. 49 Digestive system

1,2-dihydroxyethane *See* ethane-1,2-diol.
dilate To make wider, i.e. to increase the diameter of an aperture or tube. The contraction of the iris muscles causes the pupil of the eye to dilate.
dilead(II) lead(IV) oxide *See* lead oxides.
diluent *See* diluting agent.
dilute 1. To reduce the strength of a solution by the addition of water or other solvent. 2. Describes a solution which has been diluted.
diluting agent Any liquid which reduces the strength of a solution when mixed with it. Usually the diluting agent is the same as the solvent in the particular solution it dilutes.
dilution law A mathematical expression which, at a constant temperature, determines how the degree of dissociation, α (alpha), varies with the total concentration of a given weak electrolyte in dilute solutions. The expression is:
$$\frac{\alpha^2 \times c}{1-\alpha} = K$$
where c is the molar concentration of the electrolyte. K is called the dissociation constant. The expression was first put forward in 1889 by the German chemist Ostwald and is therefore sometimes called Ostwald's dilution law. From the expression it is seen that α increases when the electrolyte is diluted, i.e. when c decreases.
Example: in a 0,1M (0.1 M) solution of ethanoic acid (acetic acid), CH_3COOH, at 25°C, α is 0,0133 (0.0133) or 1,33% (1.33%); therefore
$$K = \frac{(0,0133)^2 \times 0,1}{1 - 0,0133} = 1,79 \times 10^{-5}$$
$$\left(K = \frac{(0.0133)^2 \times 0.1}{1 - 0.0133} = 1.79 \times 10^{-5} \right)$$
This value of K can now be used to calculate α for ethanoic acid at any given concentration at 25°C by using the expression for the dilution law. In the case of a very weak electrolyte, α is so small that $1 - \alpha$ is almost 1 and the expression
$$\frac{\alpha^2 \times c}{1-\alpha} = K$$
can then be written:
$$\alpha^2 \times c = K \quad \text{or} \quad \alpha = \sqrt{\frac{K}{c}}$$

dimer A compound obtained from the combination of two identical molecules called monomers. The dimer molecule may contain twice the number of atoms of the monomer molecules from which it is formed; e.g. the dimer dinitrogen tetroxide, N_2O_4, exists in equilibrium with the monomer nitrogen dioxide, NO_2:
$$2NO_2 \rightleftharpoons N_2O_4$$

This form of dimerisation is called addition dimerisation. When two identical monomers combine to form a dimer and another molecule, e.g. water, the process is called condensation dimerisation. *Example:* two molecules of monosaccharide, $C_6H_{12}O_6$, combine to give a disaccharide, $C_{12}H_{22}O_{11}$, and water:
$$2C_6H_{12}O_6 \rightleftharpoons C_{12}H_{22}O_{11} + H_2O.$$
See also polymer.

dimerisation *See* dimer.
dimethylbenzene $C_6H_4(CH_3)_2$ (xylene) Any of three isomers which occur in coal-tar. They are aromatic hydrocarbons. The isomers are distinguished from one another according to where the two methyl groups, $-CH_3$, are placed in the benzene molecule, e.g. 1,2-dimethylbenzene. Dimethylbenzenes (xylenes) are colourless, insoluble, flammable liquids widely used as solvents and in organic synthesis.
dinitrogen oxide *See* nitrogen oxides.
dinitrogen pentoxide *See* nitrogen oxides.
dinitrogen tetraoxide *See* nitrogen oxides.
dinitrogen trioxide *See* nitrogen oxides.
diode A common type of diode is the thermionic valve, consisting of a glass vacuum tube containing a hot filament (the cathode) surrounded by a cylinder or plate (the anode). Both electrodes are usually made of nickel. The cathode is coated with a mixture of barium and strontium oxides, increasing the rate of electron emission. A principal use of a diode is for rectifying a.c.
dioecious 1. (unisexual) Having male and female flowers on different plants of the same species. 2. In animals: having male and female reproductive organs on different individuals of the same species.
diol (glycol) The general name for a group of dihydric alcohols. The lower members are colourless, water-soluble, viscous, sweet, hygroscopic liquids with rather high boiling-points. They are good solvents for many dyes.
dioptre A unit of refractive power of a lens. It is expressed as the reciprocal of the focal length in metres. By convention a convex lens has a positive power (dioptre), whereas a concave lens has a negative power.
dioxin An extremely poisonous substance with the formula

[structural formula of dioxin showing two chlorine-substituted benzene rings connected by two oxygen atoms, with Cl groups at four positions]

It is a manufacturing impurity in certain types of herbicides. Dioxin accidentally polluted an area in Seveso, Italy, in 1976, causing death and disfigurement to many people and animals. Since dioxin is environmentally stable it may enter food chains.

dip *See* angle of dip.

dip circle (inclinometer) An instrument for measuring the angle of dip. It consists of a thin, magnetised needle pivoted to swing about its centre of gravity in a vertical plane. The inclination of the needle to the horizontal is read on a vertical, circular scale divided into degrees.

dipeptide A compound made from two molecules of amino acids joined together with a peptide bond. *See also* peptide *and* protein.

diphtheria An acute, contagious, bacterial disease which starts as a throat infection and produces a membrane constricting the air passages. The bacillus responsible for the disease produces a toxin which may damage the heart and the nerves. Formerly, it was one of the most fatal childhood diseases. It is treated with penicillin and a vaccine is now available for immunisation.

diphyodont Having two dentitions. This is characteristic of human beings.

diploblastic Describes animals which have two cell layers, ectoderm and endoderm, making up their body wall. Coelenterates are diploblastic. *See also* mesogloea.

diploid Describes the nucleus of a cell which has its chromosomes in homologous pairs. A cell with a diploid nucleus is called a diploid cell. Almost all the cells of higher plants and animals are diploid except the gametes, which are haploid.

diploid number The total number of chomosomes in a diploid cell. The diploid number is twice the haploid number. *See also* chromosome.

Diplopoda A class of arthropods comprising millipedes.

diplotene *See* meiosis.

dip needle An instrument similar in function to a dip circle. The scale over which the magnetised needle moves is a quadrant (quarter circle) and not a complete circle.

dipolar ion *See* zwitterion.

dipole 1. Two point charges (poles), of equal magnitude but of opposite sign, separated by a small distance. The product of either of the charges and the distance between them is the dipole moment. *See also* magnetic dipole.
2. A molecule such as hydrogen chloride, HCl, is a dipole because the electrons in the bond joining the two atoms are nearer to the electronegative chlorine atom than to the hydrogen atom, thus giving the molecule a dipole moment. Water is a common dipole. The dipole moments of molecules provide important information concerning their molecular structure. *See also* van der Waals forces.

dipole moment *See* dipole.

Diptera A large order of insects whose members include the flies and the mosquito. Characteristics include a single pair of membranous wings, the hind wings being reduced to small club-shaped halteres, and complete metamorphosis.

direct current (d.c.) An electric current flowing in one direction only.

disaccharide One of a group of soluble carbohydrates (sugars). A disaccharide molecule, $C_{12}H_{22}O_{11}$, may be obtained by the reaction of two molecules of monosaccharide, $C_6H_{12}O_6$, with the elimination of water:
$2C_6H_{12}O_6 \rightleftharpoons C_{12}H_{22}O_{11} + H_2O$
On hydrolysis, a disaccharide gives the two molecules of monosaccharide from which it is obtained. A little acid must be present to assist the reaction. Saccharose (sucrose), lactose and maltose are examples of naturally occurring disaccharides. Disaccharides are hydrolysed into monosaccharides during catabolism; this is brought about by enzymes called carbohydrases.

disc A plate-like structure consisting of cartilage and fibre separating the bones of the spinal column. Discs act as shock absorbers and allow the spine to bend.

discharge To remove or reduce the electric charge from a body, e.g. a capacitor or a cell.

discharge tube There are several types of discharge tubes. Generally they consist of a glass vessel either totally or partially evacuated. Metal electrodes are fitted at each end of the vessel and a potential difference is applied to these electrodes causing a stream of electrons to pass from cathode to anode. If a gas is present in the vessel, a luminous glow is produced, the colour of which depends on the nature of the gas, e.g. air gives a pink colour, neon gives orange-red.

disinfectant A chemical used to destroy micro-organisms.

disintegration *See* decay.

disodium hydrogenphosphate *See* sodium phosphates.

disodium tetraborate-10-water *See* borax.

dispersion The distribution of small particles (colloidal particles) in a fluid.

dispersion of light The separation of light of mixed wavelengths into a spectrum. A beam of white light may be separated into coloured beams of light by passing the beam through a prism.

displacement 1. A chemical reaction in which one element or group of elements takes the place of another element or group of elements. *Example:* if iron, Fe, is placed in copper(II)

sulphate solution, $CuSO_4$, copper, Cu, is displaced by the iron and iron(II) sulphate, $FeSO_4$, is left in solution:
$Fe + CuSO_4 \rightarrow Cu + FeSO_4$
2. The amount of fluid displaced by a floating or submerged body. The size of a ship is often given in terms of its displacement of water; e.g. a ship with a displacement of 20 000 tonnes.

displacement can (eureka can, overflow can) A metal can, fitted with an overflow tube, used for the indirect measurement of the volume of irregular solids. The can is filled with a liquid (usually water) until it overflows. When an object is immersed in the liquid, the volume of liquid which overflows is equal to the volume of the object. See Fig. 50.

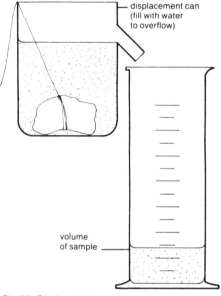

Fig. 50   Displacement can

displacement series See electrochemical series.

displayed formula See formula.

disproportionation The chemical change of a substance in which it is simultaneously oxidised and reduced. Example: two molecules of hydrogen peroxide, H_2O_2, undergo disproportionation when they react to give water and oxygen:
$H_2O_2 + H_2O_2 \rightarrow 2H_2O + O_2$
See also redox reaction.

dissect To cut open an organism in order to display its parts, structure, etc.

disseminated sclerosis See multiple sclerosis.

dissociation The breakdown of molecules into smaller molecules, atoms or ions. This process is often reversible and it may be either electrolytic or thermal. See also electrolytic dissociation and thermal dissociation.

dissociation constant See dilution law.

dissolve 1. To treat a solid, liquid or gas (solutes) with water or another suitable solvent so that the solute forms a uniform solution with the solvent. 2. (of a solute) To form a uniform solution with a solvent.

distal Situated away from the place of attachment or origin. Example: the foot is a distal part of the leg. Compare proximal.

distance ratio See velocity ratio.

distillate The liquid produced during distillation.

distillation The process in which a solution is heated to produce a vapour which is then condensed by cooling it to become liquid (the distillate) again. The process is commonly used in separating a pure liquid from one or more dissolved, solid substances which are non-volatile; and in separating liquid mixtures containing constituents with different boiling-points. See Fig. 51; see also fractional distillation.

distilled water Pure water obtained by the distillation of an aqueous solution containing non-volatile solutes.

distribution of charge The distribution of electric charge on the surface of a conductor depends on the nature of its surface. The charge density is greatest at the most highly curved convex parts of the surface. Charge is especially concentrated on sharp points.

distributor A device for sending (distributing) the current from the induction coil of a car's ignition system to the sparking-plugs in the correct firing order of the engine's cylinders.

diuresis Increased flow of urine.

diuretic A substance which causes an increased flow of urine in humans. An increased intake of liquids such as water, tea, coffee or beer leads to an increased production of urine and these liquids are therefore diuretics. However, certain drugs also have a diuretic effect and may be given in treating oedema caused by heart or kidney disorders.

diurnal 1. Describes an event which happens daily. 2. A diurnal animal is one which is active during the day. Compare nocturnal.

diurnal rhythm (circadian rhythm) A rhythm of activity in plants and animals which takes place in a cycle of approximately 24 hours, e.g. the daily leaf movements of plants, and sleeping and waking in animals. These diurnal rhythms continue even after the organism is isolated from its environment; they are independent of external factors, but arise inside the organism.

divalent (bivalent) Having a valency of two.

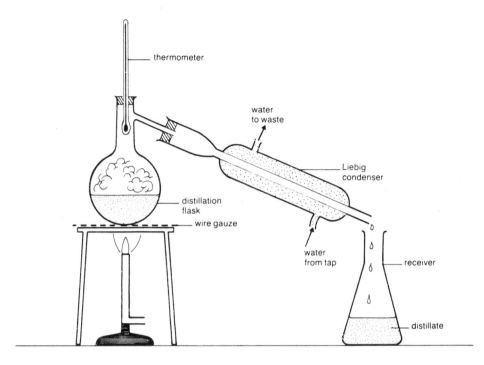

Fig. 51 Distillation

diverging lens See lens.
division See phylum.
dizygotic twins (fraternal twins) Twins formed from two ova fertilised simultaneously. They may be of different sexes and are no more alike than any other offspring of the same parents. Compare monozygotic twins.
DNA See deoxyribonucleic acid.
dolomite A naturally occurring, whitish, sedimentary rock consisting of calcium magnesium carbonate, $CaMg(CO_3)_2$. It is an important source of magnesium; it is also used as a building stone and as a refractory material.
domain See domain theory.
domain theory A theory of magnetism. In some magnetic materials, e.g. iron and steel, there are millions of small, naturally magnetised areas called domains. These point in all directions at random and so the material has no polarity. When the material is magnetised, e.g. by stroking, the domains turn round and all point in one direction. This means there are free poles at each end of the material, giving rise to the north and south poles of the magnet. See Fig. 52.
dominant Describes a characteristic (or gene) possessed by one parent which appears in the offspring and masks the corresponding characteristic derived from the other parent, which is said to be recessive, Example: in Mendel's classic experiment, he found that peas having yellow seeds crossed with peas having green seeds gave rise to progeny all having yellow seeds, i.e. yellow was the dominant characteristic and green was the recessive characteristic.
donor An individual giving blood to be transfused, tissue or organs to be transplanted, or semen to be inseminated.
Doppler effect The apparent change in the frequency of a wave motion (sound or electromagnetic radiation) caused by the relative motion between the source and the observer. Example: the apparent change in frequency of an approaching ambulance's siren is heard as a sudden fall in pitch when the ambulance passes an observer.
dormancy A resting period in which the metabolic rate of an organism is greatly reduced. The organism is said to be dormant.
dormant See dormancy.
dormin See abscisic acid.
dorsal Pertaining to or situated on or near the back, e.g. dorsal fin.
dosemeter An instrument or device used to measure the amount of radiation received by a person working with radioactive materials. A common type is the film dosemeter or film

double blind trials

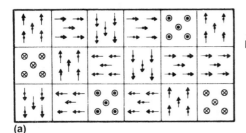

Fig. 52 Domain theory
(a) Magnetic material in demagnetised condition
(b) Magnetised state

badge, where the amount of blackening of the film is a measure of the radiation received.

double blind trials *See* placebo.

double bond *See* covalent bond.

double decomposition (metathesis) A chemical process in which two compounds simultaneously decompose to form two new compounds. *Example:* when silver nitrate solution, $AgNO_3$ is added to a sodium chloride solution, NaCl, silver chloride, AgCl (insoluble), and sodium nitrate, $NaNO_3$, are formed:
$AgNO_3 + NaCl \rightarrow AgCl + NaNO_3$
In double decompositions, the driving force is often the insolubility or the volatility of one of the products.

double helix *See* deoxyribonucleic acid.

double refraction (birefringence) The phenomenon in which a single ray of light is split into two rays which travel with different velocities and which are plane-polarised at right angles to one another. This happens when a ray of light passes through certain materials such as a crystal of calcite. The ray that obeys the ordinary laws of refraction is called the ordinary ray, whereas the other is called the extraordinary ray. The two rays emerge from different points of the crystal. *See also* Nicol prism *and* calcite.

double salt *See* salt.

down feather One of the small, fine feathers situated beneath the contour feathers of adult birds. They trap air and therefore provide insulation. A down feather is also called a plumule. *See* Fig. 67.

Down's syndrome (mongolism) A congenital disorder which may be due to the presence of an extra chromosome in the nuclei of body cells (47 chromosomes present instead of the normal number 46), or to the normal number of chromosomes being arranged in an unusual manner. Down's syndrome is more often seen in babies born to older mothers than in those born to young women. A baby suffering from Down's syndrome has a short, broad face, slanted eyes, short fingers, weak muscles, and shows some degree of mental retardation. Occasionally such a baby is born with heart disorders.

DPD *See* diffusion pressure deficit.

dragonfly *See* Odonata.

drone A fertile male bee with reduced mouth parts. *Compare* worker.

dropping funnel *See* separating funnel.

dropsy *See* oedema.

Drosophila melanogaster A small fruit fly, about 2 mm long, of the order Diptera. It has been and still is widely used in genetic research because it breeds easily and quickly. Another advantage is that wild types produce many different mutations when exposed to radiation.

drug Any substance which is used to treat disease or relieve symptoms.

drupe A fleshy fruit containing one or more seeds. The pericarp is differentiated into the inner hard portion (the endocarp) enclosing the seed, the middle fleshy region (the mesocarp) and the outer skin-like covering (the epicarp). Examples include the plum and the mango. The coconut is a drupe with a fibrous mesocarp. *Compare* berry.

dry cell A form of Leclanché cell (a primary cell) consisting of a zinc can (the negative pole) in the centre of which is placed a carbon rod (the positive pole) surrounded by a core of a mixture of manganese(IV) oxide, MnO_2, and powdered carbon. The remaining space is filled with an ammonium chloride jelly, NH_4Cl, (the electrolyte) and the cell is sealed to prevent the electrolyte from drying up. One dry cell has an e.m.f. of 1,5 volts (1.5 volts). The mixture of chemicals in the cell may vary depending on its manufacturer. *See* Fig. 53; *compare* wet cell.

dry ice *See* carbon dioxide.

drying agent A substance capable of absorbing moisture from another substance. It is either hygroscopic or deliquescent; e.g. silica gel, anhydrous calcium chloride, $CaCl_2$, and concentrated sulphuric acid, H_2SO_4.

drying tower A glass vessel containing a drying agent through which a moist gas can pass to be dried. *See* Fig. 54.

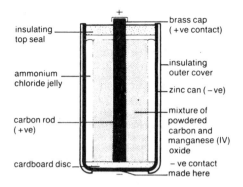

Fig. 53 Dry cell

DTs *See* delirium tremens.
duct A channel or tube carrying liquids (secretions) or gases.
ductile Describes a material which can be drawn out into a wire.
ductless glands (endocrine glands) These are glands, without ducts, whose secretions (hormones) are poured directly into the blood, e.g. the adrenal and pituitary glands.
Dulong and Petit's law A law stating that the relative atomic mass of a solid element multiplied by its specific heat capacity is equal to approximately 6,4 (6.4). This product is known as the atomic heat of the element. The figure 6,4 (6.4) is obtained by averaging the figures for different elements and there are several exceptions for which the law does not hold. However, at high temperatures many metals and several non-metals approximate to the law. The chief use of the law (put forward in 1819) has been to determine the valency of a solid element, of which the equivalent mass has been determined accurately. *Example:* the equivalent mass of an element is determined to be 32,69 (32.69). By using Dulong and Petit's law, its approximate relative atomic mass is determined to be 69. This gives

$$\text{valency} = \frac{69}{32{,}69} \left(\frac{69}{32.69}\right)$$
$$= 2 \text{ (nearest whole number)}$$

A more accurate relative atomic mass of the element is then equal to
$2 \times 32{,}69 = 65{,}38$
$(2 \times 32.69 = 65.38)$
indicating that the element is zinc.

duodenum The first part of the small intestine, situated immediately below the stomach. In humans it is about 30 cm long and about 2,5 cm (2.5 cm) in diameter. Into it run the bile and the pancreatic ducts. Digestion of fats begins in the duodenum and starch is also acted upon here. *See* Fig. 49.
Duralumin® A very light, hard, strong alloy containing more than 90% aluminium, Al; some magnesium, Mg; copper, Cu; manganese, Mn, and silicon, Si. It is used for making aircraft and the moving parts of certain machinery.
dura mater The tough, protective outer membrane enveloping the brain and spinal cord. *See also* meninges.
duramen *See* heartwood.
dwarfism A condition in which not enough growth hormone is secreted from the pituitary gland. Pituitary dwarfs are normally proportioned, which serves to differentiate them from those whose limited growth is due to other causes. *Compare* gigantism.
dye A natural or synthetic substance (usually of organic origin) which is used in colouring different materials such as textiles, paper, plastics and leather. *See also* azo compound.
dynamic equilibrium *See* equilibrium.
dynamic friction *See* friction.
dynamics The branch of mechanics concerned with the motion of bodies and the forces which change or produce motion.
dynamic viscosity *See* viscosity.
dynamite An explosive invented by the Swedish chemist Alfred Nobel. Dynamite consists of nitroglycerine absorbed in kieselguhr. Nitroglycerine is made by a reaction between glycerine (glycerol) and concentrated nitric and sulphuric acids. It is a pale yellow, oily liquid which explodes violently on slight shock. It is absorbed in kieselguhr to make it less sensitive to shock.

Fig. 54 Drying tower

dynamo *See* generator.

dynode *See* photomultiplier

dysentery An infectious disease affecting the intestines, caused by bacteria. Symptoms are diarrhoea, sometimes accompanied by a little blood and mucus, fever, vomiting and stomach pains. The disease is common in areas where there is poor sanitation. Dysentery is treated with antibiotics and plenty of liquid to avoid dehydration. The protozoan *Entamoeba histolytica* also causes dysentery, especially the type known in tropical areas. The protozoan eventually passes out of the body in the faeces. Dysentery caused by bacteria is more common in Europe.

E

e The charge on an electron. It is equal to $1{,}60219 \times 10^{-19}$ coulombs (1.60219×10^{-19} C).

ear The sense organ of hearing and balance in vertebrates. In mammals, the ear is divided into three areas: the external or outer ear, the middle ear or tympanic cavity and the inner ear or membranous labyrinth. *See* Fig. 55.

ear drum (tympanic membrane) A concave membrane made up of skin and fine collagen fibres, situated at the innermost end of the external ear canal, separating the outer ear from the middle ear. It transmits vibrations to the small bones of the middle ear. *See* Fig. 55.

early clocks *See* clock.

earpiece *See* telephone.

earth An electrical connection made between an electrical appliance and the Earth. This connection is made to protect the user from an electric shock should the appliance develop a fault which allows electricity to flow through the outer case of the appliance. The potential of the Earth is considered to be zero.

Earth One of the nine planets in our Solar System, orbiting the Sun in a slightly elliptical orbit between Venus and Mars once in $365\frac{1}{4}$ days approximately. The Earth is a sphere which is somewhat flattened towards the poles. Its age is about 4 500 million years and it rotates on its axis once in 23 hours, 56 minutes, 4 seconds. About 70% of the Earth's surface is covered with water.

earthenware A material made from coarse, baked clay. It is used in the manufacture of pots, sanitary pipes, tiles, etc.

earthquake A sudden, natural vibration within the Earth's crust. It may be caused by movements of molten rock below or within the crust or by the development of faults in the crust. *See also* Richter scale.

Earth's crust *See* crust.

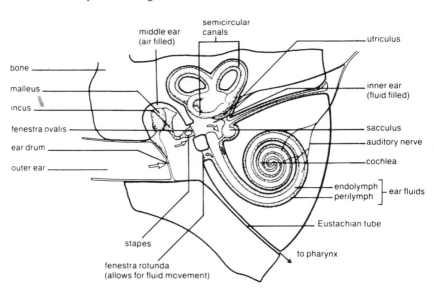

Fig. 55 Ear in section: arrows represent movements of ear drum, ossicles and ear fluids

Earth's magnetism *See* terrestrial magnetism.

ebonite (vulcanite) A hard, black, insulating material manufactured by vulcanising rubber with 25–50% of sulphur. Today ebonite is to a large extent replaced by different types of plastic. Ebonite was first prepared by C. Goodyear in 1850.

ecdysis In insects and reptiles, the periodic casting off (moulting) of the cuticle or epidermis to enable growth to occur. *See also* endysis.

ECG *See* electrocardiogram.

Echinodermata A phylum of marine animals comprising starfish, sea-urchins, sea-cucumbers, sea-lilies, etc.

echo A sound image produced when sound waves are reflected by a hard surface such as a wall or cliff.

echolocation The location of objects by the determination of sound waves or radio waves reflected from them. Echolocation is used by mammals such as the bat and the porpoise, and in radar.

echo-sounder (asdic, sonar) An apparatus used to determine the depth of sea beneath a ship, and to locate shipwrecks, submarines and shoals of fish. This is done by measuring the time taken between sending out a sound signal from the bottom of the ship and receiving the echo. The signal is reflected from the sea-bed or from objects under the ship. *See* Fig. 56.

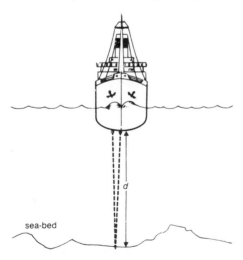

Fig. 56 Echo-sounder

eclipse The passage of one heavenly body through the shadow of another. (1) A solar eclipse occurs when the Moon passes between the Sun and the Earth. The Moon casts a shadow on the Earth. *See* Fig. 57 (a). (2) A lunar eclipse occurs with the Earth passes between the Sun and the Moon. The Earth casts a shadow on the Moon. *See* Fig. 57 (b).

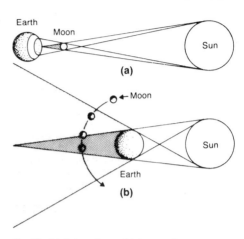

Fig. 57 (a) Solar eclipse (b) Lunar eclipse

E. coli *See Escherichia coli.*

ecology A branch of biology concerned with the relationships of animals and plants to their natural surroundings.

ecosystem (biosystem) A community of organisms consisting of producers, consumers and decomposers interacting with one another and with the environment: e.g. a pond or a sea-shore.

ecto- Prefix meaning outer or outside.

ectoderm In multicellular animals, the outer layer of the embryo. The epidermis, hair, nails and nerve tissue develop from the ectoderm. *Compare* endoderm.

ectomycorrhiza *See* mycorrhiza.

ectoparasite A parasite living on the outside of the host, e.g. flea. *Compare* endoparasite.

ectoplasm The outer layer of cytoplasm in an animal cell. It is usually clear and contains only a few granules. *Compare* endoplasm.

ectotrophic *See* mycorrhiza.

eczema A non-contagious skin disorder characterised by a red, often itchy rash and blisters.

edaphic conditions Environmental conditions determined by the biological, chemical and physical characteristics of the soil.

eddy currents Induced currents circulating throughout the cores of transformers, electromagnets or other electrical equipment as a result of changing magnetic fields. Eddy currents cause energy losses and, in transformers, can be greatly reduced by the use of a laminated core.

Edison accumulator *See* Ni–Fe cell.

EDTA *See* edta.

edta (EDTA) Ethylenediaminetetra-acetic acid. Edta is also known under the trade names Sequestrol and Versene. It is a white solid which is often used quantitatively in titrations to determine the concentration of certain cations such as calcium, Ca^{2+}, and magnesium, Mg^{2+}, in a sample. Edta forms very stable, soluble complexes with many other metal ions, especially with those of the transition elements. In titrations, a solution of the disodium salt of edta is often used as it is more soluble in water than edta itself. Since a complex is formed using edta as a titrant, such titrations are called complexometric titrations. Edta is also used in industrial water softening as it 'locks up' calcium and magnesium ions, preventing them from forming a scum or precipitate with soap or any other substance. *See also* chelate.

EEG *See* electroencephalogram.

eelworm *See* Nematoda.

effective value *See* root-mean-square value.

effector A cell, tissue or an organ responding to a stimulus; cilia, muscles and glands are effectors.

efferent Conducting away; efferent nerves conduct impulses away from the nervous centres. *Compare* afferent.

efferent neurone *See* motor neurone.

effervescence In a liquid, the production of bubbles which rise to the surface. Effervescence is seen when a bottle of fizzy drink is opened; when certain metals, e.g. calcium, react with water; or when an acid reacts with a carbonate.

efficiency The efficiency of a machine is the ratio of the work done by the machine to the work put into the machine. It is usually expressed as a percentage:

$$\text{efficiency} = \frac{\text{work output}}{\text{work input}} \times 100\%$$ or

$$\text{efficiency} = \frac{\text{mechanical advantage}}{\text{velocity ratio}} \times 100\%$$

Because of work done in overcoming friction in a machine the efficiency is always less than 100 per cent.

efflorescence 1. The time of flowering (blossoming). 2. The partial or total loss of water of crystallisation from the atmosphere from a hydrated salt, e.g. hydrated sodium carbonate (washing soda), $Na_2CO_3 \cdot 10H_2O$, and iron(II) sulphate, $FeSO_4 \cdot 7H_2O$. *Compare* deliquescence.

effort The force applied to a machine such as a lever or pulley system.

effusion The process in which a gas under pressure passes through a small opening, i.e. from a region of higher pressure to a region of lower pressure.

egestion The process of ridding the body of undigested material.

egg (ovum) The female gamete which, when fertilised, gives rise to a new individual. *See also* yolk.

egg albumen *See* albumen *and* albumin.

egg-tooth A small projection of the upper jaw of an embryo reptile or the beak of an embryo bird. It is used to crack the shell on hatching.

egg-white *See* albumen.

Einstein's equation (mass–energy equation) An equation, suggested by Einstein, stating that matter and energy are related according to the expression:

energy = mass × (velocity of light)^2

$E = m \times c^2$

E is measured in joules, m in kilograms and c in metres per second. For the complete conversion of 1 kg of matter to energy, one gets:

$E = (1 \text{ kg}) \times (3 \times 10^8 \text{ m s}^{-1})^2$
$= 9 \times 10^{16} \text{ kg m}^2 \text{ s}^{-2}$
$= 9 \times 10^{16} \text{ J}$

This relation between matter and energy has proved to be true experimentally (e.g. in fission and fusion processes). The very large value of c^2 indicates that it is possible to obtain large amounts of energy from a very small mass.

ejaculation The forcing of semen from the penis.

elaioplast *See* leucoplast.

elastic *See* elasticity.

elasticity The ability of a substance to return to its previous shape and size after distortion by a force. Substances with this ability are said to be elastic. *Compare* plasticity; *see also* Hooke's law.

elastic limit The value (limit) of force below which an elastic material returns to its original length after being stretched by a force. If this limit of force is exceeded, the material may be permanently stretched. *See also* Hooke's law *and* yield point.

elastin An elastic, fibrous protein found in the connective tissue of vertebrates, e.g. in the walls of arteries and in the lungs. *See also* scleroprotein.

elbow-joint The hinge joint between the distal end of the humerus and the proximal ends of the radius and ulna.

electrical appliance Any device converting electrical energy to other forms of energy. *Example:* an electric kettle converts electrical energy to heat energy.

electrical energy Energy obtained from moving electric charges. Units of electrical energy are joules (watt seconds) and kilowatt-hours.

electrical shielding (screening) *See* Faraday's cage.

electric arc A luminous, electrical discharge produced between two electrodes usually made of carbon (graphite). The discharge is characterised by a low potential difference and a high current. Heat is produced as a result of the discharge and evaporation of the electrodes maintains it. Electric arcs are used in arc welding, in arc lights, in production of steels, etc.

electric arc furnace A furnace used in the production of high quality steels, e.g. stainless steel. The furnace is cylindrical in shape and is fitted with a removable cover through which three graphite electrodes are lowered onto the surface of the charge. The charge is usually scrap steel to which lime and iron ore are added. Electric arcs between the electrodes and the charge generate heat, melting the metal. A slag forms on the surface of the molten metal and is raked off. Other metals such as chromium and nickel may be added to the furnace to obtain the required type of steel.

electric bell An electromagnetic device in which an electric current (direct current) flowing through a coil produces a magnetic field which attracts an iron armature. This movement of the armature breaks contact with a contact screw and switches off the current. The armature springs back and closes (makes) the circuit again. The current flows once more and the action is repeated as long as the bell-push is pressed. Thus the armature vibrates and the hammer attached to it repeatedly strikes a gong. *See* Fig. 58.

An electric bell using alternating current does not use this type of make-and-break contact. The armature consists of a permanent magnet and the reversing poles of an electromagnet through which the current flows, alternately attract and repel the armature. This produces the vibration necessary for the hammer to strike the gong.

electric cell *See* cell.
electric charge *See* charge.
electric clock *See* clock.
electric current Symbol: *I*; unit: ampere (A). An electric current consists of moving electric charges. These charges are carried through conductors by charge carriers such as electrons or ions. In solid conductors, the electric current is considered to be a flow of electrons.

electric field The space in the neighbourhood of a charged object. When an electric charge is placed in this region, a force is exerted upon it.

electricity A term used for the phenomena caused by electric charges. Static electricity is formed by stationary charges, while current electricity is formed by moving charges.

electricity meter *See* joulemeter.

electric motor Any device for converting electrical energy into mechanical energy. A simple type of electric motor consists of a coil (armature) which can turn in a powerful magnetic field. An electric current flows through a commutator to the coil which rotates due to the mechanical force exerted on a current-carrying conductor (the coil) in a magnetic field.

electric potential *See* potential.

electric power Symbol: *P*; unit: watt (W). The power of an electrical appliance is the rate at which it converts electrical energy. The power of an appliance can be calculated by multiplying the potential difference in volts by the current in amperes.

electric spark A visible discharge of electricity between two places (e.g. electrodes) having opposite high potentials. When an electric spark is seen, a sound is also heard because the heating of air between the electrodes creates a cracking noise.

electric wind A stream of ionised and unionised gaseous molecules flowing away from a highly charged electrical conductor. The ionisation occurs when gaseous molecules collide in the intense electric field around the conductor. Ions which have the same charge as the conductor are then repelled from it, dragging uncharged, gaseous molecules with them.

electrocardiogram (ECG) A graph obtained by amplifying the minute electrical impulses generated in the heart. By studying the pattern of an ECG, a trained doctor may discover

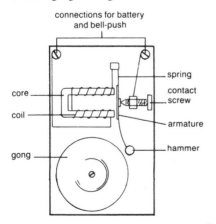

Fig. 58 Electric bell

whether the heart is healthy or whether it is damaged. An ECG is produced by an instrument called an electrocardiograph.

electrocardiograph The instrument used to record an ECG. From the instrument, electrodes lead to the skin over the patient's heart and to both arms and legs. The electrodes detect (after amplification) the minute electrical impulses generated by the beating heart. *See also* electrocardiogram.

electrochemical equivalent The mass of an element liberated during electrolysis by one coulomb. The SI unit is kilogram per coulomb, $kg\,C^{-1}$ (kg/C).

electrochemical series (of metals) (electromotive series, displacement series) Metallic elements arranged in order of their increasing electrode potentials, those highest in the series having large negative values. In this series hydrogen is included, having an arbitrary electrode potential of zero. *See also* activity series of metals *and* redox potential.

electrochemistry A branch of chemistry concerned with the interconversion of electrical energy and chemical energy.

electrode A conductor (a plate, grid or wire) for emitting, collecting or controlling the flow of electric charge carriers in an electrolyte, gas, vacuum, dielectric or semiconductor. *See also* anode *and* cathode.

electrode potential The potential difference between an element and a solution (electrolyte) containing its ions. If the ionic concentration of the electrolyte is 1 mol per dm^3 and the temperature is 25°C, then the electrode potential is called the standard electrode potential. *See also* redox potential *and* Appendix.

electroencephalogram (EEG) A graph obtained by amplifying the minute electrical impulses generated in the nerve-cells of the brain. By studying the pattern of an EEG, a trained doctor may discover whether the brain is healthy or not. An EEG is produced by an instrument called an electroencephalograph.

electroencephalograph The instrument used to record an EEG. From the instrument, electrodes are taped to the scalp of the patient. They detect (after amplification) the minute electrical impulses generated in the nerve-cells of the brain. In many countries, the instrument is used in establishing death. An EEG which is completely flat, without any peaks or valleys, for two days, indicates the death of the patient even if his heart is still beating. *See also* electroencephalogram.

electrolysis The process in which an ionic solution or fused ionic compound (an electrolyte) is decomposed by the passage of a direct current through it. The positive cations and the negative anions move towards the cathode and anode respectively, where substances are either deposited or liberated when the ions give up their electric charges by gaining or losing electrons.

electrolysis (laws of) *See* Faraday's laws of electrolysis.

electrolyte A substance which, when dissolved in a solvent (often water) or when molten, is able to conduct an electric current during electrolysis. The term also applies to the liquid itself. Strong electrolytes, such as strong acids, alkalis and most soluble salts, contain many ions. Weak electrolytes, such as weak acids and alkalis, contain only a few ions. Non-electrolytes, such as pure water, propanone (acetone), and paraffin, contain very few ions or none at all.

electrolytic Pertaining to electrolysis.

electrolytic cell *See* cell.

electrolytic corrosion Two different metals in contact with an electrolyte give rise to this type of corrosion. The more electropositive of the two metals, i.e. the one placed higher in the electrochemical series, will act as an anode and dissolve. *Example:* in galvanising, iron is coated with a layer of zinc. If this layer of zinc is scratched and the iron becomes exposed to a damp atmosphere, the zinc dissolves according to:

zinc -2 electrons $\rightarrow$ zinc ions

See also sacrificial protection.

electrolytic dissociation The formation of free positive and negative ions when an electrolyte dissolves in a liquid (usually water). *Example:* when hydrogen chloride, HCl, dissolves in water, hydrogen ions, H^+, and chloride ions, Cl^-, are formed:

$HCl \rightleftharpoons H^+ + Cl^-$

Hydrogen ions combine with water molecules to produce hydroxonium ions, H_3O^+:

$HCl + H_2O \rightleftharpoons H_3O^+ + Cl^-$

electromagnet A temporary magnet consisting of a coil of insulated wire wrapped around a soft iron core. When a current flows in the coil, the core becomes strongly magnetised.

electromagnetic induction The production of an induced e.m.f. by the use of magnets and magnetic fields. This can be done in several ways: (1) When a magnet is moved towards a coil of wire, an e.m.f. is induced in the coil. If the magnet is then moved away from the coil, the direction of the e.m.f. is reversed. The same effect is produced when the magnet is stationary and the coil is moved towards and away from the magnet. (2) When a coil, connected to an electric supply, is used instead of a moving magnet and is placed next to a second coil, an induced e.m.f. is obtained in the second coil by a changing current in the first coil. Electromagnetic induction occurs as a result of a changing magnetic field. It forms the basis of the generator (dynamo) and transformer. *See also* Faraday's law of electromagnetic induction *and* Lenz's law.

electromagnetic radiation Radiation consisting of waves of energy (electromagnetic waves), associated with electric and magnetic fields, produced by accelerating charged particles. These electric and magnetic fields vibrate at right angles to each other and to the direction of motion. Electromagnetic fields require no medium for their propagation and travel through space at $2{,}997\,925 \times 10^8$ ($2.997\,925 \times 10^8$) metres per second (the speed of light). Their properties depend on their frequency or wavelength, the range of frequency being called the electromagnetic spectrum.

electromagnetic spectrum The frequency range of electromagnetic radiation. *See* Fig. 59.

electromagnetic wave *See* electromagnetic radiation.

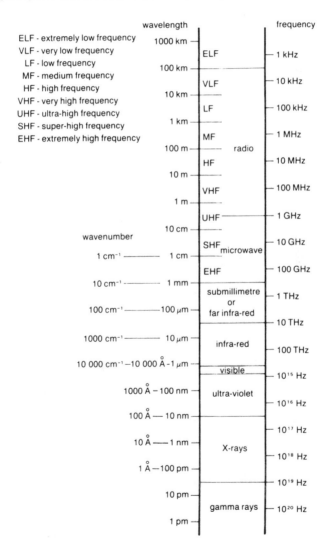

Fig. 59 Electromagnetic spectrum

electromagnetism The branch of physics concerned with the study of magnetic forces produced by electricity and electric effects produced by magnetic fields.

electromotive force (e.m.f.) Symbol: E; unit: volt (V). The driving force of a cell, generator, etc., 'pushing' electrons round a circuit. The e.m.f. of a cell is its potential difference in open circuit, i.e. when no current is flowing. It may also be expressed as the total work done in joules per coulomb of electricity conveyed in a circuit in which the cell is connected.

electromotive series *See* electrochemical series.

electron (negatron, negaton) One of the three basic particles in an atom. Electrons circle in orbits around the nucleus. The mass of an electron is 9.109×10^{-31} kg (9.109×10^{-31} kg) which is about 1/1840 of the mass of a proton. The electron carries a negative electric charge of 1.602×10^{-19} coulombs (1.602×10^{-19} coulombs). In a neutral atom the number of electrons equals the number of protons. *Compare* positron; *see also* beta ray, cathode ray, rest mass, antiparticle.

electron configuration The arrangement of electrons around the nucleus of an atom. For example, the electron configuration of nitrogen, N, is 2,5 indicating that there are two electrons in the first shell (K-shell) and five in the second (L-shell). However, the term also applies to the distribution of electrons in subshells. *See also* orbital.

electronegativity A measure of the ability of an atom to attract electrons. Fluorine is the most electronegative atom (element) known. Several attempts have been made to give a quantitative measure of electronegativity; they have all been more or less arbitrary, although they give results which agree reasonably well. The most generally accepted measure is that suggested by Linus C. Pauling who used bond energy values in his estimations. In Pauling's scale of electronegativity, fluorine is given the value 4,0 (4.0); oxygen 3,5 (3.5); chlorine 3,0 (3.0). Very electropositive elements have been given low values: e.g. 0,7 (0.7) for francium, 0,8 (0.8) for potassium and 0,9 (0.9) for sodium. *Compare* electropositivity.

electron gun A device, consisting of an arrangement of electrodes, for producing a stream of electrons. A grid controls the number of electrons emitted from a heated cathode. The electrons are focused by an anode and accelerated by a second anode. Electron guns are essential parts of electronic apparatus such as cathode-ray tubes and electron microscopes.

electronics The study, application and development of devices controlling the motion of electrons, e.g. thermionic valves and semiconductors.

electron micrograph *See* micrograph.

electron microscope A microscope producing a magnified image of an object using a beam of high-energy electrons rather than light. The electron beam passes through a magnetic or electrostatic focusing system which is similar to the optical lens system in a light microscope. Electron microscopes produce more detailed images than optical microscopes, i.e. have higher resolving powers, because the wavelengths of electrons are much smaller than the wavelength of light. Magnifications as high as 200 000 or more can be achieved with an electron microscope.

electronvolt Symbol: eV. The quantity of energy gained by an electron when passing through a potential rise of one volt. This unit of energy is commonly used in nuclear physics. One electronvolt is equal to 1.602×10^{-19} joules (1.602×10^{-19} J).

electrophile An ion, atom or molecule which is short of electrons, e.g. a cation such as H^+, H_3O^+ and NO_2^+ (nitryl cation) or $AlCl_3$ and SO_3. Electrophiles are capable of accepting an electron pair when forming a new bond with another atom or group of atoms; they react more readily at negative centres such as those found in dipolar molecules. Electrophilic reactions are common in organic chemistry. They may be either addition reactions or substitution reactions. An example of an electrophilic reaction is the formation of ethanol, CH_3CH_2OH, from ethene, $H_2C{=}CH_2$, when the latter is first treated with concentrated sulphuric acid, H_2SO_4, at 80°C and the product is then boiled with water:
$H_2C{=}CH_2 + H_2SO_4 \rightarrow H_3C-CH_2HSO_4$
and
$H_3C-CH_2HSO_4 + H_2O \rightarrow$
$\qquad\qquad\qquad CH_3CH_2OH + H_2SO_4$
The hydrogen ions, H^+, from the acid are the electrophilic reagent attacking the double bond in ethene around which there is a centre of high electron density. *Compare* nucleophile.

electrophilic reaction *See* electrophile.

electrophilic reagent *See* electrophile.

electrophoresis (cataphoresis) The movement of charged colloidal particles through a fluid when an electric field is applied to it. Electrophoresis is often carried out on specially prepared paper or in a gel on a glass slide. It is used to separate mixtures of proteins and is an important method in both qualitative and quantitative analysis.

electrophorus A simple device for generating an electric charge by electrostatic induction. It consists of a flat, insulating plate, charged by friction, and a circular metal plate with an insulating handle. The metal plate is placed on top of the charged plate and earthed momentarily by touching it. This leaves the metal plate with an induced charge of opposite sign to that on the insulating plate. This charge can be given to another conductor, e.g. an electroscope, or a spark may be drawn from it by holding a finger close to its edge.

electroplating The electrolytic process of plating a metal with a less reactive one, e.g. the plating of iron with nickel or chromium. This is done by placing the object to be plated as a cathode in an electrolyte containing ions of the metal which is used for the plating. These ions are deposited on the object in the form of a firmly adhering coat. Electroplating is carried out to protect metals from corrosion, or for decorative purposes, or both.

electropositivity A measure of the ability of an atom to lose electrons. The alkali metals are all very electropositive atoms (elements). *Compare* electronegativity.

electroscope An instrument for detecting electrical potential differences. In the common gold leaf electroscope, one or two gold leaves are attached to the end of a metal stem connected to a metal cap. The stem is placed in a metal case with glass sides and insulated from it. When an electric charge is placed on the metal cap, the gold leaves diverge because they receive like charges and so repel one another. *See* Fig. 60.

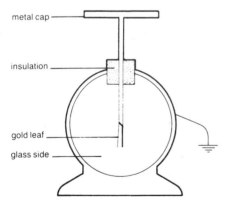

Fig. 60 Electroscope

electrostatic induction The production of an electric charge on a conductor by placing it in the electric field surrounding a charged body. If a neutral body is brought close to a positively charged body, surface electrons are attracted towards the positive charge and move towards it, so the surface nearer the positive charge becomes negatively charged. The attraction between the charges is strong enough to move a small, light object. This explains why a plastic rod charged by friction will attract and pick up small pieces of paper. *See also* charge.

electrostatics The study of static electricity (electric charges at rest).

electrovalent bond *See* ionic bond *and* covalent bond.

electrovalent compound A chemical compound existing as ions and not as molecules. Sodium chloride, NaCl (or Na^+Cl^-) is an electrovalent compound formed from its elements when an electron is transferred from sodium to chlorine:
$$Na - e + Cl \rightarrow Na^+Cl^-$$
The bond between the two atoms is ionic (electrovalent). Most acids, bases and salts are electrovalent compounds. *See also* ionic bond *and* covalent bond.

electrovalent crystal *See* ionic crystal.

element A substance which cannot be split into simpler substances by any chemical process; i.e. an element contains only one kind of atom. There are about 105 elements, of which 91 occur naturally.

elementary particles Particles found in atoms. Originally only protons, neutrons and electrons were known as elementary particles. However, many other elementary particles have been discovered, the existence and properties of which are intensively studied by scientists. Elementary particles which are not necessarily protons, neutrons or electrons may be produced when highly energetic atomic nuclei are brought to collide with each other or with elementary particles. Today there are about 200 known elementary particles.

elephantiasis (filariasis) A disease caused by a thread-like parasitic worm whose larvae are transferred from one human to another by certain mosquitoes. The worms live in the bloodstream and in the lymphatic system where they produce inflammation, swelling and obstruction of lymph flow. This causes parts of the body, especially the arms and legs, to become greatly swollen. The disease is treated surgically. *See also* filarial worms.

elevation of boiling-point An increase in the boiling-point of a solvent by the addition of a solute. Measurement of the elevation is used in relative molecular mass determinations. For pure substances, the elevation of boiling-point is proportional to the number of molecules or ions which dissolve in the solvent, and the elevation which is produced by the same molecular concentration of a substance is a constant for a particular solvent. *Compare* depression of freezing-point; *see also* Appendix.

elimination reaction A chemical reaction in which a double bond or a triple bond arises when an organic molecule loses certain atoms. E.g. when water, H_2O, is removed from ethanol, CH_3-CH_2OH, ethene, $CH_2=CH_2$, is obtained. This process can be carried out by treating ethanol with concentrated sulphuric acid at about 200°C. *Compare* addition reaction; *see also* substitution reaction.

elixir of life *See* alchemy.

eluent *See* carrier.

elytron (pl. elytra) The modified, horny fore wing of a beetle protecting the thin hind wing when the insect is at rest. *See also* Coleoptera.

embolism An obstruction of a blood vessel, which may be caused by materials such as a blood clot, a globule of fat, a clump of bacteria or a gas bubble, coming from elsewhere in the circulatory system. The object causing the obstruction is called an embolus. Embolism may cause a stroke and may lead to gangrene.

embolus (pl. emboli) *See* embolism.

embryo 1. A young plant in the first stage of development from a zygote. In seed plants, the embryo consists of the plumule, radicle and cotyledons. 2. An animal in the first stage of development from a zygote before birth or hatching. This stage of development of a human embryo lasts approximately eight weeks, after which the embryo is called a foetus.

embryology A branch of biology concerned with the study of the formation and development of embryos.

embryonic membrane One of the membranes concerned with nutrition, protection and respiration which surround the embryos of vertebrates.

embryo sac In angiosperms, a large, oval cell in the nucellus of the ovule. Fertilisation and development of the embryo take place in the embryo sac.

emergent ray *See* refraction.

emery *See* alumina.

emesis *See* vomiting.

emetic A substance (medicine) which causes vomiting.

e.m.f. *See* electromotive force.

emission of radiation *See* radiant energy.

emission spectrum *See* spectrum.

emitter *See* transistor.

empirical Describes results obtained from experimental observations rather than from theory.

empirical formula *See* formula.

emulsifier *See* emulsifying agent.

emulsifying agent (emulsifier) A substance which promotes the formation of an emulsion at the same time stabilising it. Detergents are examples of emulsifying agents.

emulsion 1. A colloidal dispersion in which small droplets of one liquid are dispersed in another, e.g. an oil in water emulsion or a water in oil emulsion. Because of the large surface area between the two phases (liquids), an emulsion is often rather unstable. However, an emulsifying agent may be added to increase the stability of the emulsion, i.e. to ensure that the emulsion does not easily separate. Emulsions are widely used, e.g. as foods, medicines, cosmetics and lubricants. 2. A light-sensitive, thin coating applied to celluloid, glass or paper used as a film in photography.

enamel 1. A form of glass fused on to a metal surface at about 700°C. Enamel is made by melting together sand, sodium carbonate, boron oxide and small amounts of other substances. When the hot molten mass is poured into water, it shatters into small solid pieces known as frit. The frit is ground to a fine powder and mixed with feldspar and water to make up a thin cream which can be applied to a metal surface. 2. The external hard layer of the crown of a tooth. It consists of more than 90% calcium and magnesium compounds.

enantiomers *See* optical isomerism.

enantiotropy The existence of two allotropes of an element or of two forms of the same substance which are directly convertible into each other by a change in temperature; each has its own temperature range of stability. Sulphur exhibits enantiotropy. *Compare* monotropy; *see also* monoclinic sulphur.

encephalon *See* brain.

endemic Pertaining to a disease or pest which persists among a certain people or in a certain region. *Examples:* cholera is an endemic disease in Asia, often rising to epidemic proportions; in tropical regions the mosquito is endemic. *See also* epidemic *and* pandemic.

endo- Prefix meaning internal.

endocardium The inner layer (membrane) of the heart, consisting of endothelial cells.

endocarp *See* drupe.

endocrine glands *See* ductless glands.

endocrinology The study of the endocrine glands and their secretions, the hormones, together with their nature and effect.

endoderm In multicellular animals, the inner layer of the embryo. The alimentary canal and its associated glands develop from endoderm. *Compare* ectoderm.

endodermis The innermost layer of the cortex, consisting of a single layer of cells surrounding the stele. It is found in all roots, in the stems of ferns and in some dicotyledons. The endodermis controls the horizontal movement of water between the cortex and the stele.

endolymph A fluid in the membranous labyrinth of the ear. See Fig. 55.

endomycorrhiza See mycorrhiza.

endoparasite A parasite living within the host, e.g. tapeworms and several fungi. Compare ectoparasite.

endoplasm The inner layer of cytoplasm in an animal cell. It is usually granular. Compare ectoplasm.

endoplasmic reticulum A network of folded membranes in the cytoplasm of a cell, forming tubes and passages for the many different chemicals used in and produced during metabolism. It is continuous with the nuclear membrane and the cell membrane. The endoplasmic reticulum may be either smooth or granulated with ribosomes on its cytoplasmic side.

endoskeleton A skeleton situated within the body of an animal, e.g. the skeleton of vertebrates. See Fig. 61; compare exoskeleton.

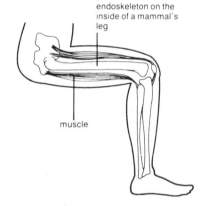

Fig. 61 Endoskeleton

endosperm In many seeds, the nutritive tissue (food store) surrounding the embryo.

endospore A resting cell formed inside a vegetative cell and normally produced under unfavourable growth conditions. This is seen in the case of a few bacteria, the *Bacillus* and *Clostridium*. Usually endospores are oval or spherical structures and are highly resistant to heat, desiccation, disinfectants and X-rays. When favourable conditions for growth return, the endospore germinates, producing a vegetative cell.

endothelium In vertebrates, a single layer of cells lining the heart, blood vessels and lymphatic vessels. See also endocardium.

endothermic process A process (system) in which heat is absorbed from the surroundings. Example: when nitrogen, N_2, and oxygen, O_2, combine to form nitrogen monoxide, NO, heat is absorbed:
$N_2 + O_2 \rightarrow 2NO - heat$
See also enthalpy.

endotrophic See mycorrhiza.

end-point (equivalence point) The point in a titration at which the number of equivalents of liquid used (from the burette) equals the number of equivalents of the sample (in the flask). The end-point is therefore the point at which the reaction is complete. It can be observed using an indicator, which changes its colour at this point.

endysis The formation of a new layer of skin (cuticle or epidermis) following ecdysis.

enema The injection of a liquid into the rectum and colon in order to rid them of their contents. This is done for several reasons, e.g. to clean out bodily waste before an operation.

energy Symbol: E; unit; joule (J). Energy is the capacity of a system to do work. There are two main forms of energy: potential energy (due to position) and kinetic energy (due to motion). However, several other forms are distinguished: e.g. chemical, electrical, heat, light, mechanical, nuclear and sound. These forms of energy are more or less interconvertible and, as stated in Einstein's equation, interconversion may occur between energy and mass. See also conservation (laws of).

energy level Electrons orbiting the nucleus of an atom possess energy depending on their distance from the nucleus. The energy level (shell) of an electron indicates its distance from the nucleus; an electron close to the nucleus has less energy than one that is further away from the nucleus. Consequently, an electron must be given energy to move it from a low energy level to a higher energy level. Conversely, energy is liberated (in the form of electromagnetic radiation) when an electron moves from a high energy level to a lower one. Energy levels are numbered outwards from the nucleus, 1, 2, 3, etc. or are given letters, K, L, M, etc. Each shell contains a certain number of electrons. In the K, L and M shells the maximum numbers of electrons are 2, 8 and 18 respectively. See also Bohr atom *and* orbital.

energy value See calorific value.

enol The general name for a group of organic compounds which contain a hydroxyl group adjacent to a carbon atom which also has a double bond:

—C=C—
　　|
　　OH

Sometimes the enol exists in an equilibrium with a carbonyl compound:

—C=C— ⇌ —C—C—
　|　　　　　|　‖
　OH　　　H　O

and the enol and the carbonyl compound are called tautomers. Unlike ordinary alcohols, enols are stronger acids than water. The word enol indicates both a double bond, C=C, by the prefix en- and an alcohol by the suffix -ol. Enols resemble phenols as the enol structure is present in a phenol. Both enols and phenols give a blue, violet or green colour with an iron(III) chloride solution and both are acidic, dissolving in aqueous solutions of strong alkalis. *See also* tautomerism.

enteric Pertaining to the alimentary canal or intestines.

enterokinase An enzyme, secreted in the duodenum, which activates trypsinogen and converts it to trypsin.

enteron The alimentary canal or corresponding body cavity concerned with digestion and absorption.

enthalpy The heat content (stored heat energy) of a substance or system. Symbol: H. H is often expressed in kJ per mole. During a chemical process it is the change in enthalpy which is important. This is called ΔH (delta H) where Δ means 'difference' or 'change'. If the reactants in a chemical process have more energy than the products, energy is released during the reaction and the process is called exothermic. Conversely, if the products have more energy than the reactants, energy will be absorbed during the reaction and the process is called endothermic. This can be expressed in the following way:
ΔH = enthalpy of products −
                    enthalpy of reactants
For the reaction:
$C_{(solid)} + O_{2(gas)} \rightarrow CO_{2(gas)}$ + heat
one can write:
$\Delta H = H_{CO_2} - H_C - H_{O_2}$
From this it is seen that in an exothermic reaction ΔH is negative, whereas ΔH is positive in an endothermic reaction. ΔH is often called the heat of reaction. *Examples:* in the process
$C_{(solid)} + O_{2(gas)} \rightarrow CO_{2(gas)}$ + heat
ΔH is $-393,6$ kJ mol^{-1} (-393.6 kJ mol^{-1}).
In the process
$H_2O_{(gas)} + C_{(solid)} \rightarrow CO_{(gas)} + H_{2(gas)}$ − heat
ΔH is $+131,5$ kJ mol^{-1} ($+131.5$ kJ mol^{-1}).
See also entropy.

entire With a smooth margin or edge. *See* Fig. 111.

entomology The study of insects.

entropy Symbol: S; unit: joules per mole per degree (J mol^{-1} K^{-1}). A measure of the atomic, ionic or molecular disorder of a system; the more disordered the system, the higher its entropy. *Example:* when ice melts and becomes water there is an increase in entropy because the molecules of water are more disordered in the liquid state than in the solid state. When water (or any other liquid) evaporates to vapour it is obvious that both liquids and vapours are disordered systems, but since the volume of a vapour is much greater than that of a liquid from which it is formed, the motions of the molecules in the vapour are even more disordered than they are in the liquid. Therefore, when a liquid forms a vapour, there is an increase in entropy. Consequently, there is a decrease in entropy when a vapour is condensed into a liquid and a further decrease when the liquid freezes and becomes a solid. From this it is seen that an increase in temperature of a substance results in an increase in its entropy (at the absolute zero of temperature the entropy is zero). The most probable states of a system are those of high entropy. The reason for this is that there are many more ways of producing disordered arrangements than ordered ones. Quantitatively it is not the absolute magnitude of S that is important, but the change in entropy. This is called ΔS (delta S), where Δ means 'difference' or 'change'. A positive value of ΔS indicates an increase in entropy.
Examples: for the melting of ice:
$H_2O_{(solid)} \rightarrow H_2O_{(liquid)}$
$\Delta S = +21,8$ joules mole^{-1} deg^{-1}
    ($+21.8$ J mol^{-1} K^{-1})
For the evaporation of water:
$H_2O_{(liquid)} \rightarrow H_2O_{(gas)}$
$\Delta S = +111,0$ joules mole^{-1} deg^{-1}
    ($+111.0$ J mol^{-1} K^{-1})
Knowing the different causes of changes in entropy it is possible to predict the sign and the size of the entropy change in a given reaction.
Example: in the following reaction:
$K_{(solid)} + \frac{1}{2}Cl_{2(gas)} \rightarrow KCl_{(solid)}$
the predicted ΔS is negative (decrease in S) and large because gas disappears and crystals are formed.

A system in a state of equilibrium results from a balance of two factors. The first is the tendency of the system to reach minimum enthalpy; the second is the tendency of the system to reach maximum entropy. In many states of equilibria these two factors oppose each other or are in competition, i.e. the equilibrium state is a compromise between them. The interplay of the enthalpy and the entropy of a system can be described by using a concept called free energy, which has the symbol G. G is expressed as
$G = H - TS$
under conditions of constant temperature and pressure, where H is the enthalpy, T is the absolute temperature and S the entropy. For a change in G at constant temperature and pressure, one gets:
$\Delta G = \Delta H - T(\Delta S)$

When equilibrium is reached,
$\Delta G = 0$
i.e. the free energy of the products and reactants in a chemical process are equal. If ΔG is negative, the free energy of the products is less than that of the reactants, i.e. the process may take place of its own accord. If ΔG is positive, the free energy of the products is greater than that of the reactants and the process will not take place unless something is done, e.g. the temperature is raised causing ΔG to become negative. From this it is seen that
$\Delta G = G_{products} - G_{reactants}$
The equation
$\Delta G = \Delta H - T(\Delta S)$
states that at low temperature ΔG depends mainly on ΔH because $T(\Delta S)$ is then small in comparison with ΔH. At high temperature ΔG depends mainly on $T(\Delta S)$ since $T(\Delta S)$ is then large in comparison with ΔH. G is also called Gibbs's function and generally expresses that part of a system's energy content which can be used for external work.

environment A collective term describing the conditions surrounding an organism. It includes air, light, soil, temperature, water and the presence or absence of other organisms, i.e. the conditions for development or growth.

enzyme A molecule of biological origin (produced by living cells) which increases the rate of a specific biochemical reaction; i.e. an enzyme is an organic catalyst. All enzymes are made up of proteins, sometimes related to metallic ions, carbohydrates or other organic molecules. Usually enzymes are named by adding the suffix -ase to the word indicating the nature of the substrate on which they act, e.g. amylase (acting on starch) and urease (acting on urea). Enzymes can be extracted from animals and plants and produced industrially using micro-organisms. It is characteristic that many of them are highly specific. Most enzymes work best within a narrow pH range and at temperatures of 35–40°C; at higher temperatures they are rapidly destroyed. *See also* coenzyme.

enzymology The study of enzymes and their functions.

Eocene A geological period of approximately 16 million years which ended approximately 38 million years ago. A sub-division of the Tertiary period.

eosin A red, fluorescent, crystalline dyestuff used as a stain in microscopy.

eosinophil *See* granulocyte

epi- Prefix meaning upon or above.

epicardium The outer layer (membrane) of the heart, consisting of collagenous and elastic fibres.

epicarp *See* drupe.

epicotyl The part of the embryonic stem above the attachment of the cotyledons. *Compare* hypocotyl.

epidemic An outbreak of an infectious disease affecting a large number of people at the same time in any region. *See also* endemic *and* pandemic.

epidermis 1. The outer layer of tissue, seldom more than one cell thick, covering the leaves, stems and roots of plants. In higher plants, cutin covers aerial epidermis. 2. The outer layer of the skin of animals. It is several cell-layers thick in terrestrial vertebrates, the outermost layer consisting of dead cells forming a protective, waterproof coat. These dead cells are continually worn away and replaced by cells beneath them. In invertebrates, the epidermis is one cell thick. *See* Fig. 208; *see also* cuticle.

epididymis In birds, mammals and reptiles, a long, greatly coiled tube receiving spermatozoa from the testis and conveying them, via the vas deferens, to the urethra.

epigeal 1. Describes germination in which the cotyledons appear above ground. 2. Describes insects living on the surface of the soil. *Compare* hypogeal.

epiglottis In mammals, a flap of cartilage situated at the back of the tongue. When food is swallowed, the epiglottis closes the glottis, preventing food from entering the trachea.

epigynous Describes a flower in which the ovary is fused to the receptacle and is inferior to (below) the sepals, petals and stamens, e.g. apple. *See* Fig. 62; *see also* perigynous *and* hypogynous.

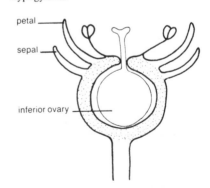

Fig. 62   Epigynous flower in section

epilepsy A brain disease which is often chronic and believed to be heritable. However, epilepsy may also be caused by brain damage or by high fever in infancy. Symptoms include fits which may occur hours, days, weeks or months apart. Epilepsy may in some patients cause loss of consciousness and violent movements. Often the eyes roll back. In mild cases the patient may experience a sudden 'black out' during which he or she makes strange movements or behaves oddly. Certain drugs may help in preventing fits, but there is still no real cure for epilepsy.

epineurium A tough sheath of connective tissue surrounding a nerve.

epiphysis (pl. epiphyses) *See* ossification.

epiphyte A plant growing on another, but not parasitic upon it, merely using it for support. Examples include mosses, lichens and orchids.

epithelial Pertaining to epithelium.

epithelium (pl. epithelia) A sheet of tissue covering the surface of the body and the organs within it. There are several main types of epithelial tissue, depending on the structure and function of their cells. Epithelia may provide protection from abrasion, smooth surfaces for the passage of fluids, ciliated surfaces to trap foreign particles, etc. Sometimes epithelial tissue also has a secretory function. In this case it may be called glandular tissue.

EPNS Electroplated nickel silver. *See* nickel silver.

epoxyethane C_2H_4O or $H_2C\!-\!CH_2$
$\backslash/$
O
(ethylene oxide) A flammable, gaseous, colourless ether with a heterocyclic structure. It is soluble in water and in many organic solvents. M.p. $-111°C$, b.p. $11°C$. Epoxyethane may be prepared by oxidation of ethene, $CH_2\!\!=\!\!CH_2$, with air at about 250°C using silver as a catalyst:
$2CH_2\!\!=\!\!CH_2 + O_2 \rightarrow 2C_2H_4O$
Because of strain in the three-membered ring, it is a very reactive compound. It easily polymerises and is useful in the organic synthesis of detergents, fumigants, disinfectants, etc.

Epsom salt Hydrated magnesium sulphate, $MgSO_4.7H_2O$. A bitter, white, soluble crystalline salt used in the leather and textile industries and as a laxative.

equation *See* chemical equation.

equations of motion Three equations connecting velocity, acceleration, time and distance for a body moving with a uniform acceleration:
(1) $v = u + at$
(2) $s = ut + \frac{1}{2} at^2$
(3) $v^2 = u^2 + 2as$
where v is the final velocity (m s^{-1}), u is the initial velocity (m s^{-1}), a is the acceleration (m s^{-2}), t is the time in seconds (s) and s is the distance in metres (m).

equatorial plate A plane midway between the poles and at right angles to a line joining the poles of a cell. During the metaphase of cell division, chromosomes become arranged across the equatorial plate. The equatorial plate is also the plane across which the original cell separates into two new cells.

equilibriant A single force which neutralises a set of forces acting on a body; i.e. the equilibriant is equal and opposite to the resultant of the set of forces.

equilibrium constant *See* mass action (law of).

equilibrium (dynamic) 1. A balanced state of constant (continual) change of a system. *Example:* if water is placed in a closed vessel at constant temperature, molecules of water evaporate at the same rate as molecules of water vapour condense. A reversible chemical reaction may reach the state of dynamic equilibrium, i.e. when the rate of the forward reaction equals that of the backward reaction. *See also* neutral equilibrium, stable equilibrium *and* unstable equilibrium. **2.** *See* chemical equilibrium.

equivalence point *See* end-point.

equivalent The combining mass (weight) of a substance relative to oxygen as a standard. For an element, one equivalent is equal to its relative atomic mass divided by its valency. *See also* equivalent mass *and* chemical combination (laws of).

equivalent mass The equivalent mass (formerly the equivalent weight) of an element is the number of parts by mass of it which will combine with or replace eight parts by mass of oxygen. The equivalent mass expressed in grams is called the gram-equivalent mass. The above definition can therefore be further expanded to: the gram-equivalent mass of an element (substance) is the number of grams of it which will combine with or replace eight grams of oxygen or the known gram-equivalent mass of any other element (substance). In redox reactions, the equivalent mass of one or more of the reactants is dependent upon the change in oxidation numbers of the atoms in these substances. *See also* Dulong and Petit's law.

equivalent weight *See* equivalent mass.

erect Standing upright.

erectile Describes tissue which can be made rigid when blood vessels in it become distended.

erecting prism or lens A prism or concave lens used in optical instruments in order to obtain an erect image from an inverted image. *See* Fig. 63.

erector muscle (of hair) A smooth muscle which, on contraction, raises the hair, making

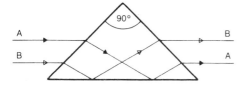

Fig. 63 Erecting prism

it more upright, and at the same time compressing the sebaceous gland.

erepsin In vertebrates, a mixture of enzymes which act on proteins in the small intestine, breaking them down to amino acids.

Erlenmeyer flask *See* conical flask.

erosion The slow destruction of a solid by the influence of chemicals, water, wind, etc. *See also* weathering.

errors of parallax *See* parallax.

erythroblast A cell, found in the bone marrow, from which an erythrocyte (red blood corpuscle) develops.

erythrocyte A red blood corpuscle (cell). *See* blood corpuscle.

Escherichia coli (*E. coli*) A bacterium inhabiting the mammalian colon. It has been and still is widely used in genetic research, and as an indicator bacterium for faecal pollution of water or foodstuffs.

essential amino acid *See* amino acid.

essential hypertension *See* hypertension.

ester The general name for a large group of organic compounds formed by the reaction of acids with alcohols. *Example:* ethanoic acid (acetic acid), CH_3COOH, and ethanol (ethyl alcohol), C_2H_5OH, react to form ethyl ethanoate (ethyl acetate), $CH_3COOC_2H_5$:
$$CH_3COOH + C_2H_5OH \rightleftharpoons$$
$$CH_3COOC_2H_5 + H_2O$$
This is a reversible reaction. The formation of an ester from an acid and an alcohol is called esterification. Important esters include cellulose esters, used in making fibres and plastics, and phosphoric acid esters, used in making insecticides. Many esters are liquids with a pleasant smell, used for flavouring essences. Most esters are insoluble in water, but soluble in ethanol and ether (diethyl ether). Several esters occur naturally in fruit. *See also* fats and oils.

esterification *See* ester.

estuarine Describes a plant or animal which inhabits the estuary of a river.

estuary The mouth of a wide river which is tidal and in which salt and fresh water mix.

ethanal CH_3CHO (acetaldehyde) A water-soluble liquid aldehyde. It is now manufactured industrially from ethene, $CH_2\!\!=\!\!CH_2$, which is treated (oxidised) with palladium(II) chloride, $PdCl_2$, in the presence of water at a temperature of about 100°C and a pressure of 10 atmospheres:
$$CH_2\!\!=\!\!CH_2 + PdCl_2 + H_2O \rightarrow$$
$$CH_3CHO + Pd + 2HCl$$
The palladium, Pd, is then oxidised back to palladium(II) ions, Pd^{2+}, by a reaction with copper(II) ions, Cu^{2+}:
$$Pd + 2Cu^{2+} \rightarrow Pd^+ + 2Cu^+$$
In the presence of air (oxygen), Cu^+ is oxidised back to Cu^{2+}:
$$4Cu^+ + O_2 + 4HCl \rightarrow 4Cu^{2+} + 4Cl^- + 2H_2O$$
The whole process is called the Wacker process. Ethanal is used in the organic synthesis of other compounds.

ethanamide *See* amide.

ethane C_2H_6 A saturated hydrocarbon belonging to the alkanes. It is a flammable gas, occurring in natural gas from which it is extracted. Ethane is used in organic synthesis. M.p. $-183°C$, b.p. $-89°C$.

ethanedioate (oxalate) A salt of ethanedioic acid (oxalic acid), $(COOH)_2$.

ethanedioic acid $(COOH)_2$ (oxalic acid) A poisonous, colourless, solid organic acid containing two carboxyl groups, —COOH. It is present in many plants such as spinach, cabbage, tomatoes and rhubarb in the form of its salts, ethanedioates (oxalates). An aqueous solution of ethanedioic acid crystallises to form the dihydrate, $(COOH)_2 \cdot 2H_2O$. Ethanedioic acid can be prepared by heating sodium methanoate (sodium formate), HCOONa:
$$2HCOONa \rightarrow (COONa)_2 + H_2$$
and then treating the sodium ethanedioate, $(COONa)_2$, with a strong, dilute acid:
$$(COONa)_2 + 2H^+ \rightarrow (COOH)_2 + 2Na^+$$
Ethanedioic acid is used in dyeing, bleaching and metal polishes.

ethane-1,2-diol CH_2OH—CH_2OH (glycol, ethylene glycol; 1,2-dihydroxyethane) A dihydric alcohol. It is a colourless, water-soluble, viscous, sweet, hygroscopic liquid used as antifreeze.

ethanoate (acetate) A salt of ethanoic acid (acetic acid), CH_3COOH.

ethanoic acid CH_3COOH (acetic acid) A weak organic acid which can be formed by the oxidation of ethanol. Ethanoic acid is the main ingredient in vinegar. Salts of acetic acid are called ethanoates (acetates). Concentrated ethanoic acid (99,4% (99.4%) W/W, m.p. 16,6°C (16.6°C), b.p. 118°C) is called glacial ethanoic acid. *See also* fermentation.

ethanol C_2H_5OH (ethyl alcohol or 'alcohol') A volatile, flammable, colourless liquid, miscible with water. It is the alcohol found in intoxicating drinks. Ethanol can be prepared by the fermentation of certain carbohydrates

or by reacting ethene, C_2H_4, with sulphuric acid, H_2SO_4, and water.
$H_2C{=}CH_2 + H_2SO_4 \rightarrow$
$\qquad H_3C{-}CH_2{-}O{-}SO_2OH$
and
$H_3C{-}CH_2{-}O{-}SO_2OH + H_2O \rightarrow$
$\qquad CH_3{-}CH_2OH + H_2SO_4$
It is used as a solvent and as a raw material for the manufacture of other organic compounds. M.p, $-117°C$, b.p. $78,5°C$ ($78.5°C$).

ethanoyl chloride *See* acyl chloride.

2-ethanoyloxybenzoic acid *See* aspirin.

ethene C_2H_4 (ethylene) An unsaturated hydrocarbon belonging to the alkenes. It is a flammable gas manufactured by cracking petroleum gases. Ethene is used in the manufacture of ethanol and certain polymers. M.p. $-169°C$, b.p. $-102°C$.

ethenyl $CH_2{=}CH{-}$ (vinyl) An organic radical. Ethenyl compounds undergo polymerisation easily. *See also* PVC *and* phenylethene.

ethenylbenzene *See* phenylethene.

ether 1. *See* ethoxyethane. 2. The general name for a group of organic compounds with the formula $R{-}O{-}R'$, where R and R' stand for organic radicals. In simple ethers such as ethoxyethane (diethyl ether), $C_2H_5{-}O{-}C_2H_5$, the two radicals are the same; in mixed ethers such as methoxyethane (methyl ethyl ether), $CH_3{-}O{-}C_2H_5$, the two radicals are different. *See also* Williamson synthesis.

ethoxyethane $C_2H_5{-}O{-}C_2H_5$ (ether, diethyl ether). A volatile, colourless and highly flammable liquid made by dehydrating ethanol. Ethoxyethane is immiscible with water. It is used as an anaesthetic and as a solvent. M.p. $-116°C$, b.p. $35°C$.

ethyl $C_2H_5{-}$ An organic radical present in thousands of organic compounds.

ethyl alcohol *See* ethanol.

ethylbenzene *See* phenylethene.

ethylene *See* ethene.

ethylene glycol *See* ethane-1,2-diol.

ethylene oxide *See* epoxyethane.

ethyne C_2H_2 (acetylene) An unsaturated hydrocarbon belonging to the alkynes. It is a flammable gas which may be made by the action of water on calcium carbide, CaC_2:
$CaC_2 + 2H_2O \rightarrow C_2H_2 + Ca(OH)_2$
Ethyne is used in organic synthesis and in welding. M.p. $-80,8°C$ ($-80.8°C$), b.p. $-84,0°C$ ($-84.0°C$).

etiolation A condition exhibited by green plants when they are grown in darkness. The plants become pale yellow because of the lack of chlorophyll, their stems become long and weak due to abnormal lengthening of the internodes and their leaves remain small.

Eucaryota One of the main groups into which all living organisms are divided. This group includes all animals and plants except blue-green algae and bacteria. Their cells are characterised by having a nucleus separated from the cytoplasm by a nuclear membrane, genetic material carried on chromosomes present in the nucleus, nuclear division by mitosis and organelles such as mitochondria and chloroplasts present in the cytoplasm. Eucaryotic cells are much bigger than procaryotic ones. *See also* Procaryota *and* Protista.

eudiometer A long, graduated glass tube fitted with a pair of electrodes at its closed end. It is used for measuring volume changes in gases on sparking. *See* Fig. 64.

Fig. 64 Eudiometer

Euglena A genus of unicellular micro-organisms found in both animal and plant classifications, as its members possess both animal and plant characteristics. These include chloroplasts, a contractile vacuole, a flagellum, a non-rigid cell wall and the presence of granules containing a polysaccharide called paramylum, $(C_6H_{10}O_5)_n$.

euphotic zone The upper water layer of a sea. In the euphotic zone there is sufficient light for active photosynthesis.

eureka *See* constantan.

eureka can *See* displacement can.

Eustachian tube The tube or canal connecting the middle ear with the pharynx in land vertebrates. Normally the tube is closed, but during swallowing and yawning it opens to allow air from the throat to pass through it, thus equalising air pressure on each side of the ear drum. *See* Fig. 55.

eutectic alloy An alloy of two or more metals having a composition which gives the lowest melting-point of any composition of the metals. *Example:* the alloy containing 36% lead and 64% tin is the eutectic alloy of these two metals. It has a melting-point of 181°C, which is also well below the melting-point of either pure metal.

Eutheria *See* Placentalia.

eutrophication Over-fertilisation of rivers and lakes resulting from an increase in the water of nitrates and phosphates. This gives rise to a very rapid growth of algae whose bacterial decomposition after death requires a great amount of oxygen. The water is thus depleted of dissolved oxygen, and fish and other animals die by suffocation. Eutrophication is mainly caused by the washing out of nitrates from fertilised soil and by discharge of treated sewage. The phosphates mainly come from certain detergents in the sewage.

eV *See* electronvolt.

evaporation The process in which a liquid is changed into vapour. This is usually brought about by heating the liquid and it occurs at temperatures below and at the liquid's boiling-point. Evaporation is carried out to concentrate a solution, either to form crystals or to adjust the solution to a required strength, or to evaporate it to dryness.

evergreen Describes plants (trees and shrubs) which do not shed all their leaves at the same time and thus appear green all the year round. *Compare* deciduous.

evolution 1. (of species). The gradual development of organisms from their related ancestors. *See also* Darwin's theory. 2. (of gases). The production of a gas, e.g. during a chemical reaction. *Examples:* when an acid is added to a carbonate carbon dioxide, CO_2 is evolved; when an acid is added to a sulphite, sulphur dioxide, SO_2 is evolved.

excited state *See* Bohr atom.

excreta Waste material discharged from the body of an animal, e.g. carbon dioxide, water, urea and uric acid.

excretion The removal of the waste products of metabolism, e.g. carbon dioxide, water and nitrogenous substances, from the body of an animal.

exhalation *See* expiration.

exo- Prefix meaning external.

exocrine glands Glands which have ducts to carry their secretions, e.g. salivary glands, tear glands and sebaceous glands.

exodermis A layer of cells, derived from the cortex, which replaces dead epidermis in the piliferous layer in older parts of roots.

exoskeleton (dermoskeleton) A skeleton situated in the skin or covering the outside of the body of an animal, e.g. the skeletons of insects. See Fig. 65; *compare* endoskeleton.

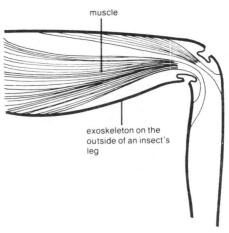

Fig. 65 Exoskeleton

exosphere The outermost layer of the Earth's atmosphere, situated beyond the ionosphere.

exothermic process A process (system) in which heat is evolved to the surroundings. *Example:* when carbon, C, and oxygen, O_2, combine to form carbon dioxide, CO_2, heat is evolved:

$C + O_2 \rightarrow CO_2 + $ heat

See also enthalpy.

expansion The process in which the length, area or volume of a substance (solid, liquid, gas) is increased.

expansion, coefficient of (expansivity) The expansion of unit length (linear), area (superficial), or volume (cubic), per degree rise in temperature.

expansion of water *See* anomalous expansion of water.

expansivity *See* expansion, coefficient of, linear expansivity *and* cubic expansivity.

expiration (exhalation) The process of giving out air or water from the respiratory organs.

explantation *See* tissue culture.

explosion A rapid increase in pressure in a confined space as a result of violent chemical or nuclear reactions. Explosions produce large amounts of hot gases.

explosive Any substance which can cause an explosion when heated or struck.

exposure meter (light meter) A photoelectric cell used in photography to measure the intensity of light available so that the correct shutter speed and aperture of a camera may be chosen. *See also* selenium cell.

extensor A muscle which extends or straightens part of the body, e.g. the triceps muscle. *Compare* flexor.

external auditory meatus A tube extending from the auricle to the middle ear. It serves to direct sound waves to the ear drum.

external ear *See* outer ear.

external respiration *See* breathing.

exteroceptor A receptor which senses stimuli from outside the body of an animal. Exteroceptors are concerned with environmental factors such as sound, light, heat and chemicals. *Compare* interoceptor.

extinct Describes a species which no longer exists. *Example:* the dodo, a large bird which lived in Mauritius, is now extinct.

extinction coefficient *See* transmittance.

extinguishing agent Any substance capable of putting out fire.

extra- Prefix meaning outside.

extract A product obtained by extraction.

extraction 1. The process in which a desired substance is removed from a solid or liquid mixture. This may be brought about by treating the mixture with a suitable solvent. **2.** The process in which pure metals are obtained from their ores.

eye The sense organ of sight in animals. It is a highly specialised organ in vertebrates and Cephalopoda. In human beings, the eyeball is an almost perfect sphere with a diameter of about 2,5 cm (2.5 cm), held in place by three pairs of muscles attached to the sclera. *See* Fig. 66.

eye lens (crystalline lens) A transparent, biconvex body (lens) situated immediately behind the iris and in front of the vitreous humour in the eyes of vertebrates and Cephalopoda. It is held in place by the suspensory ligament. *See* Fig. 66; *see also* ciliary muscle.

eyelid A folded, muscular layer of skin of the eye of some vertebrates. Eyelids are found above and below the eye. The upper lid is movable, whereas the lower lid only moves slightly in human beings. When closed, the eyelids protect the eye. The movement of the eyelids can be a reflex or voluntary action. *See also* nictitating membrane.

eyepiece (ocular) In optical instruments, such as microscopes and telescopes, the lens or lens system nearest to the eye of the observer. It is used to view the image formed by the objective.

eye spot In lower plants and animals, and in some vertebrates, a small mass of light-sensitive pigment.

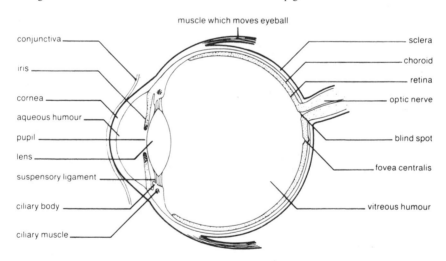

Fig. 66 Eye (horizontal section)

F

F Chemical symbol for fluorine.

F_1 Abbreviation denoting the first filial generation of offspring arising from crossing the animals or plants of parental generation (P_1).

F_2 Abbreviation denoting the second filial generation of offspring arising from crossing the members of the F_1.

facet 1. The flat side of a crystal. **2.** A segment of a compound eye. **3.** A small surface on a bone at which another bone articulates.

facultative Describes organisms possessing the ability to live under different conditions. *Examples:* a facultative anaerobe is able to grow under either aerobic or anaerobic conditions; certain saprophytic fungi are able to exist as parasites. *Compare* obligate.

faeces Undigested food, together with the residue of secretions and bacteria expelled from the body of an animal.

Fahrenheit scale A scale of temperature in which the melting-point of ice is 32°F and the boiling-point of water, at standard atmospheric pressure (101 325 Pa), is 212°F. This scale is no longer in use for scientific purposes.

Fajans's rules Rules describing whether a covalent or an electrovalent (ionic) bond will be formed between atoms. Bonds between atoms are more likely to be covalent when the atoms have large electrical charges or when the cation is small and the anion large. The bonds are more likely to be electrovalent (ionic) when the atoms have small electrical charges or when the cation is large and the anion small.

Fallopian tube (uterine tube, oviduct) In female mammals, one of a pair of muscular tubes for the transport of the ovum from ovary to uterus.

fallow Describes land left uncultivated so that plant nutrients may be replaced naturally, thus improving soil fertility.

false fruit (pseudocarp) A fruit whose structures include floral parts, e.g. the receptacle or calyx, in addition to the gynaecium. Examples include the strawberry and mulberry.

false rib A rib attached indirectly to the sternum by fused strips of cartilage. In humans the 8th, 9th and 10th ribs are false ribs.

family A group, consisting of a number of related genera, used in the classification of living organisms. Similar families are grouped into the same 'order' and orders into a class.

fang 1. The root of a tooth. **2.** A long, pointed tooth, especially the canine tooth of a carnivore. **3.** The poison tooth of a snake.

farad Symbol: F. The SI unit of capacitance. A capacitor has a capacitance of 1 farad when a potential difference of 1 volt exists across its plates holding a charge of 1 coulomb.

$$\text{capacitance} = \frac{\text{charge}}{\text{potential}} \quad \text{or}$$

$$\text{farads} = \frac{\text{coulombs}}{\text{volts}}$$

The farad is a large unit; for practical purposes the microfarad is employed.

Faraday constant Symbol: F. The quantity of electricity required to liberate 1 mole of a monovalent ion during electrolysis. The constant has the value $9,648\,670 \times 10^4$ coulombs per mole ($9.648\,670 \times 10^4$ C mol^{-1}).

Faraday cylinder A device similar to the Faraday's cage. It is used to collect a stream of charged particles, e.g. electrons, and is shielded by an earthed cylinder. An insulated conductor inside the cylinder may be connected to a detecting apparatus, e.g. a gold leaf electroscope. *See* Fig. 165.

Faraday's cage A device in which electrical instruments are placed to screen (shield) them from external electric fields. Screening (shielding) was first demonstrated by Michael Faraday. Originally Faraday's cage was a box covered with an earthed wire mesh.

Faraday's disc A simple model of a d.c. generator consisting of a copper disc which can be rotated between the poles of a permanent horseshoe magnet. When the disc is rotated, an e.m.f. is induced between the centre of the disc and its circumference.

Faraday's law of electromagnetic induction An electromotive force is induced in a circuit whenever there is a change in the magnetic flux linked with the circuit. The strength of the induced electromotive force is proportional to the rate of change of the magnetic flux. *See also* electromagnetic induction.

Faraday's laws of electrolysis Two laws stating: (1) The mass of a substance liberated at an electrode during electrolysis is proportional to the quantity of electricity passing through the electrolyte. (2) When the same quantity of electricity is passed through different electrolytes, the masses of the substances liberated at the electrodes are in the ratio of their chemical equivalents.

farmyard manure A natural manure obtained when faeces and urine from domesticated animals are collected on a bed of grass and then allowed to decompose with the aid of certain bacteria. This form of manure contains all essential substances required by any crop. It is usually spread on the soil or ploughed into it.

far point The farthest point from the eye at which objects can be seen clearly. For normal vision, the far point is at infinity. *Compare* near point.

fat *See* fats and oils.

fatal Ending in death.

fat cell A cell which accumulates drops of fat in its cytoplasm, eventually forming a large globule of fat in the middle of the cell. This pushes the nucleus to one side and the cytoplasm into a thin layer.

fatigue 1. (of metals) The failure of metals as a result of repeated stresses, e.g. bending. Continual stressing may distort the structure of a metal causing cracks to develop. **2.** (of muscles) The accumulation of lactic acid in muscles causing loss of irritability, so that they fail to respond to stimuli. Muscle fatigue is caused by an oxygen debt due to hard, sustained exercise.

fats and oils General names for organic substances which consist of mixtures of different glycerides, i.e. esters of glycerol (propane-1,2,3-triol) and fatty acids (alkanoic acids). Those glycerides with melting-points at room temperature are called oils; those with melting-points above room temperature are called fats. Unsaturated fatty acids usually give lower melting-points to the glycerides they form with glycerol than do the saturated fatty acids with the same number of carbon atoms. Sometimes the term lipid(e) is used as a synonym for a fat. Fats and oils occur in animals and plants. Of the natural fats, unsaturated fat is most often of plant origin whereas saturated fat is of animal origin. *See also* esters, saponification *and* polyunsaturates.

fatty acid (alkanoic acid) A straight-chain, saturated or unsaturated monocarboxylic acid with the formula $R\text{---}COOH$, where R is hydrogen or a radical consisting of carbon and hydrogen. The higher members of the fatty acids occur in nature, where they are combined with glycerol forming esters called fats or oils.

fauna All the animals present in a certain region or period.

Fe Chemical symbol for iron.

fear A certain behaviour in animals produced by a sudden unpleasant stimulus. The adrenal glands produce an increased amount of the hormone adrenalin(e) which speeds up the beating of the heart, increases the blood pressure and enables the muscles to work faster and longer than usual, i.e. the increased secretion of adrenalin(e) prepares the body for emergency actions such as flight or fight.

feather A structure growing from the skin of birds. There are several kinds of feathers, such as the flight feathers of the wing and tail, contour feathers, coverts, down feathers and filoplumes. Feathers form an insulating, waterproof layer around a bird's body, and some feathers are modified for flight. *See* Fig. 67.

Fehling's solution An aqueous solution obtained by mixing a solution of copper(II) sulphate, $CuSO_4$, with an alkaline solution of potassium sodium tartrate. When a reducing agent, e.g. a reducing sugar (carbohydrate), is added to Fehling's solution and then heated for a few minutes, a red precipitate of copper(I) oxide, Cu_2O, is formed. *See also* Benedict's solution *and* Barfoed's solution.

feldspar General name for a group of very abundant rock-forming minerals. Chemically, the feldspars are silicates of aluminium and one or two other metals such as potassium, sodium, calcium or barium. They are used in abrasives and in the manufacture of porcelain.

female Symbol: ♀. An individual whose sex organs produce only ova. *Compare* male.

femoral Pertaining to the femur or thigh, e.g. the femoral artery.

femto- Symbol: f. A prefix meaning 10^{-15}.

femur 1. The thigh-bone or proximal bone in the hindlimb of vertebrates. *See* Fig. 162. **2.** The third joint of the leg in insects.

fenestra cochleae *See* fenestra rotunda.

fenestra ovalis (fenestra vestibuli, oval window) An opening in the wall of the bony labyrinth, between the middle ear and the inner ear. It is closed by a thin membrane and transmits vibrations of the stapes to the perilymph in the membranous labyrinth. *See* Fig. 55.

fenestra rotunda (fenestra cochleae, fenestra tympani, round window) An opening in the wall of the bony labyrinth connecting the cochlea with the tympanic cavity. It is closed by a thin membrane and allows free movements of the contents of the inner ear. *See* Fig. 55.

fenestra tympani *See* fenestra rotunda.

fenestra vestibuli *See* fenestra ovalis.

ferment An enzyme causing fermentation, e.g. zymase.

fermentation A form of anaerobic respiration of organic substances (usually carbohydrates) brought about by micro-organisms or enzymes. When glucose, $C_6H_{12}O_6$, is fermented, ethanol (ethyl alcohol), C_2H_5OH, and carbon dioxide, CO_2, are formed:

$$C_6H_{12}O_6 \rightarrow 2C_2H_5OH + 2CO_2$$

The process is catalysed by yeast containing the enzyme zymase. If the ethanol is oxidised,

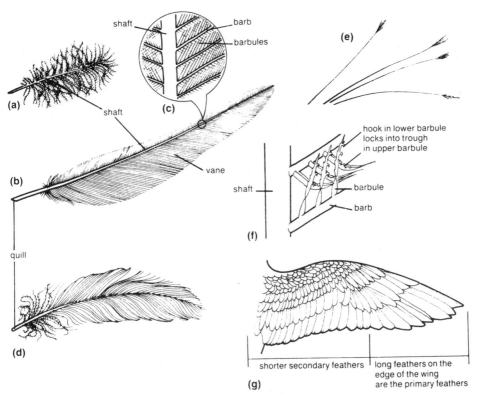

Fig. 67 (a) Down feather (b) Flight feather (c) Flight feather as seen under a microscope
(d) Covert feather (e) Filoplumes (f) Simplified diagram of barbules
(g) Arrangement of feathers on a wing

ethanoic acid (acetic acid), CH_3COOH, is produced:
$C_2H_5OH + O_2 \rightarrow CH_3COOH + H_2O$
See also respiration.

fern A vascular, non-flowering plant with feathery fronds usually carrying spores on their undersides. See Fig. 68.

ferric Term formerly used in the names of iron compounds to indicate that the iron is trivalent; e.g. ferric oxide, Fe_2O_3, now called iron(III) oxide. Compare ferrous.

ferric chloride See iron chlorides.
ferric hydroxide See iron hydroxides.
ferric oxide See iron oxides.
ferric sulphate See iron sulphates.
ferricyanide ion See cyanoferrates.
ferrocyanide ion See cyanoferrates.

ferromagnetic material All materials show some kind of magnetic effect. This is because magnetic fields are produced when electrons are moving around their nuclei and during the spin they exhibit while moving in their orbits. These magnetic fields usually interact with one another in such a way that they tend to cancel each other out. However, substances like iron, steel, nickel and cobalt are capable of forming strong magnets. Their domains line up when subjected to a strong magnetic field so that at one end are the north-seeking poles of domains, and at the other end are the south-seeking poles. Such strongly magnetic materials are called ferromagnetic. Other materials exhibit such a small magnetic effect that it is almost non-existent; these materials are called paramagnetic. Still other materials will, when placed in a magnetic field, align themselves at right angles to the lines of force rather than along the lines of force as ferromagnetic and paramagnetic materials do. Such materials are called diamagnetic and they only have a very minute effect. See also domain theory.

ferrous Term formerly used in the names of iron compounds to indicate that the iron is divalent (bivalent); e.g. ferrous oxide, FeO, now called iron(II) oxide. Compare ferric.

ferrous chloride

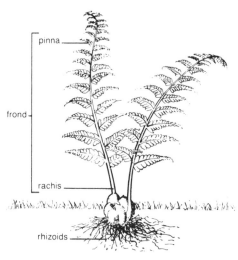

Fig. 68 Fern

ferrous chloride *See* iron chlorides.
ferrous hydroxide *See* iron hydroxides.
ferrous oxide *See* iron oxides.
ferrous sulphate *See* iron sulphates.
ferrum The Latin name for iron.
fertile 1. Describes female organisms capable of producing offspring by sexual reproduction. 2. Describes eggs or seeds capable of developing into a new individual. 3. Describes soil containing the essential ingredients for plant growth.
fertilisation 1. The union of a male and female gamete to form a zygote. 2. Adding fertilisers to soil to make it fertile.
fertiliser A natural or artificial manure used to replace plant nutrients removed from the soil by crops. Farmyard manure is a natural manure; artificial manures are chemical substances such as superphosphate or ammonia.
fertility rate This is defined as the number of live births in a year divided by the number of women aged 15–45 years in that year, and multiplied by one thousand:

$$\text{fertility rate} = \frac{\text{number of live births in a year} \times 1000}{\text{number of women aged 15–45 in that year}}$$

fever (pyrexia) An abnormal rise in the body temperature of an animal, e.g. above 37,5°C (37.5°C) in Man. Fever is most often caused by bacterial or viral infection.
fibreglass *See* glass fibres.
fibre optics An optical system consisting of long, thin strands of glass or plastic which transmit light by total internal reflection. Light is carried from one end of the fibre to the other; in a group of fibres, this produces a spray of light at the tips. This system is used by the telephone service. It is also used in medicine and industry for viewing inaccessible parts of the body or of machines.
fibrin An insoluble protein produced in blood plasma from the soluble protein fibrinogen, by the action of the enzyme thrombin and the presence of calcium ions. *See also* blood clotting.
fibrinogen A soluble protein present in blood plasma. When activated by the enzyme thrombin and in the presence of calcium ions, it is converted into the insoluble protein fibrin. Fibrinogen is synthesised in the liver. *See also* blood clotting.
fibroblast *See* areolar tissue.
fibrous roots Many roots of a similar size forming no main root. These are seen in most grasses, including maize and bamboo.
fibula 1. The smaller of the two distal bones of the hindlimb in tetrapods, the bone posterior to the tibia. In Man it is the smaller and outer shin bone, located between the knee and the ankle. *See* Fig. 162. 2. In some insects, the fibula is a structure that holds the forewings and hindwings together.
field The region around an electrically charged object, a magnetised body or a mass where an effect can be noticed. *See* electric field, magnetic field *and* gravitational field.
field capacity A property of a soil expressed as the state in which it is full of water, i.e. it holds the maximum amount of capillary water. Any extra water drains away as gravitational loss. *Compare* permanent wilting point.
field coil A coil of wire used for magnetising an electromagnet, e.g. in a generator or electric motor.
field magnet The magnet, usually an electromagnet, which provides the magnetic field in a generator or electric motor. In small electrical machines it may be a permanent magnet.
filament 1. A fine wire of high resistance which is heated by the passage of an electric current through it. In an electric light bulb the filament is usually made of tungsten (wolfram) or carbon and acts as the source of light. In thermionic valves the filaments emit electrons. 2. Any long thread-like structure such as the stem (stalk) of a stamen or the fibre produced by the silk-worm. *See also* bacterium.
filarial worms Thread-like worms belonging to the phylum Nematoda. The worms are parasitic to mammals and they live in the bloodstream or are found in the lymphatic vessels. Elephantiasis is a disease caused by filarial worms.

filariasis Infection caused by filarial worms. Filariasis is a major economic and social problem in the tropics. *See also* elephantiasis.

filial generation *See* F_1 and F_2.

film 1. Any thin layer of a substance, e.g. a layer of oil on a water surface. **2.** In photography, a strip of cellulose acetate or polyester coated with a light-sensitive emulsion of silver bromide, AgBr.

film badge *See* dosemeter.

film dosemeter *See* dosemeter.

filoplume A hair-like feather. Filoplumes are found all over a bird's body between the coverts. *See* Fig. 67.

filter 1. A device for removing suspended solid particles from a fluid by passing the fluid through a porous material such as paper or sand, which retains the solid particles. **2.** A device placed in the path of a beam of light or other electromagnetic radiation to alter the component wavelengths in the beam. *Example:* a blue filter placed in the path of a beam of white light only allows wavelengths in the blue region of the spectrum to be transmitted.

filterable (filtrable) Describing a microorganism (usually a virus) which can pass through a filter that retains most bacteria.

filter bed A tank or tube filled with a filtering material such as sand, stones or gravel and used for filtering purposes, e.g. the removal of mud from large quantities of water.

filter flask A thick-walled conical flask with a side tube for attachment to a filter pump. A tightly-fitting funnel (usually a Büchner funnel) is placed in the mouth of the flask and the attached filter pump creates a vacuum in the flask, thus speeding up filtration. *See* Fig. 69.

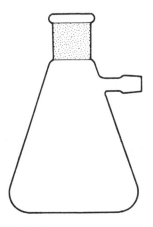

Fig. 69 Filter flask

filter funnel A funnel used in filtration. A specially folded filter paper is placed in the funnel. *See* Fig. 70.

Fig. 70 Filter funnel

filter paper A porous paper made of cellulose, used in filtration. Certain specially treated filter paper leaves almost no ash when burnt and is therefore used for quantitative purposes in gravimetric analysis.

filter pump A type of vacuum pump commonly used to assist filtration. It is made of metal, glass or plastic. A jet of cold water is forced through a narrow nozzle, thus removing air from the system to which the pump's side tube is connected (usually a filter flask). *See* Fig. 71.

filtrable *See* filterable.

filtrate The clear liquid obtained after filtration.

filtration The process in which solid particles are separated from a fluid by passing it through a filter.

fin A fold of skin, with a skeletal support of fin-rays, in aquatic vertebrates. The main function of the fins is to control stability and to steer. They are divided into median fins and paired fins. The dorsal, ventral and caudal fins are median fins; the pectoral and pelvic fins are paired fins. The dorsal and ventral fins help the fish not to veer to one side, and at the same time reduce rolling and yawing movements. The paired fins are used by the fish to control upward and downward movements. The pectoral fins have a special function as brakes to stop forward movement. By using only one pectoral fin, the fish can make sharp turns. *See* Fig. 73.

fine chemical

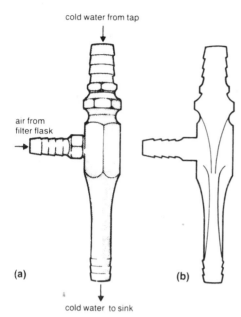

Fig. 71 Filter pump
(a) Outside view (b) Longitudinal section

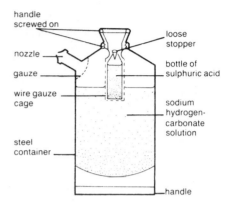

Fig. 72 Fire extinguisher (soda-acid type)

fine chemical A chemical which is manufactured to be as pure as possible. Fine chemicals are expensive. They are often used in chemical analysis. Chemicals which are not fine chemicals may be called technical chemicals or coarse chemicals. Such chemicals are much cheaper to manufacture than fine chemicals and they are often used for industrial or agricultural purposes.

fineness of gold The purity of gold in an alloy expressed as the number of parts per thousand that are gold. *See also* carat.

fin-ray A supporting rod of tissue for a fin, made from bone, cartilage or a collagen-like substance.

fire-clay Clay which is low in the oxides of iron, calcium and magnesium. It softens only at high temperatures and is therefore used as a refractory in lining high temperature furnaces.

fire extinguisher A steel vessel containing chemicals capable of putting out fires. There are several types. Their contents depend upon the type of fire that they are intended to extinguish. A classic fire extinguisher is the soda-acid type in which the extinguishing agent is water which is forced out by the pressure of carbon dioxide produced when sulphuric acid reacts with a solution of sodium hydrogencarbonate. *See* Fig. 72. Other types of fire extinguishers may use a foam, a dry powder or carbon dioxide as their extinguishing agents.

firing order The order in which the cylinders of a multicylinder internal combustion engine fire: e.g. 1, 3, 4, 2 for a four-cylinder engine.

Fischer–Tropsch process A catalytic hydrogenation of carbon monoxide, CO, to produce mixtures of hydrocarbons such as alkanes and alkenes. More hydrogen is added to water gas and the mixture is passed over a catalyst at a temperature of about 200°C and a pressure of 200–300 atmospheres. Through further development this method can also be used to produce aromatic compounds and alcohols such as methanol, CH_3OH.

fish A cold-blooded vertebrate representing the most ancient form of vertebrate life; all other vertebrates may have developed from fish. Fish are adapted to an aquatic life: they have gills (for breathing) and appendages called fins used for locomotion, balancing, etc. The body of a fish is usually streamlined and covered with a tough skin often containing scales and slime glands. The eyes are paired and are usually situated on either side of the head. *See* Fig. 73.

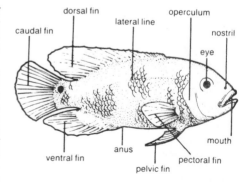

Fig. 73 Fish

The fins are divided into median fins and paired fins. The main function of the fins is to control stability and to steer. The dorsal, ventral and caudal fins are median fins; the pectoral and pelvic fins are paired fins. The dorsal and ventral fins help the fish not to veer to one side, at the same time reducing rolling and yawing movements. The paired fins are used by the fish to control upward and downward movements. The pectoral fins have a special function as brakes to stop forward movement. By using only one pectoral fin, the fish can make sharp turns.

fission 1. The cleavage of a cell either into two equal parts (binary fission) or into more than two equal parts (multiple fission). **2.** A process in which a heavy nucleus of an atom breaks down into two lighter nuclei which have a total mass slightly less than that of the total initial material. The difference in mass is radiated as energy, the amount of which is stated by Einstein's equation. Fission may take place when the uranium isotope $^{235}_{92}U$ absorbs a neutron, $^{1}_{0}n$, and first forms an unstable uranium isotope which then splits into nuclei of tellurium, Te, and zirconium, Zr:

$$^{235}_{92}U + ^{1}_{0}n \rightarrow ^{236}_{92}U \rightarrow ^{137}_{52}Te + ^{97}_{40}Zr + 2^{1}_{0}n$$

The two liberated neutrons can produce other fissions. The fission shown here is just one of the many ways in which a $^{236}_{92}U$ nucleus can split. The process of fission is made use of in atomic bombs (A-bombs) and in the production of electricity in nuclear power stations. *Compare* fusion.

fixation *See* staining.

fixation of nitrogen *See* nitrogen fixation.

fixed point A standard, reproducible temperature used to define a thermometer scale. Commonly used fixed points are those of pure melting ice (ice-point) and steam from pure boiling water at a pressure of 101 325 Pa (steam-point). On the Celsius scale these fixed points are 0°C and 100°C respectively.

fixed stars *See* star.

fixing *See* photography.

flagella *See* flagellum.

flagellum (pl. flagella) A fine, hair-like extension of cytoplasm projecting from the body of a micro-organism. Flagella are long, usually few in number and are used for locomotion. They are much longer than cilia but have the same internal structure.

flame A burning, glowing mass of gas giving out heat and light. Various, often complex, chemical reactions take place in a flame.

flame cell In animals belonging to the phylum Platyhelminthes and the phylum Annelida, a hollow club-shaped cell with a bunch of cilia inside its lumen. Flame cells are usually interconnected by means of canals which ultimately open to the exterior. Water and nitrogenous waste material diffuse into a flame cell from which they are passed on to the exterior by the flicking movement of the cilia. *See also* nephridium.

flame test A qualitative test to determine the presence of a particular element (usually a metal), by heating it or its compounds on a clean platinum or nichrome wire, moistened with concentrated hydrochloric acid, in a non-luminous bunsen flame. Different elements emit different characteristic wavelengths of light when they are heated or applied to a high voltage. *Examples:* sodium atoms emit a yellow light, potassium atoms a violet light, copper atoms a blue-green light. Not all elements emit any one wavelength intensively enough to be seen in a flame test. However, all elements emit a characteristic wavelength of light when their atoms are excited.

flammable (inflammable) Describes a substance which is easily set on fire. Note that flammable and inflammable have the same meaning. *Compare* non-flammable.

flash-point The lowest temperature at which an organic liquid at 101 325 Pa liberates so high a concentration of vapour at or near its surface that it forms an ignitable mixture with air. Some liquids, such as ethoxyethane (diethyl ether) and propanone (acetone), have flash-points below 0°C, whereas others, such as vegetable and mineral oils, have flash-points of several hundred degrees Celsius.

flatulence An accumulation of air (wind) in the stomach or the intestines. The condition is relieved when excess air, called flatus, passes out through the mouth or anus.

flatus *See* flatulence.

flatworms *See* Platyhelminthes.

flavour A sensation arising from the simultaneous stimulation of the taste buds of the tongue and the smell organs of the nose when a substance is eaten or drunk. Flavour is lost when the nasal cavity is congested, e.g. during a heavy cold, although the taste is unaffected.

Fleming's rules Two rules relating directions of motion, magnetic field and current in generators and electric motors. The thumb, first and second finger are held at right angles to one another and represent the direction of motion of the conductor, magnetic field and conventional current respectively. The left-hand rule is used for electric motors and the right-hand rule for generators. *See* Fig. 74.

flexor A muscle which bends (flexes) part of the body, e.g. the biceps muscle. *Compare* extensor.

flight feathers

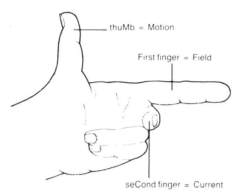

Fig. 74 Fleming's left hand rule

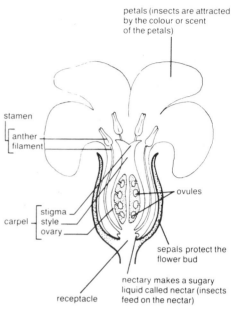

Fig. 75 Parts of a flower

flight feather See contour feather.

flint A very hard form of the mineral silicon(IV) oxide (silica), SiO_2, derived from material in the bodies of sea-living creatures. It has a dull grey or black colour. Flint was, for ages, the chief raw material used by early Man for making tools and weapons. This was mainly because of its hardness and the sharp edges formed when it breaks.

floating rib A rib which is free at its anterior end (not joined to the sternum). In humans the 11th and 12th ribs are floating ribs.

flocculating agent See flocculation.

flocculation The process in which small particles in a dispersed phase of a colloidal system, under certain conditions, clump together forming a precipitate which may float in the liquid. The clumping of clay in water can be brought about by an alkali such as lime. Flocculating agents used in the purification of water and sewage include aluminium sulphate, $Al_2(SO_4)_3$, and iron(III) sulphate, $Fe_2(SO_4)_3$.

flora 1. All the plants present in a certain region or period. **2.** A list describing the plant species of a particular region, arranged in families and genera, together with an identification key.

floret One of the small individual flowers making up the crowded inflorescence of a composite flower.

flotation (law of) A law stating that a floating body displaces its own weight of the fluid in which it floats. See also Archimedes' principle.

flower A reproductive organ in higher plants (basically a leafy shoot) generally consisting of sepals, petals, and stamens or pistils or both. See Fig. 75.

fluid Any substance which can flow, i.e. a liquid or a gas.

fluidity The fluidity of a fluid is the reciprocal of its dynamic viscosity. The SI unit of fluidity is $Pa^{-1}s^{-1}$.

fluke A common name for worms belonging to the Trematoda.

Fluon® Alternative name for Teflon.

fluorescein An organic, cyclic (aromatic), yellow compound having a brilliant greenish fluorescence. It is used to detect leakages in water systems and for colouring liquids in various instruments.

fluorescence An instantaneous kind of luminescence whereby light is emitted from certain substances when these are irradiated by light or certain other radiations. The absorbed light is usually emitted at a greater wavelength than the incident light. Ultraviolet light striking a fluorescent substance gives rise to emission of visible light. Several substances, such as fluorescein and uranyl(VI) sulphate, exhibit fluorescence. See also phosphorescence.

fluorescent lamp A lamp generating light by fluorescence. It consists of a glass tube, the inside of which is coated with a fluorescent substance. The lamp is filled with mercury vapour which emits ultraviolet radiation when an electric current is passed through it. The ultraviolet radiation is converted to visible light by the fluorescent substance coating the inside of the lamp.

fluoridation The addition of minute amounts of fluorides (one part per million of fluoride ion, F^-) to drinking water supplies with the intention of reducing dental decay (caries).

fluoride A salt of hydrofluoric acid, HF.

fluorine An element with the symbol F; atomic number, 9; relative atomic mass, 19,00 (19.00); state, gas. It is a non-metal, a halogen and the most electronegative element known. Fluorine is produced by the electrolytic oxidation of molten fluorides, with fluorine being discharged at the anode. It is a pale yellow gas which is extremely corrosive, reactive and poisonous. Fluorine causes severe burns on the skin. It is used in making certain organic compounds such as fluorocarbons.

fluorite (fluorspar) A widely distributed mineral consisting mainly of calcium fluoride, CaF_2. It has a variety of colours. It is used in making hydrofluoric acid, HF, as a flux in the smelting of metals, as an enamel and also in the glass industry. Fluorite has given its name to the phenomenon of fluorescence; however, it shows this effect only weakly.

fluorocarbon See Freon and Teflon.

fluorspar See fluorite.

flux 1. A substance capable of combining with another substance, usually an oxide, to form a compound with a lower melting-point than the oxide. Such a substance is used in the extraction of metals from their ores, in soldering and in welding. **2.** The rate of flow of mass, volume or energy per unit area normal to the direction of flow.

flux density See magnetic flux density.

fly ash Particles of ash arising from the burning of a fuel such as coal or oil. The fly ash is carried along with the smoke and is one of the causes of air pollution from factories, conventional power stations, etc. To reduce the amount of fly ash pollution, most of the fly ash is removed by making the smoke pass great electrostatic filters. The fly ash obtained from these filters is used in the manufacture of cement, in which it replaces some of the clay, in road-making as a replacement for gravel, and as a backfilling material.

FM Frequency modulation. See modulation.

f-number A number used in photography. It represents the diameter of the aperture (the effective diameter of the lens) and is expressed as a fraction of the focal length f of the lens. Example. $f/4$ means that the diameter of the aperture is one-quarter of the focal length of the lens.

foam A dispersion of small bubbles of gas in a liquid or a solid.

focal length Symbol: f. The distance from the pole of a spherical mirror or the centre of a lens to the focal point (F). See Fig. 76 (a), (b), (c) and (d).

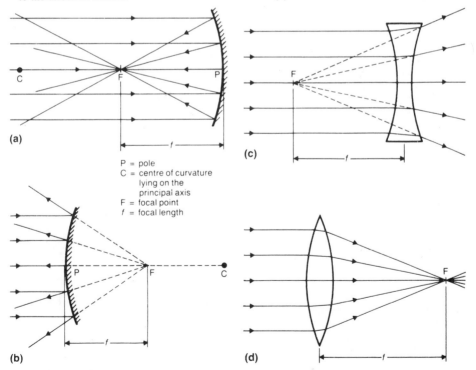

Fig. 76 Focal length
   (a) Concave mirror (b) Convex mirror (c) Concave lens (d) Convex lens

P = pole
C = centre of curvature lying on the principal axis
F = focal point
f = focal length

focal plane The plane through the principal focus, perpendicular to the principal axis of a lens or spherical mirror.

focal point (principal focus) Symbol: F. The focal point of a lens or spherical mirror is the point on the principal axis through which rays of light parallel to the principal axis are refracted or reflected. *See also* focal length.

focus (pl. foci) The point to which rays of light converge (real focus) or from which they appear to diverge (virtual focus) after passing through a lens or being reflected from a mirror.

fodder Plants which are grown as food for domesticated animals. The plants may be dried, e.g. hay.

foetal membrane One of the membranes protecting and nourishing the foetus, e.g. the amnion and the chorion.

foetus 1. The name given to the mammalian embryo after the appearance of recognisable features. The human embryo is called a foetus after about eight weeks of development. **2.** The unborn or unhatched offspring of vertebrates.

fog Tiny water droplets suspended in the atmosphere thus reducing visibility to less than one kilometre: i.e. fog is low-level cloud. Fog may develop where warm, moist winds blow over cold surfaces. Particles of dust and soot promote the formation of fog as water vapour can condense onto them. *Compare* mist.

foil A thin sheet of metal produced by hammering or rolling, e.g. aluminium foil.

folic acid A vitamin belonging to the B-complex group. It occurs in green leaves, liver, kidney, milk, yeast, etc. Deficiency of folic acid may lead to anaemia and mental disturbance.

follicle 1. A dry dehiscent fruit formed from a single carpel and splitting on one side only, e.g. the fruit of the marsh marigold. **2.** A cavity or sheath, e.g. a hair follicle.

fontanelle Soft areas of membranous tissue on an infant's skull where the bones are not fused. In humans there are usually six fontanelles which are all closed by the time the infant is eighteen months old.

food chain A sequence of organisms in which each organism feeds on the one preceding it and may in turn be eaten by the one following it. At the beginning of a food chain are green plants. *Example:* green plant (grass) → small herbivore (goat) → small carnivore (jackal) → large carnivore (lion).

food cycle (food web) All the interconnected food chains present in a community (ecosystem).

food vacuole A vacuole in unicellular organisms such as Protozoa, containing ingested food particles in various stages of digestion, including remains of indigestible food.

food web *See* food cycle.

fool's gold *See* iron(II) disulphide (pyrites).

foot-and-mouth disease An acute, contagious disease of cloven-footed animals caused by a virus. Symptoms include vesicular eruption of the mucous membranes and skin, particularly in the mouth and in the clefts of the feet. To prevent further spread of the disease, infected animals and those which have had contact with the disease are generally slaughtered. The whole area around the infected farm is put in quarantine. The slaughtered animals are buried in quicklime. A vaccine is available against the disease.

foramen (pl. foramina) Any small perforation (opening).

foramen magnum The opening in the occipital bone of the skull through which the spinal cord passes to connect with the brain.

force Symbol: *F*; unit: newton (N). Any action, e.g. a pull or a push, which causes a change of a body's shape or movement.

force meter (newtonmeter) An instrument for measuring force, e.g. a spring balance.

forceps A pincer-like instrument for picking up or holding small objects.

force pump A pump for raising water. It differs from the lift pump in that there is no valve in its piston and it can raise water to a greater height. *See* Fig. 77.

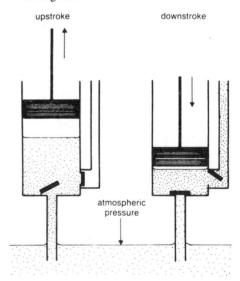

Fig. 77 Force pump

force ratio *See* mechanical advantage.
forebrain (prosencephalon) In vertebrates, the part of the brain consisting of the diencephalon and the telencephalon. The forebrain receives stimuli from the olfactory organ (nose).
foreskin *See* prepuce.
forest An area of land consisting virtually entirely of densely growing trees, shrubs and shade-tolerant herbs.
formaldehyde *See* methanal.
formalin *See* methanal.
formate *See* methanoate.
formic acid *See* methanoic acid.
formula (pl. formulae) The indication of the chemical composition of a compound or molecule using internationally accepted symbols for the atoms of the elements it contains. *Example:* the formula of water is H_2O, showing that one molecule of water consists of two atoms of hydrogen and one atom of oxygen. The molecular formula corresponds to the correct mass of the molecule. Strictly speaking this formula should only be used for substances consisting of discrete molecules. A formula showing the sequence and arrangement of the atoms present in a molecule is called a structural formula. *Example:* CH_3COOH is the structural formula of ethanoic acid (acetic acid). A formula showing the spatial arrangement of the atoms or groups of atoms in a substance is called the displayed formula or the graphic formula of that substance. *Example:*

is the displayed formula of ethanoic acid. The simplest possible formula showing the stoichiometric proportion of the atoms in a compound is called the empirical formula. *Example:* CH_2O is the empirical formula of ethanoic acid. For water and many other compounds, the molecular and the empirical formula are the same.

When identifying an unknown chemical compound, the empirical formula is first determined, then the molecular formula and lastly the displayed formula. For a series of compounds (usually organic), a general formula is sometimes used. This allows the writing of the molecular formula of any member of the series. *Example:* the alkanes have the general formula C_nH_{2n+2}, where $n \geq 1$. $n=1$ gives CH_4, the molecular formula of methane.

Fortin's barometer A type of mercury barometer used for the accurate measurement of atmospheric pressure. *See* Fig. 78.

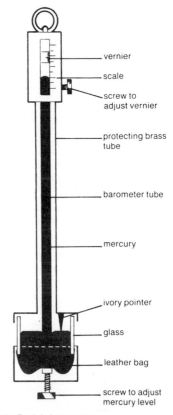

Fig. 78 Fortin's barometer

fossil Remains or evidence of an ancient animal or plant preserved in the Earth's crust.
fossil fuels Coal, petroleum and natural gas. These are derived from the remains of living organisms which existed in the remote past.
fossil wood *See* petrifaction.
Foucault's pendulum A simple pendulum consisting of a heavy metal ball suspended on a very long wire, used to demonstrate the Earth's rotation. The plane in which the pendulum swings twists gradually round in a clockwise direction.
fovea A small pit or depression.
fovea centralis A shallow depression in the centre of the yellow spot (macula lutea) in the retina of many vertebrates. It is the region of clearest vision and contains no rods, but numerous cones. *See* Fig. 66.
foxglove *See* digitalis.
fraction A certain portion of a distillate collected when fractional distillation is performed, e.g. the fraction collected between 65°C and 70°C, or the fraction collected at 75°C.

fractional crystallisation The process in which a mixture, containing substances with different solubilities, is separated by a series of recrystallisations involving the separation of crystals from the mother liquor. At a certain temperature, one component of the mixture crystallises out whilst others remain in solution.

fractional distillation The process of separating (fractionating) a mixture of several liquids which have different but often close boiling-points, by collecting the separate liquids (fractions) at different temperatures. A simple fractionating column for laboratory use consists of a long glass tube packed with small pieces of glass tubing, glass beads or porcelain rings, providing a large surface area. Fractions of lower volatility condense on this surface and only the fraction of highest volatility (of those remaining in the mixture) reaches the top as a vapour. At the top of the column is a side tube for collecting and condensing the vapours. In industry, specially constructed fractionating towers are used. See Fig. 79.

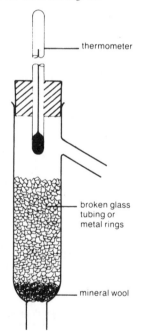

Fig. 79 Fractional distillation: laboratory fractionating column

fractionating See fractional distillation.
fractionating column See fractional distillation.
fractionating tower A large tower used in the refining of crude oil. The crude oil is pre-heated in a furnace and then sprayed into the bottom of the tower. The more volatile fractions (i.e. those with lower boiling-points) migrate to the cooled top of the tower and the less volatile to the heated base. The tower is divided into a number of compartments by means of trays containing holes covered by bubble caps. These enable vapour to pass up the tower; overflow pipes allow liquid to drop down. Pipes at different heights on the tower make it possible to drain different fractions distilling within a certain temperature range. See Fig. 80.

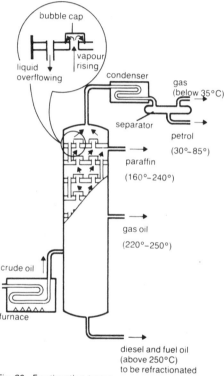

Fig. 80 Fractionating tower

Frasch process (pump) A process used in extracting sulphur from underground deposits. Three concentric pipes are sunk down to the level of the sulphur deposit. Superheated steam (at 175°C and 10 atmospheres) is forced down the outer pipe and hot, compressed air is forced down the rather narrow middle pipe. The heat of the steam melts the sulphur which is then forced up the third pipe together with air and water. This mixture is then run into large tanks where the sulphur solidifies and separates out. See Fig. 81.

fraternal twins See dizygotic twins.
Fraunhofer lines See chromosphere.

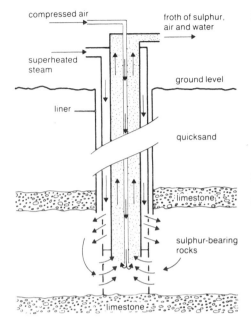

Fig. 81 Frasch process (pump)

free electron An electron which is not attached to a specific atom or molecule, but moves freely under the influence of an electric field.

free energy *See* entropy.

free fall *See* acceleration of free fall.

free radical *See* radical.

freezing The change of state from liquid to solid, taking place at a constant temperature called the freezing-point.

freezing-mixture A mixture of two or more substances, e.g. crushed ice and sodium chloride or solid carbon dioxide and ether, which absorbs heat and therefore produces a lower temperature than the original temperature of the constituents of the mixture. Freezing-mixtures are used to produce low temperatures for chemical reactions, for freezing-point determinations, etc.

freezing-point (melting-point) The temperature at which a liquid freezes or a solid melts at standard atmospheric pressure (101 325 Pa). At this temperature, there is an equilibrium between the liquid and the solid phases.

Freon® Halogenated methane, CH_4, or ethane, C_2H_6, in which some or all of the hydrogen is replaced by fluorine or by fluorine and chlorine. Freons without chlorine are called fluorocarbons. All Freons are non-toxic and non-flammable. They are widely used as refrigerants and as propellants in aerosols.

frequency Symbol: f or ν (nu). **1.** The number of complete cycles or oscillations in unit time (usually one second) of a vibrating system. Frequency is measured in hertz (Hz) and is related to the wavelength and velocity of a wave as follows:

$$f = \frac{c}{\lambda}$$

where c is the velocity and λ (lambda) is the wavelength. **2.** The frequency of an alternating current is the number of times the current passes through its zero value in the same direction in unit time.

frequency modulation (FM) *See* modulation.

friction The force tending to oppose the relative motion between surfaces in contact. The force of friction opposing the motion of a body over a surface is equal to the applied force attempting to move the body, up to a value called limiting friction. The frictional force reaches a maximum at this point and is called static friction. Any increase in the applied force causes sliding to occur. Once sliding occurs, the force needed to keep the body just moving is known as kinetic or dynamic friction and this is rather less than the static friction. The ratios of the static friction and dynamic (sliding) friction and the normal reaction between the surfaces in contact are called the coefficients of static and dynamic friction respectively:

$$\text{coefficient of friction} = \mu \text{ (mu)} = \frac{F}{R}$$

where F is the force of friction and R is the normal reaction.

frictional electricity (triboelectricity) Static electricity produced by rubbing unlike materials together. *Example:* when a glass rod is rubbed with a piece of silk, the glass rod becomes positively charged and the silk negatively charged, i.e. electrons move from the glass to the silk.

friction (coefficient of) *See* friction.

Friedel–Crafts reaction *See* substitution reaction.

fringes *See* interference.

frit *See* enamel.

frond **1.** A leaf-like organ, especially of ferns and palms. **2.** The thallus of certain seaweeds and liverworts.

fructose (fruit sugar, laevulose) A carbohydrate with the formula $C_6H_{12}O_6$. Fructose is a monosaccharide (a ketohexose) and one of the many isomers having this formula. It is a white, crystalline solid occurring naturally in fruit and honey. It forms the basic unit of inulin. Like glucose, fructose is a reducing sugar.

fruit The structure, containing the seed or seeds, which develops from the ovary of a flower after fertilisation.

fruit sugar *See* fructose.

fuel Any material capable of producing large amounts of heat, either by combustion or by nuclear fission. Fuels may be solids (e.g. coal), liquids (e.g. fuel oil) or gases (e.g. natural gas).

fuel cell A type of primary cell in which chemical energy in certain fuels is directly converted to electrical energy by using a suitable oxidising agent. The difference between a fuel cell and a dry cell or an accumulator is minor. Whilst a dry cell or an accumulator produces electricity from a built-in supply of chemicals until it has to be replaced or recharged, the fuel cell receives fuel from outside tanks and produces an electric current as long as it is fed. In simpler fuel cells, the fuel is hydrogen and the oxidising agent is oxygen. Each gas is led to porous metal electrodes, made of platinum or nickel, placed in an alkaline electrolyte or in a molten salt. An electron transfer takes place at the electrodes, producing water as a by-product. Fuel cells are quite efficient, delivering a high power.

fuel oil A brownish-black petroleum fraction used for heating in industry. Fuel oil is left as a residual substance during fractional distillation of crude oil (petroleum).

fulcrum The pivot of a lever; i.e. the point of support, about which it turns.

full-wave rectification Rectification of an alternating current to produce a unidirectional current by reversing the direction of alternate half cycles of the alternating current. This can be done using a double diode. *See* Fig. 82; *compare* half-wave rectification.

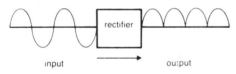

Fig. 82 Full-wave rectification

fume A visible cloud of small airborne particles either arising from condensation of a vapour or from a chemical reaction. *Example:* when ammonia gas, NH_3, and hydrogen chloride gas, HCl, combine, white fumes of ammonium chloride, NH_4Cl, are produced.

fume cupboard A cupboard, with a sliding glass door, in which laboratory experiments involving the use or production of harmful gases are performed. It is fitted with a ventilator for the extraction of these gases.

fumigant A substance which, when evaporated, is capable of killing pests. Examples of fumigants are methanal (formaldehyde), epoxyethane (ethylene oxide) and hydrogen cyanide.

fumigation The process in which pests are killed by exposure to poisonous gases (fumigants).

fuming nitric acid Concentrated nitric acid, HNO_3, containing nitrogen dioxide, NO_2.

fuming sulphuric acid *See* oleum.

functional group An atom or a group of atoms which determines the characteristic properties of a compound. The term is almost exclusively used for organic compounds in which a functional group is attached to the carbon skeleton. Examples of functional groups are the hydroxyl group, —OH; the aldehyde group, —CHO; the amino group, —NH_2; and the carboxyl group, —COOH.

functional group isomerism *See* structural isomerism.

fundamental The component of a complex vibration (note) having the greatest intensity and the lowest frequency. *See also* quality of sound.

fungi *See* fungus.

fungicide A chemical which may destroy fungi such as mildew and moulds.

fungus (pl. fungi) An organism, e.g. the mushroom, which is neither animal nor plant, but is generally regarded as a plant because its cells possess walls. Fungi are not able to perform photosynthesis and are therefore either parasites or saprophytes. There are about 100 000 species of fungus. Fungi are important in nature, e.g. in soil where they help in the decomposition of dead plants and animals. Various fungi are used in industrial processes, e.g. in baking and brewing. *See also* Mycophyta *and* Protista.

funicle In angiosperms, a stalk attaching the ovule to the placenta.

funnel *See* filter funnel, separating funnel *and* thistle funnel.

fur (scale) A deposit in kettles, boilers, water-pipes, etc., consisting mainly of calcium carbonate, $CaCO_3$, magnesium carbonate, $MgCO_3$, and sometimes iron(II) carbonate, $FeCO_3$. This is formed from soluble hydrogen carbonates of calcium, magnesium and iron when temporarily hard water is heated; e.g. $Ca(HCO_3)_2 \rightarrow CaCO_3 + CO_2 + H_2O$
Fur is a bad conductor of heat and therefore wastes fuel.

furanose *See* sugar.

furfural C_4H_3OCHO. A heterocyclic five-membered ring containing oxygen as the hetero atom and also containing an aldehyde group, —CHO. It is a colourless, pleasant-smelling liquid, completely miscible with most organic solvents, but only a little soluble in water. Furfural can be prepared from polysaccharides in which the building unit is a pentose by acid hydrolysis. When strong acids act on pentoses or hexoses, furfural or furfural derivatives are formed. This is the basis of a number of laboratory tests, one of the commonest being Molisch's test.

furrow A long narrow groove cut in the ground by a plough.

fuse A safety device used in an electrical circuit to prevent an excessive current flowing through the circuit. It consists of a short length of wire or a strip of metal which has a low melting-point. It is connected in series and is designed to melt ('blow') when the current exceeds a safe value.

fused Describes a substance in the molten state or a substance which has been melted and has solidified.

fused ring Two or more cyclic units joined to one another by sharing two atoms. Naphthalene consists of two fused six-membered rings.

fusel oil A mixture of alcohols formed as impurities in ethanol (ethyl alcohol), C_2H_5OH, when it is prepared by fermentation of molasses or other sugar-rich materials. The major components of fusel oil are different pentanols, all with the formula $C_5H_{11}OH$. These are toxic and give the ethanol a bad smell and taste.

fusion 1. The change of state from solid to liquid, taking place at a constant temperature called the melting-point.
2. A process in which two light nuclei of atoms combine to produce a heavier single nucleus having a total mass slightly less than that of the total initial material. The difference in mass is radiated as energy, the amount of which is stated by Einstein's equation. A fusion process requires enormous temperatures (several million degrees Celsius) before it takes place. Such high temperatures may be produced by fission processes. In the Sun (and other stars), the temperatures are high enough for fusion to take place, thus producing the energy the Sun radiates into space. This involves the conversion of hydrogen, H_2, to helium, He, in three stages:

(1) $^1_1H + ^1_1H \rightarrow ^2_1H + ^0_{+1}e$
(2) $^1_1H + ^2_1H \rightarrow ^3_2He$
(3) $^3_2He + ^3_2He \rightarrow ^4_2He + 2^1_1H$

In (1), two hydrogen nuclei (protons), ^1_1H, combine to form a nucleus of heavy hydrogen (deuterium), ^2_1H, and one positron, $^0_{+1}e$. In (2), a 'lightweight' helium nucleus is formed from the combination of a proton and the heavy hydrogen nucleus from (1). In (3), two 'lightweight' helium nuclei combine to give a normal helium nucleus, ^4_2He, and two protons. Nuclear fusion is uncontrolled. World-wide research is being carried out to find ways to control it so that the vast amounts of energy it liberates can be usefully applied. The fusion principle is made use of in hydrogen bombs which are ignited by an atomic bomb. Fusion reactions are sometimes called thermonuclear reactions. *Compare* fission.

G

g *See* acceleration of free fall.

galactose A carbohydrate with the formula $C_6H_{12}O_6$. Galactose is a monosaccharide (an aldohexose) and one of the many isomers having this formula. It is a white crystalline solid and occurs naturally in certain plant polysaccharides. In lactose it is combined with glucose. Like glucose, galactose is a reducing sugar.

galaxy A large star system also containing gas and dust. Usually galaxies contain between 10^6 and 10^{12} stars. They may be lens-shaped, elliptical, spiral or irregular. *See also* 'big bang' theory.

Galaxy, the (Milky Way, Home Galaxy) The disc-shaped, spiral galaxy which may appear as a faint band of starlight across the sky. It is made up of the combined light coming from millions of stars, most of which cannot be seen individually with the naked eye. The Galaxy contains 10^{11} stars and is therefore one of the largest of all galaxies. It is 10^5 light-years in diameter and 2000 light-years thick. The Sun is one of the stars of the Galaxy, lying about 30 000 light-years from its centre.

galena PbS (lead glance) The most important lead ore. Galena may occur in all types of rocks. It is a soft grey mineral, rather like lead itself. *See also* cerussite.

Galilean telescope *See* telescope.

gall *See* bile.

gall-bladder A pear-shaped, membranous, muscular sac attached to the underside of the liver. In humans it contains about 40 ml of fluid. The gall-bladder functions as a storage organ for bile (gall) which is constantly secreted by the liver. During digestion, the gall-bladder contracts and expels its bile into the duodenum. The presence of fatty foods in the small intestine stimulates the production of a hormone, cholecystokinin, which makes the gall-bladder contract. *See* Fig. 49.

gallstone A small, pebble-like formation sometimes found in the gall-bladder. Gallstones may cause acute, intense pain when they become lodged in the bile duct. This often occurs after a meal rich in fatty foods. The reason for the formation of gallstones is not known. Some types of gallstones may be dissolved by the administration of drugs, others may need to be removed surgically along with the gall-bladder.

galvanic cell An obsolete name for a primary cell. The name voltaic cell is more commonly used.

galvanised iron Plates of iron (steel) coated with a thin layer of zinc in order to prevent rusting. This is generally carried out by dipping the plates into a bath of molten zinc or by an electrolytic method. The process of coating iron with zinc is called galvanising.

galvanising *See* galvanised iron.

galvanometer (galvo) A sensitive instrument for detecting and measuring very small electric currents. *See also* ammeter.

galvo *See* galvanometer.

gamete (germ-cell, sex cell) A reproductive cell which unites with another during sexual reproduction. In animals, a male gamete is called a spermatozoon and a female gamete is called an ovum. Gametes are usually haploid.

gametogenesis The formation of gametes (sex cells).

gametophyte The phase of the life cycle of a plant which produces gametes. The gametophyte has haploid nuclei. Fusion of gametes produced by the gametophyte produces a diploid zygote which develops into a sporophyte.

gamma (γ) The third letter of the Greek alphabet.

gamma globulin A protein present in blood plasma constituting the antibodies and therefore concerned with immunity.

gamma rays Electromagnetic radiation of very short wavelength emitted by certain radioactive atoms. Gamma rays can penetrate a thick metal sheet and are only stopped by thick layers of concrete and lead, e.g. a layer of approximately 15 cm of lead. When gamma rays pass through gases, the gases are ionised. Controlled gamma radiation is used in the treatment of cancer.

ganglion (pl. ganglia or ganglions) A small mass of nerve-cell bodies appearing as a swelling and giving rise to nerve-fibres.

gangrene Death of body tissue caused by lack of oxygen due to obstructed blood circulation. *See also* necrosis.

ganja *See* cannabis.

gas A state of matter in which the particles of a substance move freely throughout the whole of the space in which it is placed.

gas chromatography *See* chromatography.

gas constant *See* universal gas constant.

gas equation An equation derived from the combination of Boyle's law and Charles's law: $pV = nRT$ where p is the pressure, V is the volume, n is the number of moles of gas, R is the universal gas constant and T is the temperature measured in kelvin.

gas laws *See* Boyle's law, Charles's law *and* gas equation.

gas mask A device worn over the nose and mouth for protection against poisonous gases. The mask contains a substance, e.g. activated carbon, to adsorb such gases, and a filter pad to retain solid particles, e.g. smoke.

gas oil *See* diesel fuel.

gasoline *See* petrol.

gassing The evolution of gas from the plates of an accumulator when it nears the end of its charging period.

gas thermometer A device for measuring temperature. Some gas thermometers measure changes in the volume of a gas caused by temperature changes in the gas while the pressure is kept constant (the constant pressure gas thermometer). Others measure changes in the pressure of a gas caused by temperature changes in the gas while the volume is kept constant (the constant volume gas thermometer). Such thermometers can be used for very accurate temperature measurement. A simple constant volume gas thermometer consists of a glass bulb containing dry air, attached by a narrow tube to a U-tube containing mercury. The bulb is immersed in a water bath and changes in the temperature of the bath cause pressure changes in the air in the bulb. The mercury levels are adjusted to keep the volume of the air constant and the pressure is measured by the difference in the heights of the mercury levels. In order to calculate a temperature using this apparatus, the following formula is applied:

$$t = \frac{p_t - p_0}{p_{100} - p_0} \times 100$$

where t is the temperature (°C), p_0, p_{100} and p_t are the pressures of the gas at the ice-point, the steam-point and the temperature t respectively. In accurate work, corrections are made for the non-ideal behaviour of the gas, expansion of the bulb, etc. *See* Fig. 83.

gastric Pertaining to the stomach.

gastric gland One of the numerous, tube-like glands in the lining of the stomach. They secrete gastric juice.

which permits the entry of ionising radiations. The gas is a noble gas, e.g. argon, at low pressure, mixed with a small amount of a halogen, e.g. bromine. A potential difference (as high as 1000 volts) is applied between the electrodes. When an ionising radiation enters the tube it ionises the gas in the tube, causing electrons to move to the anode and positive ions to move to the cathode. In doing so the electrons and positive ions cause further ionisation and an avalanche occurs. This can be detected as a pulse of electric current. The purpose of the halogen is to quench the avalanche and make the tube ready to react again within a very short time, about 10^{-4} seconds—the recovery period. The electric pulses are amplified and used to operate a counter, a loudspeaker or ratemeter. The Geiger–Müller counter is a type of ionisation chamber. See Fig. 84.

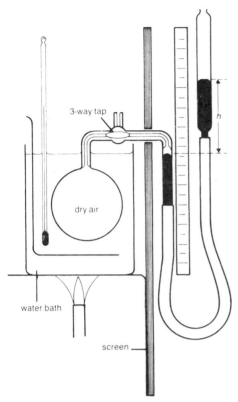

Fig. 83 Gas thermometer

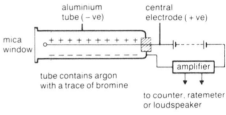

Fig. 84 Geiger–Müller counter

gastric juice A digestive juice containing hydrochloric acid, pepsinogen, pepsin and mucus.

gastrin A hormone which stimulates the production of gastric juice.

gastritis An inflammation of the mucous membrane of the stomach.

Gastropoda A class of molluscs. Members of this class are marine, freshwater or terrestrial. They have a distinct head bearing tentacles and eyes. They also have a flat, muscular foot and often a single (univalve) shell. Examples include snails, slugs and limpets.

Gay-Lussac's law When gases react they do so in volumes that bear a simple ratio to one another and to the volume of the gaseous product, provided temperature and pressure remain constant.

Gay-Lussac's law of heat See Charles's law.

Ge Chemical symbol for germanium.

Geiger–Müller counter An instrument for detecting and counting ionising radiations. It consists of a gas-filled aluminium tube (the cathode) with a wire (the anode) down its centre. At one end of the tube is a mica window

Geissler tube A discharge tube used to demonstrate the luminous effects of a discharge of electricity through rarefied gases.

gel A jelly-like, colloidal solution which may be produced from a sol when this coagulates, e.g. by heating. In a gel the solid is arranged in the liquid as a fine network. Examples of gels are gelatin(e), jellies, agar and slimy precipitates such as aluminium hydroxide, $Al(OH)_3$.

gelatin(e) A water-soluble, rather complex, colourless and odourless protein formed by the hydrolysis of collagen. It can be made by boiling skin or bones with water. Gelatin(e) is used in photography, in foods, in making pharmaceutical capsules and in making adhesives.

gelatinous Describes a substance which has a jelly-like consistency.

gem (gemstone) A cut and polished precious stone derived from a mineral and used for decorative purposes, e.g. diamond, ruby and sapphire.

gemma (pl. gemmae) An organ of asexual reproduction in organisms such as mosses, liverworts and coelenterates. A gemma is a cellular outgrowth which eventually separates from the parent organism.

gemmation Reproduction by means of gemmae. In animals, gemmation is commonly called budding.

gemstone *See* gem.

gene A part of a chromosome, consisting of DNA, with the ability to transmit hereditary characteristics from one generation to another. Since chromosomes occur in pairs (homologous chromosomes), genes also occur in pairs. Two genes occupying the same relative position (locus) on homologous chromosomes are called alleles or allelomorphs. If these genes are identical, they are referred to as homozygous genes. If they are not identical, i.e. when one is recessive and the other dominant, then they are referred to as heterozygous genes. Genes may be altered by mutation.

gene frequency The frequency of occurrence of any given gene in relation to all its alleles at the same locus.

gene mutation *See* mutation.

gene pool The collective term for all the genes of a certain population.

general formula *See* formula.

generator (dynamo) A machine for converting mechanical energy into electrical energy. In its simplest form it consists of a coil of wire (armature) rotated in a magnetic field produced by a field magnet. An electric current is produced in the coil by electromagnetic induction. The ends of the coil may be connected to two slip rings on the coil's driving shaft and the current is drawn off by brushes pressing against the slip rings. In this form the generator produces an alternating current. If the slip rings are replaced by a commutator, then the generator produces a direct current. Power stations produce electricity from large a.c. generators called alternators. *See* Fig. 85 (a) *and* (b); *see also* van de Graaff generator *and* Wimshurst machine.

genetic code In a cell, the sequence of purine and pyrimidine bases in messenger RNA which is produced from a length of DNA present in the nucleus. This sequence determines the sequence of amino acids in a protein. Three bases code for each particular amino acid.

genetics The study of heredity and variation in living organisms.

genetic variation A variation inherited by an organism and caused by a change in the genes.

genital Pertaining to the region of the reproductive organs.

genital duct *See* gonoduct.

genitalia (genitals) The reproductive organs, the gonads and all associated accessory organs, especially the external organs.

genitals *See* genitalia.

genotype The nature and arrangement of genes in an individual organism. *Example:* the height of a human being is partly determined by the genotype inherited from his or her ancestors. *Compare* phenotype.

genus (pl. genera) A group of organisms having common structural characteristics, usually a group of species. Related genera are further grouped into a family.

geo- Prefix meaning Earth.

geographic meridian The geographic meridian at any place is the plane containing the place and the axis of rotation of the Earth. *See also* Fig. 2.

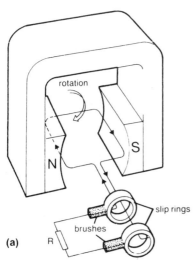 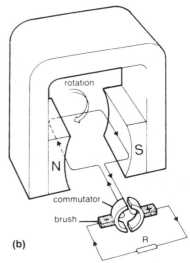

Fig. 85 (a) Simple a.c. generator   (b) Simple d.c. generator

geology The science of the origin, structure and history of the Earth and its inhabitants as recorded in the rocks of the Earth's crust.

geomagnetism *See* magnetism, terrestrial.

geometrical isomerism A type of stereo-isomerism which may occur in compounds with at least one bond about which the molecule cannot rotate. In compounds (a) and (b) below, the molecules cannot rotate about the carbon–carbon double bond. This means that (a) and (b) are different isomers.

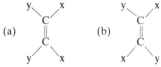

In (a) similar atoms (or groups of atoms) are on the same side of the double bond whereas they are on opposite sides in (b); i.e. the relative positions of these atoms in space are different for the two isomers. The isomer represented in (a) is called the *cis*-isomer and (b) is called the *trans*-isomer. Geometrical isomerism is not limited to compounds containing a double bond, but appears also in some derivatives of certain ring (cyclic) compounds and in some complex inorganic compounds. Geometrical isomers differ in both chemical and physical properties. Examples of compounds exhibiting geometrical isomerism are *trans*-butenedioic acid (fumaric acid), $C_2H_2(COOH)_2$, and *cis*-butenedioic acid (maleic acid), $C_2H_2(COOH)_2$.

geotropism Plant growth in response to gravity. *Example:* main roots grow towards the centre of the Earth and main stems normally grow away from it.

germ 1. A unicellular micro-organism. 2. Part of an organism capable of developing into a new organism, e.g. a bud, a fertilised egg, the embryo of a seed.

germanium An element with the symbol Ge; atomic number 32; relative atomic mass 72,59 (72.59); state, solid. Germanium is found in several sulphide ores of silver and zinc, and in coal. It is a hard, brittle metal used as a semiconductor in transistors.

German measles *See* rubella.

German silver An obsolete name for nickel silver.

germ-cell *See* gamete.

germicide A chemical which may destroy germs.

germination The beginning of growth in a spore or seed. In higher plants, this process may be said to last from the uptake of water to the development of leaves capable of performing photosynthesis.

germinative layer *See* Malpighian layer.

gerontology The study of old age in organisms and diseases related to this period of life. *Compare* paediatrics.

gestation period In mammals, the length of time the embryo spends in the uterus, i.e. from conception to birth.

geyser The ejection of boiling water and steam from underground sources. This is seen in active or recently active volcanic areas. Geysers are found in Iceland, New Zealand and a few other places. The heat energy from geysers may be used for domestic or industrial purposes.

GH *See* growth hormone.

giant star A bright star with dimensions much larger than those of an average star like the Sun. Its brightness depends on its surface temperature and size. The very biggest and brightest stars are called super giants.

gibberellin One of a group of plant hormones which have a variety of effects on plant growth and development, e.g. accelerating the growth in stems of certain plants (i.e. they may cause a dwarf plant to grow like a normal plant). Gibberellins are sometimes used in association with auxins.

Gibbs' function *See* entropy.

giga- Symbol: G. A prefix meaning 10^9.

gigantism A condition in which there is an overproduction of growth hormone from the pituitary gland, resulting in heights of about 2,10 metres (2.10 metres) or more in humans. *Compare* dwarfism.

gill bar A curved bone supporting a gill. Gill bars separate adjacent gill slits. Gill filaments are attached to the gill bars.

gill cover *See* operculum.

gill filament A thread-like structure arising from a gill bar. Gill filaments are divided and sub-divided, providing a large surface area for gaseous exchange. They are supplied with a dense network of capillaries.

gill raker In certain fish, a small rod-like structure made of bone attached to a gill bar. Gill rakers serve to filter water entering the gills, thus trapping food particles which are eventually swallowed.

gills 1. The vertical, radial, plate-like structures under the caps of mushrooms. 2. Respiratory organs of aquatic organisms. Gills consist of vascular skin occurring as plate-like outgrowths with large surface areas. Through the gill surface there is an interchange by diffusion of dissolved oxygen and carbon dioxide between water and blood.

gill slit An opening between adjacent gill bars. Water passes through the gill slits over the gill filaments.

girdle A skeletal structure supporting the appendicular skeleton. *See* pectoral girdle *and* pelvic girdle.

girdle scars A ring of scars on a twig where scale leaves of the previous year's buds were attached. *See* Fig. 86.

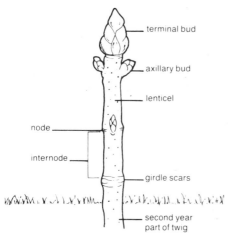

Fig. 86 Twig showing girdle scars

gizzard In birds and many invertebrates, a sac-like enlargement of the alimentary canal situated immediately after the crop. It has muscular walls which are sometimes covered with toothlike plates, or it may contain small stones, for the breaking up and grinding of food. *See also* proventriculus.

glacial Describes substances which form ice-like crystals on freezing, e.g. concentrated ethanoic acid (acetic acid), CH_3COOH.

glacial acetic acid *See* glacial ethanoic acid.

glacial ethanoic acid (glacial acetic acid) Concentrated ethanoic acid, CH_3COOH, which forms ice-like crystals on freezing (m.p. 16,6°C (16.6°C)).

glacier A large, compact, growing mass of ice formed from frozen snow in regions with heavy snowfalls where the snow remains throughout the year, i.e. in the polar regions and on high mountains. When sufficient ice has accumulated, the glacier will begin to move slowly from its point of origin due to the force of gravity. This movement may be from a few centimetres to thirty metres per day.

gland An organ in which one or more specific chemical compounds are synthesised and later on secreted from it. In animals some glands are big, e.g. the liver, and some are small, e.g. tear glands. In plants the glands are always small. Glands are either endocrine or exocrine.

gland cell An isolated secreting cell of a gland.

glandular With, or pertaining to, glands.

glandular tissue *See* epithelium.

glans A glandular structure.

glans clitoridis *See* clitoris.

glans penis The tip of the mammalian penis.

glass An amorphous, hard and brittle mixture of sand (silicon(IV) oxide), sodium carbonate and limestone fused together. There are several types of glass, some of which may contain lead, barium, potassium or other metals in place of sodium. Boron oxide may be used in place of some of the silicon(IV) oxide. Glass is considered to be supercooled liquid. *See also* Pyrex *and* Vitreosil.

glass electrode An electrode used in conjunction with a calomel electrode in the electrometric measurement of pH. The electrode consists of a silver wire coated with silver chloride placed inside a thin-walled bulb made of a special type of glass. The bulb also contains 0,1 molar (0.1 molar) hydrochloric acid into which the wire dips. The potential difference between the glass surface and the solution in which the electrode is placed is proportional to the pH of the solution; that is, the potential of the glass electrode depends solely on the pH in the solution (sample).

glass fibres Melted glass which has been drawn out by steam through tiny holes to form fibres of very small diameter, approximately 10^{-6} m. They may be spun into threads, woven into tapes and cloth, formed into pads or embedded in a matrix of, for example, polyester. *See also* fibre optics.

glass wool Very thin glass threads, resembling cotton wool. Glass wool is used as a filter agent and for packing and insulation.

Glauber's salt Crystalline sodium sulphate, $Na_2SO_4 \cdot 10H_2O$, occurring naturally. It is used in the manufacture of glass and paper and also in medicine as a laxative.

glaucoma An eye disease caused by increased pressure of fluid within the eyeball. If not treated, it may lead to blindness.

glaze A smooth, shiny coating on various ceramics making them waterproof. Glazes may be used for decorative purposes. The glaze used on porcelain contains the same ingredients as porcelain, but in different proportions.

glenoid cavity (glenoid fossa) A socket-like structure in the scapula into which fits the head of the humerus.

glenoid fossa *See* glenoid cavity.

globulin A water-insoluble protein which is soluble in dilute salt solution and coagulates when heated. Globulins are found in animal and plant tissue, e.g. fibrinogen and legumin.

glomerulonephritis *See* nephritis.

glomerulus (pl. glomeruli) A knot of capillaries in a Bowman's capsule of the kidney. It is concerned with blood filtration. *See* Fig. 149.

glottis In vertebrates, the opening of the trachea into the pharynx and also the space between the vocal cords. During swallowing the glottis is closed by the epiglottis.

glucagon A polypeptide hormone secreted by the α cells of the islets of Langerhans of the pancreas. It stimulates the breakdown of glycogen in the liver, causing an increase in blood sugar; i.e. glucagon has the opposite effect to insulin. *See also* glycogenolysis.

glucosamine $CH_2OH—(CHOH)_3—CH(NH_2)—CHO$. An organic compound (an amino sugar) which may be obtained by hydrolysis of chitin. Glucosamine is one of the constituents of a molecule of streptomycin.

glucose (dextrose, grape-sugar) A carbohydrate with the formula $C_6H_2O_6$. Glucose is a monosaccharide (an aldohexose) and one of the many isomers having this formula. It is a white, crystalline solid occurring naturally in plants and animals. In green plants, glucose is formed during photosynthesis and stored as starch. In animals, glucose is obtained when disaccharides and polysaccharides are digested and is stored as glycogen. Glucose is a reducing sugar.

glume A membrane bract situated at the base of a grass spikelet. *See* Fig. 217.

glumella *See* palea.

gluten An almost equal mixture of the proteins gliadin and glutenin occurring in cereals, e.g. wheat.

glyceride An ester of glycerol and one or more organic acids (saturated or unsaturated). Fats are mainly composed of triglycerides of fatty acids, i.e. compounds in which the three hydroxyl groups in glycerol have reacted with three similar or different fatty acids.

glycerin(e) An obsolete name for glycerol.

glycerol (propane-1,2,3-triol; glycerin(e)) A trihydric alcohol with the formula $CH_2OH—CHOH—CH_2OH$.
It is a colourless, water-soluble, viscous, sweet, hygroscopic liquid occurring in an esterified form in fats. Glycerol is formed as a by-product in the manufacture of soap. It is used in the manufacture of explosives (*see* dynamite) and plastics, and in pharmacy, etc.

glyceryl trinitrate *See* nitroglycerine.

glycine (aminoacetic acid, aminoethanoic acid, glycocoll) The simplest of the aliphatic amino acids, having the formula $CH_2(NH_2)COOH$. It is derived from ethanoic acid (acetic acid) CH_3COOH, and therefore has the IUPAC name of aminoethanoic acid (aminoacetic acid). It is a constituent of several proteins such as collagen and elastin.

glyco- Prefix meaning sweet.

glycocoll *See* glycine.

glycogen (animal starch) A soluble polysaccharide built up of many units (molecules) of glucose; i.e. glycogen is a polymer. It is formed by animals and fungi which store carbohydrate as glycogen. In vertebrates the manufacture, storing and breakdown of glycogen mainly take place in the liver and muscles. *See also* glycogenesis *and* glycogenolysis.

glycogenesis The formation of glycogen from glucose, stimulated by insulin.

glycogenolysis The breakdown of glycogen to form glucose, stimulated by glucagon and adrenalin(e).

glycol 1. *See* ethane-1,2-diol. 2. *See* diol.

glycolysis The process in living organisms in which glycogen or starch is first converted into glucose, then into pyruvic acid by a series of reactions catalysed by various enzymes and coenzymes. The chemical energy released during these processes is stored in ATP. This energy is small in comparison with that liberated in the Krebs cycle. All the stages up to the formation of pyruvic acid can proceed in the absence of oxygen. In aerobic respiration the Krebs cycle follows glycolysis. In anaerobic respiration pyruvic acid is converted into ethanal (acetaldehyde) and carbon dioxide and then to ethanol (in plants) or to lactic acid (in animals) for elimination or storage. Glycolysis takes place in the cell cytoplasm (mitochondria).

glycoprotein *See* mucoprotein.

GMT *See* Greenwich Mean Time.

gneiss Coarse-grained, banded, metamorphic rock consisting of quartz, feldspar and mica. *See also* granite.

gnomon *See* clock.

goblet cell A pear-shaped or wineglass-shaped cell which secretes mucus and is present in some epithelia, e.g. in the lining of the alimentary canal.

goitre An enlargement of the thyroid gland which produces the hormone thyroxine. This causes a swelling at the front of the throat. An enlarged thyroid gland may be caused by under-production or over-production of thyroxine and may be due to lack of iodine in the food and water. *See also* cretinism.

gold An element with the symbol Au; atomic number 79; relative atomic mass 196,97 (196.97); state, solid. It is a very inert, heavy, soft, yellow transition metal. However, it reacts with fluorine and chlorine. Cyanide ions, CN^-, form stable complex ions with gold ions. Gold dissolves in aqua regia, but is unattacked by all pure acids. It occurs as the free metal (native gold) and is used in dentistry, jewellery and medicine.

gold leaf electroscope *See* electroscope.

Goldschmidt process A process in which aluminium powder reacts with a metal oxide when ignited, to form pure metal and aluminium oxide. This process is used to produce metals of high purity such as chromium, nickel, vanadium, manganese and uranium. The process is also used in thermit(e) welding.

Golgi apparatus (body) An organelle situated in the cytoplasm of a cell. It consists of a series of membranous, smooth-surfaced sacs. Its function is uncertain but it is thought that one of its functions is to gather up chemicals manufactured in the cell and wrap them in a membrane. The chemicals may then be moved out of the cell or to another part of it.

gonad In animals, a sex gland (organ), ovary or testis, producing gametes and sometimes hormones.

gonococcus The bacterium causing gonorrhoea.

gonoduct (genital duct) A duct carrying genital products from a gonad to the exterior.

gonorrhoea A venereal disease caused by the gonococcus bacterium. The disease is almost always contracted from an infected person during sexual intercourse. Symptoms may include a painful, burning sensation on urination (especially in men) and a discharge of pus from the penis or vagina. If gonorrhoea is detected in its early stages, it may easily be cured by antibiotics.

gout See uric acid.

Graafian follicle (Graafian vesicle) A small fluid-filled sac in the mammalian ovary containing an oocyte attached to its wall. The oocyte may develop from an oogonium. It is surrounded by follicle cells. As it develops a cavity appears which enlarges, and eventually the follicle bursts releasing the oocyte (ovulation). The oocyte gives rise to an ovum. Hormones from the pituitary gland influence the growth of the follicle. See also ovarian follicle.

Graafian vesicle See Graafian follicle.

graduated flask See volumetric flask.

graduation The marking of a scale on an instrument. Example: a measuring cylinder is graduated in millilitres.

graft A part of a living organism transplanted into a larger part of another or the same organism: e.g. a skin graft, a scion inserted into a stock. See Fig. 87.

Graham's law of diffusion Under constant conditions of temperature and pressure, the rate of diffusion of a gas is inversely proportional to the square root of its density:

$$R \propto \sqrt{\frac{1}{d}}$$

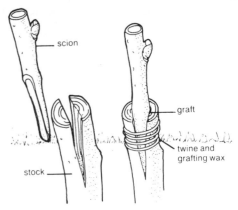

Fig. 87 Graft

where R is the rate of diffusion and d is the density. By comparing the rate of diffusion of two gases, A and B, the density of one can be found if the density of the other is known:

$$\frac{\text{rate of diffusion of A}}{\text{rate of diffusion of B}} = \sqrt{\frac{\text{density of B}}{\text{density of A}}}$$

Graham's pendulum See compensated pendulum.

gram (gramme) A unit of mass equal to $\frac{1}{1000}$ of the kilogram.

gram-atom See mole of atoms.

gram-equivalent mass See equivalent mass.

gram-formula The mass in grams of one mole of a chemical compound, calculated from its formula. Examples: one gram-formula of sodium chloride, NaCl
= 22,99 g + 35,45 g = 58,44 g
(= 22.99 g + 35.45 g = 58.44 g)
Two gram-formulae of copper(II) oxide, CuO
= 2(63.55 g + 16,00 g) = 159.10 g
(= 2(63.55 g + 16.00 g) = 159.10 g)

Gramineae A very large family of monocotyledons, consisting of the grasses.

gram-ion See mole of ions.

gramme See gram.

gram molecular volume See molar volume.

gram-molecule See mole of molecules.

Gram-negative See Gram's stain.

Gram-positive See Gram's stain.

Gram's stain A widely used technique of staining bacteria. Gram's stain is used to distinguish between two kinds of bacteria, the Gram-negative and the Gram-positive. However, some bacteria are Gram-variable; young cultures are positive, but become negative as they age. In performing the technique, four solutions are used: a basic stain (e.g. crystal violet), a mordant (e.g. iodine solution), a decoloriser (e.g. a mixture of propanone and ethanol) and a counterstain (e.g. safranin). At the end of the procedure, the Gram-negative will be stained red and the Gram-positive violet. *See also* murein.

Gram-variable *See* Gram's stain.

granite An igneous rock consisting of quartz, feldspar and mica. Its colour varies between grey, white and pink and combinations of these colours. It is mainly the feldspar that gives colour to granite, while specks of mica give it its sparkle. The great hardness of granite makes it suitable as a building material. Granite can be smoothed and polished easily and for this reason it is also used in making monuments. *See also* gneiss.

granulocyte (polymorph) A leucocyte formed in the red bone marrow. Granulocytes have large lobed nuclei and an irregular shape, and possess granules in their cytoplasm. About 70% of the leucocytes are granulocytes. Granulocytes are capable of changing shape and can ingest and therefore destroy pathogens; i.e. they are phagocytic. They may be divided into three groups: neutrophils (60–70%), basophils (0,5%; 0.5%) and eosinophils (1–3%). Neutrophils do not stain with acidic and basic dyes and they respond quickly to any bacterial infection in the body. Basophils stain with basic dyes such as methylene blue and they contain histamine. They are involved in allergic and stress conditions which develop in the body. Eosinophils stain with acidic dyes such as eosin. They are active towards parasitic infections and are (like basophils) involved in allergic reactions.

granum (pl. grana) Grana are clusters of regularly arranged, parallel membranes situated in the amorphous, colourless substance (the stroma) of chloroplasts. The enzymes involved in photosynthesis are spread over the surface of the grana, which also contain photosynthetic pigments such as chlorophyll.

grape-sugar *See* glucose.

graphic formula *See* formula.

graphite (black lead, plumbago) A naturally occurring allotrope of carbon which has a high melting-point (3650°C). It consists of hexagonal crystals arranged in flat planes which can slide over each other when a force is applied. Graphite is a rather soft mineral with a black colour exhibiting a dull metallic lustre. It readily marks paper, is greasy to the touch and is a good conductor of heat and electricity. Graphite is used in making electrodes, crucibles and lubricants. Mixed with clay it forms the 'lead' in pencils. *See* Fig. 88.

grass 1. *See* Gramineae. **2.** *See* cannabis.

graticule A network of fine wires placed at the focal point of the eyepiece of a telescope or microscope. It can be used as a reference system or for measurement.

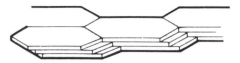

Fig. 88 Hexagonal plates of graphite 'sliding' over one another

grating *See* diffraction grating.

gravid Describes a female animal which is pregnant or carrying fertilised eggs.

gravimetric analysis A very accurate chemical analysis of a material by the precipitation of one or more of its constituents and the subsequent accurate measurement of their masses.

gravitation The mutual attraction between masses. The magnitude of this force depends upon the masses of the objects and the distance separating them. *See also* Newton's law of gravitation.

gravitational constant Symbol: G. The universal constant of gravitation as it appears in Newton's law of gravitation. Its value is $6{,}67 \times 10^{-11}$ newton metre squared per kilogram squared ($6.67 \times 10^{-11}\,N\,m^2\,kg^{-2}$).

gravitational field The region around a body in which it attracts another body as a result of their masses.

gravity The effect of gravitational attraction between the Earth or some other heavenly body and an object in its gravitational field. The Earth's gravity is responsible for the weight of an object near its surface and for the acceleration of free fall.

green algae *See* Chlorophyceae.

greenhouse effect The effect experienced inside a greenhouse: infra-red radiation of short wavelength from the Sun passes through glass and is absorbed by the contents which in turn radiate infra-red rays with a longer wavelength because of their relatively low temperature. These rays cannot pass through the glass, so the temperature inside the greenhouse rises. The same effect is seen when short wave infra-red rays from the Sun pass through a layer of carbon dioxide in the Earth's atmosphere, whereas long wave infra-red rays radiated back from the Earth do not; i.e. the solar energy is trapped by the Earth's atmosphere. It is possible that the increasing concentration of carbon dioxide in the Earth's atmosphere may cause a permanent temperature rise which could lead to catastrophes such as the melting of the polar ice-caps.

green sickness *See* chlorosis.

Greenwich Mean Time (GMT) Mean solar time at the Greenwich meridian, the line of longitude passing through the observatory at Greenwich in the UK. GMT is the basis for all measurements of time used in scientific and navigational work.

grey matter Nervous tissue consisting of cell bodies, found in the central nervous system of vertebrates. Grey matter forms the cortex enclosing the white matter in the brain, but in the spinal cord the order is reversed and white matter forms the cortex enclosing the grey matter. Grey matter is concerned with co-ordination.

grid 1. An electrode for controlling the flow of electrons in a thermionic valve. 2. A system of overhead or underground cables through which a country's electricity is distributed.

Grignard reagent *See* Grignard synthesis.

Grignard synthesis The chemical process in which an alkylmagnesium halide or an arylmagnesium halide (a Grignard reagent), e.g. ethylmagnesium chloride, C_2H_5MgCl, is brought to react with another compound, e.g. water, H_2O:
$$C_2H_5MgCl + H_2O \rightarrow CH_3—CH_3 + Mg(OH)Cl$$
In this reaction ethane, $CH_3—CH_3$, is produced. Grignard reagents are very reactive compounds; as well as water, they react with many compounds, both inorganic and organic, and they are therefore important in synthesis. A Grignard reagent can be formed by the reaction between an alkyl halide or aryl halide and magnesium, e.g.
$$C_2H_5Cl + Mg \rightarrow C_2H_5MgCl$$
The process is carried out in dry ethoxyethane (ether).

gristle *See* cartilage.

ground state *See* Bohr atom.

growth The increase in cellular mass (protoplasmic mass) of an organism. Growth is characterised either by an increase in the size of an individual cell or by the division of a cell into two daughter cells. However, cellular growth processes usually involve a cyclic repetition of an increase in the cell mass followed by a division of this mass into daughter cells.

growth hormone 1. In plants: *See* auxin, gibberellin *and* cytokinin. 2. (GH; somatotrophic hormone, somatotrophin) In animals, a polypeptide hormone promoting growth of the whole body and secreted by the anterior lobe of the pituitary gland. An excessive production of GH may cause gigantism, whereas a deficiency of the hormone may cause dwarfism.

growth ring *See* annual ring.

guanine A nitrogenous organic base (purine base) which is part of the genetic code in DNA molecules, where it pairs with cytosine. It is also a part of RNA.

guard cell One of a pair of cells surrounding a stoma. The turgidity of these cells controls the opening and closing of the stoma. *See* Fig. 223.

gullet 1. *See* oesophagus. 2. In Protozoa, a cavity through which food enters the body.

gum *See* gum arabic *and* mucilage.

gum arabic A water-soluble powder obtained from certain varieties of acacia. It is used in pharmacy and in making adhesives.

gun-cotton *See* cellulose trinitrate.

gunpowder An explosive mixture of 75% potassium nitrate, KNO_3, 15% powdered charcoal, C, and 10% powdered sulphur, S. When it is ignited, several chemical reactions occur. These produce different gases, the volumes of which are many times that of the original mixture; thus if gunpowder is ignited in a confined space an explosion occurs.

gut *See* alimentary canal.

gutta-percha A natural product obtained from the sap of certain tropical plants. Gutta-percha has the same constitution as natural rubber, but is a *trans*-isomer, whereas natural rubber is a *cis*-isomer. Gutta-percha is a tough, hornlike substance.

guttation The secretion of drops of water from the surface of leaves through structures called hydathodes. This occurs in conditions of high humidity in which transpiration proceeds slowly, and may be due to root pressure. The drops of water are often mistaken for dew.

gymnosperm A member of a major division of the plant kingdom, comprising the conifers and their allies. Gymnosperms are distinguished from angiosperms by having their ovules exposed and not protected by an ovary. *See also* Spermatophyta.

gynaeceum *See* gynoecium.

gynoecium (pistil, gynaeceum) The collective name for the female reproductive organs (carpels) of a plant.

gypsum $CaSO_4 \cdot 2H_2O$ (calcium sulphate-2-water) A naturally occurring, hydrated mineral, slightly soluble in water. It loses three-quarters of its water of crystallisation on heating to around 120°C and becomes plaster of Paris (calcium sulphate-$\frac{1}{2}$-water), $CaSO_4 \cdot \frac{1}{2}H_2O$:
$$CaSO_4 \cdot 2H_2O \rightarrow CaSO_4 \cdot \tfrac{1}{2}H_2O + 1\tfrac{1}{2}H_2O$$
Gypsum is one of the principal causes of permanent hardness in water. It is used in making plaster of Paris, paper and paint. *See also* calcium sulphate.

H

H Chemical symbol for hydrogen.

Haber–Bosch process A process used in the industrial manufacture of ammonia, NH_3. Nitrogen, N_2, obtained from the air and hydrogen, H_2, obtained from methane, CH_4, are dried, mixed and reacted together at 250 atmospheres and 500°C in the presence of a catalyst, forming ammonia:
$$N_2 + 3H_2 \rightleftharpoons 2NH_3$$
Approximately 15% of the gases are converted to liquid ammonia under these conditions. Unreacted nitrogen and hydrogen are recycled for further reaction. *See* Fig. 89.

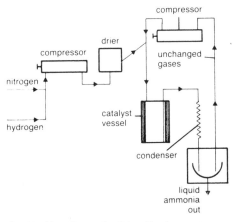

Fig. 89 Flow diagram for Haber–Bosch process

habitat The locality and environment in which an organism lives, e.g. a pond or sea-shore.

haem A red, respiratory pigment containing iron and forming the non-protein part of haemoglobin in the blood.

haematite A red or black mineral consisting of iron(III) oxide, Fe_2O_3. It is the most important ore of iron.

haematology The study of blood and its diseases.

haematuria The presence of blood in urine.

haemocyanin A blue respiratory pigment containing copper, found in the blood of several invertebrates such as molluscs and crustaceans.

haemoglobin (Hb) In the blood of vertebrates and a few invertebrates, a red respiratory pigment consisting of haem and the protein globin, found in the red blood cells (erythrocytes). It combines readily with oxygen forming an unstable compound, oxyhaemoglobin, HbO_2, which carries oxygen from the lungs to all body tissues. *See also* blood corpuscles *and* blood plasma.

haemolysin A substance causing haemolysis. Haemolysins can be formed by the spleen.

haemolysis The process in which haemoglobin is lost from the erythrocytes. This may be caused by a haemolysin, e.g. the venom of certain snakes. Haemolysis takes place when red blood corpuscles are placed in a salt solution weaker than 0,9% (0.9%); they swell and burst.

haemophilia A hereditary disease occurring almost exclusively in males. The blood lacks a substance needed in the formation of thrombokinase, which is active in blood clotting. This means that the blood clots very slowly. The disease is transmitted only by females and is due to a genetic defect. *See also* blood clotting.

haemorrhage Severe loss of blood.

haemorrhoids (piles) Enlarged veins in the wall of the region of the anus. This causes severe pain and is often accompanied by bleeding. The disorder is treated either medically or surgically.

hahnium *See* transuranic elements.

hail (hailstones) Frozen water droplets formed when raindrops are blown upwards by strong wind into regions where the temperature is below 0°C. When hailstones fall they grow, because water from the warm moist air through which they pass condenses on them and then freezes. Hailstones are usually less than 10 mm in diameter, but occasionally they can grow to the size of a cricket ball.

hailstones *See* hail.

hair 1. A thread-like outgrowth from plant epidermis. *See also* root hair. **2.** In mammals, a thread-like outgrowth from the skin. A hair consists of dead cells which have grown from a hair follicle. Hairs are strengthened by keratin.

hair cells *See* Corti's organ.

hair follicle A cavity (sheath) in the epidermis and dermis, holding a hair by surrounding its root and shaft. Muscles, capable of erecting the hair, are attached to the hair follicles, and ducts from sebaceous glands open into them.

hair hygrometer *See* hygrometer.

hairspring *See* clock.

half-cell An electrode dipping into a solution of an electrolyte. The electrode potential of a half-cell is measured against a hydrogen electrode.

half-life The time taken for one-half of any quantity of radioactive nuclei to decay (disintegrate). It is also the time taken for the radioactivity of the substance to be reduced to half its initial value. Half-lives can vary from millions of years to fractions of seconds. *Examples:* the half-life of the $^{232}_{90}$ thorium isotope is 1.40×10^{10} years (1.40×10^{10} years), whereas that of the $^{212}_{84}$ polonium isotope is 3×10^{-7} seconds. *See* Fig. 90; *see also* radiocarbon dating.

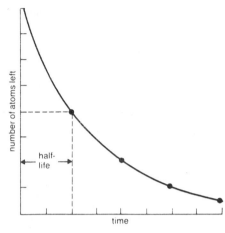

Fig. 90 Half-life

half-wave rectification Rectification of an alternating current, to produce a unidirectional current, by allowing only alternate half cycles of the alternating current to flow. This can be done using a diode. *See* Fig. 91; *compare* full-wave rectification.

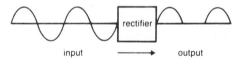

Fig. 91 Half-wave rectification

halide Any chemical compound containing halogen atoms or ions, e.g. bromoethane (ethyl bromide), C_2H_5Br, and sodium chloride, NaCl.

halite *See* rock salt.

Halley's comet A famous comet orbiting the Sun once every 76,1 years (76.1 years). It is named after the English astronomer, Edmund Halley (1656–1742), who stated that the comets observed in 1531, 1607 and 1682 were one and the same, returning at intervals of between 75 and 76 years. He therefore predicted that it would return in 1758 and so it did. It was last observed in 1910 and is therefore due to reappear in 1986.

hallucinogen A drug, such as cannabis, which causes hallucinations.

hallux (pl. halluces) In tetrapod vertebrates, the innermost digit of the hind foot. It is the big toe in humans and consists of two phalanges.

halo A luminous ring appearing round the Sun or Moon. It is caused by reflection and refraction of light by ice crystals in the Earth's atmosphere. *See also* corona.

halogen The general name for the non-metals (elements) found in Group VIIA of the Periodic Table of the Elements. These are: fluorine, chlorine, bromine, iodine and astatine. They are electronegative elements and will combine directly with most metals to form salts. *Example:* when heated, chlorine, Cl_2, combines with iron, Fe, to form iron(III) chloride, $FeCl_3$:
$3Cl_2 + 2Fe \rightarrow 2FeCl_3$
See also redox potential.

halogenation The process in which a halogen is introduced into an organic molecule which is either aliphatic or cyclic. *Examples:* methane, CH_4, reacts with chlorine, Cl_2, when heated or struck by ultraviolet radiation to form chloromethane (methyl chloride), CH_3Cl:
$CH_4 + Cl_2 \rightarrow CH_3Cl + HCl$
Benzene, C_6H_6, reacts with bromine, Br_2, in the presence of a catalyst to form bromobenzene, C_6H_5Br:
$C_6H_6 + Br_2 \rightarrow C_6H_5Br + HBr$
Halogenation can also take place using compounds containing a halogen atom. When chlorine is introduced, the process is termed chlorination; using bromine or a bromine compound it is called bromination, etc.

halteres The modified hind wings of Diptera, which are reduced to small, knotted projections. They are believed to assist in maintaining balance during flight.

hammer The common name for the malleus of the ear.

haploid Describes the nucleus of a cell which has a single set of unpaired chromosomes. The gametes of an organism have haploid nuclei.

haploid number The total number of chromosomes in a haploid nucleus. The haploid number is half the diploid number.

hapteron *See* holdfast.

haptonasty A plant movement in response to contact. *Example:* when leaf hairs of the insectivorous plant Venus's fly-trap are touched, the leaves close as a result of the rapid loss of water from cells along the midrib.

haptotropism *See* thigmotropism.

hardness *See* Mohs' scale of hardness.

hard solder *See* solder.

hardware The mechanical and electronic components of a computer. *Compare* software.

hard water Water which will not readily form a lather with soap. This hardness is due to the presence of dissolved calcium, magnesium and iron compounds in the water. When soap is added to hard water, an insoluble scum is

formed, consisting of these metals and the acid radicals of the fatty acids present in the soap. Once these metals are precipitated as scum, the addition of more soap will produce a lather. There are two types of hardness: temporary hardness and permanent hardness. Temporary hardness is due to the presence of soluble hydrogencarbonate compounds of calcium, magnesium and iron. These compounds enter the water when rain-water, containing dissolved carbon dioxide, CO_2, passes over rock containing carbonates, e.g. calcium carbonate, $CaCO_3$:
$CaCO_3 + CO_2 + H_2O \rightarrow Ca(HCO_3)_2$
Temporary hardness can be removed by heating the water because insoluble carbonates of the metals are thus formed, e.g.
$Ca(HCO_3)_2 \rightarrow CaCO_3 + CO_2 + H_2O$
Permanent hardness is largely caused by the presence of calcium sulphate, $CaSO_4$, which enters the water when rain-water passes over rock containing calcium sulphate. It cannot be removed by heating, but may be removed by the addition of chemicals such as sodium carbonate, Na_2CO_3:
$CaSO_4 + Na_2CO_3 \rightarrow CaCO_3 + Na_2SO_4$
Both types of hardness can be removed by passing the hard water through a layer of ion exchange material or by distillation. Water without hardness is called soft water. See also stalactite and stalagmite.

hardwood See deciduous.

Hardy–Weinberg equilibrium A term based on the Hardy–Weinberg principle describing the particular state of a population in which the gene frequencies are stable.

Hare's apparatus An apparatus used to compare or measure the relative densities of liquids. It consists of two vertical glass tubes connected by a T-piece fitted with a clip. The two tubes are dipped into the liquids A and B whose relative densities are to be compared. When air is removed through the T-piece, the liquids are pushed up the tubes by atmospheric pressure. The relative densities are inversely proportional to the heights to which the liquids rise in the tubes. If the relative density of one of the liquids is known, the relative density of the other may be calculated from the expression:
$$\frac{\rho_1}{\rho_2} = \frac{h_2}{h_1}$$
where ρ_1 and ρ_2 are the relative densities of the two liquids and h_1 and h_2 the heights risen. See Fig. 92.

harmonic A component of a complex vibration (note) which is a simple multiple of the frequency of the fundamental.

hashish (hash) See cannabis.

haustorium (pl. haustoria) An outgrowth of the stem, root or hypha of certain parasitic plants. The haustorium penetrates cells and absorbs food from the host.

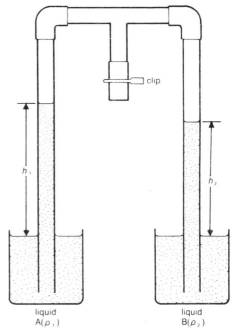

Fig. 92 Hare's apparatus

Haversian canal A small canal passing through compact bone in vertebrates. Haversian canals are interconnected and branch out throughout the bone. They carry blood vessels, nerves and lymphatic vessels. See also osteoclast.

Hb See haemoglobin.

HbO₂ See haemoglobin.

H-bomb Hydrogen bomb. See fusion.

He Chemical symbol for helium.

hearing-aid A device worn by people with impaired hearing. It consists of three parts: a microphone, a battery-powered amplifier and an earpiece. Modern hearing-aids can be made so small that they can be hidden in the frame of a pair of spectacles.

heart A hollow, muscular organ pumping blood through the circulatory system. The human heart has four chambers and contracts rhythmically. The four chambers are: the right and left atria (auricles) and the right and left ventricles. The atria receive blood and the ventricles pump blood. See Fig. 93; see also diastole.

heartwood (duramen) Xylem tissue found in the central region of a tree trunk. It consists of dead cells and no longer conducts water; the vessels are usually blocked by ingrowths. Heartwood has lignin in the cell walls and is rather hard, giving the tree mechanical

'heat'

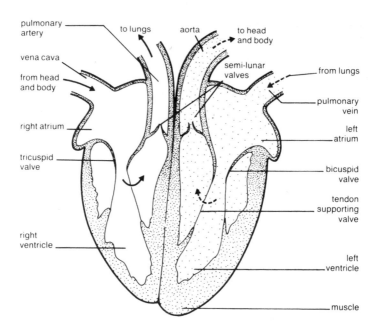

Fig. 93 Heart (longitudinal section)

support. Often heartwood is deeply coloured; this is caused by deposition of tannins and resins. *See also* sapwood.

'heat' *See* oestrus.

heat capacity Symbol: C; unit: joule per kelvin ($J K^{-1}$). The heat required to raise the temperature of a substance (solid, liquid or gas) by one kelvin. The heat capacity of an object can be calculated by multiplying its mass by its specific heat capacity.

heat energy Symbol: Q; unit: joule (J). Energy possessed by a system because of the kinetic energy of its atoms or molecules due to translation, rotation and vibration. Heat energy is transferred by conduction, convection and radiation.

heat engine A device which converts heat energy into mechanical energy, e.g. the diesel engine.

heat exchanger A device for transferring heat from one fluid to another without the fluids being in direct contact with one another, e.g. radiators in cars (used for cooling) and domestic radiators (used for heating). Heat exchangers are made of metal and usually have coiled tubes or fins to increase the surface area between the fluids.

heat of combustion The heat change observed when one mole of a compound is completely burnt in oxygen; reactants and products being at 25°C (298 K) and one atmosphere, e.g. the heat of combustion of graphite is
$\Delta H = -393$ kJ mol^{-1}:
$C_{(graphite)} + O_{2(gas)} \rightarrow CO_{2(gas)}$
ΔH is negative when heat is given out.

heat of formation The heat change observed when one mole of a compound is formed from its elements at 25°C (298 K) and one atmosphere, e.g. the heat of formation of water is
$\Delta H = -286$ kJ mol^{-1}:
$H_{2(gas)} + \frac{1}{2}O_{2(gas)} \rightarrow H_2O_{(liquid)}$
ΔH is negative when heat is given out.

heat of neutralisation The heat change observed when one mole of an acid or base is fully neutralised. The overall reaction is:
$H^+ + OH^- \rightarrow H_2O$
If the acid and base are strong (fully ionised), $\Delta H = -57$ kJ mol^{-1}
ΔH is negative when heat is given out. For weak acids or bases the heat of neutralisation may be more or less than 57 kJmol^{-1} because molecules of acid or base have to ionise before neutralisation can take place and the energy change involved in this will affect the total heat change.

heat of reaction *See* enthalpy.

heat of solution The heat change observed when one mole of a solute dissolves in a sufficiently large volume of solvent so that further dilution (addition of solvent) causes no more heat change.

heat radiation *See* infra-red radiation.

heat transfer *See* conduction, convection *and* radiation.

heavy hydrogen *See* deuterium.

heavy water D$_2$O (deuterium oxide) Water containing the heavy hydrogen isotope, deuterium, D, in place of the normal hydrogen isotope, H. Heavy water is present in natural water to a small extent and may be extracted from it by electrolysis, because the normal hydrogen isotope is discharged before deuterium, leaving the remaining water richer in heavy water. Heavy water has a boiling-point of 101,4°C (101.4°C), a freezing-point of 3,8°C (3.8°C) and a relative density of 1,11 (1.11) at 4°C. It is used as a moderator in certain types of nuclear reactors. If only one of the hydrogen atoms in water is replaced by deuterium, semi-heavy water, HDO, is obtained.

hecto- Symbol: h. A prefix meaning one hundred; e.g. one hectometre (hm) is equal to 100 metres.

helices *See* helix.

helio- Prefix meaning sun (from Greek, helios).

heliotropism *See* phototropism.

helium An element with the symbol He; atomic number 2; relative atomic mass 4,00 (4.00); state, gas. Helium was first detected on the Sun (1868) by spectrographic analysis of sunlight, but later, in 1895, it was also detected on Earth. It is a very inert gas found in the atmosphere in very small amounts. Helium belongs to the noble gases in the Periodic Table of the Elements. It is used to provide an inert atmosphere for welding operations (like argon), to fill airships and balloons, and in some fluorescent lamps. A mixture of 80% helium and 20% oxygen is used as an artificial atmosphere for divers and others working under pressure.

helix (pl. helices) A spiral shape. *Example:* the DNA molecule is in the form of a double helix.

Helmholtz coils A pair of identical coils whose centres are separated by a distance equal to their radius. They are connected in series and produce a uniform magnetic field between them when a current flows through them. They are normally used in conjunction with a cathode-ray deflection tube to demonstrate the electromagnetic deflection of an electron beam. *See* Fig. 165.

Hemiptera A large order of insects, commonly called bugs. Characteristics include two pairs of wings, piercing and sucking mouth-parts and incomplete metamorphosis. *See also* lac insect.

Henle's loop The hairpin-shaped part of the uriniferous tubule lying in the medulla of the vertebrate kidney. *See* Fig. 149.

henry Symbol: H. The SI unit of mutual inductance and self inductance. A conductor has an inductance of 1 henry if an e.m.f. of 1 volt is induced in it when the current flowing through it changes at a rate of 1 ampere per second.

Henry's law The mass of a gas dissolved by a given volume of solvent at a constant temperature is directly proportional to the pressure of the gas. This law only holds for gases which do not react with the solvent.

heparin A complex organic compound usually extracted from the liver or lung and used in the treatment of thrombosis, in which it acts as an anticoagulant. *See also* mast cell.

hepatic Pertaining to, or associated with, the liver.

Hepaticae A class of bryophytes comprising the liverworts. They are small, leaf-like plants with a prostrate plant body or a creeping central stem, growing in damp conditions and attached to the substratum by unicellular rhizoids. These are hair-like structures functioning as roots. Antheridia and archegonia (the sex organs) are grouped in various places. Male gametes are motile by flagella. After fertilisation has occurred a spore-containing capsule develops. Each spore is capable of giving rise to a protonema, from which new plants develop.

hepatic artery In vertebrates, the artery which conveys blood to the liver. *See* Fig. 27.

hepatic portal system In vertebrates, the part of the circulatory system which conveys blood to the liver. It consists of the hepatic artery and hepatic portal vein.

hepatic portal vein In vertebrates, the vein which conveys blood, rich in the products of digestion, from the alimentary canal to the liver. *See* Fig. 27.

hepatic vein In vertebrates, the vein which conveys blood from the liver to the heart. *See* Fig. 27.

hepatitis Inflammation of the liver. It is generally caused by a virus, but can also be caused by poisonous chemicals such as alcohol.

heptavalent Having a valency of seven.

herb A seed plant with a green, non-woody stem. Parts of it may be used in food, medicine, etc. A herb is usually smaller than a shrub.

herbaceous Pertaining to or being a herb.

herbarium A collection of dried plants, or the place where such a collection is stored.

herbicide A chemical which may destroy harmful plants, e.g. a selective weed-killer.

herbivore A plant-eating animal.

herbivorous Pertaining to animals that are plant-eaters.

hereditary Describes a characteristic which is passed on from parent to offspring, e.g. the colour of the eyes or hair. *See also* gene.

heredity

heredity The tendency to transmit characteristics of living organisms from parent to offspring.

hermaphrodite (bisexual) **1.** A plant which has both stamens and carpels in the same flower. **2.** An animal which has both male and female reproductive organs.

heroin A narcotic drug derived from morphine and with similar effects. It is more potent than morphine and acts more quickly. Heroin, like morphine, produces addiction, but to a much greater extent; it is therefore seldom prescribed by doctors.

hertz Symbol: Hz. The SI unit of frequency, defined as the frequency of a periodic process with a period of one second. This unit replaces the cycle per second.

Hess's law A law stating that the heat evolved or absorbed during a chemical change is the same whether the change is brought about in one stage or through several intermediate stages. *Example:* the heat change in a chemical reaction in which A becomes C is the same whether the reaction takes place in one stage as A → C or in two stages as A → B → C.

hetero- Prefix meaning other or different.

heterocyclic compound An organic compound containing one or more atoms other than carbon as part of the ring structure, e.g. pyridine, C_5H_5N, which is a six-membered ring with nitrogen as the heteroatom. *See also* homocyclic compound *and* carbocyclic compound.

heterodont Having teeth of various shapes with various functions. This is characteristic of human beings, dogs, rabbits, etc. *Compare* homodont.

heterogeneous Describes a system (e.g. a mixture) which is not uniform in composition. *Compare* homogeneous.

heterogeneous reaction A chemical reaction taking place in more than one phase. *Example:* the reaction:
$CaCO_{3(s)} \rightleftharpoons CaO_{(s)} + CO_{2(g)}$
is a heterogeneous reaction because the mentioned substances are not all in the same state. *Compare* homogeneous reaction.

heterosexuality Having a sexual relationship with a person of the opposite sex. *Compare* homosexuality.

heterosis *See* hybrid vigour.

heterotroph(e) An organism using organic material as its main source of carbon. *Compare* autotroph(e).

heterozygote An individual which has two different alleles in the two corresponding loci on a pair of homologous chromosomes. *Compare* homozygote.

heterozygous Pertaining to a heterozygote.

hexa- Prefix meaning six.

hexacyanoferrate(II) *See* cyanoferrates.
hexacyanoferrate(III) *See* cyanoferrates.
hexadecanoate *See* palmitate.
hexadecanoic acid *See* palmitic acid.
hexamethylenetetramine *See* hexamine.

hexamine $(CH_2)_6N_4$ (hexamethylenetetramine, urotropine) A colourless crystalline solid which is soluble in water. It can be prepared when methanal (formaldehyde), HCHO, reacts with ammonia, NH_3:
$6HCHO + 4NH_3 \rightarrow (CH_2)_6N_4 + 6H_2O$
In a molecule of hexamine, the four nitrogen atoms and the six CH_2- groups are at the four corners and along the six edges of a regular tetrahedron respectively. Hexamine is used in the plastics industry and in the manufacture of certain explosives. Formerly it was extensively used as an antiseptic in the treatment of infections of the urinary tract, but it has now been almost completely replaced by sulphonamides or antibiotics.

Hexapoda *See* Insecta.

hexavalent Having a valency of six.

hexose A carbohydrate which is a monosaccharide with six carbon atoms in its molecule, e.g. glucose, fructose, galactose, $C_6H_{12}O_6$. *See also* sugar.

Hg Chemical symbol for mercury.

hibernation A period of dormancy in animals during winter.

hiccup An involuntary contraction of the diaphragm followed by the sudden closure of the glottis which produces the characteristic sound. One common cause of hiccups is eating too quickly.

hidrosis The formation and excretion of sweat (perspiration).

hi-fi *See* high-fidelity.

high blood pressure *See* hypertension.

high-energy bond The bond between the phosphate groups in ATP. The chemical bond energy stored here can readily be liberated when necessary. It is more than the energy contained in the bond between the adenosine-ribose part of ATP and one of the phosphate groups.

high-fidelity (hi-fi) Describes equipment for high quality sound-reproduction.

high tension (HT) *See* high voltage.

high voltage Any voltage in excess of 550 volts.

hilum (pl. hila) **1.** The scar on the testa of a seed marking its former point of attachment to the ovary wall. **2.** A depression in the surface of an organ marking the entrance or exit of blood vessels, nerves, etc.

hindbrain (rhombencephalon) In vertebrates, the part of the brain consisting of the medulla oblongata, the cerebellum and the pons. The hindbrain receives stimuli from the ears and skin.

hinge joint A joint in the body allowing movement in one plane, e.g. the elbow and the knee joints.

hip-girdle *See* pelvic girdle.

hip-joint The ball and socket joint providing articulation between the femur and the pelvic girdle.

histamine A hormone which is a cyclic organic base, present in all body tissues and produced from the cyclic amino acid histidine by removal of its carboxyl group. Histamine is normally only released in small amounts. It stimulates the production of gastric juice and enlarges capillaries to increase the flow of blood. Sometimes large amounts of histamine may be released; this happens in shock and allergy. To counteract the release of such large amounts, antihistamine drugs are used. *See also* mast cell.

histidine A cyclic amino acid from which histamine is produced.

histology The study of animal and plant tissues, including their structure and function.

histone One of a group of basic proteins which contain large amounts of the amino acids arginine and lysine and are attached to DNA molecules in the chromosomes. Histones are believed to have a gene-regulating function.

Hofmann voltameter An apparatus for collecting gases evolved during electrolysis. It consists of two inverted burettes fitted with platinum electrodes at their bases. The two burettes are connected by a cross tube carrying an upright tube and a reservoir. If water, containing a little acid, is placed in the reservoir and a current is passed through the solution, volumes of hydrogen and oxygen in the ratio 2:1 can be collected. *See* Fig. 94.

holdfast (hapteron) In algae, a flattened sucker or disc situated at the end of the stalk or thallus and capable of attaching the plant to almost any solid support.

hole, electric *See* semiconductor.

hologram *See* holography.

holography A photographic technique for making three-dimensional pictures (holograms). The light source is usually a laser beam and a possible arrangement is shown in Fig. 95 (a). Here the laser beam illuminates the object to be holographed and a photographic film is placed so that it receives reflected light from the object and direct light from the source, which has been reflected from a mirror. Interference between these two beams of light produces an interference pattern which is recorded on the film. To project an image, laser light is sent through the developed film and two images are formed as shown in Fig. 95 (b). Holography may in the future be used to produce three-dimensional television.

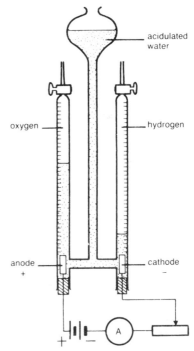

Fig. 94 Hofmann voltameter

holophytic Feeding like green plants, i.e. manufacturing food by photosynthesis. *Compare* holozoic.

holozoic Feeding like animals, i.e. food material is broken down during digestion into simpler substances which are then absorbed by the body or oxidised to obtain energy. *Compare* holophytic.

Home Galaxy *See* Galaxy, the.

homeopathy (homoeopathy) A system of medicine founded by Samuel Hahnemann and based on the principle that a disease, characterised by definite symptoms, can best be treated by administering minute quantities of a certain drug which, in a healthy person, would produce the symptoms of the disease involved if taken in a larger dose.

homeostasis The maintaining of the internal environment of an animal. *Examples:* the skin helps to regulate blood temperature; the lungs regulate the carbon dioxide concentration.

homo- Prefix meaning the same.

homocyclic compound An organic, cyclic compound containing atoms of only one element, usually carbon, in the ring structure, e.g. benzene, C_6H_6. *Compare* heterocyclic compound.

homodont Having teeth which are all the same shape and have the same function. This is characteristic of frogs. *Compare* heterodont.

homoeopathy

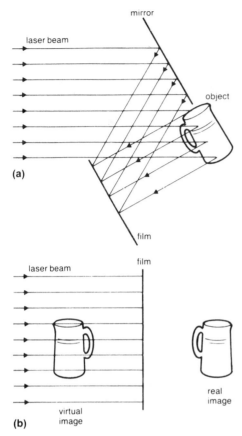

Fig. 95 Holography

homoeopathy See homeopathy.
homogeneous Describes a system (e.g. a mixture) which is uniform in composition. Compare heterogeneous.
homogeneous reaction A chemical reaction taking place in one phase only. Example: the reaction:

$$N_{2(g)} + 3H_{2(g)} \rightleftharpoons 2NH_{3(g)}$$

is a homogeneous reaction because all the substances are in the same state (the gas state). Compare heterogeneous reaction.
homoiothermic Warm-blooded.
homologous Describes animal or plant parts which have the same structure and origin. Example: the proboscis of a butterfly and the mouthparts of a cockroach, although very different in appearance, have the same function and are of similar origin. Compare analogous.
homologous chromosomes See chromosome.

homologous series A series of related organic compounds in which the formula of each member (homologue) differs from those of its preceding and succeeding members by a —CH_2— group. Example: the alkanes form a homologous series.
homologue See homologous series.
Homo sapiens The specific name of modern man.
homosexuality Having a sexual relationship with a person of the same sex. Compare heterosexuality.
homozygote An individual which has identical alleles in corresponding loci on a pair of homologous chromosomes. Compare heterozygote.
homozygous Pertaining to a homozygote.
honey-bee The common hive-bee, Apis mellifera, a four-winged, stinging, social insect which collects nectar and pollen and produces wax and honey. There are three anatomically and functionally different types of honey-bee. These are: fertile females called queens; sterile females called workers; and fertile males called drones. See also Hymenoptera.
honey guides (nectar guides) A series of markings on the petals of flowers, thought to guide insects to nectar.
Hooke's law A law stating that the extension of an elastic material is directly proportional to the force producing the extension provided the elastic limit is not exceeded.
hookworm A disease caused by the roundworm Ancylostoma duodenale. The larvae enter the bloodstream through direct penetration of the skin, usually of the legs and feet. From the blood the larvae penetrate the wall of the intestine and fasten onto the wall by means of hooks. Eggs from adult hookworms are passed out with faeces and may then infect food. Drugs are available in the treatment of the disease. Symptoms include severe anaemia and tiredness. See also Nematoda.
Hope's apparatus An apparatus to demonstrate that the maximum density of water is at 4°C. It consists of an upright metal cylinder fitted with two thermometers, A and B, and having a container surrounding its middle into which a freezing-mixture is placed. The thermometers measure the temperature of water in the cylinder. After some time, thermometer A reads 0°C whilst thermometer B reads a constant temperature of 4°C. See Fig. 96.
horizon A layer of soil possessing a more or less well-defined character, e.g. topsoil.
hormone In animals, a substance produced in small amounts by a ductless gland or by special tissues, which has a specific function. Hormones are secreted from the place of production directly into the blood and then

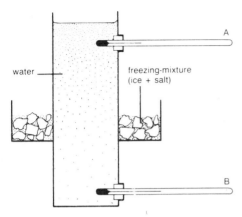

Fig. 96 Hope's apparatus

transported to another place where they exhibit their effect. Examples include adrenalin(e), insulin and thyroxine. In plants, hormones are secreted internally by certain cells. *See also* cytokinin.

horn silver *See* silver.

hornwort *See* Bryophyta.

horticulture The science or practice of garden cultivation.

host 1. An organism which is infected by a parasite. The parasite spends part or the whole of its life in or on the host, deriving nourishment and protection from it. **2.** An organism receiving grafted or transplanted tissue.

hot-wire ammeter An ammeter whose operation depends on the thermal expansion of a wire or strip. When an electric current passes through it, the wire or strip increases in length. This movement causes a pointer to move over a circular scale. The hot-wire ammeter can be used to measure alternating currents as well as direct currents. *See also* moving iron instrument.

HT High tension. *See* high voltage.

humerus The bone located between the shoulder and the elbow in the forelimb of vertebrates. *See* Fig. 162.

humid Damp and warm.

humidity The measure of the amount of water vapour in air, expressed either as absolute humidity or relative humidity. *See also* hygrometer.

humour A body fluid, e.g. the aqueous humour in the eye.

humus A complex, colloidal and usually dark material derived from the decomposition of animal and plant material in soil. Humus can hold water and contains mineral nutrients. It is therefore essential for plant growth. Humus is found in topsoil.

Hund rule *See* orbital.

hurricane A wind of force 12 (121 km/h or more) on the Beaufort scale.

husk The outer layer (coating) of some seeds.

hybrid The offspring of genetically unlike parent organisms.

hybrid vigour (heterosis) The tendency of a hybrid to exhibit increased growth, fertility, etc.

hydathode A specialised water pore (gland) from which water is secreted. Hydathodes occur on the edges or tips of the leaves of many plants. *See also* guttation.

Hydra An aquatic animal belonging to the phylum Coelenterata.

hydrargyrum The Latin name for mercury.

hydrate A salt containing water of crystallisation, e.g. copper(II) sulphate-5-water (hydrated copper(II) sulphate), $CuSO_4 \cdot 5H_2O$.

hydrated Describes a salt containing water of crystallisation. *Compare* anhydrous.

hydration A type of solvation in which molecules or (more frequently) ions are attached to water molecules, most often by coordinate bonds or by some kind of electrostatic force. *Example:* in aqueous solution the proton, H^+, is generally written as H_3O^+, indicating that it is associated with a water molecule. *See also* oxonium ion.

hydraulic press (Bramah's press) A machine whose operation depends on the transmission of pressure by a fluid. It consists of two connected cylinders filled with a liquid. The cylinders have different diameters and each is fitted with a piston. When a small force is applied at P to piston C (Fig. 97), a large force is obtained at the bottom of piston D. *Example:* if the downward force at C is 100 N and the cross-sectional area of C is 40 cm² then the pressure on the liquid is
$100 \div 40 = 2,5 \, N \, cm^{-2}$ ($2.5 \, N \, cm^{-2}$)
If D has a cross-sectional area of 2000 cm², then the upward force on D is equal to
$2,5 \times 2000 = 5000 \, N$ ($2.5 \times 2000 = 5000 \, N$)
The large upward force can be used to compress bales of material, steel plates, etc. Similar hydraulic systems working on this principle are hydraulic jacks and the hydraulic brakes of motor vehicles. *See* Pascal's principle.

hydrazine N_2H_4 A colourless, water-soluble liquid which may be prepared by treating ammonia, NH_3, with sodium chlorate(I) (sodium hypochlorite), NaOCl:
$2NH_3 + NaOCl \rightarrow N_2H_4 + NaCl + H_2O$
Hydrazine is a weaker base than ammonia. It is a very powerful reducing agent. With acids, hydrazine forms salts, e.g.
$N_2H_4 + HCl \rightarrow N_2H_4HCl$ ($N_2H_5{}^+Cl^-$)

hydric

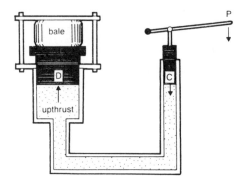

Fig. 97 Hydraulic press

Hydrazine has been used as a rocket propellant (fuel). Its reaction with hydrogen peroxide, H_2O_2:
$N_2H_4 + 2H_2O_2 \rightarrow N_2 + 4H_2O$
which is catalysed by copper(II) ions, Cu^{2+}, is strongly exothermic and is accompanied by a large increase in volume. Hydrazine is used to remove dissolved oxygen from boiler-feed water, e.g. in power-stations.

hydric Of or containing hydrogen. The term is usually applied to alcohols. *Examples:* methanol, CH_3OH, is a monohydric alcohol; ethane-1,2-diol (glycol), CH_2OH—CH_2OH, is a dihydric alcohol; and propane-1,2,3-triol (glycerol), CH_2OH—$CHOH$—CH_2OH, is a trihydric or polyhydric alcohol.

hydride A chemical compound formed from hydrogen and another element. However, strictly only those compounds in which hydrogen is present as the hydride ion, H^-, should be called hydrides. With sufficiently electropositive metals, hydrogen forms ionic hydrides, e.g. lithium hydride, LiH, containing the negative hydrogen ion (the hydride ion) H^-, which in electrolysis of such hydrides is liberated as hydrogen at the anode. Non-metals form covalent hydrides, e.g. methane, CH_4, and ammonia, NH_3. Most hydrides are covalent.

hydride ion H^- The negative hydrogen ion, formed in hydrides in which hydrogen combines with strongly electropositive metals, e.g. lithium hydride, LiH.

hydro- Prefix meaning water.

hydrocarbon An organic compound containing hydrogen and carbon only. Hydrocarbons are either aliphatic molecules, e.g. alkanes, alkenes and alkynes, or cyclic molecules, e.g. benzene, C_6H_6.

hydrochloric acid HCl A strong acid obtained when the gas hydrogen chloride, HCl, is dissolved in water. Salts of hydrochloric acid are called chlorides. It is widely used in the chemical industry. An old name for HCl is spirits of salt.

hydrocyanic acid *See* hydrogen cyanide.

hydroelectric power-station A power-station in which the kinetic energy of falling water is converted into electrical energy. The falling water is made to drive water-turbines which in turn drive generators.

hydrofluoric acid HF A rather weak acid obtained when hydrogen fluoride is dissolved in water. Salts of hydrofluoric acid are called fluorides. The acid attacks most metals and glass and is therefore stored in polythene containers. Hydrofluoric acid is very poisonous, penetrating the skin and destroying both tissue and bones.

hydrogen An element with the symbol H; atomic number 1; relative atomic mass 1,01 (1.01); state, gas. It is the lightest of all the elements, and is a colourless and highly flammable substance. Hydrogen may be produced industrially in the Bosch process. In the laboratory it can be produced by treating zinc, Zn, with dilute sulphuric acid, H_2SO_4:
$Zn + H_2SO_4 \rightarrow H_2 + ZnSO_4$
Hydrogen is a reducing agent used in the manufacture of ammonia, NH_3, in the Haber–Bosch process, in petroleum refining, in welding, etc.

hydrogenation The process in which hydrogen is added to another molecule, usually organic. This may involve the addition of hydrogen to an unsaturated molecule. *Example:* ethene, C_2H_4, and hydrogen form ethane, C_2H_6:
$C_2H_4 + H_2 \rightarrow C_2H_6$ or
$CH_2{=}CH_2 + H_2 \rightarrow CH_3$—$CH_3$
In the manufacture of margarine, unsaturated oils are converted into saturated fats using a catalytic hydrogenation.

hydrogen bomb (H-bomb) *See* fusion.

hydrogen bond A weak bond which forms between covalently bonded hydrogen atoms and electronegative atoms. This is the result of the attraction of positive and negative charges between molecules; i.e. hydrogen bonds are electrostatic. In water, H_2O, there are hydrogen bonds between the electronegative oxygen atoms and the electropositive hydrogen atoms in different molecules. *See* Fig. 98.

Hydrogen bonds are able to exist because of the small size of the hydrogen atom. This allows negative poles of other molecules to approach close to the positive hydrogen nucleus, thus giving rise to an attraction. DNA molecules consisting of double helices are united with their bases over a hydrogen atom. Hydrogen bonds can be intramolecular as well as intermolecular. A hydrogen bond is written A—H---B, where A—H is the normal

hydrogen peroxide

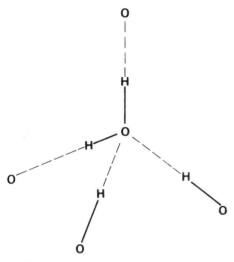

Fig. 98 Hydrogen bond

chemical bond and H---B is the hydrogen bond.

hydrogen chloride HCl A colourless gas with a pungent smell. It can be prepared by reacting sulphuric acid, H_2SO_4, and sodium chloride, NaCl:
$H_2SO_4 + 2NaCl \rightarrow 2HCl + Na_2SO_4$
or by heating hydrogen gas, H_2, with chlorine gas, Cl_2:
$H_2 + Cl_2 \rightarrow 2HCl$
Hydrogen chloride is very soluble in water, producing the strong acid hydrochloric acid, HCl.

hydrogen cyanide HCN An extremely poisonous, colourless gas evolved when cyanides are treated with acids. Hydrogen cyanide condenses at about 26°C to a liquid. In aqueous solution, hydrogen cyanide is a very weak acid called hydrocyanic acid or prussic acid, HCN, whose salts are called cyanides. Hydrogen cyanide is manufactured industrially from methane, CH_4, and ammonia, NH_3, in a catalytic oxidation:
$2CH_4 + 2NH_3 + 3O_2 \rightarrow 2HCN + 6H_2O$
Hydrogen cyanide is used in the production of certain synthetic chemicals and as a fumigant. Both hydrogen cyanide and hydrocyanic acid have a smell similar to the smell of bitter almonds, which contain cyanides.

hydrogen electrode An electrode consisting of a thin piece of platinum foil coated with a special form of precipitated platinum called platinum black and immersed in an acid solution with a hydrogen ion concentration of 1 mole per dm^3 at 25°C. Hydrogen gas at a pressure of 1 atmosphere is bubbled over the platinum and some of it is adsorbed on the platinum black. The electrode potential obtained in this way is arbitrarily taken as zero and the electrode is then called a standard hydrogen electrode against which other electrode potentials may be measured. See also Appendix.

hydrogen fluoride HF A colourless gas which can be prepared by treating calcium fluoride, CaF_2, with sulphuric acid, H_2SO_4:
$CaF_2 + H_2SO_4 \rightarrow 2HF + CaSO_4$
When dissolved in water, hydrogen fluoride forms the rather weak acid hydrofluoric acid, HF. Like this acid, moist hydrogen fluoride attacks glass and it is therefore used in the etching of glass.

hydrogen ion H^+ The positive ion of hydrogen, obtained when a hydrogen atom loses its electron; i.e. a proton is a hydrogen ion. Hydrogen ions are formed in aqueous solutions of acids. However, the hydrogen ion always combines with water molecules to form the oxonium ion (hydroxonium ion), H_3O^+:
$H^+ + H_2O \rightleftharpoons H_3O^+$
In metallic hydrides, hydrogen is negatively charged, H^-, and is therefore liberated at the anode during electrolysis of such hydrides. See also hydride ion.

hydrogen ion concentration The concentration of hydrogen ions in a solution expressed as molar concentration. The hydrogen ion concentration of a solution determines its acidity. In dealing with hydrogen ion concentration, the pH scale is useful. pH represents 'power of hydrogen ions'. This scale runs from 0 to 14, where 0 represents a very acidic solution and 14 a very alkaline solution. 7 on the pH scale is the neutral point. In calculating the pH of a solution, the expression
$pH = -\log_{10}[H^+]$
may be used, where $[H^+]$ is the hydrogen ion concentration. From this expression it is seen that
$[H^+] = 10^{-pH}$
Examples: the pH of 0,01 M (0.01 M) HCl is 2, as
$pH = -\log_{10}[10^{-2}] = 2;$
in a solution having a pH of 3,
$[H^+] = 10^{-3} = 0,001 \ (0.001)$
Since $[H^+]$ in pure water is 10^{-7} gram-ions per litre, the pH of pure water is $-\log_{10}10^{-7} = 7$, i.e. pure water contains one ten-millionth of a gram of hydrogen ions per litre. If acid is added to pure water, its pH will decrease because the hydrogen ion concentration increases. If alkali is added to pure water, its pH will increase because the hydrogen ion concentration decreases. See Fig. 99.

hydrogen peroxide H_2O_2 A colourless, rather viscous liquid which can act both as an oxidising agent and as a reducing agent, depending on the substance it reacts with.

hydrogen spectrum

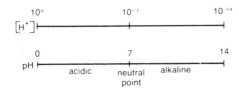

Fig. 99 Hydrogen ion concentration

Examples: when reacting with lead(II) sulphide, PbS, it is an oxidising agent forming lead(II) sulphate, $PbSO_4$, and water:
$PbS + 4H_2O_2 \rightarrow PbSO_4 + 4H_2O$
When reacting with lead(IV) oxide, PbO_2, it is a reducing agent forming lead(II) oxide, PbO, oxygen, O_2, and water:
$PbO_2 + H_2O_2 \rightarrow PbO + O_2 + H_2O$
Hydrogen peroxide may be prepared by treating barium peroxide, BaO_2, with sulphuric acid, H_2SO_4:
$BaO_2 + H_2SO_4 \rightarrow H_2O_2 + BaSO_4$
H_2O_2 is a very weak acid whose salts are called peroxides. Hydrogen peroxide gives off oxygen readily and is therefore used as a disinfectant, as a bleaching agent in the textile and paper industries and as a rocket propellant.

hydrogen spectrum *See* Balmer series.

hydrogen sulphide H_2S (sulphuretted hydrogen) A colourless, poisonous, flammable gas which smells of rotten eggs. In aqueous solution it is a weak acid whose salts are called sulphides. Hydrogen sulphide can be prepared by the action of strong, dilute acid on certain sulphides, e.g. iron(II) sulphide, FeS:
$FeS + H_2SO_4 \rightarrow H_2S + FeSO_4$
It is used as a reducing agent and in chemical analysis.

hydrolase An enzyme catalysing hydrolysis.

hydrolysis The reaction of a compound with water in which the hydroxyl group of the water remains intact. *Example:* the carbonate ion, CO_3^{2-} is hydrolysed in water to produce an alkaline solution:
$CO_3^{2-} + H_2O \rightleftharpoons HCO_3^- + OH^-$

hydrometer An instrument for measuring the densities or relative densities of liquids. A common type of hydrometer consists of a weighted glass bulb with a long graduated stem. The hydrometer floats vertically in the liquid under test and the length of stem immersed is proportional to the density of the liquid. Some hydrometers are fitted with a thermometer. *See* Fig. 100.

hydronasty A plant movement in response to changes in atmospheric humidity. *Example:* the flowers of the dandelion open and close in response to such changes.

hydronium ion *See* oxonium ion.

hydrophilic Having an affinity for water.

Fig. 100 Hydrometer

hydrophobia *See* rabies.

hydrophobic Having no affinity for water.

hydrophyte A plant adapted to live in or on water, or in very wet places. Its adaptations may include very little cuticle on the epidermis, no stomata, a reduced root system, very little xylem and supporting tissue, and leaves offering a large surface area to water, etc. Examples of hydrophytes are water-lily, duckweed and water-lettuce. *Compare* mesophyte *and* xerophyte.

hydroponics The cultivation of plants without soil. The plant roots are dipped into water containing nutrients or are rooted in sand irrigated with a nutrient solution. *See also* water cultures.

hydrosphere The water on the Earth's surface.

hydrotropism Plant growth in response to the stimulus of water.

hydrous Containing water.

hydroxide A chemical compound containing the hydroxide ion, OH^-, e.g. sodium hydroxide, NaOH.

hydroxide ion (hydroxyl ion) The negative ion OH^-, as found in alkalis. It takes up a hydrogen ion (proton), H^+, to form water and is therefore a strong base:
$OH^- + H^+ \rightleftharpoons H_2O$.

hydroxonium ion *See* oxonium ion.

2-hydroxybenzenecarboxylic acid *See* 2-hydroxybenzoic acid.

2-hydroxybenzoate (salicylate) A salt of 2-hydroxybenzoic acid (salicylic acid), $C_6H_4(OH)(COOH)$.

2-hydroxybenzoic acid $C_6H_4(OH)(COOH)$ (salicylic acid 2-hydroxybenzenecarboxylic acid) An aromatic, colourless, crystalline acid. The ester formed between 2-hydroxybenzoic acid (salicylic acid) and ethanoic acid (acetic acid) is called 2-ethanoyloxybenzoic acid (acetylsalicylic acid) or aspirin.

hydroxyl group A monovalent, functional group consisting of an oxygen atom and a hydrogen atom: —OH. It is found in many organic compounds such as alcohols and carbohydrates. In hydroxides it exists as the hydroxide ion, OH^-.

hydroxyl ion *See* hydroxide ion.

2-hydroxypropanoic acid *See* lactic acid.

hygro- Prefix meaning moisture.

hygrometer An instrument for measuring the humidity of air and other gases. Two common types are the wet and dry bulb hygrometer and the hair hygrometer. The wet and dry bulb hygrometer consists of two thermometers placed side by side. The bulb of one is in air and the bulb of the other is wrapped in muslin which dips into water. The wet bulb is cooled due to the cooling effect of evaporation, the amount of which depends on the humidity of the surrounding air. The relative humidity of the air can be obtained from tables of dry bulb temperatures and wet bulb depressions. The hair hygrometer's action depends on the changing length of a human hair caused by changes in the humidity of the air. The hair is mounted in a glass-fronted metal case and the movements caused by changes in the hair's length are transmitted to a pointer moving over a scale. This instrument measures relative humidity directly. *See* Fig. 101.

hygroscopic Describing a substance which absorbs moisture from the air, e.g. silica gel and anhydrous calcium chloride, $CaCl_2$. If a substance absorbs so much moisture that it dissolves in it, it is said to be deliquescent; i.e. all deliquescent substances are hygroscopic, but hygroscopic substances are not always deliquescent.

hygrotropism Plant growth in response to the stimuli of moisture and humidity.

hymen In mammals, a fold of mucous membrane partly blocking the external opening of the vagina.

Hymenoptera A large order of insects whose members include ants, bees and wasps. Characteristics include two almost equal pairs of membranous wings and complete metamorphosis. Certain members of this order exhibit different types of social organisation.

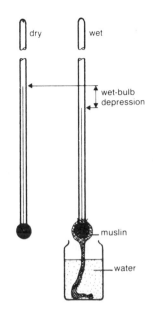

Fig. 101 Wet and dry bulb hygrometer

hyper- Prefix meaning excessive or above normal.

hyperglycaemia A condition in which the amount of glucose in the blood is too high. This occurs in diabetes and is caused by lack of insulin. *Compare* hypoglycaemia.

hypermetropia (hyperopia, long sight) A defect of the eye in which the eyeball is too short and consequently images of near objects are brought to a focus behind the retina. This is corrected by convex spectacle lenses. *See* Fig. 102; *compare* myopia; *see also* presbyopia.

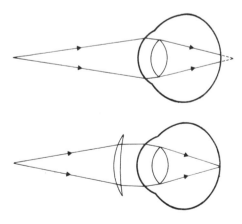

Fig. 102 Hypermetropia (long sight)

hyperopia *See* hypermetropia.

hypersonic Describes a speed in excess of Mach 5.

hypertension (high blood pressure) A blood pressure so high that a doctor decides to treat the patient. Hypertension imposes a strain on the heart and gradually damages the small blood vessels in the kidneys and eyes. This condition may lead to early death from a stroke or a coronary thrombosis. Symptoms, if any, are headache, dizziness, misty vision and breathlessness. Only in about 10% of all cases of hypertension is a specific cause found. In the remainder, the high blood pressure is described as essential, i.e. the cause can not be determined. *See also* sphygmomanometer.

hypertonic Describes a solution which has a higher osmotic pressure than that of a standard solution.

hypha (pl. hyphae) A filament of the mycelium of a fungus. *See* Fig. 219.

'hypo' The common name for sodium thiosulphate, $Na_2S_2O_3 \cdot 5H_2O$. It is used in fixing in photography.

hypo- Prefix meaning under or below normal.

hypochlorite *See* chlorate(I).

hypochlorous acid *See* chloric(I) acid.

hypocotyl The part of the embryonic stem below the attachment of the cotyledons. *Compare* epicotyl.

hypogeal 1. Describes germination in which the cotyledons remain below ground. 2. Describes insects living under the surface of the soil. *Compare* epigeal.

hypoglycaemia A condition in which the concentration of glucose in the blood is too low. This may occur in diabetes if the patient is given too high a dosage of insulin for the amount of carbohydrate ingested. *Compare* hyperglycaemia.

hypogynous Describes a flower in which the ovary is situated superior to (above) the sepals, petals and stamens, e.g. hibiscus. *See* Fig. 103; *see also* epigynous, perigynous.

hypotension Low blood pressure. This is seldom a symptom of ill health. On the contrary, low blood pressure may be a health asset since it has been proved that people having a low blood pressure seldom suffer from arteriosclerosis.

hypothalamus The region at the base of the brain beneath the floor of the third ventricle, extending downwards to contribute to the pituitary gland. It is the region of the brain controlling the autonomic nervous system. *See* Fig. 16.

hypothesis A suggestion made as a basis for further investigation.

hypotonic Describes a solution which has a lower osmotic pressure than that of a standard solution.

hypsometer An instrument used to determine the upper fixed point of a thermometer. In a hypsometer, the thermometer is surrounded by steam. A manometer attached to the hypsometer measures the pressure of the steam. Corrections must be made if the steam pressure is different from atmospheric pressure and if the atmospheric pressure is different from 101 325 Pa. *See* Fig. 104.

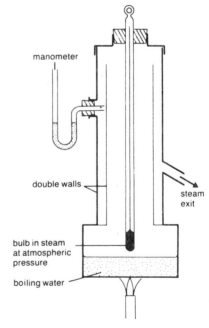

Fig. 104 Hypsometer

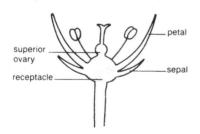

Fig. 103 Hypogynous flower in section

hypophysis *See* pituitary gland.

I

I Chemical symbol for iodine.
IAA *See* indoleacetic acid.
-ic 1. Suffix meaning that a multivalent element in a compound is in the higher of two possible valency states. *Examples:* ferric compounds contain the iron(III) ion, Fe^{3+}, while ferrous compounds contain the iron(II) ion, Fe^{2+}; cupric compounds contain the copper(II) ion, Cu^{2+}, while cuprous compounds contain the copper(I) ion, Cu^+. 2. Suffix meaning that the central atom of an acid is in the higher of two possible valency states. *Example:* In sulphuric acid, H_2SO_4, the sulphur has a valency of six; in sulphurous acid, H_2SO_3, it has a valency of four. *See also* -ous.
ice Water, H_2O, in the solid state, formed when water freezes at 0°C. The density of ice is less than that of water, so ice floats on water. When water freezes to form ice, it expands.
ice-age A period of time during which vast areas of land which are not normally covered by ice are buried beneath ice sheets. Four ice-ages occurred during the Pleistocene epoch. During these periods the northern part of North America and Europe lay under thick ice sheets and the weight of the ice caused the Earth's crust to be depressed.
iceberg A huge, floating mass of ice detached from a glacier when it reaches the sea. Only about one-tenth of an iceberg is above the surface of the water.
Iceland spar *See* calcite.
ice-point The temperature at which water and ice are in equilibrium at standard pressure, i.e. the temperature of pure melting ice at standard pressure. *See also* triple point *and* fixed point.
icterus *See* jaundice.
-ide Suffix used for most binary compounds, e.g. hydrogen sulphide, H_2S, a compound of hydrogen and sulphur only. It is also used for halides, hydroxides and cyanides, although these may not be binary.
ideal gas (perfect gas) A gas which, if it existed, would obey the gas laws exactly. It would consist of molecules of negligible or zero volume, the forces of attraction between the molecules also being negligible or zero. A real gas begins to behave like an ideal gas at extremely low pressures and temperatures.
identical twins *See* monozygotic twins.
igneous rock Rock of volcanic origin, formed when molten lava solidifies, e.g. basalt and granite.
ignition temperature The minimum temperature to which a substance must be heated before ignition can take place.

ileum The third part of the small intestine, immediately preceding the large intestine. In humans it is about 4 m long. In it digestion of fats and proteins is completed and carbohydrates are acted upon. Most absorption takes place in the ileum. *See* Fig. 49.
iliac Pertaining to the ilium, e.g. the iliac artery.
ilium The dorsal bone of the pelvic girdle in tetrapod vertebrates.
illuminance (illumination) Symbol: E; unit: lux (lx) or lumen per square metre ($lm\,m^{-2}$). The luminous flux per unit area.
illumination *See* illuminance.
ilmenite *See* titanium.
image The optical appearance of an object produced by a lens, mirror or optical system. An image may be real or virtual. A real image is formed by converging rays of light and can be produced on a screen. A virtual image is formed by diverging rays of light and cannot be produced on a screen.
imago A fully developed adult insect; the final stage of both complete and incomplete metamorphosis.
immersion heater An electric heater designed to heat liquids in which it is immersed. It is often controlled by a thermostat.
immersion objective *See* oil immersion.
immersion oil *See* oil immersion.
immiscible Describes substances which do not mix, e.g. oil and water.
immune Describes an organism capable of resisting a specific infection. *See* immunity.
immunisation The process of introducing certain substances (vaccines) into the body of an animal in order to increase its ability to produce antibodies. This is generally done by injection of the vaccine into body tissue. *See also* inoculation.
immunity The ability of an organism to resist infection. In animals this is done by the production of antibodies; in plants it is effected by certain structural characteristics, e.g. a cuticle that prevents the entry of pathogens.
immunology The study of resistance to infection in animals.
impedance Symbol: Z. The total resistance to the flow of an alternating current in a circuit. It is equal to the ohmic resistance plus the reactance.
imperfect fungus A fungus which has no known sexual cycle. *Compare* perfect fungus.
implantation (nidation) The attachment of a fertilised mammalian egg to the lining of the uterus. In humans, this takes place approximately four days after the release of the egg from the ovary and precedes the formation of the placenta.
implosion The inward collapse of the walls of an evacuated vessel (e.g. cathode-ray tube) caused by the external pressure.

impulse 1. A signal (message) conducted along a nerve-fibre in the form of a travelling wave of electrical nature. The impulse is not an electric current, but a wave of chemical activity in the nerve-fibre, accompanied by a change in potential of the fibre. Energy for the impulse is not provided by the stimulus causing the impulse, but is provided along the course of the nerve-fibre. After an impulse has passed a nerve-fibre, there is a very short period (refractory), lasting a few milliseconds, in which the nerve-fibre recovers its irritability. During this recovery period the nerve-fibre will not transmit another impulse. **2.** A force acting for a very short time. The magnitude of the impulse is the product of the force and the time for which it acts, provided the force is constant.

incandescence The emission of visible light from a material which is strongly heated, e.g. the bright, white light from calcium oxide, CaO, (limelight), and the light from the filament of an electric light bulb.

incidence, angle of See angle of incidence.

incident ray A ray of light striking a reflecting or refracting surface. See Figs. 4 and 5.

incised With deeply notched margin or edge. See Fig. 111.

incisor tooth A tooth with short cutting edges used for cutting, gnawing, tearing and holding. The incisor teeth are situated next to the canine teeth at the front of the mouth. There are eight in the jaws of adult human beings. An incisor tooth has one root.

inclination, angle of See angle of dip.

inclined plane A sloping, rigid plane used to reduce the force required to raise a load.

inclinometer See dip circle.

incomplete metamorphosis See metamorphosis.

incubation 1. The period of time between the initial infection of a host by a pathogen and the appearance of the first symptoms. **2.** The process of hatching eggs using natural or artificial heat.

incubator 1. An apparatus for hatching eggs. **2.** An apparatus for developing micro-organisms. **3.** A transparent container in which children born prematurely may be isolated and kept under controlled conditions.

incus (anvil) The middle of the three small bones in the middle ear. It receives vibrations from the malleus, transmitting them to the stapes. See Fig. 55.

indehiscent Describes fruits which do not burst open to liberate seeds. Compare dehiscent.

independent assortment, law of See Mendel's laws.

index (pl. indices) A very small piece of steel which is used to indicate the maximum temperature recorded by a maximum thermometer and the minimum temperature recorded by a minimum thermometer. See also maximum and minimum thermometer.

Indian hemp See cannabis.

indicator A substance (usually organic) which, by a colour change, indicates the presence of another substance. Example: a starch solution gives a blue colour with an iodine solution. pH indicators are usually dyes which have different colours at different pH values. See also end-point and Appendix.

indifferent equilibrium See neutral equilibrium.

indigenous Describes animals and plants which belong naturally to a locality, i.e. they have not been introduced to it.

indoleacetic acid (IAA) A natural auxin which is a very common growth-regulating hormone.

induced current An electric current produced by the induced e.m.f. during electromagnetic induction.

induced e.m.f. See electromagnetic induction.

inductance Symbol: L or M; unit: henry (H). The measure of the ability of an electrical conductor to resist a change in current flow, as a result of an induced e.m.f. opposing the current flow. See also mutual inductance and self inductance.

induction A change in the electric or magnetic state in a body caused (induced) by its nearness to a field. See also electromagnetic induction, electrostatic induction and magnetic flux density.

induction coil A device for producing high, intermittent voltages from a low voltage supply. It consists of a laminated iron core around which is wound a primary coil with a few turns of thick wire, surrounded by a secondary coil with a large number of turns of thin wire. A low current applied to the primary coil is arranged to be rapidly interrupted by a mechanism similar to that used in the electric bell. This produces a high, intermittent, induced e.m.f. in the secondary coil. This e.m.f. may be sufficient to cause a spark to jump across the adjustable spark gap. An induction coil is used in a car's ignition system. It produces a sufficiently high voltage to cause a spark to jump the sparking-plug gap. The spark ignites the petrol-air mixture. See Fig. 105; see also distributor and sparking-plug.

inductive reactance The apparent resistance of a coil to a flow of alternating current. It is set up by the self-inductance of the coil and is additional to the coil's ohmic resistance.

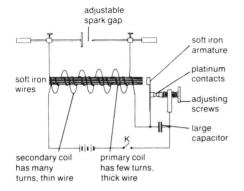

Fig. 105 Induction coil

inductor A coil connected in a circuit to introduce inductance. *See also* choke.

indusium (pl. indusia) *See* sorus.

industrial melanism An evolutionary response to the darkening of tree trunks, walls and other surfaces by soot deposits and other industrial waste, resulting in the organism being better camouflaged. *See* melanism.

inert Describes a substance which is very unreactive or does not react at all with other substances, e.g. nitrogen and the noble gases.

inert gas *See* noble gas.

inertia The tendency of a body to remain at rest or, if moving, to continue its uniform motion in a straight line.

infection 1. The invasion of an animal host by pathogens 2. A disease caused by pathogens.

inferior Describes a structure which is situated below another. *Compare* superior.

inferior vena cava *See* vena cava.

inflammable *See* flammable.

inflammation The reaction of body tissue to injury such as heat, cold, blows, cuts, infection, etc. The part involved becomes red and swollen because of wide open blood vessels and the reaction is usually painful. An inflammation may be treated with compresses or antibiotics.

inflorescence 1. A flowering shoot. 2. The arrangement of flowers on a main stalk. *See* Figs. 38 and 179. 3. The act of blossoming.

influenza An acute infectious disease, caused by a virus, which may occur in epidemics. Symptoms are fever, headache and inflammation of the mucous membranes of the nose and throat. No drugs exist for treating influenza. The patient usually recovers after a week. Sometimes influenza may develop into pneumonia because the virus lowers the body's defences against attack from bacteria.

infra-red radiation (radiant heat, heat radiation, IR radiation) The part of the electromagnetic spectrum which has a wavelength greater than the red end of the visible spectrum. It is emitted from hot bodies.

infundibulum A funnel-shaped stalk connecting the pituitary gland to the floor of the diencephalon.

infusorial earth *See* kieselguhr.

Ingenhousz's apparatus An apparatus for comparing the thermal conductivities of a number of different metals. It consists of a rectangular tank in the top of which rods of equal length and cross-section of the different metals are fitted. The rods are covered with paraffin wax and the tank is filled with hot water. The rate of melting of the wax indicates the rate of conduction of heat through each rod.

ingestion The taking in of food by swallowing or engulfing.

ingot A mass of cast metal, e.g. a gold ingot.

inhalation *See* inspiration.

inheritance Those characteristics of an offspring which are derived from its parent(s).

inhibitor A negative catalyst, i.e. a substance which slows down the rate of a chemical reaction. A well-known example is the anti-knock inhibitor tetraethyl-lead(IV), $(C_2H_5)_4Pb$, which is added to petrol to prevent premature ignition of the petrol-air mixture in the cylinder of a petrol engine.

inner ear (membranous labyrinth) The essential part of the organ of hearing. It is complex in shape, consisting of a membranous labyrinth which fits inside a bony labyrinth of the same shape. The membranous labyrinth contains a fluid called endolymph, whilst the bony labyrinth is filled with perilymph. It is in this part of the ear that sound vibrations are converted into nerve-impulses which are conveyed to the brain via the auditory nerve. *See* Fig. 55.

innervation The distribution of nerves in an organism.

inoculation The introduction of certain micro-organisms into body tissue (usually by injection) in order to produce antibodies. This leads to immunity against infectious diseases such as smallpox, diphtheria, etc. *See also* vaccination.

inorganic Pertaining to compounds which do not contain carbon. A few carbon compounds are also described as inorganic; they include carbon monoxide, CO, carbon dioxide, CO_2, carbonic acid, H_2CO_3, and the carbonates. *Compare* organic.

inorganic chemistry The study of the chemistry of inorganic compounds.

Insecta (Hexapoda) A large class of arthropods (insects). Characteristics include a body divided into three distinct parts (head, thorax, abdomen), three pairs of legs on the thorax, which may also bear one or two pairs of wings, and one pair of antennae on the head. Insects undergo complete or incomplete metamorphosis during their life histories.

insecticide A man-made chemical used to combat insects, especially those which are crop pests. Examples are DDT and BHC.

Insectivora An order of small, placental mammals feeding on insects. Members include hedgehogs, shrews and moles.

insectivorous Pertaining to animals and plants which feed on insects.

insemination The introduction of semen into the uterus.

insertion The method or place of attachment of an organ. In the case of a muscle attached to a bone, contraction of the muscle causes the bone to move. *Compare* origin.

insoluble Describes a substance which does not dissolve in a certain solvent at a given temperature. Insoluble is a relative term; normally a tiny amount of the substance will dissolve. However, this is very much less than the amount of a soluble substance which dissolves in the same amount of solute at the same temperature, and it is negligible for most purposes.

inspiration (inhalation) The process of taking in air or water into the respiratory organs.

instar A stage in the larval development of an insect between one ecdysis and the next.

instinct An inborn pattern of behaviour in animals in response to certain stimuli, e.g. nest-building in birds.

insulate To prevent the passage of energy such as electricity, heat and sound by placing non-conductors (insulators) in its path.

insulator A material which is a non-conductor of energy such as electricity, heat or sound.

insulin A polypeptide hormone secreted by the β cells of the islets of Langerhans in the pancreas. Its production is stimulated by a high concentration of glucose in the blood. It enables the oxidation of glucose to proceed in the tissues and promotes the formation of glycogen in the liver and muscles, stimulates the production of fat from glucose and stimulates protein synthesis. Insulin has the opposite effect to that of glucagon. *See also* diabetes mellitus *and* glycogenesis.

insulinase An enzyme found mainly in the tissue of the liver and kidney. Insulinase takes part in the breaking down of insulin.

integrated circuit An electronic circuit consisting of many components incorporated into a small piece of a semiconductor, usually a silicon chip. Integrated circuits are used in computers, radios, and anywhere else where a small, reliable circuit is needed.

integument A covering layer, e.g. the testa of a seed, the skin of an animal.

intelligence The degree of ability to solve unfamiliar problems and to adjust to new situations. The ability to behave intelligently may be inherited from parents, but it must be developed by learning.

inter- Prefix meaning between or among.

intercostal muscle A muscle situated in the space between two ribs. The intercostal muscles are concerned with the mechanism of breathing. *See* Fig. 118.

intercostal space The space between two ribs.

interference 1. The production of alternate light and dark bands (fringes) on a screen by the interaction of two sets of light waves which have the same frequency, phase and amplitude. Where two crests (tops) or two troughs (dips) of the wave motions coincide, constructive interference occurs causing the formation of a bright fringe. Where a crest coincides with a trough, destructive interference occurs causing the formation of a dark fringe. Coloured fringes are formed if the light is not monochromatic. This effect is seen in soap bubbles and in thin oil films. **2.** The production of beats by the interaction of two sound waves of almost the same frequency. Where two crests or two troughs of the wave motions coincide, constructive interference occurs and the resultant amplitude is high. Where a crest and trough coincide, destructive interference occurs and the resultant amplitude is almost zero. **3.** A disturbance which may be heard as background noise in a radio receiver when it receives unwanted signals such as those caused by atmospherics. *See also* static.

interferon A protein produced in an animal host in response to a viral infection. Interferon inhibits viral replication. A short time after a viral attack, interferon is produced in the infected animal in a considerable amount, thus rapidly decreasing the replication of the virus. Later on, antibodies also appear in the blood of the animal: i.e. interferon appears to be a primary response to viral infections. In humans, three or more types of interferon may be produced. The treatment of human virus diseases with interferon has proved difficult because at present it is almost impossible to obtain enough interferon in a pure form for use in therapy. However, great efforts are being made in research to develop substances that, when injected, cause the body itself to increase its production of interferon. Attempts are also being made to stimulate certain bacteria to produce interferon; this method may make large-scale production of interferon possible in the very near future.

internal combustion engine An engine in which the energy from the combustion of a gas mixed with air in a cylinder is converted to mechanical energy. The gas used is normally vaporised oil (diesel) or petrol.

internal resistance The electrical resistance of any device, e.g. a cell or a generator, which produces an e.m.f.

internal respiration See respiration.

internode 1. The part of a plant stem between two successive nodes. See Fig. 86. **2.** The part of a nerve-fibre between two adjacent nodes of Ranvier. See Fig. 150.

interoceptor A receptor which senses stimuli from inside the body of an animal, e.g. receptors sensing stimuli in the intestine and other internal organs. Compare exteroceptor.

interphase The resting phase of a cell between two mitoses or between the first and second meiotic division.

interstitial cell One of a mass of cells situated between the sperm-producing tubules of the testis or between the ovarian follicles. Those of the testis produce the male hormone testosterone. See also seminiferous tubule.

intertidal zone The area along a beach or coastal plain which lies between the highest and lowest tides and which is therefore alternately submerged by the sea and exposed.

intestinal juice (succus entericus) Digestive juice secreted by the glands in the small intestine. It contains various digestive enzymes acting on proteins, carbohydrates and fats.

intestine The part of the alimentary canal beginning with the duodenum and ending with the rectum.

intra- Prefix meaning inside or within.

intussusception In plant cells, the increase in surface area of cell walls caused by the incorporation of new cellulose into already existing cellulose. Compare apposition.

inulin A soluble carbohydrate which is a polysaccharide, $(C_6H_{10}O_5)_n$, consisting of units of fructose, $C_6H_{12}O_6$. Inulin is a storage polysaccharide which occurs in the roots and tubers of many Compositae such as dahlia and dandelion. It is also found in small amounts in many monocotyledons.

in vacuo In a vacuum.

Invar® An alloy of iron (63,8%; 63.8%), nickel (36,0%; 36.0%) and carbon (0,2%; 0.2%). It has a low coefficient of thermal expansion and for this reason is used to make balance-wheels and pendulums.

inverse square law A law stating that the intensity of an effect at a given point is inversely proportional to the square of its distance from the cause of the effect (source). This law can be applied to intensity of illumination, gravitational force, sound, etc.

invertase (saccharase, sucrase) An enzyme hydrolysing saccharose (sucrose), $C_{12}H_{22}O_{11}$, to glucose, $C_6H_{12}O_6$, and fructose, $C_6H_{12}O_6$. Invertase is found in intestinal juice. See also invert sugar.

Invertebrata A general term (not used as a scientific classification) for all animals which do not possess backbones, i.e. are not Vertebrata.

invert sugar A mixture of equal parts of glucose, $C_6H_{12}O_6$, and fructose, $C_6H_{12}O_6$, obtained when saccharose (sucrose), $C_{12}H_{22}O_{11}$, is hydrolysed using a little dilute acid:
$C_{12}H_{22}O_{11} + H_2O \rightarrow C_6H_{12}O_6 + C_6H_{12}O_6$
See also invertase.

in vitro Describes biological processes carried out experimentally in isolation from a living organism, e.g. in a test-tube.

in vivo Describes biological processes occurring in a living organism.

involuntary action A bodily action, such as digestion, the dilation of the pupil of the eye, and the heartbeat, which is not under the control of the will. The majority of involuntary actions are controlled by the autonomic nervous system. Compare voluntary action; see also reflex.

involuntary muscle (non-striated muscle, smooth muscle) A muscle composed of the simplest type of muscle tissue, forming the muscular walls of internal organs and tubular structures such as the stomach, intestines and blood vessels. Usually involuntary muscles contract slowly and rhythmically. Note: the cardiac muscle is an involuntary muscle, but it is striated. Compare skeletal muscles.

iodide A salt of hydriodic acid, HI. This acid is of no chemical importance.

iodine An element with the symbol I; atomic number 53; relative atomic mass 126,90 (126.90); state, solid. It is a rather electronegative non-metal belonging to the halogens. Iodine occurs in the form of iodides in small concentrations in sea-water and certain species of seaweed, and as sodium iodate(V), $NaIO_3$, in the sodium nitrate deposits in Chile. Iodine is manufactured industrially from sodium iodate(V) which is treated with sodium hydrogensulphite, $NaHSO_3$, in aqueous solution:
$2NaIO_3 + 5NaHSO_3 \rightarrow I_2 + 3NaHSO_4 + 2Na_2SO_4 + H_2O$
Iodine is a black, volatile solid which sublimes to produce violet vapours. It is slightly soluble in water but readily soluble in alcohol (ethanol) and in an aqueous solution of

ion

potassium iodide, forming the tri-iodide ion, I_3^-:
$I_2 + KI \rightarrow I_3^- + K^+$
Iodine solution produces a blue-black colour in the presence of starch and is used as a test for starch. It is used in the manufacture of photographic films and in the preparation of tincture of iodine. A lack of iodine in the diet may cause goitre.

ion An electrically charged atom or group of atoms: e.g. the hydrogen ion, H^+, the chloride ion, Cl^-, the ammonium ion, NH_4^+, and the sulphate ion, SO_4^{2-}. Positive ions are called cations because they are attracted by the negative electrode, the cathode, during electrolysis. Negative ions are called anions because they are attracted by the positive electrode, the anode, during electrolysis.

ion exchange See demineralisation.

ionic bond (electrovalent bond, polar bond) A type of chemical bond between atoms which form ions when electrons are transferred from one atom to another. The ions are held together by electrostatic attraction. *Example:* in sodium chloride, NaCl, a sodium atom has lost one electron to a chlorine atom thus forming Na^+ and Cl^-. The sodium atom which originally had the electron configuration 2,8,1 now has the stable configuration 2,8, and the chlorine atom which originally had the electron configuration 2,8,7, now has the stable configuration 2,8,8. In electrovalent compounds (ionic compounds) there are no distinct molecules present; they consist of aggregates of oppositely charged ions. Electrovalent compounds usually have high melting-points and boiling-points and are solids. *Compare* covalent bond; *see also* octet.

ionic compound See electrovalent compound *and* ionic bond.

ionic crystal (electrovalent crystal) A type of crystal in which ions of two or more elements occupy lattice points. The ions are held together by the electric forces (ionic bonds) between negative and positive ions. A typical example is sodium chloride, NaCl. Ionic crystals usually have high melting-points because much thermal energy is required to break the strong ionic bonds.

ionic equation A chemical equation in which some or all of the reactants or products are written as ions, e.g.
$CaCO_3 + 2H^+ \rightarrow Ca^{2+} + CO_2 + H_2O$
In this ionic equation, calcium carbonate, $CaCO_3$, reacts with hydrogen ions, H^+, (from an acid). It dissolves to form free calcium ions, Ca^{2+}, carbon dioxide, CO_2, and water.

ionic strength Symbol: I. A quantity expressing the effects of interionic attraction in an electrolyte. Interionic attraction causes electrolytes to deviate from ideal behaviour, i.e. the ions tend to become surrounded by ions of opposite charge, thus not moving as free ions. The ionic strength of an electrolyte can be calculated from the equation:
$I = \frac{1}{2}\Sigma(c_i Z_i^2)$
i.e. I is equal to half the sum (Σ, sigma) of the molar concentration, c_i, of every kind of ion multiplied by the square of the number of electrical charges on the ion, Z_i^2. *Examples:* in a 0,1 M (0.1 M) aqueous solution of potassium chloride, KCl, which is almost completely ionised,
$I = \frac{1}{2}(0,1 \times 1^2 + 0,1 \times 1^2) = 0,1$
$(I = \frac{1}{2}(0.1 \times 1^2 + 0.1 \times 1^2) = 0.1)$
i.e. I equals the molar concentration for a diluted, univalent, strong electrolyte. In an aqueous solution which is both 0,1 M (0.1 M) KCl and 0,5 M (0.5 M) HCl:
$I = \frac{1}{2}(0,1 \times 1^2 + 0,6 \times 1^2 + 0,5 \times 1^2) = 0,6$
$(I = \frac{1}{2}(0.1 \times 1^2 + 0.6 \times 1^2 + 0.5 \times 1^2) = 0.6)$
0,1 M (0.1 M) $AlCl_3$ has $I = 0,6$ (0.6);
0,2 M (0.2 M) $CaSO_4 + 0,2$ M (0.2 M) HCl has $I = 1,0$ (1.0), etc.

ionisation The process in which free ions are formed, e.g. when ionically bonded compounds dissolve in water or other solvents, or when a gas is ionised by an ionising radiation. Covalent compounds may ionise when dissolved in water or other solvents. *Example:* hydrogen chloride, HCl, dissolves in water to form hydrochloric acid, HCl. Ions form when an atom, a molecule or a group of atoms lose or gain one or more electrons.

ionisation chamber See Geiger–Müller counter.

ionisation energy Symbol: I. The minimum energy required to withdraw an electron completely from a neutral gaseous atom or ion against the attraction of the nuclear charge. It is measured in kilojoules per mole, $kJ\,mol^{-1}$, or in electronvolts, eV. Values given in $kJ\,mol^{-1}$ can be converted into eV by dividing by 96,48 (96.48). The energy required to remove the first electron from a copper atom
$Cu \rightarrow Cu^+ + e$
is 745 $kJ\,mol^{-1}$. This is called the first ionisation energy. The second ionisation energy is the energy required to remove the second electron. The full process,
$Cu \rightarrow Cu^{2+} + 2e$
requires 2699 $kJ\,mol^{-1}$, so the second ionisation energy is
$2699 - 745 = 1954\,kJ\,mol^{-1}$
This shows that the ionisation energy increases with increasing loss of electrons. This is because it becomes progressively more difficult to remove negatively charged electrons as the positive charge on the ion increases. When ionisation energy is measured in eV it is called ionisation potential.

ionisation potential Symbol: I. *See* ionisation energy.

ionosphere The region of the Earth's atmosphere situated immediately below the exosphere. It contains electrically charged particles (ions and free electrons) caused by short-wave solar radiation. It is able to reflect radio waves and is therefore important in intercontinental radio transmission.

IR Abbreviation for infra-red. *See* infra-red radiation.

iris 1. An adjustable diaphragm controlling the amount of light entering an optical instrument. **2.** The coloured part of the eye. Muscles in the iris are able to alter the size of the pupil, thus controlling the amount of light entering the eye. *See* Fig. 66.

iron An element with the symbol Fe; atomic number 26; relative atomic mass 55,85 (55.85); state, solid. Iron is a very abundant transition element occurring in the Earth's crust as haematite, Fe_2O_3; magnetite, Fe_3O_4; siderite, $FeCO_3$ and iron pyrites, FeS_2. Iron is extracted from haematite, magnetite or siderite in a blast furnace. Physical properties of iron depend on the presence of small amounts of other metals and of carbon. Iron is easily magnetised and is used in different alloys such as cast iron, wrought iron and steel. When iron rusts, hydrated iron(III) oxide is formed, $Fe_2O_3 \cdot xH_2O$; the x means that the water content is not always the same. Iron is also found in haemoglobin.

iron chlorides Iron(II) chloride (ferrous chloride), $FeCl_2$, and iron(III) chloride (ferric chloride), $FeCl_3$. Salts of iron.
Iron(II) chloride (anhydrous), $FeCl_2$, is a white solid which can be made by heating iron strongly in a stream of dry hydrogen chloride, HCl:
$Fe + 2HCl \rightarrow FeCl_2 + H_2$
When dissolved in water, $FeCl_2$ forms a pale green solution. When crystallised out of solution it forms the hydrated salt $FeCl_2 \cdot 4H_2O$.
Iron(III) chloride (anhydrous), $FeCl_3$, is a black solid which can be made by heating iron strongly in a stream of dry chlorine, Cl_2:
$2Fe + 3Cl_2 \rightarrow 2FeCl_3$
When dissolved in water, $FeCl_3$ forms a brownish-yellow solution. When crystallised out of solution it forms the hydrated salt $FeCl_3 \cdot 6H_2O$.

iron(II) di-iron(III) oxide (tri-iron tetroxide) *See* iron oxides.

iron(II) disulphide (pyrites) FeS_2 (iron pyrites, pyrite) A yellow mineral occurring in many rocks and often mistaken for gold. For this reason it is commonly called fool's gold. Iron(II) disulphide (pyrites) used to be an important source of sulphur for the manufacture of sulphuric acid, H_2SO_4. It could again be of importance if natural deposits of sulphur should become limited.

iron hydroxides Iron(II) hydroxide (ferrous hydroxide), $Fe(OH)_2$ and iron(III) hydroxide (ferric hydroxide), $Fe(OH)_3$.
Iron(II) hydroxide, $Fe(OH)_2$ is a white solid when pure. It can be prepared by adding an alkali to an aqueous solution containing iron(II) ions, Fe^{2+}:
$Fe^{2+} + 2OH^- \rightarrow Fe(OH)_2$
This produces a dirty-green, gelatinous precipitate of $Fe(OH)_2$.
Iron(III) hydroxide, $Fe(OH)_3$, is precipitated as a rust-coloured, gelatinous solid by adding an alkali to a solution containing iron(III) ions, Fe^{3+}:
$Fe^{3+} + 3OH^- \rightarrow Fe(OH)_3$

iron oxides Iron(II) oxide (ferrous oxide), FeO, iron(III) oxide (ferric oxide), Fe_2O_3, and iron(II) di-iron(III) oxide (tri-iron tetroxide), Fe_3O_4 or $Fe^{II}Fe_2^{III}O_4$.
Iron(II) oxide, FeO, is a black solid obtained by heating iron(II) ethanedioate (iron(II) oxalate), FeC_2O_4, in a vacuum:
$FeC_2O_4 \rightarrow FeO + CO + CO_2$
Iron(III) oxide, Fe_2O_3, is a rust-coloured solid occurring naturally as haematite. It may be obtained by heating iron(III) hydroxide, $Fe(OH)_3$:
$2Fe(OH)_3 \rightarrow Fe_2O_3 + 3H_2O$
Iron(II) di-iron(III) oxide (tri-iron tetroxide) Fe_3O_4, is a magnetic oxide of iron. It occurs naturally as magnetite and, as such, is a natural magnet or 'lodestone'. It is a black solid which may be formed when steam is passed over red-hot iron:
$3Fe + 4H_2O \rightarrow Fe_3O_4 + 4H_2$

iron pyrites *See* iron(II) disulphide (pyrites).

iron sulphates Iron(II) sulphate (ferrous sulphate), $FeSO_4$, and iron(III) sulphate (ferric sulphate), $Fe_2(SO_4)_3$. Salts of iron.
Iron(II) sulphate (anhydrous), $FeSO_4$, is a white solid when pure. It is made by careful heating of hydrated iron(II) sulphate, $FeSO_4 \cdot 7H_2O$. This is a green substance which is obtained when an aqueous solution containing iron(II) ions, Fe^{2+}, and sulphate ions, SO_4^{2-}, is brought to the point of crystallisation. $FeSO_4 \cdot 7H_2O$ may also be prepared by dissolving iron in dilute sulphuric acid, H_2SO_4. On evaporating the solution, green hydrated crystals of $FeSO_4 \cdot 7H_2O$ separate out:
$Fe + H_2SO_4 + 7H_2O \rightarrow FeSO_4 \cdot 7H_2O + H_2$
Iron(II) sulphate is used in the brown ring test for nitrates.
Iron(III) sulphate (anhydrous), $Fe_2(SO_4)_3$, is a yellow solid. It is made by heating an aqueous

solution of iron(II) sulphate, $FeSO_4$, sulphuric acid, H_2SO_4, and hydrogen peroxide, H_2O_2:
$2FeSO_4 + H_2SO_4 + H_2O_2 \rightarrow Fe_2(SO_4)_3 + 2H_2O$
On crystallising the solution, hydrated $Fe_2(SO_4)_3 \cdot 9H_2O$ separates out.

irradiation Exposure to any form of radiation.

irreversible reaction A chemical reaction in which the reactants are completely converted into products, and the products do not react with one another to form the original substances (reactants). *Example:* the chemical reaction between an aqueous solution of copper(II) sulphate, $CuSO_4$, and zinc, Zn, producing zinc sulphate, $ZnSO_4$, and copper, Cu, is an irreversible reaction:
$CuSO_4 + Zn \rightarrow ZnSO_4 + Cu$
Compare reversible reaction.

irrigation The artificial distribution of water to cultivated land, usually by storage or by the diversion of natural waters.

irritability The property, universal among living organisms, of responding to external changes by reacting to them, e.g. the nervous activity of animals and the growth curvature of plants.

ischium The posterior bone of the pelvic girdle in tetrapod vertebrates.

islets of Langerhans Groups of cells situated throughout the pancreas. They produce the hormones glucagon, secreted by α cells, and insulin, secreted by β cells.

iso- Prefix meaning equal.

isobar 1. A line on a map connecting points which have the same atmospheric pressure. 2. A curve which compares quantities measured at the same pressure, e.g. a graph of the volume of a gas against its temperature when the pressure is kept constant. 3. Atoms which have different atomic numbers (number of protons) but the same mass number are called isobars. *Example:* $^{40}_{18}Ar$, the isotope of argon which contains 18 protons and 22 neutrons in the nucleus, and $^{40}_{20}Ca$, the isotope of calcium which contains 20 protons and 20 neutrons in the nucleus, are isobars. Since isobars are different elements, they have different properties.

isobaric Occurring without a change of pressure in a system.

isoclinic line A line on a map joining points which have the same angle of dip (inclination).

isoelectric point The pH value of a substance or a system at which it is internally neutralised. For an amino acid, the degree of ionisation of the carboxyl group, —COOH, and that of the amino group, —NH_2, are equal at its isoelectric point. Each amino acid has a characteristic isoelectric point. *See also* zwitterion.

isogonal line A line on a map joining points which have the same angle of declination (variation).

isomer *See* isomerism.

isomerism The existence of two or more chemical compounds which have the same molecular formula, although their three-dimensional structures may be different. *Example:* butane (an alkane), C_4H_{10}, exists as two isomers:
(1) CH_3—CH_2—CH_2—CH_3 and
(2) CH_3—CH—CH_3
         |
         CH_3
To distinguish them from one another, (1) is called butane and (2) is called 2-methylpropane. The different forms occurring in isomerism are called isomers and they are not chemically and physically identical. This is because the atomic arrangement is different. Isomerism is usually divided into structural isomerism and stereoisomerism. In structural isomerism, the isomers have different structural formulae; in stereoisomerism, the isomers have the same structural formula but different spatial arrangements of their atoms or groups of atoms. Isomerism is of great importance in organic chemistry.

isomorphism The existence of two or more different substances which have the same crystal structure (form). Such substances are said to be isomorphous. *See also* Mitscherlich's law.

isomorphous *See* isomorphism.

isophthalic acid Benzene-1,3-dicarboxylic acid. *See* benzene-1,2-dicarboxylic acid.

isoprene *See* methylbuta-1,3-diene.

Isoptera An order of social insects living in large communities and commonly called 'white ants' or termites. These animals live in a system of galleries in soil or wood where they construct nests. There are four anatomically and functionally different castes. These are: primary reproductives which are fully winged; supplementary reproductives which are smaller, with reduced eyes and non-functional wings; workers; and soldiers. The last two castes are both small, sterile and wingless. The primary reproductives (king and queen) establish a new colony; the supplementary reproductives develop in a colony which has lost one or both of its primary reproductives; the workers build and maintain the nest, collect food and feed the reproductives and young termites; finally, the soldiers defend the colony.

isotonic Describes solutions which have the same osmotic pressure.

isotopes Atoms of the same element (therefore with the same atomic number) which have different relative atomic masses. This is due to different numbers of neutrons in the nuclei of the isotopes. *Example:* in the case of hydrogen, three isotopes are known: (1) ordinary hydrogen, ^1_1H, which contains 1 proton and no neutrons in the nucleus; (2) deuterium, ^2_1H, which contains 1 proton and 1 neutron in the nucleus; and (3) tritium, ^3_1H, which contains 1 proton and 2 neutrons in the nucleus. Isotopes have identical chemical properties, because they have the same electron arrangement, but different physical properties, the latter being used to separate isotopes when they are mixed. Some isotopes are radioactive. *See also* nuclide.

-ite Suffix denoting a salt of the corresponding -ous acid. *Example:* sodium sulphite, Na_2SO_3, is a salt of sulphurous acid, H_2SO_3. *See also* -ate.

IUPAC Abbreviation for International Union of Pure and Applied Chemistry. This organisation works on an international standard nomenclature (naming system) for use in chemistry. IUPAC was established in 1919.

IUPAP Abbreviation for International Union of Pure and Applied Physics. This organisation works on an international standard for the definitions and names of physical units. IUPAP was established in 1931.

J

jasper An impure, opaque form of silica. It is usually red or brown in colour because of the presence of clay and iron oxides.

jaundice (icterus) A symptom of a disease or disorder which gives the patient a yellowish skin, sometimes including a yellow colour of the whites of the eyes. This is caused by too much bilirubin in the blood. It may arise from an excessive breakdown of red blood cells; disease of the liver; or a blockage of the bile duct, usually due to a gallstone.

jejunum The second or middle part of the small intestine. In humans it is about 2,5 m (2.5 m) long, ascending from the duodenum and leading into the ileum. Its junction with the ileum is not sharply defined. Enzymes in the jejunum act on proteins and carbohydrates.

jigger A disease caused by the sandflea. The female flea burrows into the skin, usually of the feet, and lays her eggs, causing painful swelling. The flea can be removed surgically.

joint A place where two bones meet. Joints are either fixed or movable. In fixed joints no movement is possible, as in the sutures between the bones of the skull. In movable joints the bones move relative to one another, as in the ball and socket joints of the hip.

joule Symbol: J. The SI unit of energy, defined as the work done when a force of 1 newton moves 1 metre in its own direction. It may also be defined as the work done by a current of 1 ampere flowing through a resistance of 1 ohm for a time of 1 second. The joule has replaced the calorie as a unit of heat energy: 1 calorie = 4,1868 joules (4.1868 J).

Joule–Kelvin effect *See* Joule–Thomson effect.

joulemeter An instrument used to measure the amount of electrical energy supplied from the mains to an electric circuit. It is commonly called an electricity meter.

Joule's equivalent *See* mechanical equivalent of heat.

Joule's law A law stating that the heat produced by an electric current passing through a conductor for a measured time is given by the expression:
$Q = I^2 \times R \times t$
where Q is the heat produced, I is the size of the current, R is the resistance of the conductor or the effective resistance of an a.c. circuit, and t is the time for which the current flows. If the current is measured in amperes, the resistance in ohms and the time in seconds, then the heat produced is measured in joules.

Joule–Thomson effect (Joule–Kelvin effect) The drop in temperature of a real gas (except hydrogen) at ordinary temperature when it is expanded adiabatically through a throttle. The extent of the temperature change depends on the initial temperature and pressure of the gas. The Joule–Thomson effect is seen in Linde's process.

junction transistor *See* transistor.

Jupiter One of the nine planets in our Solar System, orbiting the Sun between Mars and Saturn once in 11,86 years (11.86 years). It is the largest of the planets and has thirteen satellites (moons).

Jurassic A geological period of approximately 57 million years, occurring during the middle of the Mesozoic, and which ended approximately 136 million years ago.

juvenile hormones Insect hormones regulating larval development or reproduction in insects.

K

K Chemical symbol for potassium.

kainite A mineral occurring in the upper parts of saline residues such as the Stassfurt deposits. Kainite is hydrated potassium chloride and magnesium sulphate: $KCl \cdot MgSO_4 \cdot 3H_2O$.

kala azar A disease (a form of leishmaniasis) found in East Africa, China and India. It is caused by a protozoon which is transmitted to humans by the bites of sandflies, which themselves are infected as a result of biting dogs. Symptoms include enlargement of the spleen and liver, anaemia and prolonged fever. The disease is complicated by dysentery. Kala azar is treated with drugs containing antimony.

Kaldo process A process by which pig-iron from the blast furnace is converted into steel. Molten pig-iron is poured into a pear-shaped vessel called a converter. The converter is rotated at speeds up to thirty revolutions per minute during which oxygen at low pressure is blown onto the surface of the molten metal. The impurities such as sulphur and carbon are oxidised to gaseous products. Silicon and other impurities form a slag by reacting with calcium oxide, CaO, which is added to the converter. When the iron is considered pure, measured amounts of carbon, manganese, etc. are added to produce the required type of steel. The Kaldo process was pioneered by B. Kalling. *See also* Bessemer process, L–D process *and* Open-Hearth process.

kalium The Latin name for potassium.

kaolin *See* clay.

katabolism *See* catabolism.

K-capture A form of radioactive decay in which a nucleus of an atom captures one of the electrons from the innermost shell, the K-shell. This means that the atomic number decreases by one. *Example:* an isotope of beryllium (atomic number 4) decays into one of lithium (atomic number 3).

keel 1. A boat-shaped structure formed by two joined, anterior petals of the flowers of legumes. **2.** *See* carina.

keeper *See* armature.

kelp A general term for large seaweeds.

kelvin Symbol: K. The SI unit of thermodynamic temperature, equal to $1/273,16$ ($1/273.16$) of the thermodynamic temperature of the triple point of water. One kelvin (1 K) is equal to one degree Celsius (1°C), thus
0°C = 273,16 K (273.16 K)
100°C = 373,16 K (373.16 K)
$T = (273,16 + t)$ K
($T = (273.16 + t)$ K)
where $t = $ °C

Kelvin scale A scale of temperature using a lower fixed point called absolute zero (0 K) and an upper fixed point which is the triple point of water (273,16 K; 273.16 K). In the Kelvin scale the temperature measure is based on the average kinetic energy per molecule of an ideal gas.

Kepler telescope *See* telescope.

keratin A fibrous protein containing sulphur. It has great strength, is insoluble in water and forms the basis of claws, feathers, hair, horns, etc. *See also* scleroprotein.

kerosene (kerosine, paraffin oil) A petroleum distillate mainly containing aliphatic hydrocarbons (alkanes with 10–16 carbon atoms). It is the principal constituent of jet fuels and is also used in paraffin heaters.

ketohexose *See* sugar.

ketone The general name for a group of organic substances containing the carbonyl group,
—C=O,
|
to which two carbon atoms are attached. Ketones are either aliphatic or cyclic molecules. *Examples:* propanone (acetone), CH_3COCH_3, is aliphatic; phenylethanone, $C_6H_5COCH_3$, is cyclic. The group
—CO,
|
in ketones is called the ketone group. *Compare* aldehyde.

ketopentose *See* sugar.

ketose A carbohydrate containing a ketone group,
—CO,
|
e.g. fructose, $C_6H_{12}O_6$, or
$CH_2OH—CO—(CHOH)_3—CH_2OH$.
See also sugar.

kidney One of a pair of organs whose main function is to filter waste products (urea and excess water) from the blood. The waste products are secreted from the kidney as urine, which passes down the ureter to the bladder. The kidney also regulates the concentration of salts in the blood. *See* Fig. 106; *see also* osmoregulation.

kidney machine (artificial kidney) A machine which performs the functions of human kidneys. Blood from an artery of a patient suffering from kidney failure is passed into the machine in which a semi-permeable membrane allows water and waste products to pass out of the blood, but prevents the escape of blood cells and proteins; this process is called dialysis. The filtered blood is then returned to the patient via a vein.

kidney stone A small, hard deposit sometimes formed in the kidney. Kidney stones may cause acute, intense pain when they become lodged

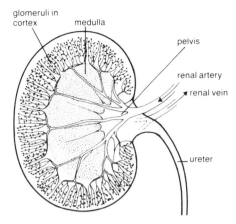

Fig. 106 Kidney in section

in the kidney or the ureter. Some of them may consist of calcium ethanedioate (calcium oxalate), which can be dissolved by the administration of drugs, others need to be removed surgically.

kieselguhr (diatomaceous earth, infusorial earth) A whitish powder made up of the skeletal remains of diatoms. It is used in filters, as an abrasive for polishing and as an absorbent of nitroglycerin in the manufacture of dynamite.

kif *See* cannabis.

kiln A furnace or oven used for firing clay, roasting sulphide ores, drying, removing carbon dioxide from limestone, etc. *See* Fig. 116.

kilo- Symbol: k. A prefix meaning one thousand; e.g. one kilogram (kg) is equal to one thousand grams.

kilogram (kilogramme) Symbol: kg. The SI unit of mass.

kilowatt-hour Symbol: kWh. A unit commonly used to measure electrical energy. It is equal to a power of one kilowatt operating for one hour and is used for calculating the cost of electrical energy taken from the mains supply: 1 kWh is equal to 3,6 megajoules (3.6 MJ).

kimberlite A type of rock occurring in volcanic pipes in South Africa. It sometimes contains diamonds.

kinase A general name for an enzyme which converts proenzymes (zymogens) into active enzymes. *Example:* enterokinase converts trypsinogen to trypsin.

kinematics The branch of mechanics concerned with the motion of bodies without reference to mass or force.

kinematic viscosity *See* viscosity.

kinetic Pertaining to motion.

kinetic energy The energy possessed by a body due to its motion. The kinetic energy (E) of a body of mass m kg moving with a velocity v m s^{-1} is given by the expression
$E = \tfrac{1}{2} m v^2$ joules

kinetic friction *See* friction.

kinetic theory (particle theory) A theory explaining some of the ways in which matter behaves, e.g. Brownian movements, diffusion, evaporation and pressure. It is especially useful in explaining the behaviour of gases because of the relatively weak forces of attraction between gas molecules. The theory is based on the assumption that all matter is made up of particles (atoms, ions or molecules) which are in a state of continual motion.

kinetin *See* cytokinin.

kingdom The largest group used in the classification of living organisms, e.g. the animal and plant kingdoms.

Kipp's apparatus (gas generator) A laboratory apparatus for the production of gases such as carbon dioxide, CO_2, hydrogen, H_2, and hydrogen sulphide, H_2S. In order to produce carbon dioxide, marble chips, $CaCO_3$, are placed on a perforated plate in bulb B (*see* Fig. 107). Dilute hydrochloric acid, HCl, is poured into bulbs A and C and rises into bulb B. There it reacts with the marble chips, producing carbon dioxide which escapes through tap T. When this tap is closed, the pressure of the gas produced in bulb B forces the acid into bulbs C and A and the reaction stops.

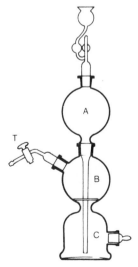

Fig. 107 Kipp's apparatus

Kirchhoff's laws for electric circuits (1) In a circuit, the algebraic sum of the currents at a junction is zero. (2) Round any closed circuit, the algebraic sum of the e.m.f.s is equal to the algebraic sum of the products of current and resistance.

The first law expresses the fact that charge does not accumulate in an electric circuit. *Example:* if a current of 2 amperes enters a junction through one wire, then a current of 2 amperes also leaves the junction through another wire. The second law expresses the law of conservation of potential (energy) and can be written:

$E = Ir_1 + Ir_2$

where E is the e.m.f., I the size of the current and r_1 and r_2 two resistors in a circuit. Together with Ohm's law, Kirchhoff's laws make up the circuit theory needed for simple electric circuits.

kiss of life The mouth-to-nose or mouth-to-mouth method of artificial respiration in which air is blown into the lungs of a patient.

Kjeldahl's method A quantitative method widely used to determine the amount of nitrogen in an organic compound (or in ammonium salts). A measured sample is boiled one at the same time with concentrated sulphuric acid and various catalysts such as mercury and copper salts. Anhydrous potassium sulphate is also added to raise the boiling-point of the acid. During this treatment the sample is 'opened', i.e. its nitrogen content is converted into ammonium ions, NH_4^+. An excess of a strong sodium hydroxide solution, NaOH, is now added to the diluted solution of ammonium ions and the ammonia, NH_3, liberated is distilled into an excess of a standard acid:

$NH_4^+ + OH^- \rightarrow NH_3 + H_2O$ and
$NH_3 + H^+ \rightarrow NH_4^+$

The acid is then titrated using a standard sodium hydroxide solution to determine the amount of acid originally neutralised by ammonia:

$H^+ + OH^- \rightarrow H_2O$

klinostat *See* clinostat.

knee-cap *See* patella.

knee-jerk A normal reflex consisting of an involuntary kick in response to a tap on the tendon below the patella.

knee-joint The hinge joint between the distal end of the femur and the proximal end of the tibia.

knocking The metallic noise heard in an internal combustion engine due to an uneven and premature ignition (pre-ignition) of the fuel-air mixture. Knocking is sometimes called pinking. *See also* inhibitor *and* octane number (rating).

Kohlrausch's law A law stating that when ionisation is complete, that is at infinite dilution where all interionic effects disappear, the conductivity of an electrolyte at a certain temperature is equal to the sum of the conductivities of the ions into which a substance dissociates.

Kolbe's reaction The formation of an alkane by the electrolysis of a strong aqueous solution of a salt of a carboxylic acid, e.g.

$2C_{17}H_{35}COONa + 2H_2O \rightarrow$
$C_{34}H_{70} + 2CO_2 + 2NaOH + H_2$

Here sodium stearate, $C_{17}H_{35}COONa$, is converted into the alkane $C_{34}H_{70}$ (tetratriacontane). This is released at the anode, while hydrogen is released at the cathode.

Krebs cycle (citric acid cycle, tricarboxylic acid cycle) A cyclic sequence of various chemical reactions providing most of the energy for biological work. It is the final stage of oxidation of organic compounds in all aerobic organisms. The cycle is operated in the mitochondria by enzymes present there. During each complete cycle molecules of ATP are released, formed from ADP and phosphate groups. The first substance entering the Krebs cycle is pyruvic acid, $CH_3-CO-COOH$, but only after it has been converted into citric acid,

$HOOC-CH_2-COH-CH_2-COOH$
$|$
$COOH$

by coenzymes. Pyruvic acid is formed during glycolysis from glucose or from products of fat and protein. During the cycle, pyruvic acid is fully oxidised to carbon dioxide and water:

$2CH_3-CO-COOH + 5O_2 \rightarrow$
$6CO_2 + 4H_2O + energy$

See also respiration.

krypton An element with the symbol Kr; atomic number 36; relative atomic mass 83,80 (83.80); state, gas. It is a colourless, odourless and very inert element found in the atmosphere in very small amounts. Krypton belongs to the noble gases in the Periodic Table of the Elements. It is used in certain lasers and in fluorescent tubes.

kurchatovium *See* transuranic elements.

kwashiorkor A deficiency disease of children caused by a lack of proteins in the diet. This often occurs in tropical countries when a child stops being breast-fed because of the birth of another child. Symptoms include cracking of the skin, swelling of the abdomen and damage of the liver. The disease often leads to death before the age of six years. *See also* marasmus.

L

labia majora The outer lips of the vulva, consisting of two thick folds of skin covered with scattered hair and containing sebaceous and sweat glands. The labia majora are similar in position and structure to the scrotum of the male.

labia minora The inner lips of the vulva, consisting of two small, thin folds of skin without hair and containing sebaceous glands.

labile Unstable, tending to undergo change or displacement.

labium (pl. labia) A lip-like structure; in insects, the name given to the lower lip.

labour *See* parturition.

labrum (pl. labra) The upper lip of an insect.

labyrinth *See* inner ear.

lachrymal Pertaining to tears.

lachrymal duct (tear-duct) One of a system of ducts draining excess tears from the eyes into the nasal cavity.

lachrymal gland (tear-gland) A small gland situated at the outer corner of the eye of land vertebrates. The gland secretes a watery, salty and slightly antiseptic liquid (tears). *See also* lysozyme.

lachrymator (tear-gas) Any of several volatile, organic substances which, even in low concentration, cause irritation of the eyes.

lac insect An insect belonging to the order Hemiptera. Lac insects excrete a resinous substance from which shellac is prepared. They obtain nourishment by piercing the outer tissue of the twigs of certain trees with their proboscis. Lac insects are orange-red; the males have membranous wings whereas the females are wingless. The female insects excrete the lac from glands which are found on all parts of their bodies. One of the centres for the manufacture of shellac is Calcutta.

lactase An enzyme which catalyses the hydrolysis of lactose into glucose and galactose. It is found in the ileum.

lactate (2-hydroxypropanoate) A salt of lactic acid, CH_3—CHOH—COOH.

lactation The secreting of milk in the mammary glands.

lacteal A blind-ended capillary found in the villi of the small intestine. Lacteals are lymphatic vessels. They absorb fatty materials which mix with lymph forming chyle. *See* Fig. 243.

lactic acid CH_3—CHOH—COOH (2-hydroxypropanoic acid) An organic acid which is a crystalline solid, soluble in water. It occurs in sour milk, where it is formed by the action of certain bacteria on the lactose in the milk. Salts of lactic acid are called lactates (2-hydroxypropanoates). Lactic acid accumulates in muscles if oxygen is not supplied to the cells fast enough (anaerobic respiration). This leads to fatigue, and rest is needed to allow the removal of lactic acid by oxidising it into pyruvic acid. Lactic acid is used in tanning and dyeing.

lactobiose *See* lactose.

lactoflavin *See* riboflavin.

lactogenic hormone *See* prolactin.

lactometer A form of hydrometer used for testing the density of milk.

lactose (milk-sugar, lactobiose) A reducing carbohydrate (sugar) with the formula $C_{12}H_{22}O_{11}$, i.e. lactose is a disaccharide. It is a white crystalline solid, occurring naturally in the milk of mammals. On hydrolysis, lactose is converted into glucose and galactose. Lactose in milk is converted into lactic acid by certain bacteria.

laevorotatory compound *See* optical isomerism.

laevulose *See* fructose.

lagging The process of wrapping water-pipes, boilers, etc., with non-conducting materials to reduce heat gain or heat loss. The term also applies to the non-conducting materials.

Lagomorpha An order of herbivorous, placental mammals including hares and rabbits.

Lamarckism A theory put forward by Jean Baptiste Lamarck, stating that evolution may have come about by the inheritance of acquired characteristics. Lamarck used the giraffe for the most often quoted example of his theory. According to Lamarck, the long neck, tongue and legs of the giraffe are evolved from a short-necked ancestor which at one time started to eat the leaves of trees and therefore stretched its neck upwards at the same time stretching out its tongue and legs as well. These longer body parts would then be passed on to the offspring and the process repeated through many generations: that is, little by little the short-necked ancestor would turn into a giraffe. Lamarck's theory of evolution is no longer accepted as it is known that acquired characteristics cause no alteration to the chromosomes, the genetic material of the cell.

lambda (λ) The eleventh letter of the Greek alphabet.

lamella (pl. lamellae) Any thin plate-like structure, e.g. a layer of calcified matrix in bone.

lamina *See* blade.

laminated core A core of a choke, transformer, relay, etc., made up of thin iron or steel plates called laminations. The

151

laminations are either oxidised or varnished to increase the electrical resistance between each one. This reduces the circulation of eddy currents in the core.

lamination *See* laminated core.
lampblack *See* carbon black.
lancelet *See* Cephalochorda(ta).
lanolin A wax which is obtained from wool, e.g. the wool of sheep. Lanolin consists of esters of higher fatty acids and complex cyclic alcohols such as cholesterol. However, aliphatic alcohols may also be present. It is used in the manufacture of ointments and cosmetics. The cholesterol obtained by saponification of lanolin is used in the manufacture of sex hormones such as testosterone and progesterone.
lanthanides *See* lanthanoids.
lanthanoids (lanthanides, lanthanons, rare earths) A series of fourteen elements starting with cerium (atomic number 58) and ending with lutetium (atomic number 71); i.e. the lanthanoids lie between lanthanum (atomic number 57) and hafnium (atomic number 72). The lanthanoids have very similar chemical properties because the outer electron structure is almost the same for them all. In the past this made them very difficult to separate. Today they are easily separated by methods such as ion exchange. The characteristic oxidation number of the lanthanoids is $+3$.
lanthanons *See* lanthanoids.
lard An animal fat. The term usually applies to fat obtained from pigs. The term 'tallow' applies to fat obtained from cattle and sheep.
large intestine Part of the alimentary canal, consisting of the caecum, appendix, colon and rectum.
larva (pl. larvae) The form which hatches from the egg of an arthropod undergoing complete metamorphosis. *Example:* the caterpillar is the larva of a butterfly. The term also applies to the immature forms of other animals undergoing some form of metamorphosis. *Example:* the tadpole is the larva of a frog.
larynx (voice-box) The organ of voice production in mammals, reptiles and some amphibians. It is an irregularly shaped, tubular structure situated at the junction of the upper end of the trachea and the pharynx. In all land vertebrates except birds, voice or sound is produced in the larynx as a result of the vibration of the vocal cords. *See* Fig. 118.
laser (Light Amplification by Stimulated Emission of Radiation) A device for producing intensified, monochromatic, electromagnetic radiation. An electron in a high energy level (in an excited atom) is stimulated by a photon to jump to a lower energy level, thus releasing a photon which has the same frequency as the stimulating photon and is travelling in the same direction. With a sufficient number of electrons in high energy levels, stimulating and stimulated (released) photons will cause further stimulated emission, resulting in a narrow beam of monochromatic radiation. The waves of this radiation are parallel and in phase and the beam has a very high energy with extremely little spread. Laser beams have many uses such as in cutting metals, in surgery and in surveying. *See also* maser.

Lassaigne test A test for nitrogen, sulphur and halogens in an organic compound. The test is carried out using a small pellet of metallic sodium which is placed in a test-tube together with a small amount of the sample. The test-tube is heated until the sodium melts and the mixture is then treated with boiling water and filtered. The filtrate will now contain cyanide ions if nitrogen was present in the sample, sulphide ions if sulphur was present and halide ions if halogens were present. These ions can then be identified by treating the filtrate with appropriate chemicals.

latent heat Symbol: L. The total quantity of heat absorbed or released during a change of state (e.g. fusion or vaporisation) at constant temperature. *See also* specific latent heat.

latent period *See* reaction time.
lateral Pertaining to or situated at a side.
lateral bud *See* axillary bud.
lateral inversion (perversion) The effect observed in a plane mirror in which the right hand side of an object appears as the left hand side of its image. *See* Fig. 108.

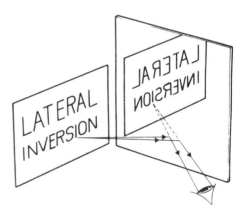

Fig. 108 Lateral inversion

lateral line In fish and some amphibians a line (canal) passing down the length of each side of the body. Lateral lines contain sense organs which detect vibrations and changes in pressure in the water.

lateral root A root growing from the side of a main root.

laterite Red or yellow tropical soil, mainly consisting of hydrated aluminium and iron oxides, produced by the weathering of rocks under hot, moist conditions.

latex A complex, milky fluid, mainly containing hydrocarbons, carbohydrates and proteins, found in many higher plants. Latex is collected from several species of these plants, e.g. the rubber tree, and is used in the manufacture of rubber. Natural rubber is a mixture of unsaturated hydrocarbons with a *cis* configuration. *See also* gutta-percha.

lattice A regular, three-dimensional arrangement of atoms, ions or molecules, as seen in a crystal.

lattice energy The energy required to separate completely the ions in an ionic crystal. This may be expressed in joules per mole, $J\,mol^{-1}$.

laughing gas *See* nitrogen oxides.

lava Molten rock flowing from a volcano. *See also* magma *and* pumice.

laxative A medicine which tends to cause emptying of the bowels. Epsom salt and Glauber's salt are traditional examples.

layering A method of artificial propagation in which a shoot from a plant stem is pegged down and covered with soil. Adventitious roots grow from nodes on the shoot, and when they are established the young plant can be detached from the parent plant. *See also* marcotting.

L–D process (Linz–Donawitz process) A process by which pig-iron from a blast furnace is converted into steel. The converter is similar to that used in the Bessemer process. However, instead of a blast of air from the base, a high-pressure blast of oxygen is applied to the surface of the molten pig-iron by a water-cooled copper tube. Impurities such as sulphur and carbon are oxidised to gaseous products. Other impurities form a slag by reacting with the lining of the converter and with added lime (calcium oxide, CaO). When the iron is considered pure, measured amounts of carbon, manganese, etc. are added to produce the required type of steel. The advantages of the L–D process are that it produces steel at a greater speed than other processes and the steel is less brittle because there is no nitrogen in the blow (blast). *See* Fig. 109; *see also* Bessemer process, Kaldo process *and* Open-Hearth process.

leaching The washing out of soluble constituents from a solid mixture. The term is usually applied to the removal of soluble ions (mineral salts) from soil during drainage.

lead An element with the symbol Pb; atomic number 82; relative atomic mass 207,19 (207.19); state, solid. It is a heavy, soft, bluish-

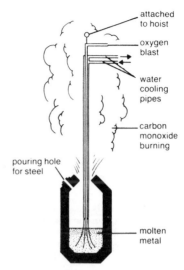

Fig. 109 L–D process

grey metal occurring naturally as galena (lead(II) sulphide, PbS) from which it is extracted by roasting and reduction with carbon:

$2PbS + 3O_2 \rightarrow 2PbO + 2SO_2$ and
$PbO + C \rightarrow Pb + CO$

Lead is used in making accumulators, as a covering for cables and in making alloys such as solder and pewter. Lead is a rather poor conductor of electricity, but it is a good absorber of sound and vibration. It is also used as a radiation shield around X-ray equipment and in nuclear reactors.

lead acetate *See* lead(II) ethanoate.

lead acetate paper Filter paper dipped in a solution of lead(II) ethanoate (lead acetate), $(CH_3COO)_2Pb$, and allowed to dry. It is used as a test for hydrogen sulphide, H_2S, which produces a brownish-black stain of lead(II) sulphide, PbS, on the paper:
$H_2S + (CH_3COO)_2Pb \rightarrow PbS + 2CH_3COOH$

lead-acid cell (lead-acid accumulator) A secondary cell in which the cathode is a lead plate, the anode is a lead plate covered with a brown deposit of lead(IV) oxide, PbO_2, and the electrolyte is dilute sulphuric acid, H_2SO_4. The plates are held in a lead-antimony alloy grid. The chemical processes taking place during the discharging (use) and the charging of a lead-acid cell can be represented by the equilibrium:
$Pb + PbO_2 + 2H_2SO_4 \rightleftharpoons 2PbSO_4 + 2H_2O$
in which the left side represents the cell when in a charged condition and the right side the cell when in a discharged condition. In a charged condition, the positive terminal (anode) con-

lead(II) bromide

tains PbO_2, the negative terminal (cathode) is pure lead and the electrolyte is dilute sulphuric acid. In a discharged condition, both electrodes contain lead(II) sulphate, $PbSO_4$, and the solution is water. This means that the relative density of the electrolyte decreases as the discharge progresses and this is a guide in determining the state of the cell. In order to charge the cell, a current of about 2 amperes is passed through the cell in the opposite direction, i.e. from the anode to the cathode. The lead(II) sulphate formed on the anode during discharge is converted back to lead(IV) oxide, the lead(II) sulphate on the cathode is removed and the relative density of the acid increases from 1,18 to 1,25 (1.18 to 1.25), i.e. sulphuric acid is produced. When fully charged, the e.m.f. of the cell is slightly more than 2 volts. A 12 volt car battery consists of six lead-acid cells connected in series. Such a battery has a low internal resistance of about 0,01 ohms (0.01 Ω), which enables it to provide a current of several hundred amperes when on short circuit.

lead(II) bromide $PbBr_2$ A white, crystalline, poisonous, solid salt. As it has a rather low melting-point (373°C), it is commonly used in experiments demonstrating electrolysis of a fused salt. It is almost insoluble in cold water, but fairly soluble in hot water.

lead(II) carbonate $PbCO_3$ A white, poisonous lead salt occurring naturally as cerussite and used as a pigment. It is insoluble in water.

lead-chamber process An obsolete process for the industrial manufacture of sulphuric acid, H_2SO_4, which takes place in large lead chambers. The process is based on the oxidation of sulphur dioxide, SO_2, by nitrogen(IV) oxide, NO_2:
$SO_2 + NO_2 \rightarrow SO_3 + NO$
The sulphur(VI) oxide produced, SO_3, is then absorbed in concentrated sulphuric acid to give oleum (fuming sulphuric acid), $H_2S_2O_7$:
$SO_3 + H_2SO_4 \rightarrow H_2S_2O_7$
When oleum reacts with water, sulphuric acid is formed:
$H_2S_2O_7 + H_2O \rightarrow 2H_2SO_4$
By the reaction of nitrogen(II) oxide, NO, with oxygen the nitrogen(IV) oxide is regenerated:
$2NO + O_2 \rightarrow 2NO_2$
The process has now been replaced almost entirely by the contact process.

lead(II) chloride $PbCl_2$ A white, crystalline, solid salt which is almost insoluble in cold water, but fairly soluble in hot water. It dissolves in concentrated hydrochloric acid, HCl, to form the complex ion $[PbCl_4]^{2-}$:
$PbCl_2 + 2Cl^- \rightarrow [PbCl_4]^{2-}$

lead(II) ethanoate $(CH_3COO)_2Pb \cdot 3H_2O$ (lead acetate) A lead salt which is a poisonous, white, crystalline solid, soluble in water. Lead(II) ethanoate may be prepared by the action of ethanoic acid (acetic acid), CH_3COOH, on lead(II) oxide, PbO:
$PbO + 2CH_3COOH \rightarrow (CH_3COO)_2Pb + H_2O$
It is used in medicine, in the textile industry and as an analytical reagent. Lead(II) ethanoate is also called sugar of lead because of its sweet taste.

lead glance See galena.

lead oxides Lead(II) oxide (lead monoxide), PbO, lead(IV) oxide (lead dioxide), PbO_2, and di-lead(II) lead (IV) oxide (tri-lead tetroxide, red lead, minium), Pb_3O_4 or $Pb^{II}_2Pb^{IV}O_4$.
Lead(II) oxide, PbO, exists in two forms, both yellow solids but with different crystalline shapes. Lead(II) oxide can be prepared by heating lead(II) carbonate, $PbCO_3$:
$PbCO_3 \rightarrow PbO + CO_2$
One of the two forms of PbO is called litharge and is obtained when PbO is heated to a temperature above its melting-point. If it is heated to a temperature below its melting-point, the other form, called massicot, is obtained. PbO is used in the manufacture of paints and glass.
Lead(IV) oxide, PbO_2, a dark-brown solid which decomposes on heating into lead(II) oxide and oxygen:
$2PbO_2 \rightarrow 2PbO + O_2$
PbO_2 can be prepared by treating di-lead(II) lead(IV) oxide with dilute nitric acid, HNO_3:
$Pb_3O_4 + 4HNO_3 \rightarrow PbO_2 + 2Pb(NO_3)_2 + 2H_2O$
It is a powerful oxidising agent used in lead-acid cells.
Di-lead(II) lead(IV) oxide, Pb_3O_4, is a red solid which can be prepared by heating lead(II) oxide in air at 400°C:
$6PbO + O_2 \rightarrow 2Pb_3O_4$
At 500°C it decomposes into lead(II) oxide and oxygen. Di-lead(II) lead(IV) oxide is used in the manufacture of paints and in the production of lead(IV) oxide for use in lead-acid cells.

lead(II) sulphate $PbSO_4$ A white, insoluble crystalline lead salt.

lead(II) sulphide PbS A brownish-black, insoluble crystalline lead salt occurring naturally as galena. See also lead acetate paper.

lead tetraethyl See tetraethyl-lead(IV).

leaf An outgrowth from a plant stem, usually consisting of a leaf base, a petiole or stalk and a thin, flattened lamina or leaf blade. A leaf is usually green and is concerned with photosynthesis and transpiration. See Fig. 110 (a) *and* (b).

leaf base The structure attaching a leaf to a stem. See Fig. 110 (a).

leaf blade See blade.

leaflet A small leaf; an individual part of a compound leaf.

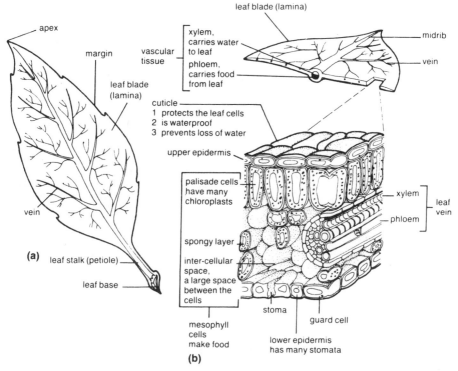

Fig. 110 (a) Parts of a leaf (b) Structure of a leaf

leaf morphology The study of the structure and shape of leaves. See Fig. 111.

leaf scar The scar remaining on a stem at the point where a leaf has fallen from it.

leaf stalk See petiole.

leakage The flow of an electric current from an imperfectly insulated conductor.

least distance (of distinct vision) The shortest distance from the eye at which an object can be seen clearly. For normal vision, this distance is about 25 cm. See also near point.

Le Chatelier's principle A principle stating that if a change is made in the conditions of any system in equilibrium, the equilibrium adjusts so as to oppose the change. The principle can be applied to the effect of changes in temperature, pressure and concentration in chemical reactions. A classic example is the Haber–Bosch process for the manufacture of ammonia, NH_3:
$$N_2 + 3H_2 \rightleftharpoons 2NH_3 + \text{heat} \; (-\Delta H)$$
Here it can be seen that the forward reaction is exothermic; increasing the temperature makes the equilibrium shift to the left because this leads to a decrease in temperature.

lecithin The general name for a group of complex substances closely related to lipid(e)s. They are mixed esters of glycerol and choline with fatty acids and phosphoric acid, and they are widely distributed in bone marrow, nervous tissue and embryonic tissue. They are extracted from natural material and can also be prepared synthetically. Lecithins are mainly used in the food industry as emulsifiers and in the manufacture of certain soaps.

Leclanché cell A primary cell consisting of a carbon rod (the positive electrode) surrounded by a mixture of powdered carbon and manganese dioxide, MnO_2, in a porous pot. Both the pot and a zinc rod (the negative electrode) stand in a glass jar containing ammonium chloride solution, NH_4Cl (the electrolyte). The cell has an e.m.f. of about 1,5 volts (1.5 V) and an internal resistance of about 1 ohm. See also dry cell *and* depolarisation.

LED See light-emitting diode.

left-hand rule See Fleming's rules.

legume 1. A dry, dehiscent fruit (a pod) formed from a single carpel. It liberates its seeds by splitting along both dorsal and ventral sutures. A pea pod is a common example. See also

legumin

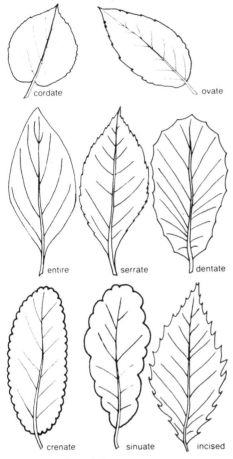

Fig. 111 Leaf morphology

Leguminosae. 2. A term used in agriculture for leguminous plants used as animal fodder.

legumin A globulin found in the seeds of Leguminosae.

Leguminosae A large family of dicotyledons ranging in size from herbs to trees. Their fruits are legumes. Many species are of great economic importance as food crops and because of the presence of nitrogen-fixing bacteria in their root nodules. Examples include peas, beans, lucernes and clovers.

leguminous Pertaining to members of the family Leguminosae.

leishmaniasis A disease caused by the protozoan *Leishmania* which lives in the red blood cells. Symptoms include anaemia, fever and enlargement of the liver and spleen. Death ensues unless the patient is treated. Infection is transmitted to human beings by the sandfly from dogs, cats and rodents. The disease is rather common in South America, Africa and the East. Effective drugs are available in the treatment of the disease. *See also* kala azar.

lemma (pl. lemmae) *See* palea.

lens A piece of transparent material (usually glass) for converging or diverging beams of light by refraction. Simple lenses are usually pieces of glass bounded by one or two spherical surfaces. They are classified according to the nature of these surfaces. *See* Fig. 112 (a). Converging or convex lenses (positive lenses) bring light to a focus. Diverging or concave lenses (negative lenses) spread light out. *See* Fig. 112 (b). The centres of the spheres of which the lens surfaces form a part are called the centres of curvature. A line joining these centres is called the principal axis (optical axis) of the lens. The optical centre is a point on the principal axis within the lens. A ray of light passing through this point emerges without deviation. A ray parallel to the principal axis is refracted through the principal focus or focal point. A ray passing through the principal focus is refracted parallel to the principal axis. The distance between the optical centre of the lens and the principal focus is called the focal length of the lens. Lenses are used in spectacles, microscopes, telescopes, binoculars, etc. *See* Fig. 112 (c) *and* Fig. 76 (c) and (d).

lens formula An equation relating the object distance (u), the image distance (v) and the focal length (f) of thin lenses. All distances are measured from the optical centre of the lens. In the 'real-is-positive' sign convention, in which real object and image distances are positive and virtual object and image distances are negative, the equation is:

$$\frac{1}{v} + \frac{1}{u} = \frac{1}{f}$$

In the 'New Cartesian' sign convention, in which distances in the same direction as the incident light are positive and distances against it are negative, the equation is:

$$\frac{1}{v} - \frac{1}{u} = \frac{1}{f}$$

lenticel A ventilating pore, found in stems and roots which occur above soil level. It contains loosely packed cells and allows the exchange of gases between the interior of the plant and the atmosphere. *See* Fig. 86.

lenticular Shaped like a biconvex lens.

Lenz's law A law stating that a current produced as a result of electromagnetic induction is in such a direction as to oppose the motion producing it.

Lepidoptera A large order of insects whose members include butterflies and moths. Characteristics include two pairs of mem-

Leyden jar

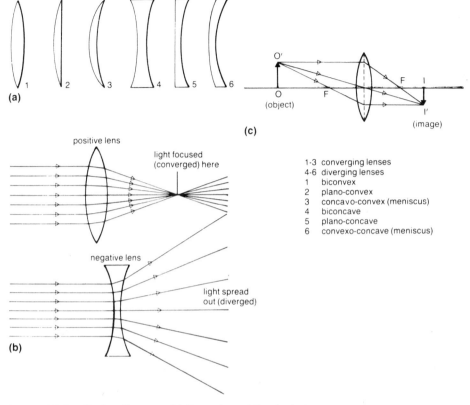

Fig. 112 (a) Classification of lenses (b) Converging and diverging lenses (c) Image formed by a lens

branous wings covered with scales, a body covered with scales, a proboscis for feeding and complete metamorphosis with caterpillars as larvae.

leprosy An infectious disease affecting skin, nerves, muscles and bones. It is caused by a bacillus and develops very slowly. The disease is common in areas with overcrowding, poor nutrition and lack of hygiene. Leprosy can be treated with drugs; if not treated it may cripple the patient.

leptotene *See* meiosis.

Leslie's cube A large, cubic metal container used to measure the magnitude of heat radiation from different surfaces. The four vertical sides are treated with different finishes, e.g. one may be polished, one may be painted matt black, etc. The cube is filled with hot water and the amount of heat radiation from each type of surface is detected with a thermopile.

lethal Causing death.

leuco- Prefix meaning white.

leucocyte A white blood corpuscle (cell). *See* blood corpuscle.

leucoplast A colourless plastid from which chloroplasts and chromoplasts arise. Leucoplasts are found in plant cells which are not normally exposed to light. There are different types of leucoplasts: amyloplasts, which store starch; aleuroplasts, which store proteins; and elaioplasts, which store fats.

lever One of the simplest machines, consisting of a rigid beam turning about a pivot (fulcrum, F). A force called the effort (E) is applied to one point on the beam in order to overcome a force called the load (L) at another point. There are three types (orders) of levers, depending upon the relative positions of the effort, fulcrum and load. *See* Fig. 113.

Lewis acid *See* acid–base theories.

Lewis base *See* acid–base theories.

Leyden jar An early type of capacitor. One of its earliest forms consisted of a glass jar (the dielectric) partly coated inside and outside with electrodes of lead foil.

liana

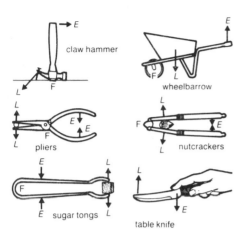

Fig. 113 Levers

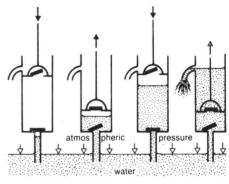

Fig. 114 Lift pump

liana (liane) A woody, climbing plant found in tropical forests.
libido Sexual desire.
lichen A partnership between an alga and a fungus living in symbiosis. Lichens are found as green, grey, bluish-green or yellow crusty growths on rocks, tree-trunks, old walls and roofs. They are the most universally distributed of all plants, being capable of growing in a variety of climates. Lichens have some economic importance in that dyes may be extracted from them, e.g. litmus. Some species are an important food source for animals, e.g. reindeer.
Liebig condenser See condenser.
life cycle The various stages through which an organism passes from being a fertilised egg (ovum) until its death. During its life cycle an organism usually produces a new generation of individuals, sexually or asexually, to repeat the process.
lift pump A pump for raising water. It differs from the force pump in that there is a valve in its piston and it can only raise water to a limited height (less than 10 metres). See Fig. 114.
ligament A strong, elastic, band of fibrous tissue connecting the ends of bones where they form joints. A ligament also forms the hinge of bivalve shells.
ligand An atom or group of atoms surrounding the central atom in a complex. Example: in the copper ammonia complex ion, $[Cu(NH_3)_4]^{2+}$ the four ammonia molecules, NH_3, are the ligands and the copper atom is the central atom. This particular complex ion has an intense blue colour. It can be formed when an excess of ammonia solution (ammonium hydroxide) is added to a solution containing copper(II) ions, Cu^{2+}:
$Cu^{2+} + 4NH_3 \rightarrow [Cu(NH_3)_4]^{2+}$.

ligases Enzymes which catalyse the linking together of two molecules (X and Y) in the presence of ATP, e.g.
$X + Y + ATP \rightleftharpoons X—Y + AMP + P—P$.
The bonds formed with ATP cleavage may be such bonds as C—O, C—C, C—S and C—N.
light-emitting diode (LED) A semiconductor diode commonly used to produce a visual display in electronic calculators, clocks, watches, etc. The colour of the emitted light depends on the material of which the diode is constructed.
light energy A form of energy consisting of electromagnetic radiation. It produces a visual sensation when it strikes the retina of the eye. The wavelength of light lies in the range of 400 (violet) – 730 (red) nanometres (nm), and the speed of light in a vacuum is $2,997\,925 \times 10^8$ metres per second ($2.997\,925 \times 10^8$ m s^{-1}). See also electromagnetic spectrum and photon.
light filter See filter.
light meter See exposure meter.
lightning An electrical discharge, producing a large spark (flash) between two charged clouds or between a charged cloud and the Earth. See also thunder.
lightning conductor A conductor of electricity, usually a thick copper strip, attached to the outside of a building. Its lower end is buried in the Earth and its upper end terminates in one or more sharp points. It serves to protect the building from lightning damage as it provides a direct path of least resistance to Earth for lightning. See Fig. 115.
light-year A unit of length expressing the distance a ray of light travels in a vacuum in one year. It is equal to 9.46×10^{12} km (9.46×10^{12} km) or 63 240 astronomical units (a.u.). It is used in stating astronomical distances.
ligneous Pertaining to or resembling wood.
lignin A complex, aromatic, organic compound found in the woody tissue of plants, often in combination with cellulose. Lignin gives strength and rigidity to the cell walls.

linear expansivity

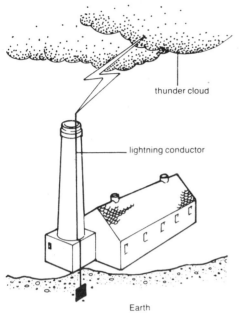

Fig. 115 Lightning conductor

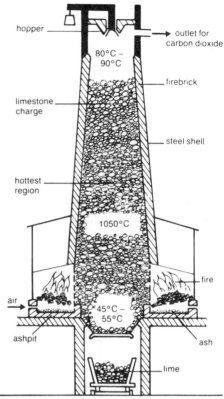

Fig. 116 Lime-kiln

lignite A dull-brown type of coal, intermediate between peat and bituminous coal.

limb A vertebrate appendage, e.g. arm, leg, wing.

lime *See* calcium oxide.

lime-kiln A furnace in which limestone, $CaCO_3$, is heated to a temperature above 825°C, at which it decomposes into lime, CaO, and carbon dioxide:
$CaCO_3 \rightarrow CaO + CO_2$
See Fig. 116.

limelight *See* incandescence.

limestone *See* calcium carbonate.

lime water *See* calcium hydroxide.

limiting friction *See* friction.

limnology The study of fresh waters and their inhabitants.

limonite A yellow or brown, naturally occurring form of hydrated iron oxide which may be represented by the formula $2Fe_2O_3 \cdot 3H_2O$. It is formed by decaying vegetation in swamps, giving the water a characteristic colour. Limonite is an iron ore. *See also* ochre.

lindane $C_6H_6Cl_6$ (1,2,3,4,5,6-hexachlorocyclohexane; benzene hexachloride, BHC). Lindane is prepared by treating benzene, C_6H_6, with chlorine, Cl_2:
$C_6H_6 + 3Cl_2 \rightarrow C_6H_6Cl_6$
The process is carried out in the presence of light of short wavelength, which provides the necessary activation energy. Lindane is an insecticide and a mixture of the possible isomers arising from the above process; the insecticidal property is due to only one of these isomers. Lindane is a colourless, crystalline solid which is insoluble in water, but soluble in several organic solvents such as benzene, propanone (acetone) and trichloromethane (chloroform).

Linde's process A process used for liquefying air. Dry air free from carbon dioxide enters a concentric tube at a pressure of about 200 atmospheres. The tube is very long and coiled; from it, the air can be suddenly allowed to expand into a chamber at a pressure of about 50 atmospheres. This cools the air, which is then returned to cool the coiled tube. Incoming, compressed air is therefore cooled before it expands. By repeating this procedure, the temperature eventually drops to around $-195°C$, the point at which the air changes into a liquid. *See also* Joule–Thomson effect.

linear expansivity Symbol: α. The linear expansivity of a substance is its increase in

linear magnification

length per unit length when its temperature rises by one degree. The unit of linear expansivity is 'per °C' or 'per K'.

$$\alpha = \frac{\text{increase in length}}{\text{original length} \times \text{temperature rise}}$$

This equation is usually expressed as follows: $l_2 = l_1(1 + \alpha\theta)$ where l_2 and l_1 are the increased length and the original length respectively and θ (theta) is the temperature rise.

linear magnification (transverse magnification) *See* magnification.

line of force An imaginary line in an electric or magnetic field whose direction at any point is the same as the direction of the field at that point.

line spectrum A type of spectrum which is composed of a number of distinct lines (spectral lines) each representing a certain wavelength of emitted or absorbed radiation. A line spectrum is produced from atoms in the gaseous state (e.g. when an electrical discharge passes through them) and the line spectrum may extend into both the ultraviolet and infrared regions. A particular element always produces a set of lines in the same position in the spectrum. Line spectra occur as a result of electrons moving between different energy levels. *See also* Balmer series.

linkage 1. The tendency of several genes not to separate from one another through several generations, except by accident, i.e. they are linked together because they are situated on the same chromosome, forming a linkage group. 2. The amount of magnetic flux passing through a coil or circuit. It is the product of the total number of lines of magnetic flux and the number of turns in the coil (or circuit) through which they pass.

linkage group *See* linkage.

linseed oil An oil extracted from the seeds of flax plants. It contains glycerides of unsaturated fatty acids and is therefore easily oxidised by oxygen in air attacking the double bonds. For this reason linseed oil is widely used in the varnish and paint industries.

Linz–Donawitz process *See* L–D process.

lipase An enzyme which hydrolyses fats to glycerol and fatty acids.

lipid(e) The general name for a group of organic substances which is difficult to define precisely. However, lipid(e)s may be considered to be esters of long chain carboxylic acids and various alcohols. Oils, fats and waxes are lipid(e)s; cholesterol and lecithin are also considered to be lipid(e)s.

lipolysis A process in which fats (lipid(e)s) are broken down during digestion. This takes place in the duodenum and ileum and is catalysed by enzymes (lipases).

lipoprotein A complex compound of protein and lipid(e). Together with mucoproteins, lipoproteins are important structural material for cell membranes.

liquation The process by which a solid mixture (usually an alloy or an ore) is separated by heating. The components with the lower melting-points will form a fused fraction which runs off.

liquefaction of a gas The change of a gas into a liquid. This can be brought about by pressure alone if the gas has a critical temperature above room temperature, otherwise some cooling must be applied.

liquefied petroleum gases (LPG) These gases are mainly alkanes, such as propane, C_3H_8, and butane, C_4H_{10}, but may include some alkenes.

liquid A state of matter, intermediate between a solid and a gas, in which the particles of a substance move relatively freely. Liquids have a low compressibility.

liquid air A pale blue liquid obtained by the liquefaction of air. It mainly contains liquid nitrogen (b.p. $-195,8°C$; $-195.8°C$) and liquid oxygen (b.p. $-183,0°C$; $-183.0°C$). It is used in the industrial preparation of nitrogen and oxygen by fractional distillation of the mixture.

liquid crystal A substance which has both liquid and solid properties. The arrangements of molecules (atoms) in liquid crystals are not completely fixed as they are in solids; they can easily be changed by an electric or magnetic field or by heat. A change in the arrangements of molecules alters the optical properties of liquid crystals. They may change from transparent to opaque, or change colour. Liquid crystals are used in digital displays of electronic calculators, watches, thermometers, etc.

liquid level apparatus *See* Pascal's vases.

litharge *See* lead oxides.

lithium An element with the symbol Li; atomic number 3; relative atomic mass 6,94 (6.94); state, solid. Lithium is a very electropositive element belonging to the alkali metals in the Periodic Table of the Elements. It occurs naturally in various minerals, e.g. spodumene, $LiAlSi_2O_6$. Lithium is the least dense of all solid elements, with a relative density of 0,534 (0.534). It is extracted from lithium ores using concentrated sulphuric acid, H_2SO_4, to form lithium sulphate, Li_2SO_4. The free element is then produced by converting lithium sulphate into lithium chloride, LiCl, which is then electrolysed when molten. Lithium is used to a slight extent in some alloys and in the form of lithium hydride, LiH, or lithium aluminium hydride, LiAlH$_4$, as a reducing agent. As lithium has the highest specific heat capacity of any solid element, it has found a use in heat transfer applications.

lithium aluminium hydride LiAlH$_4$ An important reducing agent in both inorganic and organic chemistry. It is a crystalline solid which is white when pure, but is more usually grey. It is manufactured by the action of lithium hydride, LiH, on aluminium chloride, AlCl$_3$:
4LiH + AlCl$_3$ → LiAlH$_4$ + 3LiCl
With water LiAlH$_4$ reacts violently releasing hydrogen:
LiAlH$_4$ + 4H$_2$O → 4H$_2$ + Al(OH)$_3$ + LiOH
It is able to accomplish many difficult reductions, e.g. the reduction of the carboxyl group, —COOH to a primary alcohol group, —CH$_2$OH.

lithium chloride LiCl A white, solid, soluble salt which is extremely deliquescent. It can be obtained in its anhydrous form or with one, two or three molecules of water of crystallisation. It is prepared by treating lithium carbonate, Li$_2$CO$_3$, with hydrochloric acid, HCl:
Li$_2$CO$_3$ + 2HCl → 2LiCl + CO$_2$ + H$_2$O.
Lithium chloride is used as a flux in soldering and in mineral waters.

lithium hydride LiH A colourless solid which is produced when lithium reacts with hydrogen:
2Li + H$_2$ → 2LiH.
During electrolysis of fused LiH, hydrogen is liberated at the anode. With water it releases hydrogen:
LiH + H$_2$O → H$_2$ + LiOH
Lithium hydride is seldom used except for the preparation of the more useful hydride, lithium aluminium hydride, LiAlH$_4$.

lithopone *See* zinc sulphide.

lithosphere *See* crust.

litmus A soluble, purple substance extracted from certain lichens and used as an indicator. Its colour changes to red in acid solutions (pH<5) and to blue in alkaline solutions (pH>8). Litmus is never used in titrations because its colour change is difficult to follow precisely.

litre Symbol: l. A metric unit of volume defined as 10^{-3} cubic metres (m^3). This unit is not used in high precision measurements.

litter 1. The dead leaves, twigs and other plant debris which collect on forest floors and which eventually break down to form humus. **2.** The animals produced in a single multiple birth.

live *See* alive.

liver The largest organ of the body. In humans it has a mass of about 1,5 kg (1.5 kg). It is shaped like a wedge and is dark red in colour. The liver is situated immediately beneath the diaphragm in the upper right side of the abdominal cavity. It has a number of functions which include the production of bile, the deamination of amino acids, the formation and storage of glycogen, the storage of certain vitamins, the detoxication of certain products of digestion and the regulation of the blood sugar. *See* Fig. 49.

liverwort *See* Hepaticae.

lixiviation The process by which the soluble components of a mixture are washed out with water or another solvent.

lizard *See* Squamata.

load The force applied to a body by a machine.

loam A type of soil which is one of the best for crop production. It consists of a mixture of almost equal amounts of sand and silt, a small amount of clay and a good humus content.

lobe Any roundish, flattish projection of an organ, e.g. the lobe of the ear.

local action One of the faults of a simple cell, caused by impurities in the zinc electrode. These set up small local cells on the surface of the zinc, the impurity forming one electrode and the zinc the other. This results in the flow of local currents and the zinc gradually dissolves in the electrolyte even when no current is being taken from the cell. Local action can be prevented by amalgamating the zinc electrode, i.e. rubbing its surface with mercury to form a coating of zinc amalgam, or by using pure zinc.

lockjaw *See* tetanus.

locus (pl. loci) The position occupied by a gene on a chromosome.

locust *See* Orthoptera.

lodestone *See* iron oxides.

loins The waist regions of the body, situated below the thorax and above the pelvis.

lone pair A pair of electrons in the outermost shell of an atom which are not involved in the formation of covalent bonds. Together with bonding pairs of electrons, the lone pair determines the shape of a covalent molecule. A nitrogen atom, N, has five electrons in its outermost shell, $\cdot\ddot{N}\cdot$. When forming ammonia, NH$_3$, with three hydrogen atoms, H, three of the five electrons form covalent bonds with electrons from the three hydrogen atoms, whereas the remaining two electrons form a lone pair: H:N̈:H.
                                              H

In a covalent bond between two atoms, one electron is contributed from each atom to the bond. However, in some cases both electrons (the lone pair) in an electron-pair bond are supplied by only one of the bonded atoms. Such a bond is called a coordinate bond or a dative bond and is typical of complexes. *Example:* the complex ion $[Ag(NH_3)_2]^+$ is formed when two ammonia molecules each donate their lone pair to a silver ion, Ag^+:

$2NH_3 + Ag^+ \rightarrow [Ag(NH_3)_2]^+$ or
$H_3N: + Ag^+ + :NH_3 \rightarrow [H_3N \rightarrow Ag \leftarrow NH_3]^+$

where $\rightarrow$ signifies a coordinate bond. *See also* acid–base theories.

long-day plant *See* photoperiodism.

longitudinal wave A wave motion in which the vibration or displacement is in the same direction as the direction of propagation, e.g. sound waves. *See also* transverse wave *and* travelling wave.

long sight *See* hypermetropia.

Loschmidt's number Symbol: *L*. The number of molecules present in one cubic centimetre (cm³) of an ideal gas at standard temperature and pressure. It is equal to $2,687\,19 \times 10^{19}$ ($2.687\,19 \times 10^{19}$).

loudspeaker A device for converting varying electric currents into sound energy. The most common type is the moving coil loudspeaker which consists of a small coil attached to a cone of stiff paper. The coil is free to move in the magnetic field produced by a permanent magnet. Varying currents flowing through the coil cause the cone to vibrate at the same frequencies, thus emitting sound waves. *See* Fig. 117.

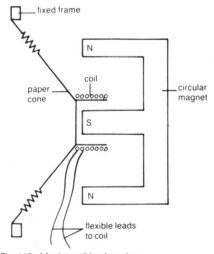

Fig. 117 Moving coil loudspeaker

low blood pressure *See* hypotension.
low tension (LT) Low voltage.
low voltage Any voltage less than 250 volts.
LPG *See* liquefied petroleum gases.
LSD (lysergic acid diethylamide, 'acid') A very potent hallucinogenic drug which is active in minute doses. It is a derivative of lysergic acid and was originally obtained from an alkaloid produced by a parasitic fungus that sometimes can be found on rye or other grain. However, LSD can now be fully synthesised in the laboratory. LSD has an unpredictable effect on the human mind. The physical effects include dilatation of the pupils, rise in blood pressure and heart-rate, nausea, tremor, fever and insomnia. LSD is not a drug of addiction and has been used in psychotherapy.

LT Low tension. *See* low voltage.

lubricant Any solid or liquid which is capable of reducing friction between sliding surfaces. Examples include graphite, talc and oils.

lucerne (alfalfa) A leguminous plant cultivated for cattle fodder.

luciferase An enzyme found in all luminescent organisms. It activates the substance luciferin to emit light. However, luciferase itself may be the light-emitting molecule.

luciferin A very complex organic substance which is capable of emitting light as a result of its oxidation brought about by luciferase. Luciferin causes bioluminescence.

lumbar Pertaining to or near the loins.

lumbar puncture The insertion of a hollow needle into the spinal cord to remove a sample of cerebrospinal fluid for examination. This procedure is very useful in the diagnosis of several diseases such as meningitis and poliomyelitis. The presence of blood in the sample indicates brain haemorrhage.

lumbar vertebrae The vertebrae situated between the thoracic and sacral vertebrae. In humans there are five; they are the largest and strongest of all the vertebrae. *See* Fig. 242.

lumen 1. (pl. lumina) The cavity within a tube, sac or gland. **2.** Symbol: lm. The SI unit of luminous flux. It is equal to the amount of light energy emitted per second in a cone of unit solid angle by a point source of one candela. 1 lumen is equal to $0,001\,471$ watts ($0.001\,471$ W).

lumen-hour A unit of quantity of light energy equal to a flux of one lumen continued for one hour.

lumen-second A unit of quantity of light energy equal to a flux of one lumen continued for one second.

luminescence The emission of light from a substance or an organism as a result of a non-thermal process. Particular cases of luminescence are fluorescence and phosphorescence.

luminous flux Symbol: Φ_v (phi$_v$); unit: lumen (lm). The amount of light energy passing through an area in one second.

luminous intensity Symbol: I_v; unit: candela (cd). The amount of light emitted per second per unit solid angle from a point source of light in a given direction.

lunar Pertaining to the moon.

lunar eclipse *See* eclipse.

lung The paired or single respiratory organ of vertebrates in which the exchange of oxygen and carbon dioxide takes place between air and blood. In humans, the lungs consist of two thin-walled, elastic sacs situated in the thorax. Inhaled air passes down the trachea (windpipe) which divides into two branches, the bronchi. These further divide forming bronchioles which terminate in alveoli where the gaseous exchange takes place. Exhaled air returns via the same route. *See* Fig. 118.

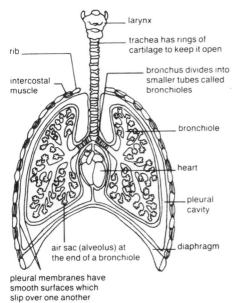

Fig. 118 Lung: section through the thorax

lung book A respiratory structure found in some arachnids such as spiders and scorpions. Lung books consist of blood-filled projections of tissue resembling the leaves of a book. Oxygen can diffuse into these projections which are found in a cavity of the body wall. This arrangement increases the surface area for gaseous exchange.

lustre The quality of a smooth or polished surface (e.g. of a metal) of shining by reflected light.

luteinising hormone In vertebrates, a hormone which is secreted by the anterior lobe of the pituitary gland. In females it stimulates the production of oestrogen in the ovaries, promotes the final development of the ovarian follicles and produces ovulation. In males it activates the production of testosterone and, in some vertebrates, it promotes the transfer of sperm from testes to ducts. In female mammals, luteinising hormone initiates the development of the corpus luteum.

lux Symbol: lx. The SI unit of illuminance. It is equal to one lumen per square metre.

lycopodium powder A fine powder obtained from the spores of plants belonging to the genus Lycopodium. It is used in fireworks.

Lyman series *See* Balmer series.

lymph A clear, colourless, alkaline fluid which is formed when tissue fluid which is not absorbed into the blood stream drains into the lymphatic vessels in which lymph circulates. Lymph is similar in composition to blood plasma. It normally contains lymphocytes and a few leucocytes, but no erythrocytes, platelets or fibrinogen. Like blood, it contains enzymes and antibodies. Lymph flows in one direction only (from tissue to heart) by the contraction of the walls of the lymphatic vessels and the movement of muscles near them. Lymph has several functions: e.g. bacteria and dead cells are engulfed in the lymph nodes; lymph carries fats away from the villi in the small intestine.

lymphatic Pertaining to or conveying lymph.

lymphatic system A system of small vessels conveying lymph. It consists of a network of blind-ending capillaries which permeate most tissue and drain into larger vessels which have valves to ensure that the lymph only flows in one direction. These larger vessels finally join veins, usually near the heart.

lymph gland *See* lymph node.

lymph node (lymph gland) A small, flattened, round or bean-shaped organ. Lymph nodes are of variable size and are situated at intervals along lymphatic vessels. They contain a network of fibres through which the lymph drains. Attached to the walls of the fibres are large phagocytes which engulf bacteria and dead cells. Lymphocytes and antibodies are also produced in the lymph nodes. In humans, large lymph nodes are found in the groin, the armpit and the neck.

lymphocyte A type of white blood corpuscle (cell) produced in the lymph nodes, red bone marrow and spleen. They are concerned in the production and carrying of antibodies.

lyophilic Having an affinity for a solvent.

lyophobic Having no affinity for a solvent.

lysin A substance capable of causing lysis (dissolution) of cells and bacteria.

lysis The dissolution of cells or bacteria caused by a lysin.

lysosome A type of membrane-bounded organelle found in the cytoplasm of animal cells. They have not yet been clearly shown in any plant cells. Lysosomes contain enzymes

lysozyme

capable of breaking down proteins and fats. The enzymes are liberated in injured cells, and assist in the digestion and removal of dead cells. They are also involved in metamorphosis.

lysozyme An anti-bacterial enzyme found in mammalian tissue secretions such as tears and saliva. It is also found in egg-white.

M

M See molarity.
MA See mechanical advantage.
machine Any of various devices for transmitting power. A small force (the effort) is applied to a machine to overcome a large force (the load). Examples of simple machines are the lever, the inclined plane and the pulley. See also efficiency, mechanical advantage and velocity ratio.

Mach number The ratio of the speed of a fluid or of an object in a fluid to the speed of sound in the fluid. Mach 1 is the speed of sound in the fluid. Speeds below Mach 1 are subsonic, speeds above Mach 1 are supersonic and speeds above Mach 5 are hypersonic.

macro- Prefix meaning large. Compare micro-; see also mega-.

macromere See megamere.

macromolecule A very large molecule such as a polymer.

macronutrient An element required in relatively large quantities by living organisms. Examples include calcium, magnesium, phosphorus, carbon, hydrogen, nitrogen and oxygen. Compare micronutrient.

macrophage A large phagocytic cell, which may be either fixed or moving. Macrophages are found in lymph nodes, bone marrow, blood and lymph of vertebrates. They remove foreign particles from blood and lymph.

macroscopic (megascopic) Visible to the naked eye. Compare microscopic.

macula (pl. maculae) A sensory area composed of supporting cells and sensory hair cells on the lining of the saccule and utricle of the ear. The hairs are embedded in a gelatinous substance containing granules of calcium carbonate. These granules are called otoliths and they are sensitive to the pull of gravity.

macula lutea See yellow spot.

Magdeburg hemispheres Two closely-fitting metal hemispheres, one of which has a tap. Air inside the sphere is pumped out through the tap to create a vacuum. When the tap is then closed, the hemispheres are difficult to separate owing to the external pressure of the atmosphere on them. See Fig. 119.

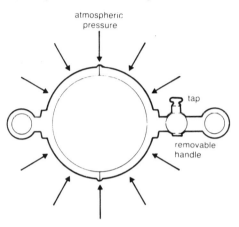

Fig. 119 Magdeburg hemispheres

magenta See colour mixing.

maggot A worm-like larva of certain insects such as the house-fly. It has no appendages and no distinct head.

magic numbers Most of the stable nuclei of elements (isotopes) contain even numbers of protons or neutrons or both. Some exceptions are deuterium, ^2_1H, lithium, ^6_3Li, boron, $^{10}_5B$, and nitrogen, $^{14}_7N$, which have odd numbers of both protons and neutrons. The numbers 2, 8, 20, 28, 50, 82 and 126 are called 'magic numbers' of protons and neutrons. It has been proved that more energy is required to remove a proton or a neutron from a nucleus with a magic number of these particles than from a nucleus with a non-magic number of them.

magma Molten rock formed in the upper mantle of the Earth. Magma may be brought to the surface of the Earth by volcanic action, when it becomes known as lava. See also igneous rock.

magnalium A magnesium-aluminium alloy containing 5–30% magnesium and 70–95% aluminium. Certain varieties of this alloy may contain small amounts of copper, nickel, tin and lead. Magnalium is a light alloy with great strength and is used in the manufacture of aircraft components.

magnesia See magnesium oxide.

magnesite A magnesium mineral consisting of magnesium carbonate, $MgCO_3$, and often occurring with small amounts of iron and calcium. It is a refractory material, used as a furnace-liner and in the cement and paper industries. Magnesite is a source of magnesium.

magnesium An element with the symbol Mg; atomic number 12; relative atomic mass 24,31 (24.31); state, solid. It is a light metal belonging to the alkaline earth metals. Magnesium is extracted from mineral ores such as magnesite, $MgCO_3$, dolomite, $CaMg(CO_3)_2$ and carnallite, $KMgCl_3 \cdot 6H_2O$, or from sea-water. The magnesium in these substances is converted to magnesium chloride which is eventually electrolysed in the fused state. When magnesium is exposed to air a protective layer of magnesium oxide, MgO, forms on its surface, thus preventing further oxidation. Magnesium is used in the production of lightweight alloys such as magnalium. As it burns with an intensive white flame, magnesium is also used in fireworks and flash tubes. Magnesium is found in chlorophyll.

magnesium carbonate $MgCO_3$ A magnesium salt occurring naturally in the minerals magnesite, $MgCO_3$, and dolomite, $CaMg(CO_3)_2$. It is only slightly soluble in water. When heated, it decomposes into magnesium oxide, MgO, and carbon dioxide, CO_2:
$MgCO_3 \rightarrow MgO + CO_2$
Like all carbonates, it dissolves in acids to give a salt, carbon dioxide and water. Magnesium carbonate is a refractory material.

magnesium chloride $MgCl_2$ A soluble magnesium salt extracted from carnallite, $KMgCl_3 \cdot 6H_2O$, or sea-water. It exists commonly as the hydrated form, $MgCl_2 \cdot 6H_2O$. The anhydrous form, $MgCl_2$, is deliquescent and is used as a source of magnesium by its electrolysis in the fused state. Hydrated magnesium chloride, $MgCl_2 \cdot 6H_2O$, loses water of crystallisation when heated. However, it is also hydrolysed by its water with the evolution of hydrogen chloride, HCl, and the formation of magnesium oxide, MgO:
$MgCl_2 \cdot 6H_2O \rightarrow MgO + 2HCl + 5H_2O$
For this reason it is not possible to obtain the anhydrous salt simply by evaporating an aqueous solution of the chloride. The anhydrous form can be made by evaporating an aqueous solution of the chloride in a stream of hydrogen chloride, which reverses the above equation. Magnesium chloride is used in fireproofing and fire-extinguishing materials and in the textile industry.

magnesium hydrogencarbonate $Mg(HCO_3)_2$ A soluble magnesium salt which cannot be isolated in the pure state as its aqueous solution decomposes on heating into magnesium carbonate, $MgCO_3$, carbon dioxide, CO_2, and water:
$Mg(HCO_3)_2 \rightarrow MgCO_3 + CO_2 + H_2O$

It is one of the causes of the temporary hardness of water: when rain-water, containing dissolved carbon dioxide from the atmosphere, attacks magnesium carbonate, which is found in most soils, the insoluble magnesium carbonate is converted into soluble magnesium hydrogencarbonate:
$MgCO_3 + CO_2 + H_2O \rightarrow Mg(HCO_3)_2$

magnesium hydroxide $Mg(OH)_2$ A crystalline, slightly soluble white powder occurring naturally in the mineral brucite. Magnesium hydroxide can be made by adding a strong alkali to a solution containing magnesium ions, Mg^{2+}:
$Mg^{2+} + 2OH^- \rightarrow Mg(OH)_2$
It is used in medicine as an antacid (milk of magnesia).

magnesium oxide MgO (magnesia) A white solid occurring naturally in the mineral periclase. It can be made by heating magnesium in air or oxygen:
$2Mg + O_2 \rightarrow 2MgO$
or by the thermal decomposition of several magnesium compounds such as magnesium carbonate, $MgCO_3$:
$MgCO_3 \rightarrow MgO + CO_2$
Magnesium oxide is slightly soluble in water, producing magnesium hydroxide, $Mg(OH)_2$:
$MgO + H_2O \rightarrow Mg(OH)_2$
It is a refractory material, used in the cement industry and in medicine as an antacid and a laxative.

magnesium silicide See silicide.

magnesium sulphate $MgSO_4$ A white, soluble magnesium salt usually occurring as the hydrated form $MgSO_4 \cdot 7H_2O$. Magnesium sulphate is found naturally in certain salt deposits. It can be made by the action of sulphuric acid, H_2SO_4, on magnesium carbonate, $MgCO_3$:
$MgCO_3 + H_2SO_4 \rightarrow MgSO_4 + CO_2 + H_2O$
It is used in the textile industry. See also Epsom salt.

magnet A piece of iron or other material possessing the property of magnetism. Magnets are either permanent or temporary. See alo electromagnet and pole.

magnetic circuit The complete path described by the lines of force of a magnetic field. See also armature.

magnetic compass See compass.

magnetic declination (magnetic variation) See angle of declination.

magnetic dip (magnetic inclination) See angle of dip.

magnetic dipole A north pole and a south pole separated by a distance, as in magnets or in a loop of wire carrying an electric current.

magnetic domain See domain theory.

magnetic equator (aclinic line) A line around the Earth joining points where the magnetic field is horizontal; the angle of dip at these points is zero. *See* Fig. 122.

magnetic field The space in the neighbourhood of a magnetic object or a wire carrying an electric current. When a magnet is placed in this region, a force is exerted upon it. A magnetic field can be represented by lines of force. *See* Fig. 120.

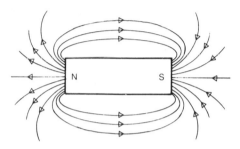

Fig. 120 Magnetic field

magnetic flux Symbol: Φ (phi); unit: weber (Wb). The total magnetic field through a small plane surface. It is the product of the magnetic flux density and the area of the surface through which the field passes.

magnetic flux density (magnetic induction) Symbol: B. The magnetic flux that passes through a unit area of a magnetic field normal to the magnetic force. The SI unit of magnetic flux density is the tesla (T). One tesla is equal to one weber per square metre.

magnetic focusing The focusing of beams of charged particles, e.g. electrons, by magnetic fields. This technique is used in cathode-ray tubes, electron microscopes, etc.

magnetic force A force of attraction or repulsion exerted by a magnetic field on a magnetic pole or an electric charge.

magnetic inclination Magnetic dip. *See* angle of dip.

magnetic induction *See* magnetic flux density.

magnetic iron ore *See* iron oxides.

magnetic lens A system of magnets used to focus beams of charged particles in cathode-ray tubes, electron microscopes, etc. Its action is similar to the action of a glass lens on beams of light. *See also* magnetic focusing.

magnetic line of force *See* line of force.

magnetic material One of the elements iron, nickel and cobalt or an alloy containing one or more of these elements, e.g. Alnico. Such materials are used in the manufacture of magnets and are classed as either 'hard' or 'soft'. Hard magnetic materials retain their magnetism (e.g. steel) whereas soft magnetic materials lose it easily (e.g. soft iron). *See also* magnetise *and* ferromagnetic material.

magnetic meridian The magnetic meridian at any place is the vertical plane containing the direction of the Earth's magnetic field at that place. *See also* Fig. 2.

magnetic pole *See* pole.

magnetic saturation The point beyond which it is not possible to make a material more magnetic. The strength of a magnet cannot be further increased beyond this limit because all the domain axes are now brought into alignment with the magnetising field. *See also* domain theory.

magnetic shielding (screening) Placing a delicate measuring instrument, e.g. a watch, in a soft iron case or Mumetal case to protect it from external magnetic fields.

magnetic storm A sudden, irregular disturbance in the Earth's magnetic field as a result of increased solar activity. This may seriously affect navigating instruments, telegraph, radio and television and may displace the positions of the Earth's magnetic poles. *See also* sunspot.

magnetic tape A strip of flexible plastic tape coated on one side with a layer of magnetic iron oxide. The tape is first demagnetised and subsequently magnetised as it passes through a tape recorder. Magnetic tape is used to store information for computers, in sound reproduction and in video equipment.

magnetic variation Magnetic declination. *See* angle of declination.

magnetise To induce magnetism. A piece of steel can be made into a magnet by placing it inside a coil (solenoid) through which a direct current is flowing for a few seconds. Another way is by 'stroking' the piece of steel with a permanent magnet using the methods of single touch or divided touch. An old method is to hammer a red-hot piece of steel whilst it is cooling and lying in a North–South direction. A weak magnet can be made by hammering the end of a piece of steel which is inclined to the horizontal at an angle equal to the angle of dip where the magnet is being made. *See* Fig. 121 (a), (b) *and* (c).

magnetism The study of magnets and magnetic fields. *See also* domain theory.

magnetism, terrestrial (geomagnetism) The Earth's magnetism. The Earth possesses a magnetic field similar to that of a powerful bar magnet. It behaves as if a bar magnet were placed at the centre of the Earth with its poles pointing in a North–South direction, the north pole of the magnet being in the Southern hemisphere of the Earth. The cause of the Earth's magnetic field is not exactly known,

magnifying glass

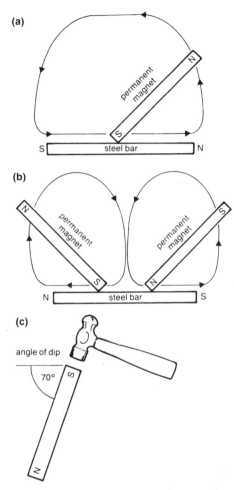

Fig. 121 Magnetising a piece of steel
(a) Single touch (b) Divided touch
(c) Hammering

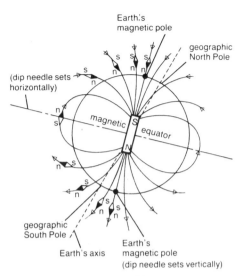

Fig. 122 Magnetism, terrestrial

but it probably involves convection currents of molten material within the Earth's core. It varies with time and locality. See Fig. 122.

magnetite See iron oxides.

magnetometer An instrument used to compare the strengths of magnetic fields at different places. The deflection magnetometer consists of a short, freely suspended magnet with a non-magnetic pointer attached at right angles across it. The magnet and pointer are pivoted to move in a horizontal plane over a circular scale graduated in degrees. The pointer measures deflections of the magnet.

magnetosphere The region around a planet in which its magnetic field is effective.

magnetotaxis The reaction of certain aquatic bacteria in response to magnetic field lines.

Such bacteria are influenced by the Earth's magnetic field. Magnetotactic bacteria (discovered in 1975) are able to synthesise tiny crystals of magnetite, Fe_3O_4. They are either north-seeking or south-seeking. Magnetotactic bacteria are found in both freshwater and seawater. They include cocci, bacilli and spirals.

magnification Symbol: m. A measure of the performance of an optical system such as a microscope, telescope or pair of binoculars. The linear (transverse) magnification of the system is the ratio of the image height and the object height:

$$m = \frac{\text{image height}}{\text{object height}}$$

or it is the ratio of the image distance and object distance from the centre of the system:

$$m = \frac{\text{image distance}}{\text{object distance}}$$

magnifying glass (simple microscope) A thick, convex lens used to view an object placed between the lens and its focal point. An erect, magnified, virtual image is obtained. See Fig. 123.

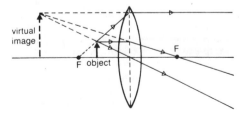

Fig. 123 Magnifying glass

mains The electricity supply from a power-station. *See also* grid.

make-and-break system A system in which an electric current is arranged to be switched on and off repeatedly. *See also* electric bell.

malachite A copper mineral (ore), varying in colour between light and dark green, which may be represented by the formula $Cu_2CO_3(OH)_2$. Some compact types of malachite are used for carvings and mosaics because of their vivid colouration and complex banding forming handsome patterns.

malaria A common, non-contagious tropical disease which is spread by mosquitoes. The disease is caused by a parasitic protozoan, of the genus *Plasmodium,* which is transmitted to Man by bites of certain female mosquitoes belonging to the genus *Anopheles.* Symptoms include acute attacks of high fever, chills and a tendency to relapse. If not treated, the disease leads to anaemia and disorders of the liver and spleen. Several drugs are used in the treatment and prevention of malaria (anti-malarial drugs). The use of insecticides against the anopheline mosquitoes has brought malaria under control in many places.

malathion $C_{10}H_{19}O_6PS_2$ An aliphatic compound which is a brownish-yellow liquid. It is sparingly soluble in water, but soluble in several organic solvents. Malathion is used as an insecticide.

male Symbol: ♂. An individual whose sex organs produce spermatozoa or corresponding gametes. *Compare* female.

malignant Describes a disorder which is likely to get worse (if not corrected). *Example:* certain tumours are said to be malignant (cancerous), i.e. they invade surrounding tissue, or cells from them spread through the blood or lymph to other parts of the body. *Compare* benign; *see also* metastasis.

malleable Describes a material which can be shaped by hammering or rolling.

malleus (hammer) The first of the three small bones in the middle ear. It is attached to the surface of the ear drum and transmits vibrations of the ear drum to the incus. *See* Fig. 55.

malnutrition A condition in which the body is deprived of a balanced diet. *See also* marasmus *and* kwashiorkor.

Malpighian body *See* Malpighian corpuscle.

Malpighian corpuscle (Malpighian body, renal corpuscle) One of numerous small, round structures situated in the cortex of vertebrate kidneys. A single corpuscle consists of a glomerulus surrounded by a Bowman's capsule and is concerned in filtration. *See also* nephron.

Malpighian layer (germinative layer) The inner layer of the epidermis, situated above the dermis. It is in this layer that skin cells actively divide producing new epidermis. The skin pigment melanin is found in this layer.

Malpighian tubule One of a number of blind-ending excretory tubules leading into the hind-gut. Malpighian tubules are found in all arthropods (except crustaceans), arachnids and myriapods. They are concerned in the excretion of uric acid and urates from the surrounding blood.

malt Grain such as barley, oats and wheat, which has been germinated under controlled conditions and then dried. It is rich in proteins, including enzymes, and is used in food manufacture and brewing.

maltase An enzyme hydrolysing maltose, $C_{12}H_{22}O_{11}$, into glucose, $C_6H_{12}O_6$. Maltase is found in intestinal juice and yeast.

Maltese cross tube A cathode-ray tube used to demonstrate that electrons travel in straight lines. It consists of a glass tube in which the metal anode is in the shape of a Maltese cross. A beam of electrons emitted from the cathode casts a sharp shadow of the cross on a fluorescent screen. *See* Fig. 124.

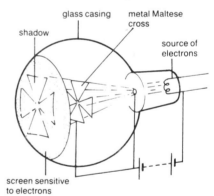

Fig. 124 Maltese cross tube

maltobiose *See* maltose.

maltose (malt-sugar, maltobiose) A reducing carbohydrate with the formula $C_{12}H_{22}O_{11}$, i.e. maltose is a disaccharide. When hydrolysed it forms two units of glucose, $C_6H_{12}O_6$. It is formed as an intermediate compound when starch is hydrolysed to glucose.

malt-sugar *See* maltose.

mamilla (pl. mamillae) A nipple or nipple-shaped organ.

mamma (pl. mammae) *See* mammary gland.

mammal *See* Mammalia.

Mammalia (mammals) A class of tetrapod vertebrates characterised by having skin covered with hair, females suckling their young which are born alive, a four-chambered heart, a constant body temperature, a diaphragm and lungs for breathing, three ossicles (small bones) in the middle ear, and usually sweat and sebaceous glands in the skin. Examples of mammals are humans, dogs, cats and whales.

mammary gland (mamma) In female mammals, a milk-secreting gland whose function is controlled by hormones secreted during pregnancy and at the birth of the offspring. It produces milk for the nourishment of the new-born progeny and is thought to be a specialised sweat gland.

mammography The examination of the breast using either X-rays or infra-red rays to reveal the possible presence of abnormal tissue such as tumours or cysts.

mandible 1. In vertebrates, the lower jaw. 2. In various arthropods, one of a pair of mouthparts used for biting and crushing. 3. In birds, the beak consists of an upper and lower mandible.

manganese An element with the symbol Mn; atomic number 25; relative atomic mass 54,94 (54.94); state, solid. Manganese is a transition metal occurring naturally in several minerals such as pyrolusite, MnO_2, and rhodochrosite, $MnCO_3$. The metal is extracted from these ores and is used in the manufacture of certain types of steel and other alloys. Large quantities of manganese nodules have been found on the floor of the oceans. The nodules are about 24% manganese, together with many other elements in lesser abundance.

manganese dioxide See manganese(IV) oxide.

manganese(IV) oxide MnO_2 (manganese dioxide) An insoluble black powder occurring naturally as the mineral pyrolusite. It can be prepared by thermal decomposition of manganese(II) nitrate, $Mn(NO_3)_2$:
$$Mn(NO_3)_2 \rightarrow MnO_2 + 2NO_2$$
Manganese(IV) oxide is a powerful oxidising agent. *Example:* it releases chlorine, Cl_2, from concentrated hydrochloric acid, HCl:
$$MnO_2 + 4HCl \rightarrow Cl_2 + MnCl_2 + 2H_2O$$
It is used in the laboratory preparation of chlorine (see above), as a catalyst and as a depolariser in dry cells.

manganin An alloy of copper (86%), manganese (12%) and nickel (2%). Its high electrical resistance varies only slightly with temperature. It is used to make resistors.

mannitol $CH_2OH—(CHOH)_4—CH_2OH$ A white, crystalline polyhydric alcohol derived from mannose or fructose. It is found in many plants, e.g. in brown algae and in the sap of various trees. *See also* sorbitol.

mannose A carbohydrate with the formula $C_6H_{12}O_6$. Mannose is a monosaccharide (an aldohexose) and one of the many isomers having this formula. It is a white, crystalline solid occurring naturally in many polysaccharides.

manometer A device for measuring pressure in fluids. A simple type consists of a glass or clear plastic U-tube half filled with a liquid such as mercury or water. One end of the tube is open to the atmosphere, the other is connected to the system to be measured. Pressure is measured by the difference in liquid levels (heights) between the two limbs of the U-tube. See Fig. 125.

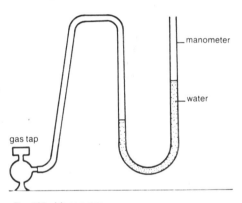

Fig. 125 Manometer

mantle 1. The layer of the Earth extending from beneath the crust (40 km below the surface) to 2900 km below the surface. 2. In molluscs, a fold of tissue covering all or most of the body. It serves to protect internal organs and its outer surface secretes the shell.

Mantoux test A test for the presence of tuberculosis by the injection of tuberculin under the skin. If the test is positive, i.e. if tuberculosis is present or has been present thus leaving the patient immune, a weal will develop on the skin near the place of injection within 48 hours. *See also* BCG vaccine *and* patch test.

manure *See* farmyard manure *and* fertiliser.

manus The hand or corresponding structure as found in mammals, amphibians and reptiles.

manyplies *See* psalterium.

marasmus A deficiency disease with symptoms similar to those of kwashiorkor, but with no abdominal swelling or skin rash. Marasmus is due to general starvation and not just lack of protein; it can occur at any age.

marble A compact calcium carbonate mineral, $CaCO_3$, formed from limestone and dolomite. When pure it is white; however, marble is often discoloured by impurities. It is used in sculpture and in ornamental building work. Marble is a metamorphic rock.

marcotting A method of artificial propagation similar to layering. A shoot from a plant stem is covered with well-manured earth and wrapped in a waterproof covering in which it is kept moist. Adventitious roots grow from nodes on the shoot, which can be detached from the parent plant and transplanted.

margarine Originally a butter substitute. It is made by emulsifying mixed vegetable and animal oils (fats) with ripened skimmed milk. The vegetable oils may be coconut, palm or sunflower oil; the animal fat is often lard. Usually vitamins and some colouring are added to margarine.

margin The edge or border of a leaf.

marijuana *See* cannabis.

Markownikoff's rule A rule stating that when an acid, HX, adds to a double bond of an alkene or a triple bond of an alkyne, the H-atom becomes attached to the carbon atom which has the larger number of H-atoms. *Example:*
$CH_3-CH=CH_2 + HCl \rightarrow CH_3-CHCl-CH_3$
Here propene gives 2-chloropropane with HCl and not 1-chloropropane,
$CH_3-CH_2-CH_2Cl$.

marrow *See* bone marrow.

Mars One of the nine planets in our Solar System, orbiting the Sun between Earth and Jupiter once in 686,98 days (686.98 days). Mars has two satellites (moons).

marsh gas *See* methane.

marsupial *See* Metatheria.

Mary Jane *See* cannabis.

maser (Microwave Amplification by Stimulated Emission of Radiation) A maser operates on the same principle as a laser, but emits radiation of a longer wavelength. Masers have several uses, e.g. as amplifiers that generate less noise than other types of amplifiers.

mass Symbol: m; unit: kilogram (kg). The quantity of matter in a body. Mass is a measure of a body's inertia and is also the property which determines the mutual attraction between one body and another. *See also* gravitation *and* Newton's law of gravitation.

mass action (law of) A law stating that the rate of a chemical reaction at a constant temperature is proportional to the product of the active masses of the reactants. For the reaction
$A + B \rightarrow AB$
the law of mass action states:
rate of reaction $\propto [A] \times [B] \times k$
where [A] and [B] are the active masses of A and B and k is called the velocity constant or rate constant. The active masses are often taken as their molecular (molar) concentrations. However, the active masses are rarely equal to their molecular concentrations because of interference, especially from ions, which leads to the expression:
active mass = molecular concentration × activity coefficient
The activity coefficient normally has a value less than one. In very dilute solutions in which there is a negligible ionic interference, the activity coefficients are almost equal to one, i.e.
active mass = molecular concentration
In electrolytes of a high concentration the activity coefficient may exceed one. The activity coefficient is a correction factor which makes the thermodynamic calculations correct.

The law of mass action applied to the reaction:
$A + 2B \rightarrow X$
gives:
rate = $k \times [A] \times [B]^2$
Applying the law of mass action to a simple type of ionic equilibrium such as:
$AB \rightleftharpoons A^+ + B^-$
one gets:
$$\frac{[A^+] \times [B^-]}{[AB]} = K$$
where K is called the equilibrium constant. K is used to calculate the extent of displacement of an equilibrium when the concentration of one or more of the reactants is changed. The value of K changes with variations in temperature because the rates of the forward and reverse reactions are unequally affected by temperature changes. The law of mass action was first put forward in 1863 by the two Norwegian chemists, Guldberg and Waage. *See also* order of reaction.

mass defect The masses of protons and neutrons (nucleons) in a nucleus of an atom are not strictly additive, because some part of their masses is converted into energy when the nucleus is formed, i.e. a slight difference exists between the actual relative atomic mass and the total masses of nucleons. This difference in mass is known as the mass defect. It means that energy must be applied to an atomic nucleus to break it up into its nucleons. This energy, equivalent to the mass defect, is called the binding energy of the nucleus. It holds the nucleus together. The greater it is, the more stable is the nucleus. *See also* Einstein's equation.

mass–energy equation *See* Einstein's equation.

mass flow A theory explaining the mechanism of transport of materials in phloem. The theory states that the materials move as a result of differences in turgor pressure at each end of a sieve-tube, resulting in a flow from regions of high to regions of low pressure.

massicot *See* lead oxides.

mass number (nucleon number) Symbol: *A*. The total number of protons and neutrons (nucleons) in a nucleus of an atom. *Examples:* the mass number of hydrogen is 1, of helium is 4 and of sodium is 23; i.e. the mass number of an atom is the integer nearest to its relative atomic mass.

mass spectrometer An instrument used in the accurate measurement of relative atomic masses, in the analysis of isotope abundance and in the qualitative and quantitative analysis of various chemical compounds (especially complex organic molecules) and mixtures. This is done by deflecting ions using magnetic and electric fields, e.g. by ionising a gaseous sample using a beam of electrons and then accelerating the positive ions produced into an evacuated chamber by an electric field. Here magnetic fields will deflect the ions into circular paths; the extent of deflection depends on the masses and charges of the ions. Both the electric and magnetic fields can be varied continuously and successive different types of ions are focused by magnetic lenses onto a detector (e.g. a photographic plate). This gives rise to a mass spectrum of the sample corresponding to particles with different ratios of charge to mass.

mass spectroscopy The study and analysis of mass spectra from different isotopes or chemical compounds.

mass spectrum *See* mass spectrometer.

mast cell A cell found in the connective tissue of vertebrates. Mast cells secrete heparin and histamine.

mastication The process of grinding (chewing) food until it is reduced to small particles or a pulp.

matrix (pl. matrices) **1.** A ground substance consisting of non-living intercellular material in which living cells are embedded. **2.** The part of a nail beneath its body (the exposed part) and its root. **3.** *See* uterus. **4.** The material on which a fungus or a lichen grows.

matt Describes a smooth surface which is dull.

matte An intermediate product, consisting of mixed metal sulphides, obtained in the smelting of sulphide ores.

matter Anything which has mass and occupies space.

maxilla (pl. maxillae) **1.** In vertebrates, the upper jaw. **2.** In various arthropods, one of a pair of mouthparts situated behind the mandibles.

maximum and minimum thermometer (Six's thermometer) A thermometer which records both maximum and minimum temperatures during a period of time. It consists of a large bulb A containing either ethanol or, more commonly, oil of creosote, connected by a U-shaped, graduated capillary tube to a second bulb almost full of the same liquid. In the bend of the capillary tube is a thread of mercury with a steel index at each end. Changes in temperature cause the liquid in A to expand or contract, thus producing a movement of the mercury thread. The indices are fitted with springs so that they stay in position in the stem and can only move when pushed by the convex mercury meniscus. Thus the lower end of the index in the right limb indicates the maximum temperature reached; the lower end of that in the left limb indicates the minimum temperature. The thermometer is reset by bringing the indices back into contact with the mercury using a magnet. *See* Fig. 126.

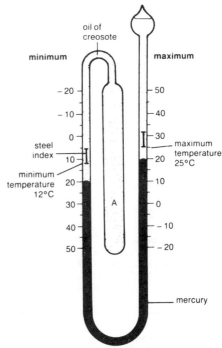

Fig. 126 Maximum and minimum thermometer

maximum density of water This is the density of water $(1{,}000\,\mathrm{g\,cm^{-3}}\ (1.000\,\mathrm{g\,cm^{-3}}))$ at 4°C (277 K). In the temperature range 0°C to 4°C, the density increases with increasing temperature. Above 4°C, the density decreases. *See also* anomalous expansion of water.

maximum thermometer A mercury thermometer designed for measuring maximum temperature during a period of time. A rise in temperature causes a steel index to be pushed, by the convex mercury meniscus, along a graduated capillary tube. When the temperature drops, the index remains in position, its lower end indicating the maximum temperature. The index can be reset by tilting the thermometer or by using a magnet. *See* Fig. 127.

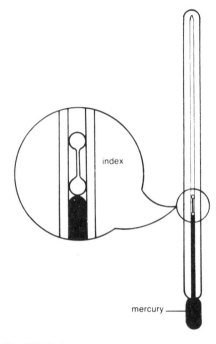

Fig. 127 Maximum thermometer

Maxwell's corkscrew rule A method for determining the direction of the magnetic field produced in a current-carrying conductor. The rule states that if a right-handed corkscrew were turned in the direction of the current (conventional current), the way in which it would turn would indicate the direction of the lines of force.

mean free path Symbol: λ (lambda). The average distance travelled by molecules between two successive collisions. In gases the mean free path between the molecules is inversely proportional to the pressure.

mean solar day The average time the Earth takes to rotate once on its axis with respect to the mean Sun.

mean solar time Time measured in terms of the Earth's rotation relative to the mean Sun.

mean Sun An imaginary Sun representing the yearly motion the true Sun would have across the sky if the Earth's orbit were circular instead of elliptical and if its axis were upright and not inclined.

measles (rubeola, morbilli) A contagious, viral disease which usually occurs in childhood. Symptoms include sore, red eyes and dislike of light, coughing and a dark-red skin rash beginning behind the ears. No drugs are available to treat the disease; however, antibiotics may be administered to suppress secondary bacterial infection. A vaccine against measles is available.

measuring cylinder A tall, graduated, glass or plastic vessel used to measure the volume of liquids. *See* Fig. 128.

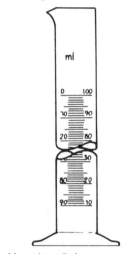

Fig. 128 Measuring cylinder

meatus A duct or channel, e.g. the external auditory meatus.

mechanical advantage (MA, force ratio) In a machine, the ratio of the load to the effort. *See also* efficiency.

mechanical clock *See* clock.

mechanical energy Energy due to the motion of a body, its position or its state of tension or compression. *See also* kinetic energy *and* potential energy.

mechanical equivalent of heat (Joule's equivalent) Symbol: J. A constant relating an old unit of heat, the calorie, to mechanical units of energy.
$J = 4.1868$ joules per calorie (4.1868 J/cal)
Experiments which were once used to calculate the value of J are now used in specific heat capacity determinations.

mechanics The study of the effect of forces on bodies. It is divided into dynamics, kinematics and statics.

median fin *See* fin.

medulla The central part of an organ or tissue, e.g. the medulla of a kidney.

medulla oblongata The posterior segment of the hindbrain, connected to the cerebellum and continuous with the spinal cord. It is concerned with the control of breathing movements, heart rate, blood pressure, reflex movements of the eye muscles, etc. *See* Fig. 16.

medullary ray A plate of tissue which runs radially from the inside to the outside of a plant stem. Medullary rays conduct and store synthesised food material in the stem and root. Primary medullary rays are situated between the primary vascular bundles and connect the pith (medulla) and the cortex. Secondary medullary rays are formed from cambium during secondary thickening and end in secondary xylem or phloem. These medullary rays are not connected with the pith and therefore are sometimes called vascular rays.

medullary sheath *See* nerve-fibre.

medulla spinalis *See* nerve-cord.

medullated nerve-fibre (myelinated nerve-fibre) *See* nerve-fibre.

medusa (pl. medusae) A stage in the life cycle of some coelenterates in which the animal is free-swimming and its body is bell-shaped or shaped like an inverted saucer with tentacles hanging vertically downwards all round the edge. A medusa has a four-lobed mouth leading to a short gullet which is connected to the gastric cavity (stomach). The tentacles are used to capture small organisms and push them into the mouth. Usually medusae drift passively for much of their time, but they may also swim actively by contractions of the bell. Medusae have male or female sex organs and reproduce sexually when sperm cells swim from the male animal to fertilise the egg cells in the female animal. The zygote develops into a polyp. In some jellyfish, the medusa is the only form of the coelenterate.

meerschaum (sepiolite) A compact, fibrous, usually white mineral of hydrated magnesium silicate, $Mg_2Si_3O_8 \cdot 2H_2O$. It is used for making tobacco pipes.

mega- 1. Symbol: M. A prefix meaning one million; e.g. one megahertz (MHz) is equal to one million hertz. 2. Prefix meaning large. *See also* macro-.

megamere (macromere) In animal embryos, a large cell produced during cleavage. Megameres collect at the vegetal pole of the blastula. *Compare* micromere.

megascopic *See* macroscopic.

megatonne A unit expressing the explosive power of nuclear weapons. One megatonne is the equivalent explosive power of one million tonnes of trinitrotoluene (TNT).

meiosis (pl. meioses) (reduction division) A type of cell division in which the number of chromosomes is halved by two successive divisions, i.e. the daughter cells have haploid nuclei. At fertilisation the fusion of two haploid gametes doubles the chromosome number to produce diploid cells once more. Thus meiosis ensures that the chromosome number of a species remains constant generation after generation. Meiosis follows a sequence of phases: prophase, which is subdivided into five stages, leptotene, zygotene, pachytene, diplotene and diakinesis; metaphase I; anaphase I; telophase I; interphase; metaphase II; anaphase II; and telophase II.

Prophase: leptotene – Chromosomes appear as uncoiled, thread-like strands.
Zygotene – Homologous chromosomes have joined together forming bivalents.
Pachytene – The bivalents become shorter and thicker and chromatids become visible. Each chromosome is now a double structure spiralled on the other.
Diplotene – A double structure of each chromosome is apparent. The bivalents are attached at connecting points called chiasmata, but repel each other at the centromeres.
Diakinesis – The nuclear membrane starts to break down and nucleoli disappear. The bivalents have contracted still further and crossing-over commences.

Metaphase I: The nuclear membrane has broken down completely and a spindle forms. The bivalents move towards the spindle equator about which they align themselves.

Anaphase I: The centromere pairs move towards opposite poles dragging the chromatids behind them. This causes the chiasmata to slip apart.

Telophase I: The groups of chromosomes form two nuclei and a new cell wall begins to form across the equatorial plate.

Interphase: The formation of a new cell wall is completed. This is the resting phase at the end of the first meiotic division.

Metaphase II: The chromosomes move towards a new spindle equator, at right angles to that of the first division, about which they align themselves.

Anaphase II: The centromeres divide and the chromatids separate, moving to opposite poles.

Telophase II: The groups of chromosomes form new nuclei on reaching the poles. A new cell wall forms in each cell. In this way four haploid cells have been formed. *See* Fig. 129.

meiotic Pertaining to meiosis.

Meker burner A laboratory burner which burns a mixture of gas and air. It has a wide,

melanin

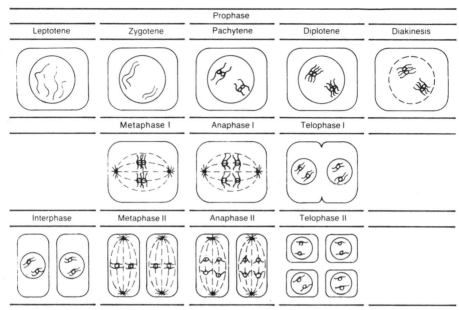

Fig. 129 Meiosis

perforated burner head which breaks up the flame into many small flames thus producing a high, homogeneous temperature. The perforated head also prevents the flame from 'striking back'. This occurs when the gas catches fire at the base of the burner instead of at its top. It can happen in Bunsen burners. *See* Fig. 130.

Fig. 130 Meker burner

melanin A pigment, usually dark-brown or black, derived from an amino acid called tyrosine, found in the Malpighian layer and in hair. It also occurs in some plants. *See also* albino.

melanism Excessive development of the pigment melanin. *See also* industrial melanism.

melting (fusion) Changing from the solid to the liquid state, usually by heating. For a pure substance which has a crystalline structure, melting occurs at a particular temperature called the melting-point (or freezing-point) which is dependent upon the external pressure. However, melting-points are usually quoted at standard atmospheric pressure (101 325 Pa). At the melting-point, there is an equilibrium between the solid and the liquid phases. Impurities will lower the melting-point of a crystalline solid. Amorphous substances have no fixed melting-points, but become progressively softer within a certain temperature range.

melting-point (m.p.) *See* melting *and* freezing-point.

membrane A thin layer of tissue covering a cell or part of an animal or plant.

membranous labyrinth *See* inner ear.

menarche The first occurrence of menstruation.

Mendelism The study of inheritance by breeding experiments with animals and plants.

Mendel's laws Two laws of inheritance proposed by Gregor Mendel. (1) The Law of Segregation states that the characteristics of an organism are determined by internal factors (genes) which always occur in pairs. Only one of a pair of these factors (genes) can be carried in a single gamete. (2) The Law of Independent Assortment states that the characteristics of the parent organisms can be inherited independently of each other; each one of a pair of alleles may combine randomly with either allele of another pair. *See also* dominant.

meninges (sing. meninx) In humans and other mammals, three membranes enclosing the brain and the spinal cord. In the order from the inside outwards, they are the pia mater, the arachnoid and the dura mater. These membranes afford protection and nourishment to the brain and spinal cord.

meningitis An inflammation of the meninges caused by several bacteria, e.g. meningococcus, pneumococcus and tubercle bacillus, or by different viruses. Both bacterial and viral meningitis may occur in epidemics and are always considered serious illnesses. Symptoms of both types of meningitis are often the same: severe headache, high fever, vomiting and stiffness of the muscles in the neck. Meningitis is treated with drugs such as antibiotics, but not all types of viral meningitis can be cured in this way. However, most clear up by themselves.

meninx *See* meninges.

meniscus The curved surface of a liquid in a narrow tube. The curvature is either concave (water in a glass tube) or convex (mercury in a glass tube). *See* Fig. 131; *see also* surface tension.

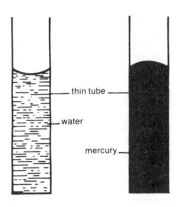

Fig. 131 Meniscus

meniscus lens A thin lens, convex on one side and concave on the other. *See* Fig. 112.

menopause *See* climacteric.

mensa The grinding or biting surface of a tooth.

menses The flow of fluids (blood, etc.) during menstruation.

menstrual cycle A modified oestrous cycle occurring, in the absence of pregnancy, in humans and some primates. In human females, it is a regular cycle of about 28 days from the beginning of one menstruation to the beginning of the next. It takes place from puberty (menarche) to climacteric (menopause).

menstrual period *See* menstruation.

menstruation (menstrual period) The periodic discharge (bleeding) from the uterus in most higher mammals. It occurs as a result of the breakdown of the lining of the uterus and usually lasts for 3–5 days. It is the sign that an ovum has not been implanted in the uterus, i.e. pregnancy has not commenced. *See also* corpus luteum.

mercaptan *See* thiol.

mercuric Term formerly used in the names of mercury compounds to indicate that the mercury is divalent (bivalent); e.g. mercuric chloride, $HgCl_2$, now called mercury(II) chloride. *Compare* mercurous.

mercuric chloride *See* mercury chlorides.

mercuric oxide *See* mercury(II) oxide.

mercuric sulphide *See* mercury(II) sulphide.

mercurous Term formerly used in the names of mercury compounds to indicate that the mercury is monovalent (univalent); e.g. mercurous chloride, Hg_2Cl_2, now called dimercury(I) chloride. *Compare* mercuric.

mercurous chloride *See* mercury chlorides.

Mercury One of the nine planets in our Solar System. It is the nearest planet to the Sun, orbiting the Sun between the Sun and Venus once in 87,97 days (87.97 days).

mercury (quicksilver) An element with the symbol Hg; atomic number 80; relative atomic mass 200,59 (200.59); state, liquid. Mercury is a heavy metal occurring naturally in the mineral cinnabar, HgS (mercury(II) sulphide), from which it is extracted by heating (roasting) it in air:
$$HgS + O_2 \rightarrow Hg + SO_2$$
Poisonous mercury vapour is liberated and is condensed to the liquid metal. Mercury forms alloys (amalgams) with many metals, but it does not attack iron, cobalt or nickel directly; hence it can be stored in iron bottles. Mercury is used as the cathode in several electrolytical processes, in barometers and thermometers, in making detonators, in the electrical industry and in dentistry in amalgams for fillings.

mercury barometer A simple mercury barometer consists of a thick-walled glass tube about a metre long, closed at one end, filled with mercury and inverted in a dish containing mercury. The vertical height (barometric height) of the mercury column that atmospheric pressure will support is a measure of atmospheric pressure and is usually measured in millimetres of mercury. The standard value of atmospheric pressure is 760 mm of mercury (101 325 N m^{-2} or Pa). The space above the mercury column is known as a Torricellian vacuum after the inventor of the simple mercury barometer, Evangelista Torricelli. See Fig. 132.

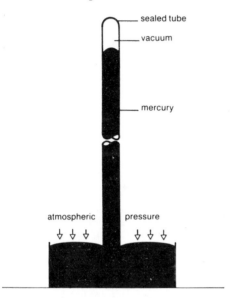

Fig. 132 Mercury barometer

mercury cathode cell See Castner–Kellner cell (process).

mercury cell A miniature primary cell used in hearing-aids, electronic calculators and watches, etc. It is capable of producing a constant discharge voltage at various temperatures. The anode consists of an amalgamated zinc plate and the cathode of mercury(II) oxide, HgO, mixed with graphite. The electrolyte is potassium hydroxide, KOH. The e.m.f. of a mercury cell is about 1,3 volts (1.3 V).

mercury chlorides Dimercury(I) chloride (mercurous chloride, calomel), Hg_2Cl_2, and mercury(II) chloride (mercuric chloride), $HgCl_2$. Salts of mercury.

Dimercury(I) chloride, Hg_2Cl_2, is a white insoluble solid which is precipitated when chloride ions, Cl$^-$, are mixed with dimercury(I) ions, Hg_2^{2+}:

$Hg_2^{2+} + 2Cl^- \rightarrow Hg_2Cl_2$

Hg_2Cl_2 can also be prepared by heating a mixture of mercury(II) chloride, $HgCl_2$ and mercury, Hg:

$HgCl_2 + Hg \rightarrow Hg_2Cl_2$

When reacting with ammonia, NH_3, Hg_2Cl_2 turns black due to the formation of free mercury. This also happens when it is exposed to sunlight. Hg_2Cl_2 is used in physical chemistry in the calomel electrode. It is also used as a fungicide and in medicine.

Mercury(II) chloride, $HgCl_2$, is a white, almost colourless, soluble solid which is rapidly formed when chlorine, Cl_2, is in contact with mercury:

$Hg + Cl_2 \rightarrow HgCl_2$

When $HgCl_2$ dissolves in water, it exists in the form of $HgCl_2$ covalent molecules. $HgCl_2$ sublimes and is very poisonous. In dilute solution it is used as an antiseptic.

mercury(II) oxide HgO (mercuric oxide) An oxide which is formed reversibly from mercury, Hg, when this is heated in air at about 350°C for a long time:

$2Hg + O_2 \rightleftharpoons 2HgO$

Made in this way, it is a red solid, but a yellow variety of HgO also exists. This can be made by adding a solution of a strong alkali to a solution containing mercury(II) ions, Hg^{2+}:

$Hg^{2+} + 2OH^- \rightarrow Hg(OH)_2 \rightarrow HgO + H_2O$

In this process, unstable mercury(II) hydroxide, $Hg(OH)_2$, is formed as an intermediate compound. The difference in colour between the two varieties of HgO is due to a difference in particle size. As seen above, HgO is unstable when heated (strongly):

$2HgO \rightarrow 2Hg + O_2$

This process was the one Priestley used to obtain oxygen. Mercury(I) oxide appears not to exist.

mercury(II) sulphide HgS (mercuric sulphide) A naturally occurring salt of mercury, also called cinnabar (a red mineral). HgS is obtained as a black solid, extremely insoluble in water, when hydrogen sulphide, H_2S, is passed through a solution containing mercury(II) ions, Hg^{2+}:

$Hg^{2+} + H_2S \rightarrow HgS + 2H^+$

The red and black forms of HgS have different crystal structures; the black form changes into the red form on heating. HgS dissolves in concentrated nitric acid, HNO_3, and in aqua regia.

mercury thermometer The most common type of thermometer. It consists of a bulb containing mercury, connected to a graduated capillary tube. The mercury expands with rising temperatures and so moves up the capillary. The melting-point of mercury is −39°C and its boiling-point is 357°C, so the thermometer can be used to measure temperatures in this range. If the space above the mercury in the capillary is filled with an inert gas such as nitrogen (thus increasing the pressure) the thermometer may be used to measure temperatures above 357°C. See also alcohol thermometer.

meridian, geographic See geographic meridian.

meridian, magnetic See magnetic meridian.

meridian, terrestrial Any imaginary circle drawn on the surface of the Earth passing through the North and South Poles.

meristem A layer of actively dividing, non-vacuolated cells found in regions of growth in plants, e.g. between xylem and phloem. Meristems are the only areas of cell division in plants, the principal meristems being situated at the tips of stems and roots (apical meristems).

mesencephalon See midbrain.

mesentery In vertebrates, thin sheets of connective tissue consisting of a double layer of peritoneum, surrounding all parts of the alimentary canal and attaching it to the dorsal wall of the abdominal cavity.

meso- Prefix meaning middle.

mesocarp See drupe.

mesoderm In multicellular animals, the middle layer of the embryo, lying between ectoderm and endoderm. From the mesoderm, muscles, blood and connective tissue develop.

mesogloea In coelenterates, the jelly-like layer separating the ectoderm and the endoderm. Tissue or specialised cells are never formed from mesogloea.

mesophile A micro-organism which grows best at temperatures between 30°C and 40°C. Most micro-organisms are mesophiles and are said to be mesophilic. See also psychrophile and thermophile.

mesophyll Specialised tissue composed of parenchyma cells and situated within the epidermis of a leaf. It is made up of the palisade mesophyll and the spongy mesophyll. The palisade mesophyll consists of tissue of cells containing large numbers of chloroplasts. Its primary function is photosynthesis. Beneath the palisade mesophyll is a layer of loosely-packed cells with fewer chloroplasts. This is the spongy mesophyll, which communicates with the atmosphere through stomata in the lower epidermis. See Fig. 110 (b).

mesophyte A plant adapted to live in conditions that are neither too wet nor too dry. Its adaptations may include a well-developed root system, large thin leaves with unprotected stomata, and plenty of chlorophyll. Most angiosperms are mesophytes. Compare hydrophyte and xerophyte.

mesosphere The region of the Earth's atmosphere situated between the stratosphere and the ionosphere.

Mesozoic The geological age (era) of middle life, estimated to have commenced approximately 225 million years ago. See Appendix.

messenger RNA (mRNA) A form of RNA concerned with the transfer of information for protein synthesis from nuclear DNA to the ribosomes in the cytoplasm of a cell. See also transfer RNA, transcription and codon.

meta- 1. Prefix meaning a change of position or condition; after or behind, among or within, between. 2. Prefix meaning that two substituents are separated by an intervening carbon atom in an aromatic ring. Example:

is meta-dichlorobenzene (m-dichlorobenzene). Its IUPAC name is 1,3-dichlorobenzene.

3. Prefix meaning a less hydrated form of an acid. Examples: metaphosphoric acid, HPO_3, is less hydrated than orthophosphoric acid, H_3PO_4; metasilicic acid, H_2SiO_3, is less hydrated than orthosilicic acid, H_4SiO_4. See also ortho- and para-.

metabolic Pertaining to metabolism.

metabolin See metabolite.

metabolism The sum total of the chemical reactions which take place in living organisms. These reactions are the result of anabolism and catabolism.

metabolite (metabolin) Any product of metabolism.

metacarpal A bone of the lower forelimb or forefoot of tetrapods. In humans there are five such bones forming the palm of the hand. The heads of the metacarpals form the knuckles on the back of the hand. See Fig. 162.

metacarpus The set of metacarpal bones forming the palm of the hand in human beings and part of the lower forelimb or forefoot in tetrapods.

metal The general name for a large group of elements. They are all solids except mercury, which is a liquid at room temperature. Metals are electropositive, are good conductors of heat and electricity, have metallic bonds, tend to form ionic compounds, have lustre when polished and are sonorous when struck. Usually metals are ductile and malleable; however, some are brittle. Oxides of metals are basic. Example: calcium oxide, CaO, forms calcium hydroxide, $Ca(OH)_2$, when treated with water:

$$CaO + H_2O \rightarrow Ca(OH)_2$$

Most metals form alloys. Those which are high in the electrochemical series are very reactive, e.g. potassium and sodium, whereas those low in the electrochemical series are only slightly reactive, e.g. gold and platinum. The above description is general; many metals do not

metal alkoxide

have all the characteristics described. *Compare* non-metal; *see also* metalloid.

metal alkoxide An organic compound formed when an alcohol reacts with a metal such as lithium, sodium or potassium. These are the metals commonly used in reactions with alcohols. *Example:*
$2CH_3OH + 2Na \rightarrow 2CH_3\text{—}O\text{—}Na + H_2$
Here methanol, CH_3OH, reacts with sodium, Na, to give the metal alkoxide sodium methoxide, $CH_3\text{—}O\text{—}Na$, and hydrogen. Alkali metal alkoxides are strong bases, soluble in alcohol. They may be compared to alkali metal hydroxides. *Example:* sodium methoxide and sodium ethoxide closely resemble sodium hydroxide in physical appearance as they are both white hygroscopic solids which have a 'soapy' feel. Metal alkoxides hydrolyse reversibly to alcohol and alkali:
$CH_3\text{—}O\text{—}Na + H_2O \rightleftharpoons CH_3OH + NaOH$
However, the equilibrium lies far to the right. Metal alkoxides are used as catalysts and as reactants in several organic syntheses. Alkoxides are also called alcoholates. *See also* Williamson synthesis.

metallic bond A type of bond occurring in metallic crystals. It cannot be described as purely ionic or covalent. In a metallic crystal, positive ions of the metal are packed in a regular pattern forming a lattice within a 'sea' of electrons which are liberated from them. This 'sea' of electrons binds the positive ions together. The electrons forming the 'sea' are relatively free to move throughout the crystal under an applied potential difference. Since there are few electrons in the outer shells of metallic atoms, each can contribute only a few electrons to the electron 'sea'. When this number is compared to the number of neighbouring atoms surrounding each atom in the lattice, it is clear that there are not enough electrons to form individual bonds between atoms. The metallic bond can also account for differences in hardness, density or melting-point between metals. The more electrons of each metallic atom which are available for the electron 'sea', the harder and denser is the structure of the metal. For instance, the alkali metals are very soft and have low densities because each atom only contributes one electron to the electron 'sea', whilst a transition metal such as chromium is much harder and denser because its atoms contribute more electrons to the electron 'sea' to take part in bonding. However, metals may be dense but still very soft, e.g. gold; this may be explained in terms of the packing structure of the metallic atoms in the crystal. *See* Fig. 133.

metallic crystal *See* metallic bond.

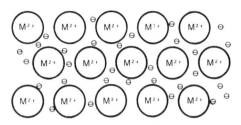

Fig. 133 Metallic bond: the 'sea' of electrons in a crystal of a metal which has two available electrons per atom

metallography A branch of metallurgy concerned with the structure and properties of metals and alloys.

metalloid An obsolete term describing an element with properties intermediate between those of metals and non-metals. Examples include arsenic, silicon and germanium. The term semi-metal has been proposed to replace the term metalloid.

metallurgy The study of metals and alloys, including their extraction, refining, fabricating, treatment, structure and properties.

metamorphic rock Rock formed from the transformation of previous rock (sedimentary or igneous) by temperature, pressure and the introduction of foreign constituents. Examples include marble, slate and gneiss.

metamorphism The process in which metamorphic rock is formed, during which existing rock undergoes physical and chemical alterations.

metamorphosis A rapid change in bodily shape and structure in the life cycle of many animals such as insects and amphibians. This change involves destruction of larval tissue, brought about by lysosomes, and growth of new tissue. Metamorphosis may be complete or incomplete. Complete metamorphosis involves a larval and pupal stage before the adult form (imago). This is exhibited by butterflies, bees, wasps, flies, etc. Incomplete metamorphosis has no pupal stage between the larval and adult stages. This is exhibited by cockroaches, grasshoppers, termites, etc. The larvae of such insects are called nymphs.

metaphase The stage of mitosis or meiosis following prophase when the chromosomes align themselves along the spindle equator.

metaphosphate A salt of metaphosphoric acid, HPO_3.

metaphosphoric acid HPO_3 A weak, glassy, solid acid whose salts are called metaphosphates. It can be obtained from orthophosphoric acid, H_3PO_4, by dehydration at around 315°C:
$H_3PO_4 \rightleftharpoons HPO_3 + H_2O$

methanoic acid

Strictly speaking, the formula HPO_3 for metaphosphoric acid is the empirical formula. The acid is a polymer with the formula $(HPO_3)_n$ where $n \geqslant 3$.

metaphysis (pl. metaphyses) *See* ossification.

metasilicate A salt of metasilicic acid, H_2SiO_3.

metasilicic acid *See* silicic acid.

metastable state A state of a system in which it seems to be in a stable equilibrium, but can undergo a rapid change if the system is disturbed. *Example:* water can be cooled very slowly to a temperature below 0°C (supercooled water); however, if a small piece of ice is added, a rapid freezing is observed and the system settles into a lower energy state. The same phenomenon is seen in supersaturated solutions.

metastasis (pl. metastases) The spread (transfer) of diseased tissue (cells) from one part of the body to another. This takes place either through the blood vessels or the lymphatic vessels. Usually the term is applied to the spread of cells from a malignant tumour, giving rise to secondary (metastatic) tumours.

metatarsal A bone of the lower hindlimb or hindfoot of tetrapods. In humans there are five such bones, forming the region of the foot between the digits (toes) and the tarsus (ankle). *See* Fig. 162.

metatarsus The set of metatarsal bones forming part of the lower hindlimb or hindfoot of tetrapods.

Metatheria A subclass of mammals containing the marsupials (pouch-bearing mammals). Marsupials are primitive in that they have only one set of teeth, there is no placenta connecting the embryo to the mother, and the young are born after a brief period of gestation in the uterus followed by a further period of development in which suckling takes place in a pouch on the mother's abdomen. Examples include kangaroos, koala bears and opossums.

metathesis Double decomposition.

Metazoa A subkingdom comprising all multicellular animals as distinct from Protozoa and Parazoa. Their bodies are composed of differentiated tissues and they possess a co-ordinating nervous system.

meteor (shooting star) Usually a small body from space which enters the Earth's atmosphere, where friction with air particles causes it to glow and streak across the sky. Millions of meteors enter the atmosphere each day, but most are too faint to be seen with the naked eye. Most meteors burn up entirely on their way through the atmosphere, but a few are big enough to fall on the Earth. *See also* meteorite *and* meteoroid.

meteorite The name given to a meteor which survives the passage through the Earth's atmosphere and reaches ground level. Meteorites are divided into three types: iron meteorites, which are almost entirely composed of iron and nickel; stone meteorites, which are primarily composed of silicates; and stony-iron meteorites, which are composed of almost equal mixtures of both. The mass of a meteorite is usually 3–15 kg. Meteorites with masses of more than 100 tonnes have been found but they are rare. The origin of meteorites is still unknown, but they are believed to originate within our Solar System and not to be fragments of stars. They are usually named after the geographical location where they fall. *See also* meteoroid.

meteorograph A collection of recording instruments carried into the upper atmosphere by kites or balloons to measure temperature, pressure, relative humidity, wind speed, etc.

meteoroid The collective name given to particles of meteoritic origin found orbiting the Sun. *See also* meteor *and* meteorite.

meteorology The study of the Earth's atmosphere with respect to weather and climate. Meteorology as a science commenced in 1643 when Torricelli invented the mercury barometer.

meter Any measuring instrument.

methanal HCHO (formaldehyde). The simplest of the aldehydes. It is manufactured by the catalytic oxidation of methanol, CH_3OH. Methanal is a colourless gas with a pungent smell. When it is dissolved in water, a solution (40%) known as formalin is formed and this is used as a preservative. Methanal is used in the plastics and textile industries and in organic synthesis. M.p. −92°C, b.p. −21°C.

methane CH_4 (marsh gas) A hydrocarbon; the simplest of the alkanes. It is a flammable gas which occurs in natural gas and coal-gas. It is used as a fuel, burning in air with a hot, non-luminous flame to give carbon dioxide, CO_2, and water:
$$CH_4 + 2O_2 \rightarrow CO_2 + 2H_2O$$
Methane is also the starting substance in the production of various organic compounds. Sometimes methane is formed from decaying organic matter. M.p. −182,5°C (−182.5°C), b.p. −164°C.

methanoate (formate) A salt of methanoic acid, HCOOH.

methanoic acid HCOOH (formic acid). A weak organic, colourless liquid with a pungent smell. It is the simplest carboxylic acid known and is found in nature in various plants and in ants. Methanoic acid is produced commercially by mixing sodium methanoate (sodium

methanol

formate), HCOONa, and sulphuric acid, H_2SO_4:
$2HCOONa + H_2SO_4 \rightarrow 2HCOOH + Na_2SO_4$
Sodium methanoate is produced in the reaction between carbon monoxide, CO, and sodium hydroxide, NaOH:
$CO + NaOH \rightarrow HCOONa$
Methanoic acid is used in textile dyeing, tanning, electroplating and in the manufacture of certain pesticides. Salts of methanoic acid are called methanoates (formates).

methanol CH_3OH (methyl alcohol, wood naphtha, wood alcohol) A volatile, flammable, colourless, poisonous liquid alcohol, miscible with water. Originally methanol was prepared by distillation of wood, but it is now produced from methane, CH_4, extracted from natural gas. Methane is passed with steam over a nickel catalyst at about 900°C under pressure:
$CH_4 + H_2O \rightarrow CO + 3H_2$
The mixture of carbon monoxide, CO, and hydrogen, H_2, known as synthesis gas, is then converted into methanol when passed at 300°C over a catalyst of zinc oxide and chromium(III) oxide under pressure:
$CO + 2H_2 \rightarrow CH_3OH$
Methanol is used as a solvent and in the manufacture of methanal (formaldehyde), HCHO.

method of mixtures A method used in calorimetry in which measured amounts of liquids or liquids and solids at different temperatures are mixed and the final temperature of the mixture is measured. The quantity of heat gained by the colder substance is equal to the quantity of heat lost by the warmer substance.

methyl CH_3—. An organic radical present in thousands of organic compounds.

methyl alcohol See methanol.

methylated spirit(s) Ethanol, C_2H_5OH, which has been made unfit for human consumption by adding to it methanol (about 9,5% (9.5%)), pyridine (about 0,5% (0.5%)) and a small amount of blue dye. Such spirit is sold without taxation for use as a solvent or fuel. See also denaturant.

methylated spirit(s), industrial A type of methylated spirit(s) which is free from pyridine and dye, but is still very poisonous because of its methanol content.

methylbenzene $C_6H_5CH_3$ (toluene) An aromatic hydrocarbon occurring in coal-tar, from which it is extracted by fractional distillation. It is a colourless, flammable, insoluble liquid which dissolves in many organic solvents. Methylbenzene is used as a solvent and in organic synthesis. See also methyl-2,4,6-trinitrobenzene.

methylbuta-1,3-diene (isoprene) A diene: i.e. an unsaturated, aliphatic hydrocarbon containing two conjugated double bonds. It has the formula
$$CH_2\!=\!C\!-\!CH\!=\!CH_2$$
$$|$$
$$CH_3$$
Methylbuta-1,3-diene is a colourless, volatile, insoluble liquid found in natural rubber and terpenes. It may be prepared from methane, CH_4, in a catalytical process. Methylbuta-1,3-diene easily polymerises and is used in the manufacture of synthetic rubber. See also buta-1,3-diene.

methylene —CH_2. An organic radical.

methylene blue A water-soluble, green organic substance which turns blue in solution. It is used in the textile industry for dyeing cotton, as a laboratory indicator and as a biological stain.

methyl orange A water-soluble, acid–base indicator often used in titrations of a weak base with a strong acid giving a weak acidic end-point. The indicator is red in solutions which have a pH below 3,2 (3.2) and yellow in solutions which have a pH above 4,4 (4.4). In the pH range 3,2–4,4 (3.2–4.4) it is orange. It is an organic substance.

methyl red An acid–base indicator which is soluble in ethanol. It is red in solutions which have a pH below 4,8 (4.8) and yellow in solutions which have a pH above 6,0 (6.0). This indicator is used in the same type of titrations as methyl orange. Methyl red is an organic substance.

methyl-2,4,6-trinitrobenzene
$C_6H_2CH_3(NO_2)_3$ (trinitrotoluene, TNT) A pale-yellow, crystalline solid which is very unstable. It is prepared by nitration of methylbenzene (toluene), $C_6H_5CH_3$:
$C_6H_5CH_3 + 3NO_2^+ \rightarrow C_6H_2CH_3(NO_2)_3 + 3H^+$
Methyl-2,4,6-trinitrobenzene (TNT) is used as an explosive.

metra See uterus.

metre Symbol: m. The SI unit of length, defined as the distance travelled by light in a vacuum in $1/299\,792\,458$ seconds.

metre bridge A type of Wheatstone bridge in which two of the four resistors are replaced by a one-metre uniform resistance wire mounted over a millimeter scale and with a sliding contact. See Fig. 134.

metric system A system of units based on the metre and kilogram.

metrology The study of weights and measures.

metronome An instrument normally used to indicate the tempo (speed) of music, but which is also used in the laboratory to assist in counting seconds or fractions of seconds. It

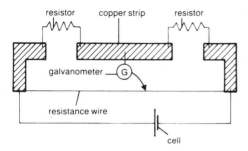

Fig. 134 Metre bridge

consists of a pivoted pendulum fitted with a sliding weight and driven by a clockwork motor. An audible click is produced at the end of each half swing.

MeV Abbreviation for mega-electronvolt: one mega-electronvolt is equal to 10^6 electronvolts (eV).

Mg Chemical symbol for magnesium.

mho (reciprocal ohm) A unit of conductance which has been replaced by the SI unit the siemens (S).

mica group A large group of silicate minerals which have complex formulae. The mica group is divided into two main types: dark mica, which is rich in iron and magnesium, and white mica, which is rich in aluminium. In addition, some micas contain lithium. Crystals of mica can easily be split into very thin, shiny elastic plates. Mica minerals are used as a dielectric in capacitors, in the heating elements of electric irons and in the manufacture of rubber tyres.

Michaelis constant The substrate concentration at which an enzyme reaction occurs at half its maximum rate.

micro- 1. Symbol: μ (mu). A prefix meaning 10^{-6}. 2. Prefix meaning small. Compare macro- and mega-.

microbalance A very sensitive analytical balance with an accuracy in the range 10^{-4}–10^{-6} grams.

microbe A micro-organism, especially a bacterium.

microbiology The study of micro-organisms.

microclimate The climate of a small area. For example, the prevailing temperatures, humidity levels and wind speeds within a forest.

micro-element See trace element.

microfarad Symbol: μF. One-millionth of a farad, 10^{-6} F.

micrograph A photograph taken through a microscope. A photomicrograph is taken through an optical (light) microscope whereas an electron micrograph is taken through an electron microscope.

micromere In animal embryos, a small cell produced during cleavage. Micromeres collect at the animal pole of the blastula. Compare megamere.

micrometer eyepiece A small glass disc on which is engraved a scale, usually of 100 divisions. It is fitted inside the eyepiece of a microscope in such a way that the scale is visible in the field of view. It is used for measuring the size of an object under the microscope. Since the scale is arbitrary, it must be calibrated for each objective with which the eyepiece is used. This is done using a micrometer slide, which is a glass slide with an accurate scale, similar to that of the micrometer eyepiece, marked on it. This slide is placed on the stage and brought into a sharp focus. The number of 'eyepiece' divisions corresponding to a certain number of 'slide' divisions is found. One slide division is usually equal to 100 μm (micrometres) and therefore the size of one eyepiece division can be calculated. See Fig. 135 (a) and (b).

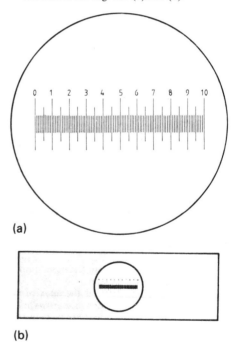

Fig. 135 (a) Micrometer eyepiece
(b) Micrometer slide

micrometer screw gauge An instrument used for measuring small or accurate distances. It consists of a U-shaped frame fitted with a screwed spindle which is connected to a graduated thimble moving over a graduated sleeve. The object to be measured is placed between the anvil and the spindle and the thimble is turned until the object is gripped between them. A spring ratchet may be fitted to the thimble to prevent overtightening. In Fig. 136, the sleeve is graduated in half millimetres. As the thimble has a scale of 50 divisions and moves forward a half millimetre in one complete turn, each division represents $\frac{1}{50}$ of a half millimetre or 0,001 cm (0.001 cm). The instrument is read by combining the readings on the sleeve and thimble.

objective (object) lens and the eye lens (eyepiece), fitted at each end of a tube. The objective lens produces a real, inverted, magnified image I_1 of an object O placed just outside its focal point F'_o. Image I_1 is formed just within (or in) the focal point F'_e of the eye lens and acts as an object for this lens. A virtual, inverted, further magnified image I_2 is produced and seen by the eye. The linear magnification of a microscope is the product of the magnification of the objective lens and the magnification of the eye lens. See Fig. 137; see also electron microscope.

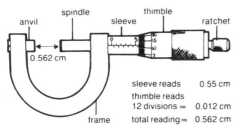

sleeve reads   0.55 cm
thimble reads
12 divisions =   0.012 cm
frame total reading = 0.562 cm

Fig. 136 Micrometer screw gauge

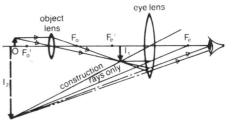

Fig. 137 Compound microscope

micrometer slide See micrometer eyepiece.
micrometre (micron) Symbol: μ. One-millionth of a meter, 10^{-6} m.
micron Symbol: μ or μm. An obsolete name for the micrometre.
micronutrient An element required in relatively small quantities by living organisms. These are vitamins and trace elements. Compare macro-nutrient.
micro-organism An organism too small to be seen with the naked eye. Examples include bacteria and viruses.
microphone A device for converting sound energy into electrical energy. See also carbon microphone.
microprocessor The controlling unit of a computer. It is a complex part and may consist of a single integrated circuit or a small number of integrated circuits. The microprocessor, together with the storage unit and input/output units, make up the computer.
micropyle 1. A tiny hole, at the apex of an ovule, through which a pollen tube grows prior to fertilisation. The micropyle can be seen as a minute pore in the testa of a mature seed. It is through this pore that water is absorbed before germination 2. In many insects, a minute pore in the egg membranes of an ovum, through which sperm enters.
microscope An instrument used to obtain magnified images of small objects. The simple microscope is just a magnifying glass. The compound microscope makes use of two convex lenses of short focal length, the

microscopic Too small to be seen by the naked eye. Compare macroscopic.
microtome An instrument for cutting thin sections of tissue to be examined under the microscope.
microtrabecular lattice Recent studies using very powerful electron microscopes have revealed that organelles such as mitochondria, ribosomes, polysomes and endoplasmic reticulum are supported in the ground substance (cytoplasm) of a living cell by a network of interlinked filaments (trabeculae). This network is known as the microtrabecular lattice. It is believed to organise the individual cell components into a functional unit, to regulate and direct transport within the cell and to control cell shape and cell movement.
microwave Electromagnetic radiation with wavelengths in the range of 1 mm–30 cm, i.e. between infra-red radiation and radio waves. Microwaves are used as carrier waves in ordinary telecommunications and satellite communications, and are the waves used by radar systems. Recently, their ability to carry energy has been applied in microwave ovens that cook food rapidly.
micturition See urination.
midbrain (mesencephalon) In vertebrates, part of the brain connecting the forebrain and hindbrain. It is the centre for visual and auditory reflexes.
middle ear (tympanic cavity) An air-filled space between the ear drum and the inner ear. It contains the three small ossicles (malleus,

incus and stapes) and is connected to the throat by the Eustachian tube. *See* Fig. 55; *see also* fenestra ovalis *and* fenestra rotunda.

midrib The largest vein of a leaf, running through the middle of the leaf blade.

migraine Severe recurrent headache, often attacking only one side of the head and usually accompanied by nausea, vomiting, visual disturbances and, occasionally, speech difficulties. In migraine, the blood vessels to the brain constrict and then dilate. The actual cause of migraine is not known. However, it has a hereditary link and tends to affect the intelligent. Attacks of migraine may be started by certain foods or alcohol. Pain-killing drugs may relieve migraine and other types of drugs may prevent it.

migration The regular, instinctive, two-way movements of animals from one habitat to another according to climate, seasonal changes, breeding habits, food supply, etc. Birds, for example, may fly thousands of kilometres to their breeding grounds.

mildew (mould) 1. A plant disease caused, in damp conditions, by a fungus producing a powdery or downy growth on the surface of the plant. The disease may be prevented by spraying the plant with a fungicide. 2. The fungus causing mildew.

mild steel Steel containing 0,04–0,25% (0.04–0.25%) carbon, with no special alloying components. It is rather soft and is used in the manufacture of car bodies, ship's plates, etc.

milk An important food produced by the mammary glands of all female mammals. Milk is used in feeding the new-born young until they are old enough to take solid foods. It contains nearly all the necessary substances required for growth and energy. These substances are proteins, fats and carbohydrate (lactose). It also contains calcium, potassium and phosphorus, together with several vitamins such as A, B and D. Milk is therefore very important for the formation of bone, teeth and blood. However, there is no iron present for the manufacture of haemoglobin. To compensate for this, the liver of the new-born infant contains plenty of iron. Breast milk also contains antibodies which give the infant some protection against certain diseases such as measles and poliomyelitis, provided the mother's body contains such antibodies.

milk of lime A thick suspension of lime, CaO, in water, thus containing calcium hydroxide, $Ca(OH)_2$.

milk of magnesia *See* magnesium hydroxide.

milk of sulphur *See* amorphous sulphur.

milk-sugar *See* lactose.

milk teeth *See* deciduous teeth *and* dental formula.

Milky Way *See* Galaxy, the.

milli- Symbol: m. A prefix meaning one thousandth, 10^{-3} (0,001; 0.001); e.g. one millimetre is equal to 10^{-3} metre.

millibar *See* bar.

Millikan's oil drop experiment An experiment (method) of measuring the charge on an electron. This is done by balancing the electric force on a charged oil drop against its weight (Fig. 138 (a)). Minute droplets of oil are allowed to fall between two parallel plates. These droplets of oil may be charged by ionising the air through which they fall, e.g. by X-rays or a little radioactive material. The potential difference between the plates, produced by a d.c. supply, is adjusted until one drop of oil, observed through a microscope, remains stationary. If the plates (Fig. 138 (b)) are separated by a distance of d metres and V volts is the potential difference across them, then the electric force on an oil drop with a charge of q coulombs is Vq/d newtons. When the oil drop is stationary this force must exactly balance its weight W; so

$$\frac{Vq}{d} = W \quad \text{or} \quad q = \frac{Wd}{V}$$

When this experiment is repeated different results are obtained, depending on the number of electrons producing the charge on the oil drop. However, these results are all multiples of one particular small value, which is taken to be the charge on a single electron. This value is $1{,}602 \times 10^{-19}$ coulombs (1.602×10^{-19} C). It is known as the charge of the electron.

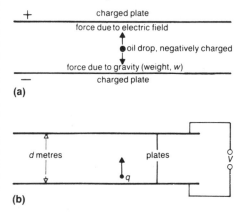

Fig. 138 Millikan's oil drop experiment

millimetres of mercury *See* atmosphere.

millipede An arthropod belonging to the class Diplopoda. It has a segmented body, each segment (apart from the first four and a few posterior ones) bearing two pairs of jointed legs. Usually there are not more than 200 legs. The head bears a pair of antennae, a pair of jaws and a group of simple eyes on each side. Millipedes are vegetarian, although large species may eat animal material.

Millon's reagent An aqueous solution obtained by mixing two volumes of a solution of mercury(II) sulphate and sulphuric acid with one volume of a solution of sodium nitrite. When heated with a substance containing protein, a pink or brick-red precipitate is formed.

mimicry A form of adaptation whereby an animal adopts the colour, habits or structure of a dangerous or inedible species in order to gain protection from predators. There are two types of mimicry: Batesian and Müllerian. In Batesian mimicry, an edible animal mimics the warning colouration of a noxious one. In Müllerian mimicry, both species are unpalatable and gain mutually since predators soon learn to avoid them after tasting a small number of each species. Mimicry occurs most amongst insects, but is also exhibited by other animals such as birds, fish and snakes.

mineral A crystalline, inorganic solid (except native mercury), occurring naturally within the Earth's crust. However, the term can also be used of organic solids, liquids and gases such as coal, oil and natural gas. Minerals are often obtained by mining.

mineral acid A term usually applied to the inorganic acids: hydrochloric acid, HCl, sulphuric acid, H_2SO_4, nitric acid, HNO_3, and phosphoric acid, H_3PO_4. They are so named because they can be prepared from minerals.

mineralisation The process in which various bacteria and fungi convert organic material from dead animals and plants into an inorganic form, e.g. organic nitrogen → ammonia → nitrite → nitrate.

mineralography The study of polished surfaces of minerals with the aid of a microscope using reflected light.

mineralogy The study of minerals.

mineral oil Petroleum and petroleum distillates, i.e. any oil of mineral origin. Compare vegetable oil.

mineral salt A salt needed by both animals and plants for the manufacture of protein, growth, tissue repair, digestion, transmission of nerve-impulses, etc. These salts are mainly chlorides, nitrates, sulphates and phosphates of metals such as potassium, sodium, calcium, magnesium, aluminium and iron. Mineral salts are formed by the weathering of rocks in the Earth's crust. However, nitrates are mainly derived from humus and nitrogen fixation. Soluble mineral salts dissolve in the soil water which plants absorb through their roots. See also trace element.

mineral wool (rock wool, slag wool) Fibres produced by blowing air or steam through molten slag. It is used as an insulating material.

minimum thermometer An alcohol thermometer designed for measuring minimum temperature during a period of time. A fall in temperature causes a steel index to be pulled, by the concave alcohol meniscus, along a graduated capillary tube. When the temperature rises the index remains in position, and therefore its upper end always indicates the minimum temperature. The index can be reset by tilting the thermometer or by using a magnet. See Fig. 139.

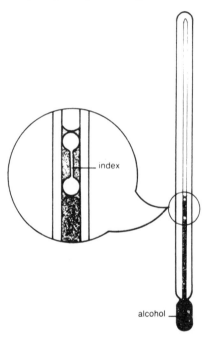

Fig. 139 Minimum thermometer

minium See lead oxides.

minor planet See asteroid.

Miocene A geological period of approximately 19 million years which ended approximately 7 million years ago. It is a sub-division of the Tertiary period.

mirage An optical illusion which usually appears as a pool of water above the surface of a hot road. It is caused by the progressive and continuous refraction of light passing through warmer layers of air of decreasing refractive index. The rays of light become parallel to the ground and then proceed to bend upwards. To an observer, the rays appear to come from the road surface. This produces an image of the sky which appears as a pool of water some distance away. See Fig. 140.

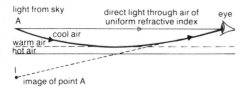

Fig. 140 Mirage

mirror A polished surface able to reflect light. Mirrors are usually made of glass coated with a tin amalgam or silver. A plane mirror is a flat mirror producing a virtual, erect image the same size as the object and situated the same distance behind the mirror as the object is in front. See Fig. 141; see also lateral inversion.

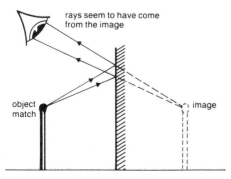

Fig. 141 Mirror

Two common types of curved mirror are the diverging or convex mirror and the converging or concave mirror. Both of these are spherical. The centre of the sphere of which the mirror surface forms a part is called the centre of curvature. A line from this point to the middle point (pole) of the mirror is the principal axis (optical axis) of the mirror. In a concave mirror, a ray of light parallel to the principal axis is reflected and passes through the principal focus or focal point. The distance between the pole of the mirror and the principal focus is called the focal length of the mirror. See Fig. 76 (a), (b); see also parabolic reflector.

mirror formula An equation relating the object distance (u), the image distance (v) and the focal length (f) of mirrors:

$$\frac{1}{v}+\frac{1}{u}=\frac{1}{f}=\frac{2}{r}$$

r is the radius of curvature. All distances are measured from the pole of the mirror. In the 'real-is-positive' sign convention, real object and image distances are positive and virtual object and image distances are negative. In the 'New Cartesian' sign convention, distances in the same direction as the incident light are positive and distances against it are negative.

mirror image An image of an object as it would appear if viewed in a mirror. The image is identical in structure to the object, but the two cannot be superimposed, e.g. a right hand is the mirror image of a left hand.

miscarriage See abortion.

mischmetall An alloy of cerium and other rare earth metals. One of its common uses is as a constituent in cigarette-lighter flints.

miscible Describes liquids which mix completely, so that they look like a single liquid; e.g. alcohol and water.

mist Tiny water droplets suspended in the atmosphere which reduce visibility to not less than one kilometre. Compare fog.

mite A small arachnid belonging to the order Acarina. They may be parasitic and some are responsible for the transmission of certain diseases in animals and plants. There are believed to be more than a million species, but little is known of the habits of many mites.

mitochondrion (pl. mitochondria) A structure, containing enzymes, bounded by a double membrane, which is found in the cytoplasm of all animal and plant cells except bacteria and blue-green algae. The inner membrane is folded inwards to form cristae. Mitochondria vary in shape and size. They are the sites at which energy transfer in cells takes place. In a single cell there may be very few or several thousand mitochondria. See also adenosine diphosphate and Krebs cycle.

mitosis (pl. mitoses) A type of cell division taking place during an organism's normal growth, in which the chromosome number is not reduced and which results in the equal division of the nucleus. Mitosis is a continuous process and follows a sequence of phases: prophase, metaphase, anaphase and telophase.

Prophase: Chromosomes appear and become shorter and thicker. Nuclear membrane breaks down and nucleoli disappear.

Metaphase: The nuclear membrane has broken down completely and a spindle forms. The chromosomes move towards the spindle equator about which they align themselves.

Anaphase: The centromeres divide and the daughter chromosomes separate and move to opposite poles.

Telophase: The chromosomes form two daughter nuclei and a new cell wall is formed between them. Two new cells have been formed, the nuclei of which contain exact replicas of the chromosomes of the parent nucleus. *See* Fig. 142.

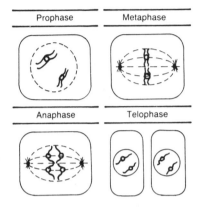

Fig. 142 Mitosis

mitral valve *See* bicuspid valve.
Mitscherlich's law (law of isomorphism) A law stating that substances with identical crystalline forms can be generally represented by similar formulae, i.e. they have similar chemical compositions. *Example:* potassium dihydrogenphosphate(V) is isomorphous with potassium dihydrogenarsenate(V) and, since the formula of the former is $KH_2PO_4 \cdot H_2O$, this implies that the formula of the latter is $KH_2AsO_4 \cdot H_2O$. This law has been used in determining the relative atomic masses of some elements. However, the law has several exceptions. *Example:* lead(II) sulphide, PbS, is isomorphous with silver sulphide, Ag_2S.
mixture A system consisting of two or more substances dispersed one within the other, in which there is no chemical bonding between the different substances. Mixtures may be heterogeneous or homogeneous and they can be solid, liquid or gaseous. A heterogeneous mixture (e.g. a mixture of powdered iron and sulphur) contains distinguishable phases, i.e. the different constituents can be distinguished. In a homogeneous mixture (e.g. air) there is only one phase, i.e. the molecules of the components mix completely. Mixtures differ from chemical compounds in the following ways: (1) they may vary widely in composition; (2) they can be separated by physical means such as distillation and crystallisation; (3) the properties of a mixture are the sum of the properties of its components, i.e. each component retains its own characteristic properties; (4) the preparation of a mixture does not usually result in a change in temperature.

MKS system A system of units based on the metre, the kilogram and the second. This system of units formed the basis of the SI system of units and has been replaced by it.
mmHg *See* atmosphere.
Mn Chemical symbol for manganese.
mobile phase *See* chromatography *and* R_F-value.
moderator A material, such as heavy water, used to retard fast neutrons in a nuclear reactor so that they are more likely to cause fission.
modulation The process of superimposing a second wave motion on a carrier wave so that a signal may be carried to a receiver. The signal may be either a sound or vision signal. Two common types of modulation are amplitude modulation (AM), in which the amplitude of the carrier wave is varied by the characteristics of an audiofrequency signal, and frequency modulation (FM), in which the frequency of the carrier wave is varied by the characteristics of the audiofrequency signal. Amplitude modulation is used for radio transmissions in the long, medium and short wavebands. It can be affected by static interference. Frequency modulation is used for radio transmissions in the very high frequency (VHF) and microwave bands. It is relatively free from static interference. *Compare* demodulation.
Mohs' scale of hardness A scale of hardness used to determine the relative hardness of solid minerals by comparing them to ten reference minerals each having a number. In ascending order of hardness, these are: (1) talc; (2) gypsum; (3) calcite; (4) fluorite; (5) apatite; (6) feldspar; (7) quartz; (8) topaz; (9) corundum; (10) diamond. The steps between minerals in the scale are not equal, e.g. corundum is about 10% harder than topaz, whereas diamond is about 10 times harder than corundum. Mineral hardness, determined after Mohs' scale, is a scratch hardness as opposed to indentation hardness. If a mineral can be scratched by quartz and not by feldspar, then its hardness lies between 6 and 7. *See* Appendix.
mol *See* mole.
molal *See* molality.
molality Symbol: m. The concentration of a solute expressed as the number of moles of solute dissolved in one kilogram of solvent. *Example:* a 0,5 molal (0.5 m) aqueous solution of sodium chloride, NaCl, contains $0,5 \times 58,44 = 29,22$ g $(0.5 \times 58.44 = 29.22$ g) NaCl per kg of water. In the tables (*see* Appendix) for molecular boiling-point elevations and molecular freezing-point depressions, the molality is used. Molality should not be confused with molarity. In dilute, aqueous solutions the molality and molarity are almost the same.

molar 1. A term usually denoting that a physical quantity is divided by moles, i.e. meaning 'per mole'. *Example:* the molar heat capacity of a substance is its heat capacity per mole, written C_m. *See also* molarity. **2.** *See* molar tooth.

molar gas constant (universal gas constant) Symbol: R. The constant used in the gas equation
$$pV = nRT$$
where p is the pressure, V is the volume, n is the number of moles of gas and T is the temperature. The value of this constant (in SI units) is 8,314 joules per kelvin per mole (8.314 J K^{-1} mol^{-1}). This can be obtained by substituting into the equation the values of p, V and T for one mole of a gas at STP:
101 325 Nm^{-2} × 0,0224 m^3 = 1 mol × R × 273 K
(101 325 Nm^{-2} × 0.0224 m^3 = 1 mol × R × 273 K)

molarity Symbol: M. The concentration of a solution expressed as the number of moles of solute dissolved in one cubic decimetre, dm^3, of solution. *Example:* a 0,5 molar (0.5 M) solution of sodium chloride, NaCl, contains 0,5 × 58,44 = 29,22 g (0.5 × 58.44 = 29.22 g) of NaCl per dm^3 of solution.

molar solution *See* molarity.

molar tooth (molar) A tooth with a flattened crown, suitable for crushing and grinding. In humans there are 8 in the milk dentition and 12 in the permanent dentition. Molars in the lower jaw have two roots, whereas those in the upper have three.

molar volume (gram molecular volume) Symbol: V_m. The volume occupied by one mole of any substance. All gases measured at the same temperature and pressure have approximately the same molar volumes. At STP the molar volume of an ideal gas is 22,4 cubic decimetres (22.4 dm^3), and the molar volumes of real gases approximate to this.

molasses The uncrystallised residue remaining after sugar crystallises out of the juice from sugar-cane and sugar-beet. It is used in the manufacture of rum and treacle, and as a food for domestic animals.

mole Symbol: mol. The basic SI unit of amount of substance, defined as the amount of a substance which contains the same number of entities as there are atoms in 0,012 kilogram (0.012 kg) of the carbon-12 isotope. The entities must be specified; they may be atoms, molecules, ions, electrons, protons, or any specified group of particles. The actual number of carbon atoms in 0,012 kg (0.012 kg) of carbon-12 is known as the Avogadro constant (number) and is 6,023 × 10^{23} (6.023 × 10^{23}). Therefore, one mole of any substance is that amount of the substance which contains 6,023 × 10^{23} (6.023 × 10^{23}) particles of that substance. From the definition of relative atomic mass, it is now seen that the relative atomic mass of an element, expressed in grams, always contains the Avogadro number of atoms. *Examples:* 1,01 grams (1.01 g) of hydrogen, 16,00 grams (16.00 g) of oxygen and 55,85 grams (55.85 g) of iron all contain 6,023 × 10^{23} (6.023 × 10^{23}) atoms. If a sample of iron has a mass of 22,34 grams (22.34 g), it will correspond to
$$\frac{22,34}{55,85} = 0,4 \text{ moles of iron atoms, containing}$$
$0,4 \times 6,023 \times 10^{23} = 2,41 \times 10^{23}$ atoms
$$\left(\frac{22.34}{55.85} = 0.4 \text{ moles of iron atoms, containing}\right.$$
$\left. 0.4 \times 6.023 \times 10^{23} = 2.41 \times 10^{23} \text{ atoms}\right)$

The mole therefore replaces the older terms gram-atom, gram-molecule, etc.

molecular crystal A type of crystal in which molecules occupy lattice points. Molecular crystals have low melting-points because the molecules are held together by weak bonds such as Van der Waal's bonds (forces) or hydrogen bonds. Most organic substances crystallise as molecular crystals. *See also* crystal.

molecular equation A chemical equation in which the reactants and the products are written as atoms or molecules, but not as ions, e.g.
$CH_4 + Cl_2 \rightarrow CH_3Cl + HCl$

molecular formula *See* formula.

molecularity *See* order of reaction.

molecular orbital *See* orbital.

molecular sieve *See* zeolite.

molecular weight *See* relative molecular mass.

molecule The smallest particle of an element or chemical compound which can exist in the free state and still have all the properties of that element or compound. A molecule is made up of one or more atoms, e.g. He is a molecule of helium; H_2 is a molecule of hydrogen; CO_2 is a molecule of carbon dioxide. Ionic substances such as sodium chloride do not form discrete molecules. Sodium chloride, NaCl, consists of aggregates of positively and negatively charged ions. *See also* atomicity.

mole fraction The mole fraction of a solute is defined as:
$$\frac{\text{moles of solute}}{\text{moles of solute} + \text{moles of solvent}}$$

mole of atoms (gram-atom) *Example:* 1,01 g (1.01 g) of hydrogen is called one mole of hydrogen atoms (formerly, one gram-atom of hydrogen). *See* mole.

mole of ions (gram-ion) *Example:* 40,08 g (40.08 g) of calcium ions is called one mole of calcium ions (formerly, one gram-ion of calcium). *See* mole.

mole of molecules (gram-molecule) *Example:* 2,02 g (2.02 g) of hydrogen is called one mole of hydrogen molecules (formerly, one gram-molecule of hydrogen). *See* mole.

Molisch's test (alpha-naphthol test) A test for all carbohydrates. A few drops of Molisch's reagent, containing alpha-naphthol in alcohol, are added to an aqueous solution of a carbohydrate. A small amount of concentrated sulphuric acid is added and a purple (violet) ring is observed at the junction of the two liquids. *See also* furfural.

Mollusca A large phylum of invertebrates. Some of its members live on land and others in water. They have a soft, unsegmented body, often covered with a shell; a muscular foot; and a distinct head bearing tentacles and eyes. Examples include snails, mussels, oysters and squids.

moment The turning effect of a force about a point when it is applied to a body. The moment of a force is the product of the force and the perpendicular distance from the line of action of the force to the point. The moment of a force is also called its torque. *See also* couple.

moments, principle of When a body is in equilibrium, the sum of the clockwise moments about any point is equal to the sum of the anticlockwise moments about the point.

momentum Symbol: p. The momentum of a body is the product of its mass and its velocity. The SI unit of momentum is the kilogram metre per second (kg m s^{-1}). *See also* Newton's laws of motion.

momentum, conservation of *See* conservation (laws of).

monatomic Describes a molecule consisting of one atom, e.g. helium, He, and all other noble gases.

monazite *See* thorium.

Mond process A process for purifying nickel, in which nickel(II) oxide, NiO, is treated with synthesis gas (a mixture of carbon monoxide, CO, and hydrogen, H$_2$) at a temperature of about 350°C:
NiO + CO → Ni + CO$_2$
The nickel so produced is impure. It is then heated to about 60°C in a stream of carbon monoxide, resulting in the formation of tetracarbonylnickel(O) (nickel carbonyl), [Ni(CO)$_4$]:
Ni + 4CO → [Ni(CO)$_4$]
Volatile [Ni(CO)$_4$] evaporates leaving impurities behind, and when heated to about 200°C it decomposes into very pure nickel and carbon monoxide:
[Ni(CO)$_4$] → Ni + 4CO

mongolism *See* Down's syndrome.

mono- Prefix meaning one.

monobasic Describes an acid which has one acidic hydrogen atom which can be replaced by a metal or metal-like group such as the ammonium group, NH$_4$, thus forming only one series of salts. Examples of monobasic acids are hydrochloric acid, HCl, nitric acid, HNO$_3$, and ethanoic acid (acetic acid), CH$_3$COOH.

monochasial cyme *See* cyme.

monochord *See* sonometer.

monochromatic radiation Light or any other electromagnetic radiation which has only one colour or wavelength, i.e. it all has the same or nearly the same frequency. *Compare* polychromatic radiation.

monoclinic sulphur (prismatic sulphur, β-sulphur) One of the two main allotropes of sulphur. It consists of needle-shaped crystals and is stable above 96°C. If monoclinic sulphur is kept at room temperature, it gradually changes to rhombic sulphur (the other main allotrope of sulphur):

monoclinic sulphur $\underset{\text{above 96°C}}{\overset{\text{below 96°C}}{\rightleftharpoons}}$ rhombic sulphur

Both monoclinic and rhombic sulphur consist of S$_8$ molecules in which the atoms are arranged in a distorted ring. *See* Fig. 143; *see also* enantiotropy *and* polymorphism.

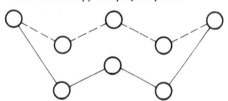

Fig. 143 Monoclinic sulphur: the S$_8$ molecule

monocotyledon A member of the smaller class of angiosperms, Monocotyledoneae, having an embryo with one cotyledon. Other characteristics include parallel venation of the leaves, and vascular tissue of the stem irregularly arranged. Examples include grasses, lilies, palms and bananas. *Compare* dicotyledon.

monocular Involving the use of one eye.

monoculture Growing the same crop on the same piece of land year after year.

monocyte The largest type of white blood corpuscle (leucocyte). Monocytes are phagocytic and have a single oval or kidney-shaped nucleus. In humans, they make up about 5% of all leucocytes.

monoecious 1. In plants: having male and female reproductive organs on the same individual; having unisexual flowers, both male and female, on the same individual. 2. In animals: having male and female reproductive organs on the same individual, i.e. hermaphrodite.

monoestrous *See* oestrous.

monogamous Having only one mate.

monohybrid The offspring of parents which are genetically different in only one characteristic. *Compare* dihybrid.

monohydric Describes a molecule containing one hydroxyl group, -OH. *Example:* methanol, CH_3OH, is a monohydric alcohol.

monomer A basic unit (substance) from which a polymer is made. *Example:* glucose may polymerise to a polysaccharide; in this case it acts as a monomer. However, the term is usually applied to a substance used in the synthesis of plastic polymers.

monophyodont Having only one dentition. This is characteristic of some whales.

monosaccharide The general name for a group of soluble carbohydrates (sugars) which may be represented by the formula $C_nH_{2n}O_n$. However, monosaccharides containing six carbon atoms are the most common, and therefore the formula $C_6H_{12}O_6$ is often used to represent all monosaccharides. A monosaccharide may be obtained by hydrolysis of a disaccharide or polysaccharide. *Example:* hydrolysis of the disaccharide saccharose (sucrose), $C_{12}H_{22}O_{11}$, yields two molecules of monosaccharide:
$$C_{12}H_{22}O_{11} + H_2O \rightleftharpoons 2C_6H_{12}O_6$$
A little acid must be present to assist the reaction. Monosaccharides cannot be further hydrolysed to simpler carbohydrates. Examples of monosaccharides are glucose and fructose.

monosilane *See* silicide.

Monotremata (Prototheria) A subclass of mammals containing the duck-billed platypus and the spiny ant-eater. Monotremes are primitive in that they possess both mammalian and reptilian characteristics. They are the only egg-laying mammals; their young are suckled in a pouch by mammary glands without nipples and they possess a cloaca.

monotropy The existence of a single form (allotrope) of a substance which is more stable than any other, regardless of the temperature. *Example:* phosphorus exhibits monotropy, the red allotrope being the stable one; i.e. white phosphorus will slowly change into red phosphorus at any temperature. However, at room temperature this change is very slow. At 250°C the change from white to red takes place in a few days. The reverse change is not directly possible. *Compare* enantiotropy.

monovalent (univalent) Having a valency of one.

monozygotic twins Genetically identical twins formed from one fertilised ovum. At some early stage in the development of the embryo it divides into two parts, each part developing into a normal embryo. *Compare* dizygotic twins.

Moon The Earth's natural satellite and its nearest celestial neighbour. Its mean distance from Earth is 384 400 km. It orbits the Earth once in a month (27,3 days; 27.3 days), always keeping the same face pointing towards the Earth. It has no atmosphere. Its gravity, which is one-sixth of that of the Earth, causes the tides on Earth.

moon *See* satellite.

morbilli *See* measles.

mordant A substance which combines with different dyes and fixes them to certain textile fibres. Most mordants are inorganic substances such as compounds of aluminium, chromium and iron.

morphine An organic, aromatic alkaloid obtained from opium. It is a narcotic and a powerful pain-killer leading to addiction. Morphine represses the cough reflex.

morphology The study of (external) structure and form of organisms.

Morse code The earliest and best known electrical system used in telegraphy and signalling. The code consists of combinations of dots and dashes representing each letter of the alphabet. The international distress call 'SOS' is simply $\cdots --- \cdots$

mortar 1. A mixture of 1 part slaked lime (calcium hydroxide), $Ca(OH)_2$, and 3–4 parts sand, made into a suitable paste with water. Sometimes mortar also contains cement. It is used in building, e.g. between bricks. On exposure to air it slowly reacts with carbon dioxide, CO_2, forming small insoluble crystals of calcium carbonate, $CaCO_3$, which effectively bind the grains of sand:
$$Ca(OH)_2 + CO_2 \rightarrow CaCO_3 + H_2O$$
Weathering of mortar is caused by rain-water containing dissolved carbon dioxide from the air, resulting in the formation of soluble calcium hydrogencarbonate, $Ca(HCO_3)_2$:
$$CaCO_3 + CO_2 + H_2O \rightarrow Ca(HCO_3)_2$$
2. A strong bowl, made of agate, glass, metal or porcelain, in which a solid may be ground to a powder using a pestle. *See* Fig. 144.

mosaic disease A virus disease in plants causing mottling or spotting of the leaves. The disease is of great economic importance as it results in considerable losses to the agricultural economy. Plants affected include tobacco, sugar-cane, cabbage, tomatoes, potatoes and other fruits and vegetables. The disease also infects some flowers. The effects of such infection on flower petals have been exploited

mosaic virus

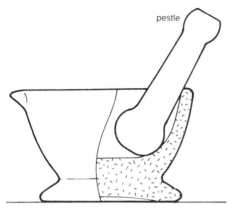

Fig. 144 Mortar

by horticulturalists to produce unusual variegations in flowers, e.g. in the tulip.

mosaic virus A plant virus which may cause mottling or spotting of the leaves (sometimes with lesions).

mosquito An insect belonging to the order Diptera. Some species are important agents in the transmission of diseases such as elephantiasis, malaria and yellow fever. In most species, the female has a sharp proboscis adapted for piercing and sucking fluids, usually blood, whereas the male does not bite and lives on plant secretions (nectar). Mosquitoes lay their eggs on the surface of water, and the larvae and pupae live in this water until the adults emerge.

moss See Musci.

moth A member of the order Lepidoptera. The vast majority of moths fly by night, have dull-coloured wings, have thread-like or feathered antennae, have thick bodies and rest with their wings held horizontally or folded around their bodies. Moths have a specialised proboscis for sucking nectar, which is coiled under the head when not in use.

mother liquor The solution remaining after the precipitation (formation) of crystals.

mother-of-pearl See nacre.

motile Capable of movement.

motion, equations of See equations of motion.

motion, laws of See Newton's laws of motion.

motoneurone See motor neurone.

motor 1. Pertaining to movement. 2. A device for converting some other form of energy into mechanical energy. See also electric motor and internal combustion engine.

motor end plate The flat termination of a branch of a motor nerve-fibre on a muscle fibre. It does not penetrate the muscle fibre. See Fig. 150.

motor neurone (efferent neurone) A nerve-cell which conducts impulses away from the central nervous system to an effector organ such as a muscle or a gland. See Fig. 150; compare sensory neurone.

mould See mildew.

moulting The casting-off or shedding of cuticle, feathers, hair, skin or horns at periodic intervals, depending on seasonal changes or on growth rate. See also ecdysis.

moving coil instrument An electrical measuring instrument consisting of a flat, rectangular coil (carrying the direct current to be measured) suspended vertically between the curved pole pieces of a permanent magnet. A current flowing through the coil causes it to rotate and its movement is restrained by a torsion wire or a flat hairspring. The deflection of the coil is proportional to the size of the current. A pointer which moves over a scale may be attached to the coil; in this form the instrument is an ammeter. If a high resistance is included in the coil's circuit, the instrument is a voltmeter. In sensitive moving coil galvanometers (mirror galvanometers) the deflection of the coil is measured by attaching a mirror to the torsion wire. See Fig. 145.

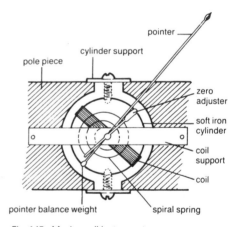

Fig. 145 Moving coil instrument

moving coil loudspeaker See loudspeaker.

moving iron instrument An electrical measuring instrument consisting of a pivoted piece of soft iron which is free to move into a fixed coil of wire (carrying the current to be measured). A current flowing through the coil causes the piece of soft iron to be attracted by the coil and its movement is restrained by a spring. The degree of attraction is proportional to the size of the current, which is measured by a pointer moving over a scale. In this form the instrument is an ammeter. If a high resistance is included in the coil's circuit, the instrument is a voltmeter. A moving iron instrument can be used to measure alternating and direct currents. There are several different types of moving iron instruments. *See also* hot-wire ammeter.

m.p. *See* melting-point.

mRNA *See* messenger RNA.

mucilage One of a group of complex, highly-branched carbohydrates which occur in the cell walls of many plants. They are hard when dry, but absorb water becoming swollen and slimy. There is no marked difference in chemical composition between mucilages and gums.

mucin A protein found in mucus.

mucopeptide A complex substance containing polysaccharide derivatives to which are attached short peptide chains, e.g. each consisting of four amino acids. An example of a mucopeptide is murein.

mucopolysaccharide A complex compound of polysaccharide and protein. On hydrolysis a mucopolysaccharide yields monosaccharides and monosaccharide derivatives, e.g. glucosamine. In mucopolysaccharides, the polysaccharide tends to predominate chemically.

mucoprotein (glycoprotein) A complex compound of protein and polysaccharide. The protein tends to predominate chemically. Together with lipoproteins, mucoproteins are important structural material for cell membranes.

Mucor A genus of saprophytic fungi. One of the most common species is *Mucor mucedo* (bread mould), appearing as a mass of white threads on damp bread.

mucosa *See* mucous membrane.

mucous Pertaining to mucus.

mucous membrane (mucosa) In vertebrates, a moist layer of tissue lining many body cavities such as the trachea and gut. It is often ciliated and is composed of a surface layer of epithelium containing goblet cells and an underlying layer of connective tissue.

mucus 1. A slimy, viscous substance containing mucin and secreted by the goblet cells of mucous membranes in vertebrates. 2. Any slimy, viscous substance secreted by animals and providing protection and lubrication.

mulching Covering the surface of a soil with a layer of leaves, straw, etc., to prevent water loss by evaporation and to provide nutrients for growing plants.

Müllerian mimicry *See* mimicry.

multi- Prefix meaning many.

multicellular Consisting of many cells. *Compare* unicellular.

multimeter An electrical measuring instrument designed to measure currents, voltages and resistances over various ranges. It has separate graduated scales for indicating the three values which are measured.

multiple bond A chemical bond in which more than two electrons (one pair) take part, e.g. a double or a triple bond between carbon atoms, as seen in alkenes and alkynes.

multiple fission *See* fission.

multiple proportions, law of *See* chemical combination (laws of).

multiple sclerosis (disseminated sclerosis) A disease in which the protective linings of nerve-fibres of the brain or spinal cord are attacked by unknown agents. The tissue may be destroyed and it is then replaced by hard scar tissue. This causes the nerves involved to cease to function and paralysis is often the result. The disease especially affects young adults of both sexes. Initial symptoms include a temporary defect of vision, 'pins and needles' and difficulty in walking. Very often the symptoms improve and disappear for a shorter or longer period of time, but they will recur. However, serious disabilities may not happen for thirty years or more. So far there is no cure for multiple sclerosis.

multivalent element An element which has more than one valency. Transition elements are multivalent elements. *Examples:* iron has the valencies two and three; copper has the valencies one and two.

Mumetal® An alloy of iron (18%), nickel (75%) and copper and chromium (7%). It is used in magnetic shielding and where changes in magnetic flux are involved, e.g. in transformers, electric motors and generators.

mumps An acute, contagious disease, mainly occurring in childhood, caused by a virus transmitted in the saliva of an infected person. Symptoms include swelling of various glands, particularly of the salivary glands and especially those situated beneath the ears, resulting in pain on chewing. The swelling often starts on one side of the head and may then later appear on the other side also. In males, mumps may affect the testicles, causing a rather painful inflammation and swelling which may lead to infertility (sterility). Mumps usually (but not always) give a life-long immunity. A vaccine against mumps is available.

murein (peptidoglycan) A mucopeptide found in the cell wall of most bacteria, which gives the cell wall strength so that it can withstand the high internal osmotic pressure. This ranges from 5 to 20 atmospheres and is a result of solute concentration. The internal osmotic pressure is 3–5 times greater in Gram-positive bacteria than in Gram-negative ones and the layer of murein in the cell walls of Gram-positive bacteria is therefore much thicker than in the cell walls of Gram-negative bacteria. In the Gram's stain, the Gram-positive bacteria are stained blue because of the large amount of murein in their cell walls. The murein combines with the dye and is not removed by the decoloriser. The difference in sensitivity of Gram-positive and Gram-negative bacteria to various penicillins also depends on the greater amount of murein in the cells walls of Gram-positive bacteria since penicillin prevents the formation of murein. It has no effect on already existing Gram-positive bacteria.

Musci A class of bryophytes comprising the mosses, of which there are thousands of species. The plants are relatively small and grow in shady, damp habitats. They have leafy stems which are either erect or prostrate and which are attached to the substratum by multicellular rhizoids, hair-like structures functioning as roots. At the tips of the stems are situated antheridia and archegonia (the sex organs). After fertilisation has occurred, a spore-containing capsule develops. Each spore is capable of giving rise to a branched thread-like structure, a protonema, from which a new moss plant develops from a lateral bud. *See* Fig. 191.

muscle A mass of contractile fibres or cells capable of contracting and relaxing to produce movement. Muscles are usually attached to the skeleton by tendons and most of them work in pairs. *See also* antagonism, involuntary muscle *and* skeletal muscles.

muscular dystrophy Any of various diseases of yet unknown cause, characterised by progressive wasting away (atrophy) of the muscles. Muscular dystrophy is an inherited disease which often begins in childhood. In the treatment of the disease, massage and physiotherapy are helpful.

mushroom A fungus belonging to the genus Agaricus. There are several species of edible mushrooms, differing largely in size, colour and form. All the species are terrestrial and saprophytic, having underground mycelia which obtain food from dead organisms in the soil. The part of the mushroom which grows above ground consists of an umbrella-shaped portion made up of a round part, the cap (sporangium), situated on a stalk. Underneath the cap there are gills, thin, plate-like structures on whose sides spores are formed. *See* Fig. 146.

mustard gas (oil)
Cl—CH$_2$—CH$_2$—S—CH$_2$—CH$_2$—Cl
A highly toxic, colourless, oily liquid, b.p. 217°C, m.p. 13°C. Both the liquid and the gas can be absorbed through the skin, forming skin blisters. Mustard gas has been used in chemical warfare. It can be destroyed by

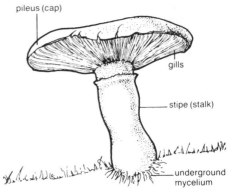

Fig. 146 Mushroom

oxidising agents. Nitrogen mustards contain nitrogen instead of sulphur. They are also very toxic and are used as powerful mutagens. They have some special medical uses.

mutagen (mutagenic agent) A substance which is capable of inducing mutation, e.g. caffeine, methanal (formaldehyde) and mustard gases (oils).

mutagenic Capable of inducing mutation, e.g. short wave radiation or certain chemicals (mutagens).

mutagenic agent *See* mutagen.

mutant Describes a gene which has undergone mutation, or an observed organism (in a population) bearing a gene which has undergone mutation.

mutation A spontaneous change in the nature of a gene during chromosome duplication. This is a result of a change in DNA. Natural mutations are rare and occur at random. They may be either beneficial or harmful to the organism. When a mutation is beneficial, it increases the ability of the organism (mutant) to survive. However, most mutations are harmful. Mutation is usually divided into gene mutation (described above) and chromosome mutation. In chromosome mutation there is a structural change in the chromosomes or a change in their numbers. Mutation in cells giving rise to gametes is more important than mutation in somatic cells, since those producing gametes can give rise to a change in the characteristics of the next generation developing from them. The rate of mutation is increased by short wave radiation such as X-rays and gamma rays or by bombardment with neutrons and by certain chemicals (mutagens). *See also* translocation.

mutual inductance Symbol: *M*. An arrangement of two coils, such that a change of magnetic flux in one causes an induced e.m.f. in the other, is said to have mutual inductance.

The mutual inductance is given by the expression
$E = M \times$ (rate of change of current)
where E is the induced e.m.f. The SI unit of inductance is the henry (H). *See also* self inductance.

mutual induction The generation of an e.m.f. in one circuit (coil) as a result of changing magnetic flux in an adjacent circuit (coil). This forms the basis on which a transformer operates. The size of the induced e.m.f. depends upon the number of turns of wire in each coil, the relative position of the coils, the type of core, etc. *See also* electromagnetic induction *and* self induction.

mutualism (symbiosis) An association between two or more organisms in which both (all) benefit and none is harmed. An example of mutualism is the association between a leguminous plant and nitrogen-fixing bacteria living in its root nodules. The bacteria require carbohydrate energy to perform their task of fixing nitrogen from the atmosphere; this carbohydrate is supplied by the plant. In return, the plant gets a continuous supply of organic nitrogen compounds liberated by the bacteria in excess of any need of their own. *See also* commensalism *and* parasitism.

mycelium (pl. mycelia) The vegetative body of a fungus, composed of a network of hyphae.

mycology The study of fungi.

Mycophyta A division comprising simple eucaryotic fungal organisms, e.g. mushrooms, rusts, moulds and yeasts. Mycophyta lack chlorophyll; they are unicellular or possess tubular filaments called hyphae; they reproduce sexually and asexually with the formation of a large number of spores. Because of the lack of chlorophyll the organisms live either as saprophytes, e.g. *Mucor* and mushrooms, or as parasites causing plant and animal diseases such as potato blight, ringworm and athlete's foot.

mycorrhiza A symbiotic association between a fungus and the roots of certain higher plants, e.g. in several species of orchids and in pine trees. This association is generally found in perennial plants and is usually divided into ectomycorrhiza and endomycorrhiza. These may also be described as ectotrophic mycorrhiza and endotrophic mycorrhiza respectively. In ectomycorrhiza the hyphae form an extensive sheath around the outside of the root. In endomycorrhiza the mycelium is embedded in the root. There is evidence that infected plants grow better; this may be caused by an improved nutrient absorption due to the greater surface area (ectomycorrhiza) or by the fact that the fungus provide channels inside the root along which absorbed nutrients may migrate (endomycorrhiza).

myelin A white, fatty substance containing protein and forming an insulating sheath around a nerve-fibre in vertebrates and crustaceans. Nerve-fibres with a thick myelin sheath conduct nerve-impulses very rapidly. *See also* Schwann cells.

myelinated nerve-fibre *See* medullated nerve-fibre.

myeloid tissue Tissue situated in the bone marrow, actively engaged in the production of red blood corpuscles and polymorphs.

myo- Prefix meaning muscle.

myocardium In vertebrates, the muscular wall of the heart.

myoglobin A pigment which is a conjugated, globular protein. It is found in muscles and has a very high affinity for oxygen. Myoglobin resembles haemoglobin in structure and function. However, it is simpler than haemoglobin, containing fewer amino acid units and carrying less oxygen.

myology The study of muscles.

myopia (short sight) A defect of the eyeball in which the eyeball is too long and consequently images of far objects are brought to a focus in front of the retina. The defect is corrected by concave spectacle lenses. *See* Fig. 147; *compare* hypermetropia.

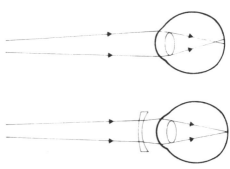

Fig. 147 Myopia (short sight)

myosin A protein found in the thick filaments in muscles. Myosin combines with actin (a protein found in thin filaments in muscles) forming actomyosin which, when stimulated, causes muscle contraction.

Myriapoda A class of arthropods comprising centipedes and millipedes.

myxoedema *See* cretinism.

N

N Chemical symbol for nitrogen.
Na Chemical symbol for sodium.
nacre (mother-of-pearl) A form of calcium carbonate secreted by specialised cells of certain molluscs. It lines the interior of the shell. In the case of oysters, a physical stimulus such as a sand grain may cause a build-up of nacre around the sand grain thus forming a pearl.
NAD Nicotinamide adenine dinucleotide (coenzyme I). A derivative of nicotinic acid. It is a coenzyme generally acting in redox reactions during energy-producing (catabolic) processes in the living cells. It catalyses the removal of a hydrogen atom (or electron) from a hydrogen-donating substrate, so that the substrate is oxidised and NAD is reduced to NADH. NADH is oxidised back to NAD by reducing a substrate.
NADH See NAD.
NADP Nicotinamide adenine dinucleotide phosphate. (coenzyme II). A coenzyme which is closely related and similar in its action to NAD. NADP is generally involved in energy-consuming (anabolic) processes in the living cells. Its reduced form is NADPH and it is oxidised back to NADP by reducing a substrate.
NADPH See NADP.
nano- Symbol: n. A prefix meaning 10^{-9}.
naphthalene $C_{10}H_8$ A fused, aromatic hydrocarbon which may be obtained during the destructive distillation of coal. It is a white, crystalline solid with a characteristic smell of moth-balls. Naphthalene is used in organic synthesis, e.g. in the manufacture of dyes and plastics, and in moth-balls. See Fig. 148.

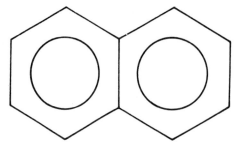

Fig. 148 Naphthalene

narcotic A substance (usually a drug) which produces drowsiness, sleep and insensibility (numbness), e.g. codeine, opium and morphine.
nares (sing. naris) External nares are the paired external openings of the nasal cavity. In animals having a nose or snout, they are called nostrils. Internal nares lead from the nasal cavity to the mouth. They are found in all vertebrates except most fish.
naris See nares.
nasal Pertaining to the nose.
nascent state ('newly-born' state) The state of certain elements, e.g. hydrogen and oxygen, in which they are more active and are probably in the form of single atoms instead of molecules. This occurs when they are first formed in a chemical reaction. *Example:* nascent hydrogen, produced by mixing hydrochloric acid and zinc or magnesium, is a much stronger reducing agent than molecular hydrogen.
nastic movement (nasty) A non-directional plant movement in response to stimuli. Such movements may be the result of changes in turgor pressure in certain cells or growth curvatures. See also chemonasty, haptonasty, hydronasty, nyctinasty, photonasty, seismonasty *and* thermonasty.
nasty (pl. nasties) See nastic movement.
native 1. Describes animals and plants which originated in the area in which they live. **2.** Describes an element occurring naturally in the free state (uncombined), e.g. copper, gold, platinum, silver and sulphur.
natrium The Latin name for sodium.
natural frequency The frequency of a freely vibrating object or body. *Example:* a low-flying aircraft often causes a window to vibrate if the natural frequency of vibration of the window corresponds to one of the frequencies making up the noise from the aircraft's engines.
natural gas A flammable mixture of hydrocarbon gases composed mainly of methane, CH_4. It is usually found in association with crude oil deposits. In this respect it was once considered a nuisance when discovered during oil drilling operations. However, huge volumes are now used as fuel and in the synthesis of organic compounds.
natural selection See Darwin's theory.
navel (umbilicus) In mammals, the place of attachment of the umbilical cord to the embryo/foetus.
Ne Chemical symbol for neon.
neap tide See tide.
near point The nearest point from the eye at which objects can be seen clearly. With increasing age, the near point becomes increasingly further away from the eye as the eye lens becomes less flexible. The standard near point distance (D) is 25 cm. *Compare* far point.

nebula (pl. nebulae or nebulas) A cloud of gas and dust found in space. Nebulae can be observed in several constellations such as Lyra, Orion and Cygnus. Some nebulae give light because their gas is glowing due to the presence of a nearby very hot star, whereas others reflect light from a nearby star. Certain nebulae absorb light from stars behind them, thus appearing as dark patches. Nebulae may be remains of exploded stars; in some of them new stars may be born.

neck vertebrae (cervical vertebrae) The first bones of the vertebral column. In humans there are seven such bones and they are the smallest of the single vertebrae. See Fig. 242.

necrosis Death of cells, tissue or part of an organ. *See also* gangrene.

nectar A sugary fluid secreted by the nectaries of flowers. It contains various carbohydrates such as saccharose (sucrose), glucose and fructose. Nectar is drunk or collected by certain insects and birds who, in return, help in pollination.

nectar guides *See* honey guides.

nectary A glandular region, secreting nectar, on the receptacle or other part of a flower. *See* Fig. 75.

negative *See* photography.

negative catalyst *See* catalyst *and* inhibitor.

negative charge *See* charge.

negative electrode *See* cathode.

negative ion *See* anion.

negative lens *See* lens.

negative pole *See* pole.

negative taxis *See* tactic movement.

negative tropism *See* tropic movement.

negaton *See* negatron.

negatron (negaton) Alternative name for the electron to distinguish it from the positron.

nekton Freely-swimming animals found at various depths in lakes and seas, e.g. fish and whales.

Nematoda A phylum of slender, cylindrical, unsegmented worms tapered at both ends, which have a body covered with a thick cuticle and a mouth and alimentary canal. Nematodes live in soil, fresh water or the sea and may be either free-living or parasitic. Members include roundworms, hookworms, threadworms and eelworms. *See also* filarial worms.

neon An element with the symbol Ne; atomic number 10; relative atomic mass 20,18 (20.18); state, gas. It is a colourless, odourless, very inert element found in the atmosphere in a very small amount. Neon belongs to the noble gases of the Periodic Table of the Elements. It is used in fluorescent lamps and neon signs.

Neoprene® The first synthetic rubber, produced in the United States in 1931. It is made by polymerisation of 2-chlorobuta-1,3-diene (chloroprene), $CH_2\!\!=\!\!CH\!-\!CCl\!\!=\!\!CH_2$. Neoprene is superior to natural rubber in its resistance to chemicals, oils, heat and sunlight, but is less elastic. It is used as a lining material in storage tanks, in the production of conveyor belts, hoses, etc.

nephridium (pl. nephridia) An excretory structure in some invertebrates such as Annelida. In most cases it consists essentially of a duct (tubule) leading from the body cavity (coelom) to the exterior. It can be closed with flame cells.

nephritis (Bright's disease, glomerulonephritis) An acute or chronic inflammation of the glomeruli of the kidneys, usually following an infection by streptococcus bacteria elsewhere in the body, e.g. a sore throat. Symptoms include puffy face and ankles, small urinations, dark brown urine because of the presence of red blood corpuscles, fever, headache and fatigue. The disease is treated with penicillin or sulpha drugs, and the patient must stay in bed.

nephron A filtration unit in the kidney. It consists of a Bowman's capsule, a glomerulus and a uriniferous tubule. A Bowman's capsule and a glomerulus make up a Malpighian corpuscle. Blood from the renal artery enters a glomerulus which is surrounded by a Bowman's capsule. High pressure in the capillaries of the glomerulus causes water, salts and various nitrogenous waste to filter out through the capillary walls and collect in the Bowman's capsule. From the Bowman's capsule, the filtrated fluid passes into the uriniferous tubule; from this certain substances are reabsorbed into a surrounding capillary network, which eventually joins to form the renal vein. *See* Fig. 149.

Neptune One of the nine planets in our Solar System, orbiting the Sun between Uranus and Pluto once in 164,79 years (164.79 years). Neptune has two satellites (moons).

nervation *See* venation.

nerve A specialised tissue consisting of a bundle of long fibres through which impulses pass to and from the central nervous system. A nerve is surrounded by a protective sheath of connective tissue, the epineurium.

nerve-cell *See* neurone.

nerve-cord (medulla spinalis, spinal cord) In vertebrates and other chordates, a solid strand of nerve-fibres forming a part of the central nervous system; the dorsally-situated spinal cord. In invertebrates, two or more nerve-cords run dorsally the length of the animal, with ganglia situated at regular intervals. *See also* cerebrospinal fluid *and* grey matter.

nerve-ending The terminal portion of a neurone in which a nerve-impulse starts or finishes.

nerve-fibre

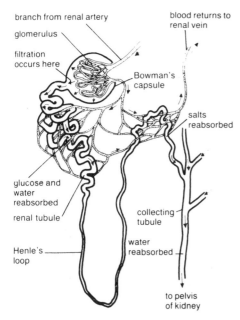

Fig. 149 Nephron

nerve-fibre The axon of a neurone. It may be surrounded by a sheath (medullary sheath) consisting of myelin. *See* Fig. 150.

nerve gas One of a group of highly toxic gases. Originally these gases were used as insecticides, but they were later produced for military use. Chemically, nerve gases are organic derivatives (esters) of fluoro-phosphoric acid. They have a serious effect on the nervous system, inhibiting the enzyme cholinesterase from deactivating acetylcholine, which is found in the majority of synapses and is concerned in the transmission of nerve-impulses.

nerve-impulse *See* impulse.

nerve-net A primitive type of nervous system, consisting of a network of nerve-cells, which is found in certain invertebrates such as coelenterates and echinoderms.

nerve-plexus *See* solar plexus.

nervous Pertaining to nerves.

nervous system The collective name for the brain, spinal cord, nerves and their branches. A nervous system is found in all Metazoa except sponges. However, in primitive animals this may be just a network of nerve-cells.

Nessler's reagent A solution containing sodium or potassium hydroxide, potassium iodide and mercury(II) iodide. When mixed with a solution containing ammonia, NH_3, a yellow or orange-brown colour is observed.

net venation *See* venation.

neural Pertaining to nerves or the nervous system.

neural arc The connection between a receptor and an effector by afferent and efferent nerve-fibres. *See also* reflex arc.

neural arch The dorsal arch of bone arising from the centrum of a vertebra. It encloses the neural canal (spinal canal) through which runs the spinal cord. *See* Fig. 242.

neural canal (spinal canal) *See* neural arch.

neuralgia Severe, intermittent pain along the course of a nerve. Neuralgia may be a symptom of a disease and is sometimes caused by injury to the nerve. However, often the cause of neuralgia cannot be explained.

neural spine A vertical, bony projection from the neural arch of a vertebra. Neural spines serve as points of attachment for muscles. *See* Fig. 242.

neuration *See* venation.

neurilemma (neurolemma) A thin sheath composed of Schwann cells, surrounding the medullary sheath of a medullated nerve-fibre and also surrounding a non-medullated nerve-fibre.

neuritis Inflammation of a nerve. It is often associated with neuralgia.

neurolemma *See* neurilemma.

neurology The study of the nervous system.

neurone (nerve-cell) A specialised cell which conducts nerve-impulses from one part of the body to another. A neurone is made up of a cell body, dendrites, an axon and a dendron. At the end of the axon a neurone may connect with another neurone by means of a synapse or it may connect with an effector such as a muscle or a gland. There are two types of neurones: sensory neurones and motor neurones. *See* Fig. 150.

Neuroptera An order of insects whose members include alder-flies, ant-lion flies and lacewings. Characteristics include four membranous wings with a profuse venation, biting mouthparts, long antennae and complete metamorphosis. Lacewing larvae feed on aphids, thus helping in the control of these pests.

neurosis (pl. neuroses) A mental or emotional disorder. Symptoms include obsessional behaviour such as excessive anger, unreasoned fear or hatred, anxiety or jealousy. Neuroses may be treated by psychotherapy. Sedatives and tranquilisers may suppress the symptoms, but do not alter the fundamental state.

neurotransmitter A chemical substance, such as acetylcholine, which is found at the synapses of neurones where it is involved in the transmission of nerve-impulses from one neurone to another.

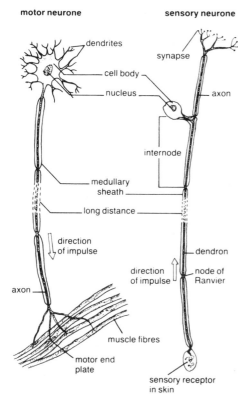

Fig. 150 Neurones

neuter An individual which is neither male nor female, e.g. a flower without pistils or stamens, or a castrated animal. The term may be applied to a non-fertile female insect.

neutral 1. Describes a solution which is neither acidic nor alkaline, i.e. one which has a pH of 7. **2.** Describes a conductor at earth potential, i.e. one which has neither negative nor positive net electric charge.

neutral equilibrium (indifferent equilibrium) The equilibrium of a body such that when it is displaced slightly, it comes immediately to rest and remains in this new position, neither moving further nor returning. *See also* stable equilibrium *and* unstable equilibrium.

neutralisation The reaction between an acid and an alkali or base in which the properties of the reactants disappear and the products are a salt and (often) water. *Example:* hydrochloric acid, HCl, can be neutralised by sodium hydroxide, NaOH, to form sodium chloride, NaCl, and water:
HCl + NaOH → NaCl + H$_2$O
The end-point of the reaction may be observed using an indicator or a pH meter.

neutral oxide *See* oxide.

neutral point 1. A point in a magnetic field where there is no resultant magnetic force, e.g. a point in the magnetic field of a magnet where the Earth's magnetic field is neutralised by the field of the magnet. A compass placed in a neutral point does not settle in any particular direction. **2.** A pH value of seven.

neutron One of the three basic particles in an atom, situated in the nucleus of the atom. Its mass is $1{,}675 \times 10^{-27}$ kg (1.675×10^{-27} kg), about the same as that of the proton, and it has no electric charge. Neutrons are found in all atomic nuclei except those of normal hydrogen, $^1_1$H. Outside the nucleus a neutron is unstable and decays with a half-life of about 12 minutes into a proton, $^1_1$H, and an electron, $^0_{-1}$e:
$$^1_0 n \rightarrow {}^1_1 H + {}^0_{-1} e$$
where $^1_0$n is the symbol of a neutron. *See also* antiparticle.

neutron bomb A specially constructed small hydrogen bomb which is ignited by a small atomic bomb. In the neutron bomb only a small proportion of its energy (about 20%) is converted into heat and pressure; the remaining energy is released as radiation, particularly as a high-energy neutron flux. The neutron bomb has a lethal effect on animals within a certain distance of the explosion centre (the distance depending on the size of the bomb), whereas the effect on buildings is limited because of the relatively small amount of released heat and pressure. Mainly because of the radiation, the explosion of a neutron bomb would cause severe damage to any ecosystem, from which it could take centuries to recover.

neutron number Symbol: N. The number of neutrons in the nucleus of an atom.

neutron star *See* supernova.

neutrophil *See* granulocyte.

'New Cartesian' convention *See* lens formula *and* mirror formula.

newton Symbol: N. The SI unit of force, defined as the force which will give an acceleration of one metre per second squared to a mass of one kilogram:
1 N = 1 kg m s^{-2}

Newtonian telescope *See* telescope.

newtonmeter *See* force meter.

Newton's disc A disc divided into seven sectors, each sector painted with one of the seven spectrum colours of white light. When the disc is rotated at speed its colour seems to be grey-white, thus demonstrating that white light is composed of these seven colours. The disc may have the seven colours arranged in any order.

Newton's law of cooling A law stating that the rate of cooling of a body under given conditions is proportional to the temperature difference between the body and its surroundings.

Newton's law of gravitation A law stating that any two particles (bodies) of matter attract one another with a force which is proportional to the product of their masses and inversely proportional to the square of the distance between them. This can be expressed by the equation:

$$F = \frac{G m_1 m_2}{d^2}$$

where F is the force of attraction, G is the gravitational constant, m_1 and m_2 are the respective masses of the particles (bodies) and d is the distance between them.

Newton's laws of motion Three laws of mechanics:
(1) Every body continues in its state of rest or uniform velocity unless acted upon by an external force.
(2) The rate of change of momentum of a body is proportional to the applied force and takes place in the direction of the force. This can be expressed by the equation:

$F = ma$

where F is the force measured in newtons, m is the mass of the body in kilograms and a is the acceleration in metres per second squared.
(3) If a force acts on one body, then an equal and opposite force acts on some other body. This is sometimes expressed as: to every action there is an equal and opposite reaction.

Newton's rings Concentric rings which may be observed when a plano-convex lens of large radius is placed on a flat glass plate and illuminated from above. If the light is monochromatic, the rings appear as alternate dark and light rings. However, if white light is used the rings are coloured. This effect is caused by interference between light waves reflected from the glass plate and those reflected from the curved surface of the lens.

Ni Chemical symbol for nickel.

niacin See nicotinic acid.

niche The particular set of environmental and biotic conditions required by an organism in order to live.

Nichrome® An alloy of nickel (80%) and chromium (20%). However, it may contain 10–20% of iron. Nichrome wire is used in heating elements because of its high resistance and high melting-point.

nickel An element with the symbol Ni; atomic number 28; relative atomic mass 58,71 (58.71); state, solid. Nickel is a transition metal. It occurs naturally as nickel(II) sulphide, NiS, in the mineral pentlandite, which is nickel-iron sulphide, $(Ni,Fe)_9S_8$. Nickel also occurs with magnesium in certain silicates. Nickel is extracted from its ores by a number of complex processes which ultimately end in the formation of nickel(II) oxide, NiO. This is then reduced to nickel by heating it with carbon: $NiO + C \rightarrow Ni + CO$
The nickel so produced is then purified by the Mond process. Nickel is a rather inert metal, resistant to corrosion. It is used as a catalyst in the hydrogenation of alkenes, in the conversion of certain oils to fats in the margarine industry, as an alloying component and in the nickel-cadmium alkaline cell. Nickel is found in most meteorites and this serves as one of the criteria for distinguishing a meteorite from other minerals.

nickel-cadmium alkaline cell (accumulator) (Alcad-accumulator) A secondary cell in which the cathode is a cadmium plate covered with cadmium hydroxide, the anode is a nickel plate covered with nickel hydroxide and the electrolyte is potassium hydroxide solution. When fully charged, the e.m.f. of the cell is 1,2 volts (1.2 V). Some of its advantages are that it is lighter than the lead-acid cell, it can be charged in a short time, and it can be inadvertently short-circuited without damage.

nickel carbonyl See tetracarbonylnickel(O).

nickel-iron cell See Ni-Fe cell.

nickel silver (German silver) Any of a number of alloys of copper, nickel and zinc, but containing no silver. Nickel silver is resistant to corrosion and is used in making tableware. EPNS (electroplated nickel silver) is nickel silver coated with a thin layer of silver.

Nicol prism A polarising prism (a polariser) consisting of two crystals of calcite made by cutting a calcite crystal in a special direction and then cementing the two pieces together with Canada balsam. When an ordinary light ray strikes the prism, the ray is double refracted, i.e. it is split into two rays. One (the extraordinary ray) passes straight forward through the balsam and leaves the prism as plane-polarised light. The other (the ordinary ray) is reflected from the balsam layer, emerging away from the extraordinary ray, i.e. it suffers total internal reflection. See Fig. 151.

nicotine $C_{10}H_{14}N_2$ An aromatic alkaloid found in tobacco leaves. It is a colourless, highly toxic liquid used as an insecticide.

nicotinic acid (niacin) C_5NH_4COOH An aromatic organic acid belonging to the B-complex group of vitamins. A deficiency of nicotinic acid causes pellagra. See also NAD and NADP.

nictitating membrane The third eyelid of amphibians, birds, reptiles and some mammals

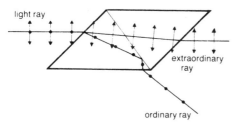

Fig. 151 Nicol prism

such as the cat and the rabbit. Its movement across the eye serves to clean the eyeball.

nidation *See* implantation.

nielsbohrium *See* transuranic elements.

Ni-Fe cell (Edison accumulator) A secondary cell in which the cathode is iron oxide, the anode is nickel hydroxide and the electrolyte is potassium hydroxide solution. When fully charged, the e.m.f. of the cell is about 1,3 volts (1.3 V). Some of its advantages are that it is lighter than the lead-acid cell, it can be charged in a short time and can be short-circuited without damage.

night-blindness (nyctalopia) A condition in which a person's night vision is poor owing to a defect in the rods of the retina. This is caused by a deficiency of vitamin A in the diet.

ninhydrin An organic chemical used in detecting amino acids, with which it gives a blue colour on heating. It is used in several types of chromatography.

nipple (mamilla) The projection at the centre of the breast of a human being. In other mammals the nipple is more often called a teat.

Nissl granule (body) A granule found in the cytoplasm of the cell body of a neurone, which can be stained with basic dyes. Nissl granules are believed to play a part in the synthesis of cytoplasmic proteins.

nitrate A salt of nitric acid, HNO_3.

nitrate bacteria Bacteria e.g. *Nitrobacter*, which convert nitrite, NO_2^-, to nitrate, NO_3^-; they are found in soil. *See also* nitrifying bacteria.

nitration The process in which a nitro-group, —NO_2, is introduced in an organic molecule. *Example:* nitration of benzene, C_6H_6, produces nitrobenzene, $C_6H_5NO_2$. Nitration is carried out with a mixture of concentrated nitric acid, HNO_3, and concentrated sulphuric acid, H_2SO_4. When these acids are mixed, the nitryl cation (nitronium ion), NO_2^+, is formed and this is the active agent of nitration:
$C_6H_6 + NO_2^+ \rightarrow C_6H_5NO_2 + H^+$

nitre *See* potassium nitrate.

nitric acid HNO_3 A colourless, fuming liquid. It is a strong acid which forms soluble salts called nitrates. Nitric acid may be prepared in the laboratory from sodium nitrate, $NaNO_3$, heated with concentrated sulphuric acid, H_2SO_4:
$NaNO_3 + H_2SO_4 \rightarrow HNO_3 + NaHSO_4$
Commercially it is prepared by catalytic oxidation of ammonia, NH_3 (Ostwald method):
$4NH_3 + 5O_2 \rightarrow 4NO + 6H_2O$
The nitrogen monoxide produced is rapidly cooled and combines with oxygen from the air to form nitrogen dioxide, NO_2:
$2NO + O_2 \rightleftharpoons 2NO_2$
In the presence of more air, the nitrogen dioxide is absorbed in water to produce nitric acid:
$4NO_2 + O_2 + 2H_2O \rightarrow 4HNO_3$
Nitric acid is a strong oxidising agent widely used in the chemical industry such as in the manufacture of fertilisers and explosives. Concentrated nitric acid is also called aqua fortis.

nitric oxide *See* nitrogen oxides.

nitride A substance composed of nitrogen and a metal or a non-metal. At red heat, nitrogen combines with several metals and non-metals to produce nitrides. *Examples:* calcium nitride, Ca_3N_2, is produced according to the reaction:
$3Ca + N_2 \rightarrow Ca_3N_2$
and boron nitride, BN, is produced according to the reaction:
$2B + N_2 \rightarrow 2BN$
The properties of nitrides depend upon the element with which nitrogen has combined. Some of them hydrolyse to ammonia, NH_3, e.g.
$Ca_3N_2 + 6H_2O \rightarrow 2NH_3 + 3Ca(OH)_2$
Others, especially those containing transition metals, are chemically very inert, hard and high melting.

nitrification The process in which nitrifying bacteria oxidise ammonium, NH_4^+, or ammonia, NH_3, to nitrite, NO_2^-; and nitrite to nitrate, NO_3^-.

nitrifier *See* nitrifying bacteria.

nitrifying bacteria (nitrifiers) Any of several aerobic bacteria which belong to the family Nitrobacteriaceae and cause nitrification; they are found in soil. Examples include *Nictobacter*, *Nitrococcus* and *Nitrosomonas*. *Compare* denitrifying bacteria; *see also* nitrite bacteria.

nitrile An organic, rather stable compound containing the cyano-group, —C≡N. Most of the common nitriles are colourless liquids which are moderately toxic and have a pleasant smell. Nitriles may hydrolyse to give acids and can be reduced to give amines.

nitrite A salt of nitrous acid, HNO_2.

nitrite bacteria Bacteria, e.g. *Nitrococcus* and *Nitrosomonas*, which convert ammonium,

NH_4^+, or ammonia, NH_3, to nitrite, NO_2^-; they are found in soil. *See also* nitrifying bacteria.

Nitrobacteriaceae A family of nitrifying bacteria.

nitrobenzene $C_6H_5NO_2$ An organic, aromatic, pale yellow oily liquid prepared by nitration of benzene. When reduced, phenylamine (aniline), $C_6H_5NH_2$, is formed.

nitro-cellulose *See* cellulose trinitrate.

nitro-compound An organic compound containing the nitro-group, $—NO_2$. The best known nitro-compounds are the aromatic ones in which the nitro-group is attached to one of the carbon atoms in the ring, e.g. nitrobenzene, $C_6H_5NO_2$. An aromatic ring may contain more than one nitro-group. Nitro-compounds may be prepared by nitration. When reduced, they are converted to amines, e.g.
$C_6H_5NO_2 + 3H_2 \rightarrow C_6H_5NH_2 + 2H_2O$
where $C_6H_5NH_2$ is phenylamine (aniline).

nitrogen An element with the symbol N; atomic number 7; relative atomic mass 14,01 (14.01); state, gas. Nitrogen is a colourless, odourless non-metal forming a diatomic molecule, N_2, with a triple bond between the two nitrogen atoms. This triple bond has a high dissociation energy, which accounts for the fact that nitrogen is a rather inert element at room temperature. Nitrogen occurs naturally in the atmosphere (about 78% by volume) from which it is extracted by fractional distillation of liquid air. In the laboratory, nitrogen may be prepared by heating an aqueous solution of ammonium nitrite, NH_4NO_2:
$NH_4NO_2 \rightarrow N_2 + 2H_2O$
Nitrogen is widely used in the chemical industry, e.g. in the manufacture of ammonia, NH_3 (Haber–Bosch process).

nitrogenase An enzyme present in nitrogen-fixing micro-organisms which assists in the transformation of atmospheric nitrogen, N_2, into ammonia, NH_3.

nitrogen cycle The circulation of nitrogen in the biosphere between organisms and their environment. This involves fixation of gaseous nitrogen, N_2, into a usable form for higher organisms, absorption of nitrogenous substances from the soil and water, and the return of such substances, including denitrification of inorganic nitrogenous material. Some nitrogen may also be lost from soil by leaching, e.g. through drainage water carried away by rivers. *See* Fig. 152.

nitrogen dioxide *See* nitrogen oxides.

nitrogen fixation 1. The process in which certain soil bacteria (e.g. *Rhizobium*), blue-green algae and fungi convert atmospheric nitrogen, N_2, into nitrogenous substances, starting with ammonia, NH_3. The micro-organisms are either free living or live in symbiosis with leguminous plants such as beans, peas and clover. *Compare* denitrifying bacteria; *see also* mutualism *and* Azotobacter. **2.** The process in which atmospheric nitrogen, N_2, is converted into nitric acid, HNO_3, by lightning. First of all nitrogen combines with oxygen to form nitrogen monoxide, NO, which is then immediately oxidised to nitrogen dioxide, NO_2. With water droplets, the NO_2 is converted into nitric acid, HNO_3, and probably a little nitrous acid, HNO_2, which are then washed into the soil. These are later converted into proteins and nucleic acids by the plants.

nitrogen monoxide *See* nitrogen oxides.

nitrogenous Pertaining to or containing nitrogen.

nitrogen oxides The four stable oxides of nitrogen are: dinitrogen oxide (nitrous oxide, laughing gas), N_2O; nitrogen monoxide (nitrogen oxide, nitric oxide), NO; nitrogen dioxide, NO_2, and dinitrogen pentoxide, N_2O_5. Dinitrogen tetraoxide, N_2O_4, is a dimer of NO_2, and dinitrogen trioxide, N_2O_3, exists but is unstable at room temperature.
Dinitrogen oxide, N_2O, is a colourless gas with a faint, sweet smell. It is used as an anaesthetic in minor surgical operations such as those performed in dentistry. Ammonium nitrate, NH_4NO_3, liberates dinitrogen oxide on heating:
$NH_4NO_3 \rightarrow N_2O + 2H_2O$
Dinitrogen oxide will rekindle (relight) a glowing splint, due to the decomposition of the gas on warming into nitrogen and oxygen:
$2N_2O \rightarrow 2N_2 + O_2$
Nitrogen monoxide, NO, is a poisonous, colourless, insoluble, very heat-stable gas. It can be prepared from dilute nitric acid, HNO_3, and copper, Cu:
$3Cu + 8HNO_3 \rightarrow 2NO + 3Cu(NO_3)_2 + 4H_2O$
or by catalytic oxidation of ammonia, NH_3:
$4NH_3 + 5O_2 \rightarrow 4NO + 6H_2O$
It can also be prepared by direct combination of nitrogen and oxygen in an electric arc:
$N_2 + O_2 \rightarrow 2NO$
In contact with air (oxygen), NO is oxidised to nitrogen dioxide, NO_2:
$2NO + O_2 \rightarrow 2NO_2$
Nitrogen dioxide, NO_2, is a poisonous brown gas. It can be prepared from concentrated nitric acid, HNO_3, and copper, Cu:
$Cu + 4HNO_3 \rightarrow 2NO_2 + Cu(NO_3)_2 + 2H_2O$
or by heating dry lead(II) nitrate, $Pb(NO_3)_2$:
$2Pb(NO_3)_2 \rightarrow 4NO_2 + 2PbO + O_2$

nitrous acid

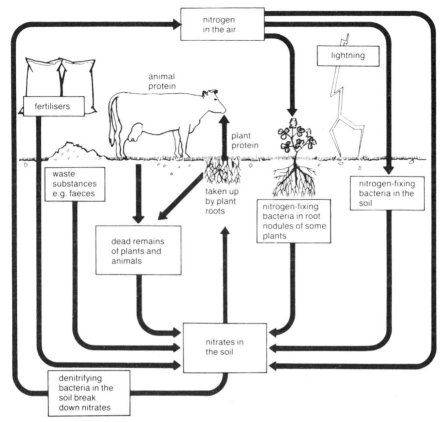

Fig. 152 Nitrogen cycle

It dissolves in water to give a mixture of nitrous acid, HNO_2, and nitric acid, HNO_3:
$2NO_2 + H_2O \rightarrow HNO_2 + HNO_3$

Dinitrogen tetraoxide, N_2O_4, is a light yellow poisonous dimer of nitrogen dioxide, NO_2:
$2NO_2 \rightleftharpoons N_2O_4$
i.e. an equilibrium exists between the two oxides. The equilibrium position depends, among other things, upon the temperature: at 22°C the colour of the mixture is orange-yellow due to a high percentage of N_2O_4, whereas it is almost black at about 145°C, i.e. N_2O_4 is more stable at low temperatures and NO_2 is more stable at high temperatures.

Dinitrogen pentoxide, N_2O_5, is the anhydride of nitric acid:
$N_2O_5 + H_2O \rightarrow 2HNO_3$
N_2O_5 is a white solid.

Dinitrogen trioxide, N_2O_3, is the anhydride of nitrous acid:
$N_2O_3 + H_2O \rightarrow 2HNO_2$
N_2O_3 is a pale blue solid. It is unstable at room temperature. M.p. −102°C.

nitroglycerine $C_3H_5(NO_3)_3$ (propane-1,2,3-triyl trinitrate; glyceryl trinitrate) An ester (not a nitro-compound) between glycerol (propane-1,2,3-triol), $C_3H_5(OH)_3$, and nitric acid, HNO_3, obtained when glycerol is treated with a mixture of concentrated nitric acid, HNO_3, and concentrated sulphuric acid, H_2SO_4:
$C_3H_5(OH)_3 + 3HNO_3 \xrightarrow{H_2SO_4} C_3H_5(NO_3)_3 + 3H_2O$
Nitroglycerine is a pale yellow, oily liquid which explodes violently on slight shock. It is the active ingredient of dynamite, but it also has a medical application in the treatment of angina pectoris.

nitro-group *See* nitration.

nitronium ion *See* nitration.

nitrous acid HNO_2 A weak acid which only exists in aqueous solution. Its salts are called nitrites and it can be prepared by treating a nitrite with a dilute strong acid:
$NO_2^- + H^+ \rightarrow HNO_2$
It is unstable when heated, liberating brown

nitrous oxide

fumes of nitrogen dioxide, NO_2 and leaving nitric acid, HNO_3:
$3HNO_2 \rightarrow 2NO + HNO_3 + H_2O$
followed by
$2NO + O_2 \rightarrow 2NO_2$
when NO reacts with oxygen in the air. HNO_2 is used in the manufacture of certain dyes.

nitrous oxide *See* nitrogen oxides.

nitryl cation *See* nitration.

noble gas One of the gases in Group VIII A of the Periodic Table of the Elements. These gases are: helium, neon, argon, krypton, xenon and radon. All the gases have complete octets of electrons in their valency shells, except helium, which has its valency shell completed with two electrons. This makes the noble gases chemically very inert. However, those with high atomic numbers, i.e. krypton, xenon and radon, have (since 1962) been brought to a chemical reaction with very electronegative elements such as fluorine, F_2, and oxygen, O_2, e.g.
$$Xe + 3F_2 \xrightarrow[\text{pressure}]{300°C} XeF_6$$
in which xenon hexafluoride, XeF_6, is formed by direct combination of its elements. XeF_6 will hydrolyse to give xenon trioxide, XeO_3:
$XeF_6 + 3H_2O \rightarrow XeO_3 + 6HF$
All the noble gases are found in air. Helium is also found as a component of some natural hydrocarbon gases. The noble gases were formerly called the rare gases and the inert gases, but both these names are now obsolete.

noble metal The transition elements silver, gold and platinum are sometimes referred to as the noble metals because of their value and resistance to corrosion.

nocturnal 1. Describes an event which happens during night. 2. A nocturnal animal is one which is active during the night. *Compare* diurnal.

node 1. The part of a plant stem (usually swollen) from which a leaf grows. *See* Fig. 86. 2. A point of zero displacement in a standing wave. The distance between two adjacent nodes is equal to half the wavelength of the wave motion. *See also* antinode.

node of Ranvier A constriction of the myelin sheath along the length of a myelinated nerve-fibre. *See* Fig. 150.

nodule Any small knob-like structure, e.g. a root nodule of Leguminosae.

noise An undesirable combination of sounds of irregular frequency.

nomenclature 1. The naming of groups of plants and animals. *See also* binomial system of nomenclature. 2. The systematic naming of chemicals. *See also* IUPAC.

non-conductor *See* insulator.

non-electrolyte A substance which, when dissolved in a solvent (often water) or when molten, does not conduct an electric current. The term also applies to the liquid itself. A non-electrolyte contains very few ions or none at all. Examples of non-electrolytes are pure water, propanone (acetone), paraffin and naphthalene.

non-flammable (non-inflammable) Describes a substance which is not easily set on fire. *Compare* flammable.

non-inflammable *See* non-flammable.

non-metal An element which is not a metal. Non-metals are electronegative, are poor conductors of heat and electricity, have no lustre and are neither ductile nor malleable. Oxides of non-metals are acidic. *Example:* sulphur trioxide, SO_3, forms sulphuric acid, H_2SO_4, when treated with water:
$SO_3 + H_2O \rightarrow H_2SO_4$
Non-metals may be minor components of alloys. The above description is general; many non-metals do not have all the characteristics described. *Compare* metal; *see also* metalloid.

non-polar molecule A molecule which has no dipole, i.e. is without a dipole moment. Carbon dioxide, CO_2, and tetrachloromethane, CCl_4, are examples of non-polar molecules. *Compare* polar molecule.

non-polar solvent A solvent which has non-polar molecules. The molecules do not possess a dipole moment and therefore the solvent usually is a poor solvent for ionic or polar solutes such as most salts. However, a non-polar solvent is usually a good solvent for non-polar solutes. *Example:* tetrachloromethane, CCl_4, is a poor solvent for sodium chloride, NaCl, but a good solvent for iodine, I_2. *Compare* polar solvent.

nonsense codon A codon coding for nothing, e.g. it may indicate the premature termination of a polypeptide chain which is then released from the ribosome.

nonsense mutation The alteration of a codon in which it is changed from one coding for an amino acid to a nonsense codon.

non-striated muscle *See* involuntary muscle.

noradrenalin(e) A hormone secreted by the adrenal gland. It is chemically closely related to adrenalin(e); noradrenalin(e) has one methyl group, $—CH_3$, fewer than adrenalin(e) and its biochemical function is different from that of adrenalin(e). Noradrenalin(e) is a neurotransmitter in the sympathetic nervous system.

normal 1. A line drawn at right angles to a surface, e.g. a normal to the point of incidence of a ray of light striking a mirror or lens. 2. An obsolete term describing an organic compound whose molecule has an unbranched chain of carbon atoms. *Example:* butane, CH_3—CH_2—CH_2—CH_3, may be called normal butane or n-butane to distinguish it from 2-methylpropane,

CH_3—CH—CH_3,
        |
       CH_3

In modern nomenclature, the name butane (in this example) is used for the isomer with the unbranched carbon chain. **3.** *See* normality.

normality Symbol: N. The concentration of a solution, expressed as the number of equivalents of solute dissolved in one cubic decimetre (dm^3) of solution. *Example:* a 1 normal (1 N) solution of sulphuric acid, H_2SO_4, contains

$$\frac{98,08}{2} = 49,04 \, g \left(\frac{98.08}{2} = 49.04 \, g \right)$$

of H_2SO_4 per dm^3 of solution.

normal salt *See* salt.

normal solution *See* normality.

normal temperature and pressure (NTP) *See* standard temperature and pressure.

north pole *See* pole.

nostrils The paired external openings of the nasal cavity in animals which have a nose or snout. *See also* nares.

note A musical sound of definite pitch (frequency) produced by a musical instrument, etc.

notochord A skeletal structure characteristic of all chordates. It consists of a thin, flexible rod situated dorsally between the central nervous system (or the nerve-cords) and the gut. The notochord is composed of tightly-packed, vacuolated cells surrounded by a firm, elastic sheath. In vertebrates, it is replaced during the development of the embryo by the vertebral column. In invertebrate chordates, except *Amphioxus* which retains its notochord throughout its life, the notochord is modified during the development of the animal.

nova (pl. novae or novas) A faint star which suddenly flares up in brightness by a factor that varies from ten thousand times to one hundred thousand times its original brightness. A nova is a member of a very close binary system in which mass (gas) is pulled by gravitation from a companion star to the secondary very dense star in the system. The material transferred to the secondary star is dramatically heated, resulting in explosive burning and giving rise to the nova outburst. The amount of material transferred is only an extremely small part of the total mass of the companion star. This process may only last for a very short time (from a few hours to a few days). *Compare* supernova.

***n-p-n* transistor** *See* transistor.

NTP Normal temperature and pressure. *See* standard temperature and pressure.

***n*-type semiconductor** *See* semiconductor.

nucellus In angiosperms, a mass of nutritional tissue surrounding the embryo sac in the centre of the ovule.

nuclear Pertaining to an atomic or biological nucleus.

nuclear energy The energy released as a result of nuclear fission or nuclear fusion.

nuclear fission *See* fission.

nuclear fusion *See* fusion.

nuclear membrane A delicate, porous membrane surrounding the nucleus of a cell, allowing exchange of material between the nucleus and the cytoplasm.

nuclear physics The branch of physics concerned with the atomic nucleus, including its structure, properties, etc.

nuclear power-station A power-station producing electricity as a result of nuclear energy released in a nuclear reactor. This energy is converted into kinetic energy of steam which drives a turbine connected to a generator. *See* Fig. 153.

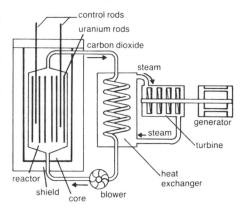

Fig. 153 Nuclear power-station layout

nuclear reaction *See* fission *and* fusion.

nuclear reactor A device in which nuclear energy (fission energy) is converted into heat energy. A reactor consists of a central region or core containing a radioactive material (usually uranium) in the form of rods; a moderator; and control rods. Nuclear fission takes place when the radioactive material is bombarded by slow-moving neutrons. Fast-moving neutrons bounce off the nuclei of the radioactive material instead of entering them. The moderator is a substance, e.g. graphite or heavy water, which slows down the fast-moving neutrons so that they cause further fission. Control rods made of a neutron-absorbing material, e.g. boron or cadmium, are inserted into or withdrawn from the core to control the number of neutrons present in the core. Fission ceases when the control rods are fully inserted. The heat produced during fission is removed by a liquid or gaseous coolant, e.g. water or carbon dioxide, which is pumped through the core. In a nuclear power-station, the heat energy produces steam in a secondary circuit and this is fed to turbines driving generators. The core of the reactor is surrounded by a very thick wall (shield) of concrete or is immersed in a deep tank of water to prevent escape of neutrons or atomic radiation. There are several types of reactors. They are described according to their function, radioactive material, moderator or coolant. *See* Fig. 153.

nuclear sap A colourless, slightly alkaline, viscous liquid found in the nuclei of animal and plant cells. *See also* nucleoplasm.

nuclear spindle *See* spindle.

nuclear weapon An explosive device, e.g. an atomic bomb, a hydrogen bomb or a neutron bomb, whose explosive power is obtained from fission or fusion.

nuclease One of a group of enzymes which hydrolyse nucleic acids to nucleotides.

nucleic acid *See* deoxyribonucleic acid (DNA) and ribonucleic acid (RNA).

nucleolus (pl. nucleoli) A small, spherical body found in the resting nuclei of animal and plant cells. There may be one or more per nucleus. A nucleolus consists of protein and RNA. It is concerned with the synthesis of ribosomes. Nucleoli disappear from nuclei during mitosis.

nucleon A proton or a neutron.

nucleon number *See* mass number.

nucleophile An ion or molecule which is 'nucleus-seeking', i.e. which is rich in electrons. *Example:* it may have one or more unshared pairs of electrons which it can donate when forming a new bond with an atom. Nucleophiles therefore react more readily at positive centres as found, for example, in dipolar molecules. They are anions, e.g. Br^-, OH^- and CN^-, or neutral molecules possessing a lone pair of electrons, e.g. NH_3 and H_2O. Nucleophilic reactions are common in organic chemistry. They may be either addition reactions or substitution reactions. In the reaction:
$CH_3Cl + NH_3 \rightarrow CH_3NH_2 + HCl$
chloromethane (methyl chloride), CH_3Cl, reacts with the nucleophilic reagent ammonia, NH_3, to produce methylamine (aminomethane), CH_3NH_2, and hydrogen chloride, HCl. NH_3 attacks the carbon atom in CH_3Cl which forms a positive centre because of its attachment to the electronegative chlorine atom. *Compare* electrophile.

nucleophilic reaction *See* nucleophile.

nucleophilic reagent *See* nucleophile.

nucleoplasm The dense protoplasm making up the nucleus of an animal or plant cell. The term is often used as a synonym for nuclear sap.

nucleoprotein One of a group of compounds made up of nucleic acids and proteins. Examples include chromosomes, histones and ribosomes.

nucleoside A compound containing a purine or pyrimidine base chemically associated with ribose or deoxyribose.

nucleotide A compound which is an ester of a nucleoside and phosphoric acid. Nucleotides are found free in cells, in nucleic acids and in various coenzymes.

nucleus 1. The chief organelle situated in the cytoplasm of plant and animal cells (that is, eucaryotic cells). It contains nuclear sap in which are found one or more nucleoli and chromosomes. The nucleus is surrounded by a nuclear membrane. *See also* meiosis *and* mitosis. **2.** *See* atomic structure.

nuclide A specific isotope of an element, containing a stated number of protons and neutrons. *Examples:* the three isotopes of hydrogen, H: ^1_1H (normal hydrogen), ^2_1H (deuterium) and ^3_1H (tritium), are all different nuclides.

nuptial flight The flight of a virgin queen bee or queen ant, during which mating takes place.

nut A dry, indehiscent fruit usually containing one seed. The pericarp is hard, woody and brittle. Examples include acorns, cashew nuts and hazel nuts. Many fruits are commonly called nuts; e.g. walnuts and coconuts, which are drupes, and ground-nuts, which are legumes.

nutation (circumnutation) The spiral rotation during growth of the apex of certain plant stems, enabling them to obtain a solid support. It also occurs in tendrils, roots, flower stalks and the sporangiophores of some fungi.

nutrient A substance serving as or providing food. For plants, inorganic salts taken up by the roots from the soil are nutrients.

nutrition The ingestion, digestion and assimilation of food (nutrients) by animals and heterotrophic plants. In autotrophic plants, the process of photosynthesis.

nyctalopia *See* night-blindness.

nyctinasty A plant movement in response to changes in light intensity and temperature.

nylon A polyamide which may be prepared from a diamine and a dicarboxylic acid. It contains a recurring unit:

$$\left[-\underset{|}{\overset{H}{N}}-(CH_2)_x-\underset{|}{\overset{H}{N}}-\underset{\parallel}{\overset{}{C}}-(CH_2)_y-\underset{\parallel}{\overset{}{C}}- \right]_n$$

Nylon was the first synthetic fibre to be manufactured by polymerisation. It is widely used in the textile industry and in the manufacture of many moulded articles. *See also* plastic.

nymph *See* metamorphosis.

O

O Chemical symbol for oxygen.

obesity The condition in which a person is overweight due to excessive deposits of fat. Obesity may be dangerous to health as it puts an extra strain on the heart and circulatory system.

object Anything observed with an optical instrument.

objective In optical instruments, such as microscopes and telescopes, the lens or lens system nearest to the object being viewed. It produces the image viewed by the eyepiece.

obligate Describes organisms which are limited to one way of life. *Examples:* obligate anaerobes can only exist in the total absence of oxygen; obligate parasites cannot exist independently of their hosts. *Compare* facultative.

occipital Pertaining to the back of the head.

occlusion 1. The closure of a structure, e.g. a duct or opening. **2.** In crystallisation, the trapping of small amounts of a substance inside a crystal. The occluded substance mostly occupies places where the crystal structure is imperfect. The term also applies to sorption of a gas by a metal, e.g. hydrogen by platinum. *See also* clathrate.

occultation The temporary concealment of one heavenly body when another passes in front of it. *Example:* the passage of the Moon in front of a star obscures light and radio waves from the star.

ocellus (pl. ocelli) *See* simple eye.

ochre One of several earthy pigments, varying from light yellow to brown, occurring naturally as hydrated iron oxides sometimes mixed with clay.

octadecanoate *See* stearate.
octadecanoic acid *See* stearic acid.
octadecenoate *See* oleate.
octadecenoic acid *See* oleic acid.

octane C_8H_{18} A saturated hydrocarbon belonging to the alkanes. It is a flammable liquid of which there exist 18 isomers. Octane (the unbranched isomer), m.p. $-57°C$, b.p. $126°C$, is found in petroleum. This isomer used to be called normal octane (n-octane).

octane number (rating) A measure of a petrol's performance in an internal combustion engine. This is expressed as a number (the octane number) which is the percentage of 2,2,4-trimethylpentane (an isomer of octane), mixed with heptane which has the same knocking characteristics as the petrol when compared in a standard engine. Heptane is given the octane number 0 (a poor fuel) and 2,2,4-trimethylpentane is given the octane number 100 (a good fuel). The octane number of a petrol may be raised by the addition of ethanol, methanol, tetraethyl-lead(IV), etc.

octavalent Having a valency of eight.

octave The interval between two frequencies of any type of oscillation (e.g. sound waves) which have the frequency ratio 2:1. In the diatonic scale, middle C has a frequency of 256 Hz and upper C has a frequency of 512 Hz.

octet A stable group of eight electrons found in the outermost shell of an atom. The noble gases (except helium) have such an octet of electrons and are therefore chemically very inert. In compounds, atoms often form bonds so that the octet is achieved either by sharing electrons or by donating and accepting electrons. *See also* covalent bond *and* ionic bond.

ocular *See* eyepiece.

Odonata An order of carnivorous insects comprising the dragonflies. Characteristics include large eyes, two pairs of membranous wings, biting mouthparts and aquatic nymphs.

odontoblast A cell which produces dentine. Odontoblasts line the outside of the pulp cavity of a tooth.

odontoclast A cell which destroys dentine and absorbs the roots of deciduous teeth.

odontoid process (dens) A tooth-like projection from the second cervical vertebra (axis). It forms a pivot around which the first cervical verebra (atlas) revolves, thus allowing rotation of the head.

odontology The study of teeth.

oedema (dropsy) Swelling of tissue due to abnormal accumulation of fluid. Oedema may be a symptom of a disease of the kidneys, liver or heart.

Oersted's apparatus The apparatus with which Oersted showed the relation between electricity and magnetism. A d.c. current is passed through a wire placed just above a compass needle. This causes the needle to take up a position nearly perpendicular to the wire and remain in that position as long as the current flows. When the direction of the current is reversed, or if the wire is placed just below the compass needle, the needle again takes up its position, but with its ends reversed. See Fig. 154.

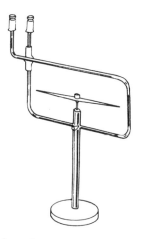

Fig. 154 Oersted's apparatus

Oersted's experiment See Oersted's apparatus.

oesophagus (gullet) The part of the alimentary canal leading from the pharynx to the stomach. It is situated behind the trachea and passes through the diaphragm. The oesophagus has a lining of mucous membrane; the secreted mucus provides lubrication for the passage of food. In vertebrates, muscles around the oesophagus cause peristaltic movements which help in the transport of food. See Fig. 49.

oestrogen The general name for a group of female sex hormones (steroids) produced in the ovaries and the placenta. Oestrogens induce oestrus in vertebrates. In mammals, they activate the womb in preparation for pregnancy and cause a thickening of its lining. They stimulate development of the mammary glands and the other secondary sexual characteristics. Oestrogens are also produced synthetically. They are used in the treatment of several female disorders such as painful menstruation, and for the relief of physical and emotional symptoms of the menopause. See also progesterone and luteinising hormone.

oestrous cycle See oestrus.

oestrus ('heat') The particular period or periods of the year in which sexually mature female animals will receive a male for the purpose of fertilisation of the ovum. Oestrus and the sex phases which immediately precede and follow it are collectively called the oestrous cycle. In the absence of pregnancy the duration of the oestrous cycle varies from five to sixty days in most mammals. It may occur only during a breeding period or throughout the year. Animals which have a single oestrous cycle during their breeding season are called monoestrous animals, e.g. dogs, whereas those which have two or more oestrous cycles during their breeding season are called polyoestrous animals, e.g. cow, mouse and rat. The menstrual cycle in a woman is analogous to the oestrous cycle and may be called a modified oestrous cycle.

offspring An animal's young or its descendants.

ohm Symbol: Ω. The SI unit of electrical resistance, defined as the resistance between two points of a conductor such that a potential difference of one volt between the points produces a current of one ampere.

ohmic Describes a resistance or any other electrical device which obeys Ohm's law.

ohmmeter An instrument for the direct measurement of electrical resistance. It is usually a modified voltmeter fitted with a dry cell or a battery.

Ohm's law A law stating that at constant temperature, the current passing though a wire is proportional to the potential difference between its ends. This may be expressed as
$V = IR$
where V is the potential difference in volts, I is the current flowing in amperes and R is the resistance of the wire in ohms. See also Kirchhoff's laws for electric circuits.

oil See fats and oils and mineral oil.

oil immersion A technique used in microscopy in which the front lens of a high-powered special objective (immersion objective) is immersed in a drop of a special oil (immersion oil), usually cedar-wood oil. The oil drop is placed either directly on the slide or on the coverslip. The working distance of the objective is very small and therefore the objective is often spring-loaded to avoid damage to the front lens or slide. The immersion oil has a greater refractive index than air, whereas the oil and the glass have the same refractive index, so there is no refraction or total internal reflection of light passing from the slide to the oil. Thus a cone of light with a very wide angle enters the objective. In this way the resolving power of the microscope improves.

oil of vitriol Sulphuric acid, H_2SO_4.

oleate (octadecenoate) A salt of oleic acid, $C_{17}H_{33}COOH$.

olecranon process In mammals, the strong bony process on the ulna. It extends beyond the elbow joint and serves as a place of attachment for the triceps muscle.

oleic acid $C_{17}H_{33}COOH$ (cis-octadec-9-enoic acid; octadecenoic acid) A liquid, unsaturated fatty acid which has one double bond between the ninth and tenth carbon atoms (counted from and including the carbon atom of the carboxyl group). Oleic acid is found naturally as a part of many fats and oils. Salts of oleic acid are called oleates (octadecenoates). It is used in the manufacture of soap, polishes and lubricants.

oleum (fuming sulphuric acid, pyrosulphuric acid). A solution of sulphur(VI) oxide, SO_3, in concentrated sulphuric acid, H_2SO_4. Oleum is also given the formula $H_2S_2O_7$.

olfaction The sense of smell.

olfactory Pertaining to the sense of smell.

olfactory lobes In vertebrates, two outgrowths of the forebrain which are concerned with the sense of smell.

olfactory nerve A cranial, sensory nerve connected to the nerve-cells of the nose. It is the first cranial nerve.

oligo- Prefix meaning few.

Oligocene A geological period of approximately 12 million years, which ended approximately 26 million years ago. It is a sub-division of the Tertiary period.

oligosaccharide A carbohydrate molecule consisting of a few units of monosaccharide, i.e. from two to five units. The disaccharides are therefore oligosaccharides.

omasum See psalterium.

omega (ω or Ω) The last letter of the Greek alphabet.

ommateum See compound eye.

ommatidium (pl. ommatidia) See compound eye.

omnivore An animal which eats both plants and animals (flesh).

omnivorous Pertaining to animals which eat both plants and animals (flesh).

onchocerciasis See river blindness.

onium ion An ion such as the ammonium ion, NH_4^+, and the oxonium ion, H_3O^+, in which a hydrogen ion (proton), H^+, is taken up by a molecule:
$NH_3 + H^+ \rightleftharpoons NH_4^+$ and
$H_2O + H^+ \rightleftharpoons H_3O^+$

oocyte A cell which forms an ovum after undergoing meiosis. An oocyte develops from an oogonium. See also Graafian follicle.

oogenesis The process in which ova are produced in the ovary.

oogonium (pl. oogonia) 1. A reproductive cell formed when a gamete divides by mitosis. This takes place in the ovary. Oogonia give rise to oocytes. 2. In certain algae and fungi, a female reproductive organ containing one or more oospheres.

oosphere A large, spherical, non-motile, naked female gamete which is formed within an oogonium.

OP 1. See osmotic potential. 2. See osmotic pressure.

opal An amorphous form of hydrated silica, $SiO_2 \cdot nH_2O$. Opals are used in jewellery.

opaque Describes a substance which absorbs a particular type of incident radiation.

Open-Hearth process (Siemens–Martin process) A process in which pig-iron, scrap iron, some iron ore and limestone are added to a furnace where the charge lies in a shallow hearth. The charge is heated by burning gases passing over it. Oxygen is blown through the charge and oxidises unwanted impurities, some of which form a floating slag with the lime. When the iron is considered pure, measured amounts of carbon, manganese, etc., are added to produce the required type of steel. The process is slow; although it makes use of large amounts of scrap, its use is declining. See also Bessemer process, Kaldo process and L–D process.

operculum (pl. opercula) A cover or lid, e.g. the cover over the gill slits of fish and the fold of skin that grows over the external gills of tadpoles.

ophthalmic Pertaining to the eye.

opium A mixture of alkaloids extracted from poppy seeds, containing morphine, codeine, etc.

opsonin See bacteriotropin.

optic Pertaining to vision.

optical activity The ability possessed by certain substances to rotate the plane of polarisation of plane-polarised light either to the right or to the left. It can be measured using a polarimeter. This ability is usually due to dissymmetry of a molecule. Substances containing one or more so-called asymmetric carbon atoms are optically active. An asymmetric carbon atom is one which is joined to four different atoms or groups of atoms. Example: one of the pentanols, $C_5H_{11}OH$ contains an asymmetric carbon atom. The structure is

$$CH_3-\overset{C_2H_5}{\underset{H}{C^*}}-CH_2OH$$

The asymmetric carbon atom is indicated by the asterisk.

optical antipodes See optical isomerism.

optical axis (optic axis, principal axis) See lens and mirror.

optical bench A rigid metal or wooden bench fitted with a scale, on which optics experiments are performed; i.e. lenses, mirrors, light sources, etc., may be arranged on it.

optical centre See lens.

optical isomerism A type of stereoisomerism in which two compounds only differ from one another in their behaviour towards polarised light. The two isomers are not superimposable, i.e. they are related as object and mirror image. This form of isomerism can only be explained by assuming a tetrahedral arrangement for the molecules involved. Molecules containing one or more asymmetric carbon atoms exhibit this form of isomerism. One of the isomers rotates the plane of polarised light to the right and is therefore said to be dextrorotatory (the (+) form), and the other rotates it to the left and is therefore said to be laevorotatory (the (−) form); the magnitude of rotation is the same. Molecules containing a single asymmetric carbon atom exist in only two forms. Generally a compound containing n asymmetric carbon atoms exists in 2^n optic forms. The dextrorotatory isomer and the laevorotatory isomer are described as enantiomers or optical antipodes. *See also* optical activity *and* racemic mixture.

optical pyrometer *See* pyrometer.

optic disc *See* blind spot.

optic lobe In vertebrates, one of two or four outgrowths of the midbrain which are concerned with the sense of sight.

optic nerve A cranial, sensory nerve connected to the nerve-cells of the retina. It is the second cranial nerve. *See* Fig. 66.

optics The study of light.

oral Pertaining to the mouth.

orbit 1. The path of a heavenly body, e.g. a planet circling (orbiting) the Sun. 2. The curved path of a particle, e.g. an electron orbiting the nucleus of an atom. 3. In vertebrates, one of two depressions in the skull in which the eye is situated.

orbital The region in the space around the nucleus of an atom (atomic orbital) or the nuclei of a molecule (molecular orbital) in which there is a high probability of finding one or more electrons. In the case of molecular orbitals, they may be thought of as being formed by an overlapping of atomic orbitals which gives rise to bonding between the atoms of the molecule. Orbitals of atoms have characteristic energies and shapes, e.g. spherical, dumb-bell shaped, etc., and they are given the symbols s, p, d and f. Atomic orbitals are divided into shells and sub-shells which successively are farther and farther away from the nucleus. The first shell (the K-shell) is the closest to the nucleus and consists of only one orbital, called the $1s$ orbital. The second shell (the L-shell) has four orbitals which are divided into two sub-shells; the first sub-shell consists of the $2s$ orbital, the other of three $2p$ orbitals. Each orbital may contain a maximum of two electrons, provided they are spinning in opposite directions. This fact is called the Pauli exclusion principle. All the orbitals of any one sub-shell are at the same energy state, i.e. an electron has the same tendency to occupy any one of the orbitals in a given sub-shell as any other. Because electrons repel each other, they tend to distribute themselves in different orbitals (Hund rule), e.g. if an atom has four p-electrons, then they will not be found as two pairs, even though the total number of electrons is even, but as one pair and two unpaired electrons.

order *See* family.

order of reaction When discussing how the concentration influences the rate of a chemical reaction, one distinguishes between its molecularity and its order. If a reaction takes place by a spontaneous decomposition of one molecule it is called unimolecular, if it takes place between two molecules it is called bimolecular, etc. Since most chemical reactions are not so simple, it is difficult in practice to determine their molecularities. However, their rates of reaction often follow the mathematical expression for a unimolecular or bimolecular process, and they are therefore called first-order and second-order reactions. Generally we can write the reaction
$$nA + mB \rightarrow pC + rD$$
where capital letters represent substances and small letters represent numbers. This is called a reaction of the order $n + m$. *Example:* the reaction between nitrogen monoxide, NO, and hydrogen, H_2:
$$2NO + H_2 \rightarrow N_2O + H_2O$$
is a third-order reaction. The order of a reaction can be defined as the sum of the powers of the concentrations as they appear in the law of mass action for the reaction.

Ordovician A geological period of approximately 70 million years, occurring between the Cambrian and the Silurian. It ended approximately 430 million years ago.

ore A mineral from which a metal or non-metal may be extracted at a profit. *Example:* galena is an ore of lead.

organ A part of an animal or plant which is specialised to perform a particular function or functions, e.g. the skin, the heart, the kidney, a leaf.

organelle A part of the protoplasm of a cell specialised to perform a particular function or functions, e.g. the nucleus (the largest organelle of a cell), a ribosome, a mitochondrion.

organic 1. Pertaining to or derived from living organisms. 2. Pertaining to compounds containing carbon. *Compare* inorganic.

organic chemistry The study of the chemistry of organic compounds.

organism Any animal or plant. The term is also applied to living things such as bacteria and fungi which are neither plants nor animals.

organometallic compound An organic compound in which there is a chemical bond between a carbon atom and a metal, e.g. tetraethyl-lead(IV), $(C_2H_5)_4Pb$:

$$CH_3-CH_2-\underset{\underset{\underset{CH_3}{|}}{\underset{CH_2}{|}}}{\overset{\overset{\overset{CH_3}{|}}{\overset{CH_2}{|}}}{Pb}}-CH_2-CH_3$$

orgasm The climax of the sexual act. In men it is accompanied by ejaculation of semen and in women by reaching a height in excitement during which there are contractions of the vagina. In both cases an orgasm is followed by a decline in tension.

orifice The opening of a duct or tube, etc.

origin The place of attachment of a muscle to a bone. When the muscle contracts, the origin does not move. *Compare* insertion.

origin of life It is believed that life originated on Earth between 3500 and 4000 million years ago. At that time, the atmosphere is thought to have consisted of several gases such as water vapour, H_2O, ammonia, NH_3, methane, CH_4, nitrogen, N_2, and carbon dioxide, CO_2. Laboratory experiments have shown that, when a mixture of some of these gases is circulated in a closed system and exposed to intense energy released from an electric spark, complex organic molecules characteristic of living matter are produced. These include amino acids and carbohydrates. A similar process is thought to have taken place in the Earth's atmosphere with the aid of energy from the Sun and lightning. The way in which these organic compounds organised and developed into living things is uncertain.

origin of species *See* Darwin's theory.

ornithine An amino acid concerned in the ornithine cycle. It is formed from the hydrolysis of arginine, which also results in the formation of urea.

ornithine cycle (urea cycle) In the liver of ureotelic animals, the formation of urea as a result of a complex cyclical series of reactions. Ornithine, carbon dioxide and ammonia react to form the intermediate compound arginine, which, in the presence of the enzyme arginase, breaks down to urea and ornithine. The ammonia is produced as a result of deamination.

ornithology The study of birds.

ortho- 1. Prefix meaning that two substituents are next to each other in an aromatic ring, e.g.

is ortho-dichlorobenzene (o-dichlorobenzene) or 1,2-dichlorobenzene **2.** Prefix meaning the most hydrated form of an acid. *Examples:* orthophosphoric acid, H_3PO_4, is more hydrated than metaphosphoric acid, HPO_3; orthosilicic acid, H_4SiO_4 is more hydrated than metasilicic acid, H_2SiO_3. *See also* meta- *and* para-.

orthoboric acid *See* boric acid.

orthophosphate *See* phosphate(V).

orthophosphoric acid *See* phosphoric(V) acid.

orthophosphorous acid *See* phosphonic acid.

Orthoptera An order of insects whose members include grasshoppers, locusts and crickets. Characteristics include thickened protective fore-wings, membranous hind-wings, biting mouthparts, powerful hind-legs modified for jumping, stridulatory (sound-producing) organs and incomplete metamorphosis.

orthosilicate *See* silicic acid.

orthosilicic acid *See* silicic acid.

oscillation *See* vibration.

oscilloscope *See* cathode-ray oscilloscope.

-ose Suffix denoting a carbohydrate, e.g. glucose, saccharose, cellulose.

osmometer An instrument for measuring osmotic pressure.

osmoregulation The control of osmotic pressure within an organism, i.e. the control of water content and dissolved substances such as mineral salts in the cells. This is especially important in animal cells, as they do not possess the strong cellulose cell walls found in plant cells and are therefore more liable to rupture of the cell membrane when exposed to hypotonic solutions. Osmoregulation can be carried out either by the removal of water, thus increasing the concentration of solutes (increasing the osmotic pressure), or by the removal of solutes (decreasing the osmotic pressure). The kidneys are associated with osmoregulation.

osmosis The process in which a solvent passes through a selectively permeable membrane from a dilute solution to a more concentrated one, continuing until an equilibrium is reached, i.e. until both solutions have the same concentration. The pressure which is just

osmotic potential

enough to stop the movement of the solvent through the membrane is known as the osmotic pressure. All living cells are surrounded by selectively permeable membranes and therefore osmosis plays an important part in the movement of water into and out of animal and plant cells. See Fig. 155; see also Pfeffer's pot.

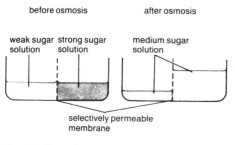

Fig. 155 Osmosis

osmotic potential (OP) The potential pressure that a solution is capable of developing when separated from pure water by a semi-permeable membrane. In a cell it is equal to the net osmotic pressure, i.e. the difference between internal and external osmotic pressures.

osmotic pressure (OP) See osmoregulation and osmosis.

ossicle Any small bone, e.g. the malleus, incus or stapes of the middle ear.

ossification The process of bone formation. In mammals, most bones are formed from cartilage, e.g. long bones and vertebrae. When ossification begins, it does so at three centres: in the shaft (diaphysis) and in the two heads of the bone (epiphyses). Chondroblasts begin to precipitate calcium salts in the matrix (calcification), osteoclasts penetrate the cartilage to form channels and, via the blood vessels extending into these channels, osteoblasts are brought and deposit bone material. Ossification proceeds until only a layer of cartilage (the metaphysis) remains separating the shaft from the heads of the bone. New cartilage produced in the metaphyses continues to be ossified, and in this manner the bone grows in length. Growth is completed when the metaphyses are completely ossified.

ossify To turn into bone.

osteoblast A bone-forming cell. See also ossification.

osteoclast A cell which penetrates calcified cartilage, forming channels into which blood vessels extend. The channels develop into Haversian canals. See also ossification.

osteocyte An osteoblast which has been trapped between the lamellae of bone as the bone developed. Osteocytes have ceased to form bone matrix.

osteology The study of bones.

osteolysis The dissolution of bone, as performed by osteoclasts. See also ossification.

osteomalacia A deficiency disease in adults caused by a lack of vitamin D, resulting in the softening of bones which may lead to deformities. Compare rickets.

Ostwald method See nitric acid.

Ostwald's dilution law See dilution law.

otic Pertaining to the ear.

otolith See macula.

-ous 1. Suffix meaning that a multivalent element in a compound is in the lower of two possible valency states. Example: ferrous compounds contain the iron(II) ion, Fe^{2+}, while ferric compounds contain the iron(III) ion, Fe^{3+}. **2.** Suffix meaning that the central atom of an acid is in the lower of two possible valency states. Example: in sulphurous acid, H_2SO_3, the sulphur has a valency of four; in sulphuric acid, H_2SO_4, it has a valency of six. See also -ic.

outer ear (external ear) The part of the ear consisting of the external auditory meatus and the auricle (pinna).

oval window See fenestra ovalis.

ovarian follicle In many Metazoa, a sac of cells situated in the ovary which surrounds a developing oocyte. The ovarian follicle nourishes the growing oocyte, and in vertebrates it also produces female sex hormones such as oestrogens. In mammals, the ovarian follicle is called the Graafian follicle.

ovary 1. In plants, the part of the gynoecium containing the ovules. See Fig. 75. **2.** In vertebrates, one of a pair of female reproductive organs producing ova and sex hormones.

ovate Shaped like an egg. See Fig. 111.

overflow can See displacement can.

overtone (upper partial) A component of a complex vibration (note) which has a lower intensity and a higher frequency than the fundamental. Overtones have frequencies which are simple multiples of the fundamental frequency. Example: the first overtone (second harmonic) has twice the frequency of the fundamental.

oviduct See Fallopian tube.

oviparous Describes animals which lay undeveloped eggs. These may be fertilised either before they leave the female's body, e.g. in birds, or after they leave the body, e.g. in amphibians. Compare ovoviviparous and viviparous.

ovipositor In female insects, the tubular, egg-laying organ situated at the end of the abdomen. The ovipositor of ants, bees and wasps is modified into a sting.

ovotestis (pl. ovotestes) The reproductive organ of certain hermaphrodite animals, e.g. the snail. It produces both ova and spermatozoa.

ovoviviparous Describes animals whose fertilised eggs are covered by membranes and develop within the animal until hatching takes place. This may occur either within the animal's body or outside it. Ovoviviparous animals include many insects, some fish, some reptiles and snails. Compare *oviparous* and *viviparous*.

ovulation See Graafian follicle.

ovule In seed plants, a structure consisting of a nucellus which contains an embryo sac and is surrounded, except at the micropyle, by one or two integuments (an integument is not present in the ovules of conifers). An ovule develops into a seed after fertilisation. See Fig. 75.

ovum (pl. ova) *See* egg.

oxalate *See* ethanedioate.

oxalic acid *See* ethanedioic acid.

oxidant A term usually applied to an oxidising agent which supplies oxygen in a combustion process. *Examples:* liquid oxygen or hydrogen peroxide, H_2O_2, are oxidants in rockets.

oxidase An enzyme catalysing various redox reactions in the presence of free oxygen.

oxidation Originally, oxidation was considered to be a process in which oxygen was taken up by another substance. *Example:* when heated in air (or oxygen), copper combines with oxygen to form copper(II) oxide, CuO, and the copper is then said to be oxidised:
$$2Cu + O_2 \rightarrow 2CuO$$
The removal of hydrogen from a substance also causes oxidation. *Example:* when hydrogen sulphide, H_2S, reacts with chlorine, Cl_2, the H_2S is said to be oxidised:
$$H_2S + Cl_2 \rightarrow S + 2HCl$$
However, the concept of oxidation has broadened and the term is now used to mean a process in which one or more electrons are lost from an atom, ion or molecule, thus producing oxidation of these particles. *Example:* when copper dissolves in nitric acid, the copper loses two electrons (2e) and becomes a copper(II) ion:
$$Cu - 2e \rightarrow Cu^{2+}$$
Oxidation may also take place electrically at the anode during electrolysis. *Compare* reduction; *see also* redox reaction, oxidising agent, reducing agent and oxidation number.

oxidation number (oxidation state) A theoretical number given to an element which is helpful in solving problems such as balancing certain chemical equations. Oxidation numbers are either positive, negative or zero and they follow a set of simple rules: (1) The oxidation number of an atom in a molecule of a free element is zero. *Example:* the oxidation number of both hydrogen atoms in H_2 is zero. (2) In neutral molecules, the sum of the oxidation numbers of the elements (atoms) is zero. *Example:* in water, H_2O:
(2 × oxidation number of hydrogen) +
(oxidation number of oxygen) = 0
(3) Hydrogen has the oxidation number +1 in chemical compounds. (4) Oxygen has the oxidation number −2 in chemical compounds, except in peroxides where it is −1. (5) The oxidation number of a simple ion is equal to its charge. *Example:* the oxidation number of the iron(III) ion, Fe^{3+}, is +3 and of the sulphide ion, S^{2-}, is −2. In the case of more complex ions, the oxidation number of the different elements (atoms) present must be taken into consideration. *Example:* in the manganate(VII) ion, MnO_4^-, the oxidation number of each oxygen atom is −2, giving
$$(-2) \times 4 = -8$$
for all four of them; the oxidation number of manganese is therefore +7, giving
$$+7 + (-8) = -1$$
Knowledge of oxidation numbers leads to a definition of oxidation as a process in which there is an increase in oxidation number, and reduction as a process in which there is a decrease in oxidation number. If fractional values do not arise, the oxidation numbers are sometimes written using capital Roman numerals. *Example:* the calcium ion, Ca^{2+}, has the oxidation number +II. *See also* equivalent mass.

oxidation–reduction process *See* redox reaction.

oxidation state *See* oxidation number.

oxide A binary compound of oxygen and another element. Oxides are either covalent or ionic. They may be classified into the following four groups:
(1) *acidic oxides.* Many oxides of non-metals are acidic, i.e. they produce an acidic solution with water. *Example:* sulphur(VI) oxide, SO_3 forms sulphuric acid, H_2SO_4, with water:
$$SO_3 + H_2O \rightarrow H_2SO_4$$
Acidic oxides also react with bases to form salts and water.
(2) *basic oxides.* Many oxides of metals are basic, i.e. they produce a basic (alkaline) solution with water. *Example:* calcium oxide, CaO, forms calcium hydroxide, $Ca(OH)_2$, with water:
$$CaO + H_2O \rightarrow Ca(OH)_2$$
They also react with acids to form salts and water.
(3) *neutral oxides.* These are oxides which are neither acidic nor basic, e.g. carbon monoxide, CO, and dinitrogen oxide, N_2O.

(4) *amphoteric oxides*. These are oxides which have both acidic and basic properties, e.g. aluminium oxide, Al_2O_3, and zinc oxide, ZnO. These react with both alkalis and acids to produce salts and water:
$Al_2O_3 + 2NaOH \rightarrow 2NaAlO_2 + H_2O$ (acidic)
$Al_2O_3 + 6HCl \rightarrow 2AlCl_3 + 3H_2O$ (basic)
The two salts formed are sodium aluminate, $NaAlO_2$, and aluminium chloride, $AlCl_3$. *See also* peroxide.

oxide ion O^{2-} This ion does not exist in aqueous solution as it is a stronger base than the hydroxyl ion, OH^-; soluble metal oxides form hydroxyl ions in aqueous solution and not oxide ions:
$Na_2O + H_2O \rightarrow 2Na^+ + 2OH^-$
The oxide ion may exist in a fused metal oxide, e.g. in molten aluminium oxide, Al_2O_3:
$Al_2O_3 \rightarrow 2Al^{3+} + 3O^{2-}$

oxidising agent (oxidant) A substance which can oxidise another and at the same time be reduced itself. *Example:* in the reaction between iron, Fe, and chlorine, Cl_2, iron(III) chloride, $FeCl_3$, is formed:
$2Fe + 3Cl_2 \rightarrow 2FeCl_3$ ($Fe^{3+}, 3Cl^-$)
Here the chlorine is the oxidising agent, oxidising Fe to Fe^{3+} and being reduced itself to Cl^-. An oxidising agent is therefore an acceptor of electrons. *Compare* reducing agent; *see also* redox reaction.

oxonium ion (hydroxonium ion, hydronium ion) The hydrated hydrogen ion, H_3O^+, formed when a hydrogen ion, H^+, combines with water:
$H^+ + H_2O \rightleftharpoons H_3O^+$

2-oxopropanoate *See* pyruvate.

2-oxopropanoic acid *See* pyruvic acid.

oxygen An element with the symbol O; atomic number 8; relative atomic mass 16,00 (16.00); state, gas. Oxygen is a colourless, odourless, chemically rather active, diatomic gas. It occurs naturally as free oxygen in the atmosphere (about 21% by volume) and is the most abundant element in the Earth's crust, where it is found chemically combined in the majority of rocks and minerals. Oxygen exists in the form of two allotropes, the diatomic oxygen molecule O_2 (dioxygen) and the triatomic ozone molecule O_3 (trioxygen). Ozone is much less stable than oxygen. Oxygen can be prepared in the laboratory using several methods, e.g. by heating dry potassium chlorate, $KClO_3$, with manganese dioxide, MnO_2, as a catalyst:
$2KClO_3 \rightarrow 3O_2 + 2KCl$
or by heating dry potassium manganate(VII), $KMnO_4$:
$2KMnO_4 \rightarrow O_2 + MnO_2 + K_2MnO_4$

or by heating a strong solution of hydrogen peroxide, H_2O_2:
$2H_2O_2 \rightarrow O_2 + 2H_2O$
Oxygen is manufactured on a large scale by fractional distillation of liquid air. It forms oxides with other elements, e.g. calcium oxide, CaO, with calcium and sulphur dioxide, SO_2, with sulphur. Oxygen is used in respirators, in welding and metal-cutting, in the extraction of certain metals from their ores, etc.

oxygen cycle The circulation of oxygen between organisms and their environment. Oxygen used in respiration is returned to the surroundings by photosynthesis. Some oxygen is also used in combustion.

oxygen debt The state in an aerobic organism in which there is a demand for oxygen greater than can be met. Oxygen debt may be seen during hard, sustained physical exercise; an increased activity needs an increased supply of oxygen, i.e. heartbeat and breathing rates increase. However, there is a limit to the quantity of oxygen that can be passed to the cells in a given time, so if the exercise continues oxygen debt occurs. Strenuous exercise leads to an accumulation of lactic acid in the cells, causing fatigue. The additional oxygen which is required to remove the lactic acid, by oxidising it into pyruvic acid, constitutes the oxygen debt. The pyruvic acid is quickly broken down to carbon dioxide and water.

oxygen mixture A mixture of very pure chemicals used in the laboratory preparation of oxygen; e.g. potassium chlorate, $KClO_3$, mixed with manganese dioxide, MnO_2, or potassium chlorate mixed with potassium nitrate, KNO_3.

oxyhaemoglobin HbO_2 *See* haemoglobin.

oxytocin A polypeptide hormone secreted by the hypothalamus and released by the posterior lobe of the pituitary gland. It induces contraction of smooth muscles. In mammals, it stimulates contractions of the uterus during birth (produces labour) and promotes secretion of milk during suckling. Oxytocin is used to induce labour artificially.

ozone O_3 (trioxygen) A bluish gas with a characteristic penetrating smell. It is one of the two allotropes of oxygen. Ozone is a strong oxidising agent and is unstable. On warming it decomposes to oxygen:
$2O_3 \rightarrow 3O_2$
Ozone is a much stronger oxidising agent than oxygen. It oxidises lead(II) sulphide, PbS, to lead(II) sulphate, $PbSO_4$:
$PbS + 4O_3 \rightarrow PbSO_4 + 4O_2$
and potassium iodide, KI, to iodine, I_2:
$2KI + O_3 + H_2O \rightarrow I_2 + 2KOH + O_2$
This last reaction is often used as a test for ozone, the iodine produced being detected by

the blue colour it forms with a starch suspension. Other oxidising agents may give the same reaction.

Ozone occurs naturally in the stratosphere where it is formed from atmospheric oxygen by the energy of sunlight. This ozone layer protects the surface of the Earth from excessive ultra-violet radiation. Ozone can be prepared by passing an electrical discharge through oxygen:

$3O_2 \rightarrow 2O_3$

It is used in water purification and as an oxidising and bleaching agent. *See also* ozonosphere.

ozone layer *See* ozonosphere.

ozoniser An apparatus used for producing ozone by maintaining an electrical discharge in an atmosphere of oxygen:

$3O_2 \rightarrow 2O_3$

ozonosphere (ozone layer) A layer of ozone in the stratosphere lying between altitudes of 12 and 50 kilometres. This layer of ozone absorbs a large proportion of the ultra-violet radiation from the Sun, thus shielding the Earth's surface from the harmful effects of this radiation, e.g. skin cancer. It has been claimed that the ozone layer is self-generating; that is, the ozone concentration in the ozone layer remains more or less constant over a period of time.

P

P Chemical symbol for phosphorus.

P$_1$ Abbreviation for the first parental generation.

P$_2$ abbreviation for the second parental generation (the grandparents).

Pa *See* pascal.

pace-maker The region of the vertebrate heart which consists of a small nodular mass of specialised muscle cells situated on the posterior wall of the right atrium (auricle). It is supplied with nerves from both the sympathetic and parasympathetic nervous systems. The pace-maker is thought to initiate the stimuli that produce contraction of the heart. The term is also applied to a small electrical device powered by batteries which is implanted in the chest of a patient whose natural pace-maker fails to work properly. The batteries are constructed to provide power for about five years, after which they have to be replaced.

pachytene *See* meiosis.

paediatrics The study of childhood and the diseases related to this period of life. *Compare* gerontology.

pain *See* phantom pain *and* referred pain.

paired fins *See* fin.

pairing The process in which homologous chromosomes are attracted to one another during the zygotene stage of meiosis.

Palaeocene A geological period of approximately 10 million years which ended approximately 54 million years ago. A subdivision of the Tertiary period.

palaeontology The study of fossil animals and plants.

Palaeozoic The geological age (era) of ancient life, estimated to have commenced approximately 570 million years ago. *See* Appendix.

palate In vertebrates, the roof of the mouth. In mammals, it is divided into two parts, a bony anterior portion called the hard palate and a posterior portion called the soft palate which is made up of soft tissues.

palea (pl. paleae) (glumella) In grasses, the upper of two bracts. Together with the lower bract (lemma), it encloses the young flower. *See* Fig. 217.

palisade cells Elongated cells making up the palisade mesophyll (layer) of a leaf. *See* Fig. 110 (b).

palisade mesophyll *See* mesophyll.

palmate Describes a compound leaf which has several lobes or leaflets attached to the apex of the petiole. *See* Fig. 111.

palmitate (hexadecanoate) A salt of palmitic acid (hexadecanoic acid), $C_{15}H_{31}COOH$.

palmitic acid $C_{15}H_{31}COOH$ (hexadecanoic acid) A solid, saturated fatty acid (alkanoic acid). It is found naturally as a part of many fats and oils. The salts of palmitic acid are called palmitates (hexadecanoates). It is used in the manufacture of soaps and lubricants.

palp *See* palpus.

palpus (pl. palpi) (palp) **1.** In insects, a sensory organ situated at the mouth. Palps are sensitive to touch. In some insects they are concerned with the sense of smell. **2.** In crustaceans, the distal parts of appendages which carry mandibles and which are used for feeding and locomotion.

palynology Pollen analysis.

pancreas In vertebrates, a glandular organ of the digestive system. In humans it is situated in the abdominal cavity, behind the stomach and extending from the duodenum to the spleen. The pancreas is concerned with both endocrine and exocrine secretions. It secretes glucagon and insulin into the bloodstream from the α-cells and β-cells of the islets of Langerhans respectively (endocrine secretions). It also

pancreatic juice

secretes pancreatic juice which drains into a duct system emptying into the duodenum (exocrine secretion). *See* Fig. 49.

pancreatic juice A digestive juice secreted by the pancreas into the duodenum. Pancreatic juice is an alkaline mixture of the enzymes trypsin, amylase and lipase. The secretion of pancreatic juice is stimulated by nerve-impulses or by the hormone secretin.

pandemic Describes an epidemic disease which is widespread over a country, a continent or the whole world. *Example:* certain types of influenza have been pandemic. *See also* endemic.

pantothenic acid An acid belonging to the B-complex group of vitamins. It is widely distributed in animals and plants. Pantothenic acid stimulates the growth of yeast. In dogs, a deficiency leads to anaemia; in Man, no deficiency has been observed.

paper chromatography *See* chromatography.

papilla (pl. papillae) A small projection of tissue, e.g. the numerous projections on the surface of the tongue or on the lining of the alimentary canal.

pappus (pl. pappi) A ring of hairs, scales or teeth making up the calyx of composite flowers. It acts as a parachute and aids wind dispersal of the fruit. *See* Fig. 156.

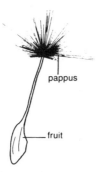

Fig. 156 Pappus

para- Prefix indicating that two substituents lie directly across from each other in an aromatic ring. *Example:*

is para-dichlorobenzene (p-dichlorobenzene) or 1,4-dichlorobenzene. *See also* meta- *and* ortho-.

parabolic mirror *See* parabolic reflector.

parabolic reflector A concave reflector whose section is a parabola. A light source placed at the focal point results in the production of a parallel beam of light. Conversely, a parallel beam of light is converged to the focal point. Parabolic reflectors are used in car headlamps, solar power devices, etc. When light is the reflected radiation, the reflector is often called a parabolic mirror. However, reflectors of this shape are also used to reflect radio waves in radio telescopes. *See* Fig. 157.

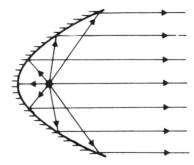

Fig. 157 Parabolic reflector

parachute A structure, e.g. a pappus, assisting the wind dispersal of a fruit or seed.

paradox 1. A statement which is contrary to an accepted theory. **2.** An apparently absurd or self-contradictory statement which nevertheless is true or may be true.

paraffin oil *See* kerosene.

paraffin wax A solid mixture of higher alkane hydrocarbons melting in the range 50–60°C. It is obtained from the distillation of petroleum and is used in the manufacture of candles, polishes, etc.

parallax The apparent relative change of position of two objects when viewed from different places. There is 'no parallax' between the two objects when they are in the same position. 'No parallax' methods are used in optics experiments to locate the position of an image. When reading a measuring instrument in which the pointer is a distance away from the scale, errors of parallax can occur if the eye is not at right angles to the scale. Such errors are eliminated if the instrument is fitted with a plane mirror behind the pointer. When the pointer covers its image in the mirror, there is no error of parallax.

parallel Components of an electric circuit are 'in parallel' when each is connected between the same two points A and B in the circuit. The effective (total) resistance R of three resistors connected in parallel as in Fig. 158 is given by the expression:

$$\frac{1}{R} = \frac{1}{R_1} + \frac{1}{R_2} + \frac{1}{R_3}$$

Compare series.

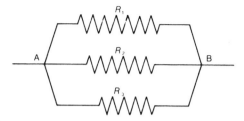

Fig. 158 Resistors connected in parallel

parallelogram of forces See parallelogram of vectors.

parallelogram of vectors If two inclined vectors F_1 and F_2 are represented in magnitude and direction by the adjacent sides of a parallelogram, then their resultant vector **F** is represented in magnitude and direction by the diagonal of the parallelogram passing through the point of intersection of the two sides. The parallelogram of vectors can be applied to forces and velocities. See Fig. 159.

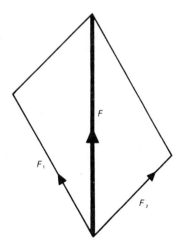

Fig. 159 Parallelogram of vectors

parallelogram of velocities See parallelogram of vectors.

parallel venation See venation.

paramagnetic material See ferromagnetic material.

Paramecium A genus of Protozoa living in fresh water which contains decaying organic matter. *Paramecium* is shaped like a flattened slipper and is covered with cilia for locomotion. It reproduces sexually by conjugation and asexually by binary fission.

paramylum See Euglena.

parasite An organism living in or on another living organism (the host). The parasite obtains food from the host and is usually harmful to it. Compare saprophyte.

parasitism An association between two organisms in which one organism (the parasite) benefits and the other (the host) is harmed. *Example:* the tapeworm is parasitic in the alimentary canal in humans and other vertebrates. See also commensalism *and* mutualism.

parasitology The study of parasites, especially those of animals.

parasympathetic nervous system A division of the autonomic nervous system which consists of the vagus nerve and certain other cranial and spinal nerves. Acetylcholine is produced at the ends of the nerve-fibres of the parasympathetic nervous system. This system works in opposition to the sympathetic nervous system. It controls such involuntary actions as constricting the iris and coronary arteries, slowing the heartbeat, lowering the blood pressure, increasing peristalsis, etc.

parathion $C_{10}H_{14}NO_5PS$. An aromatic nitro compound. It is a yellow liquid which is soluble in most organic solvents, but only slightly soluble in water. Parathion is used as an insecticide.

parathormone (parathyrin) See parathyroid gland.

parathyrin See parathormone.

parathyroid gland In tetrapod vertebrates, a ductless gland secreting the polypeptide hormone parathormone. In humans there are four such glands, arranged in pairs, lying near the thyroid gland. Parathormone regulates the metabolism of calcium and phosphorus in the body.

paratyphoid See typhoid.

Parazoa A subkingdom of invertebrates comprising the phylum Porifera (the sponges). They are multicellular animals, but differ so much from metazoans that they are considered a separate subkingdom. See also Metazoa *and* Protozoa.

parenchyma 1. Plant tissue made up of relatively large, loosely-packed cells which have thin cellulose walls and variable shapes. The cells have no specific function, but are able to change their nature according to a plant's needs. *Examples:* they can develop chloroplasts and begin to photosynthesise; they can store starch. 2. In animals: the specific tissue of an organ as opposed to the connective tissue, blood vessels, etc.

parental care A pattern of behaviour exhibited by many animals, particularly mammals and birds, in which they provide for their young. *Example:* birds build nests, incubate their eggs and protect and feed their young.

parietal Pertaining to, situated in, or forming, part of the wall of a structure.

Parkinson's disease A chronic, slowly progressive disease which affects that part of the brain controlling voluntary actions. It may be caused by arteriosclerosis or carbon monoxide poisoning. Symptoms include stiffness and weakness of the muscles, including those of the face, which leads to a mask-like appearance; and body tremors, especially trembling of the hands and nodding of the head. These symptoms may be relieved by medical treatment by drugs or surgery.

parotid gland In mammals, one of a pair of salivary glands whose ducts open into the mouth. In humans they lie in front of and below the external ear. Mumps may cause viral infection of the parotid glands.

parthenocarpy The formation of fruit without pollination or fertilisation, resulting in seedless fruit. Parthenocarpy may occur naturally, e.g. in the banana, but it can be induced in certain plants by spraying them with auxins after removing the stamens. Tomatoes, apples and pineapples are examples of fruits which may be produced in this way.

parthenogenesis The development of a new individual without fertilisation of the ovum (egg). This occurs in certain animals and plants such as aphids and dandelions. There are two kinds of parthenogenesis: diploid and haploid. In diploid parthenogenesis the eggs are formed by mitosis and the individuals developing from such eggs are diploid like the parent. In haploid parthenogenesis, the eggs are formed by meiosis and develop into haploid individuals. The factors causing parthenogenesis are unknown. However, in several animals parthenogenesis may be induced artificially, e.g. by pricking the eggs with a needle or exposing them to an acid.

partial A component of a complex vibration (note) which has a higher frequency than the fundamental. Partials include overtones as well as any non-harmonic components which may be present in the vibration.

partial pressure *See* Dalton's law of partial pressures.

particle theory *See* kinetic theory.

parturition (labour) In viviparous animals, a term given to the process of birth at the end of pregnancy. This is not due to a single factor, but to several, some of which are not yet understood. However, hormones play an important part in the onset of parturition. They stimulate the uterus to undergo powerful, rhythmic contractions, eventually forcing the foetus and, a little later, the afterbirth out of the mother. *See also* oxytocin.

pascal Symbol: Pa. The SI unit of pressure, defined as a pressure of one newton per square metre ($N\,m^{-2}$).

Pascal's principle When pressure is applied to any part of the surface of a fluid which completely fills a vessel, that pressure is transmitted equally throughout the whole of the fluid. *See also* hydraulic press.

Pascal's vases (liquid level apparatus) An apparatus used to demonstrate that a liquid finds its own level, i.e. the pressure at any particular level in the liquid is dependent only on depth and not the surface area or shape of the container. *See* Fig. 160.

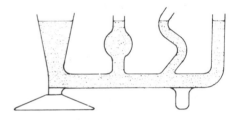

Fig. 160 Pascal's vases

Paschen series *See* Balmer series.

passive Describes a metal which is unreactive because of the formation of a thin, protective film (layer) of oxide on its surface. *Examples:* pure aluminium quickly forms a thin layer of aluminium oxide, Al_2O_3, on its surface when exposed to air; iron treated with concentrated nitric acid or any other very strong oxidising acid is passivated by a thin layer of oxide, probably iron(II) di-iron(III) oxide, Fe_3O_4. Other metals such as chromium, molybdenum, cobalt and nickel also become passive when treated with very strong oxidising acids.

pasteurisation A process of partial sterilisation, usually of milk, in which the milk is heated to a certain temperature and kept there for some time, e.g. to 72°C for 15 seconds, and then rapidly cooled to 3°C and bottled. This kills most bacteria, but not spores. It does not affect the quality of the milk. However, milk may also be sterilised by steam-heating the pasteurised, bottled milk. Such milk will keep well for several weeks even without refrigeration, but the flavour is affected. Pasteurisation is also applied to wine, fruit juices, etc.

pasture Land which has vegetation such as grasses and legumes growing on it, suitable for grazing animals.

patch test A test for the presence of tuberculosis. A small area of skin is rubbed in tuberculin. If the test is positive, i.e. if tuberculosis is present or has been present thus leaving the patient immune, a weal will develop on the skin within 48 hours. *See also* Mantoux test *and* BCG vaccine.

patella (pl. patellae) In the hind-limbs of most mammals, some birds and some reptiles, a small, rather flat bone situated in the tendon of extensor muscles in front of the knee-joint, with its apex pointing downwards. The patella articulates with the femur and forms the knee-cap.

pathogen A micro-organism which causes disease.

pathology The study of the changes in the body causing or caused by disease, including the nature of the disease.

Pauli exclusion principle *See* orbital.

paunch *See* rumen.

Pavlovian conditioning *See* conditioned reflex.

Pb Chemical symbol for lead.

p.d. *See* potential difference.

peak value The maximum positive or negative value reached by an alternating current or voltage. *See also* amplitude *and* root-mean-square value of an alternating current or voltage.

pearl *See* nacre.

peat A dark brown or black type of coal. It is the youngest stage in coal formation. Peat is composed of decayed and partly carbonised vegetable matter.

pectin The general name for a group of complex organic substances which are closely related to the carbohydrates (polysaccharides). Pectins are found in the cell wall of plants, especially in their fruits. They form gels in the presence of acid and sugar (saccharose or sucrose) and are used commercially in the food industry, e.g. in making jams and mayonnaise.

pectoral Pertaining to the breast or chest.

pectoral fin *See* Fig. 73.

pectoral girdle (shoulder girdle) In vertebrates, a skeletal structure which supports the front appendages (front limbs or fins). In humans two bones, the scapula and the clavicle, form the pectoral girdle which is attached to the thoracic wall by muscles and to the sternum by the clavicle.

pedicel The stalk of an individual flower of an inflorescence. *See* Fig. 161.

pedipalpus (pl. pedipalpi) (pedipalp) In arachnids, a head appendage which may be sensory. In the scorpion, the pedipalpi are modified for gripping; in king-crabs, they are used for locomotion; in male spiders, they are modified for fertilisation.

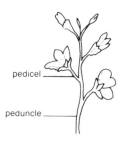

Fig. 161 Pedicel and peduncle of an inflorescence

pedology The study of soil.

peduncle The stalk of an inflorescence. *See* Fig. 161.

pelagic Describes organisms (plankton and nekton) living in the middle or surface levels of lakes and seas. *Compare* benthic.

pellagra A deficiency disease caused by lack of nicotinic acid in the diet. Symptoms include dark, inflamed areas of the skin, diarrhoea and mental depression. The treatment is to give the patient a balanced diet.

pellicle The thin, protective covering of Protozoa.

Peltier effect When an electric current is passed through a junction between two different metals or semiconductors, heat is produced or absorbed at the junction, depending on the direction of the current. The quantity of heat is proportional to the current. The effect is used in heating or cooling elements. *Compare* Seebeck effect.

pelvic Pertaining to or situated near the pelvis.

pelvic fin *See* fin.

pelvic girdle (hip-girdle) In vertebrates, a skeletal structure which supports the hind appendages (hind-limbs or fins). In humans it consists of the ilium, the ischium and the pubis, which are fused to form a bowl-shaped cavity, the acetabulum.

pelvis 1. In Primates, the basin-shaped, bony cavity formed by the pelvic girdle together with the sacrum and coccyx. **2.** In the mammalian kidney, the expanded portion of the ureter where it enters the kidney and into which urine drains.

pencil lead *See* graphite.

pendulum A simple pendulum consists of a small mass (bob) swinging freely at the end of a light string or wire. For small amplitudes of swing, the time (T) for one complete swing (the period) is given by the expression:

$$T = 2\pi \sqrt{\frac{l}{g}},$$

where l is the length of the string and g is the acceleration of free fall. Pendulums can be used to determine g, to regulate clocks, to

penicillin

demonstrate the Earth's rotation, etc. See also compensated pendulum *and* Foucault's pendulum.

penicillin 1. An antibiotic with a bactericidal effect, discovered in 1928 by Alexander Fleming and marketed in 1941. It is obtained from moulds of the genus *Penicillium* and is used in the treatment of many infectious diseases caused by Gram-positive bacteria. Penicillin has no effect on viruses. Some bacteria may become resistant to penicillin, either by the production of enzymes, such as penicillinase, which break down the penicillin, or by the development of a tolerance to penicillin. Allergic reactions to penicillin are not uncommon in humans. *See also* murein. **2.** One of a group of semi-synthetic drugs in which the structure of penicillin is altered by the addition of different molecular side-chains to the basic penicillin structure. This produces drugs which have a wider application.

penicillinase *See* penicillin.

Penicillium A genus of moulds which includes the species *Penicillium notatum*, from which Alexander Fleming extracted penicillin. It also includes *Penicillium camemberti* and *Penicillium roqueforti*, which are used in cheese-making.

penis The male copulatory organ in mammals, some reptiles and some invertebrates. In humans, it consists of three cylindrical masses of erectile tissue through which urine and semen pass via the urethra. The tip of the penis is called the glans and has an orifice which is the external opening of the urethra. During erotic stimulation, the penis becomes enlarged, hard and erect. This is caused by blood flowing into the spaces in the erectile tissue.

penta- Prefix meaning five.

pentadactyl limb The characteristic limb ending in five digits, found in all vertebrates except fish. In some species of animals, the basic arrangement of the limb is modified by loss or fusion of some of the bones. *See* Fig. 162.

pentavalent Having a valency of five.

pentlandite *See* nickel.

pentose A carbohydrate which is a monosaccharide with five carbon atoms in its molecule, e.g. ribose, $C_5H_{10}O_5$. *See also* sugar.

penumbra *See* shadow.

pepsin An enzyme formed from pepsinogen in gastric juice. It catalyses the partial breakdown (hydrolysis) of proteins to smaller peptide molecules.

pepsinogen A zymogen secreted by the gastric glands. In the stomach it is activated by hydrochloric acid to form pepsin.

peptide An organic substance containing one or more peptide bonds. Peptides are composed of units of amino acids:

$$2\begin{pmatrix} CH_2-COOH \\ | \\ NH_2 \end{pmatrix} \rightarrow H-N-\underset{\underset{NH_2}{|}}{\overset{\overset{CH_2-COOH}{|}}{\underset{\|}{C}}}-CH_2 + H_2O$$

Here a dipeptide is formed from two molecules of glycine with the release of one molecule of water. The bond between a nitrogen atom and a carbon atom connected to an oxygen atom is called the peptide bond. Proteins are composed of many units of amino acids forming

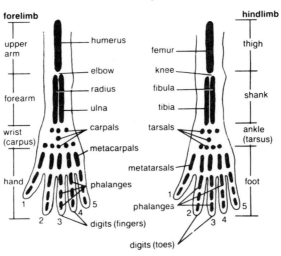

Fig. 162 Pentadactyl limb

polypeptides. However, there is no strict distinction between polypeptides and proteins. It has been suggested that polypeptides with relative molecular masses of 5000 or more are proteins.

peptide bond *See* peptide.

peptidoglycan *See* murein.

peptisation The process in which a solid substance formed in an aqueous solution changes into a colloidal form. *Example:* part of a precipitate of silver chloride, AgCl, may revert to a colloidal form, especially if pure water is used as a wash liquid:

$$AgCl_{(colloidal)} \underset{peptisation}{\overset{coagulation}{\rightleftharpoons}} AgCl_{(s)}$$

The colloidal particles will pass through a filter medium, which means that part of the precipitate is lost. Peptisation may be induced by the action of an electrolyte. *Example:* if freshly prepared iron(III) hydroxide is treated with a small quantity of iron(III) chloride solution, it immediately forms a colloidal solution with a dark reddish-brown colour.

peptone The general name given to a peptide which is formed by a partial, enzymatic breakdown of a protein. Peptones are soluble in water. They are used as a nitrogen source for bacterial cultures grown in the laboratory.

perchlorate *See* chlorate(VII).

perchloric acid *See* chloric(VII) acid.

perennation The survival of plants by vegetative means for a number of years or seasons.

perennial plant A plant which lives for a number of years or seasons.

perfect conductor *See* superconductivity.

perfect fungus A fungus which has a known sexual cycle. *Compare* imperfect fungus.

perfect gas *See* ideal gas.

perhydrol A 30% aqueous solution of hydrogen peroxide, H_2O_2.

perianth A collective term for the calyx and corolla of a flower. When the calyx and corolla are not clearly divided, the perianth refers to the outer whorl of floral parts.

pericardium The membrane lining the cavity in which the heart lies. The term is often applied to the cavity itself.

pericarp The structure which develops from a ripened ovary wall and contains the seeds of a fruit. *See also* drupe.

perichondrium A membrane made up of fibrous connective tissue, surrounding cartilage. It provides nourishment to the cartilage by diffusion.

periclase *See* magnesium oxide.

pericycle In plants, a thin layer of parenchyma lying immediately within the endodermis and surrounding the stele. It gives rise to lateral roots.

periderm A collective term for cork cambium (phellogen), cork (phellem) and phelloderm; i.e. periderm means cork cambium and its products.

perigee The point in the orbit of a moon, planet or artificial satellite when it is nearest to the Earth. *Compare* apogee.

perigynous Describes a flower in which the ovary is superior and is situated at the centre of a concave receptacle with the sepals, petals and stamens attached to its rim, e.g. rose. *See* Fig. 163; *see also* epigynous *and* hypogynous.

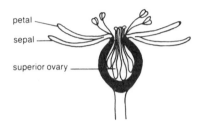

Fig. 163 Perigynous flower in section

perihelion The point in a planet's orbit when it is nearest to the Sun. *Compare* aphelion.

perilymph A fluid in the bony labyrinth of the ear. The perilymph provides a shock-absorbing protection and conducts vibrations transmitted from the middle ear. *See* Fig. 55.

period 1. *See* menstruation.   2. (periodic time) Symbol: T. The time for one complete cycle of any regularly repeating process, e.g. a wave motion or an oscillation. *See also* pendulum.

periodic law A law stated by the Russian chemist Dmitri Ivanovich Mendeléev in 1869, on which the present Periodic Table of the Elements is based. Mendeléev arranged the elements known at that time in ascending order of their relative atomic masses and noticed that elements which have similar chemical properties appeared at fixed intervals (the periodic law). There were several gaps in his table but, by considering the properties of elements above and below these gaps, he was able to predict many chemical and physical properties of these undiscovered elements. Today, the Periodic Table of the Elements is based on an ascending order of atomic numbers. Up to uranium (atomic number 92), this arrangement only differs in four places from Mendeléev's system: argon, Ar, comes before potassium, K; cobalt, Co, comes before nickel, Ni; tellurium, Te, comes before iodine, I; and thorium, Th, comes before protactinium, Pa. In each case, Mendeléev placed the pair in the opposite order using relative atomic masses. *Example:* the atomic number of argon is 18 and its relative atomic mass

Periodic Table of the Elements

is 39,95 (39.95), whereas potassium has a higher atomic number, 19, but a lower relative atomic mass, 39,10 (39.10).

Periodic Table of the Elements A table of the elements arranged in an ascending order of atomic numbers, thus demonstrating the periodic law. The table is given in the Appendix. It consists of eight main groups (the vertical columns labelled A) and seven periods (the horizontal rows). Similarities in chemical properties are found among the elements placed in a particular group because they have the same number of electrons in their outer shells. Both chemical and physical properties may change gradually or be similar throughout a group. Therefore, knowing the position of an element in the table makes it possible to predict its properties. The main group number indicates the greatest possible positive oxidation number and the main group number minus eight indicates the smallest possible negative oxidation number that the element in the group may have (Abegg's rule of eight). The halogens belonging to Group VIIA have a greatest possible positive oxidation number of $+7$, e.g. chlorine in chloric(VII) acid (perchloric acid), $HClO_4$, and a smallest possible negative oxidation number of $7-8 = -1$ e.g. chlorine in hydrochloric acid, HCl. This rule is particularly useful for the elements in the main groups IV, V, VI, and VII. Moving from one period to the next, within the same main group, we see that the number of electron shells increases by one. This means that the elements increase in atomic size, and therefore become more and more metallic. *Example:* in Group IVA, the first element is the non-metal carbon, the second element is the semi-metal silicon and the fifth element is the metal lead. Elements in the same period have the same number of electron shells. Six of the periods end in a noble gas. In the seven periods there are 2, 8, 8, 18, 18, 32 and 19 elements respectively. However, the table is still incomplete; new elements are still artificially made. To each main group except group eight, there is a sub-group, labelled B; e.g. Group IA (the alkali metals) has the sub-group IB. Chemical similarities between elements in a main group and a sub-group are less distinct than between the elements in the main group.

periodic time *See* period.

periodontal membrane A membrane made up of connective tissue, surrounding the root of a tooth. It forms a suspensory ligament serving to anchor the tooth to the jaw-bone.

periodontitis A progressive inflammatory disease of the gums. The disease slowly spreads into the underlying tissue where it causes destruction of the periodontal membrane and the supporting bone. If left untreated it leads to looseness of the teeth and ultimately to loss of the affected teeth. The disease is most often due to poor oral hygiene. Periodontitis may be treated surgically.

periosteum (pl. periostea) A tough membrane made up of connective tissue containing osteoblasts and surrounding a bone. *See* Fig. 14.

peripheral nervous system The part of the nervous system connecting all parts of the vertebrate body with the brain and spinal cord.

periscope An optical instrument for looking over or around obstacles. (1) The mirror periscope consists of two parallel mirrors placed at an angle of 45° to the horizontal at each end of a long tube. *See* Fig. 164 (a). (2) The prism periscope consists of two right-angled prisms placed at each end of a long tube. They reflect light by total internal reflection. Periscopes of this type are used in submarines. *See* Fig. 164 (b).

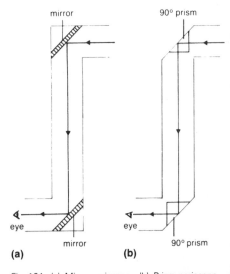

Fig. 164 (a) Mirror periscope (b) Prism periscope

Perissodactyla An order of herbivorous, placental mammals commonly called the odd-toed ungulates. Members include the horse, the rhinoceros and the tapir. They have either one toe (horse) or three toes (rhinoceros). *Compare* Artiodactyla.

peristalsis The process by which the contents of tubular organs, e.g. the oesophagus, are made to pass through them. This is brought about by a wave-like motion produced by contractions of smooth muscles in the walls of the tubes. The muscular contractions are controlled by the autonomic nervous system.

peritoneum The membrane lining the walls of the abdominal cavity.

permanent gas A gas which cannot be liquefied by pressure alone, but must also be cooled. Gases such as hydrogen, helium, nitrogen and oxygen are permanent gases.

permanent hardness See hard water.

permanent magnet A piece of magnetic material which retains its magnetism (if not heated or roughly treated) after being magnetised by electric currents or by stroking, etc. Permanent magnets have many uses, such as in loudspeakers and small electric motors. Compare temporary magnet.

permanent set The extension remaining in an elastic material, e.g. a spring, when it has been stretched beyond its elastic limit.

permanent teeth In most mammals, the teeth replacing the deciduous teeth (milk teeth). See also dental formula.

permanent wilting point A property of a soil: the state in which so much water has dried away that plants can no longer benefit from the remaining water attached to the soil particles and therefore begin to wilt. Unless water is added to the soil, the plants will not recover. Compare field capacity.

permeable Describes a material which allows the passage of a fluid through itself.

Permian A geological period of approximately 55 million years, which ended approximately 225 million years ago.

Permutit® A type of synthetic or natural zeolite used in the removal of calcium and magnesium ions in water-softening.

pernicious anaemia A type of anaemia caused by the failure of red blood corpuscles (cells) to develop in the normal manner due to a deficiency of vitamin B_{12}. The disease is treated by giving the patient injections of vitamin B_{12}. See also cobalamin(e).

peroxidase An enzyme which splits hydrogen peroxide, often found in higher plants but (with a few exceptions) rare in animal cells. Peroxidase catalyses the reduction of hydrogen peroxide, H_2O_2, to water, H_2O, and oxygen, O_2. The oxygen is then used in the oxidation of organic material. See also catalase and dehydrogenase.

peroxide A metallic, electrovalent compound containing the peroxide ion, O_2^{2-}, in which oxygen has the oxidation number -1 and in which there is a direct oxygen–oxygen bond. With water or dilute acids, peroxides liberate hydrogen peroxide, H_2O_2, e.g.
$Na_2O_2 + 2H_2O \rightarrow H_2O_2 + 2NaOH$
Here sodium peroxide, Na_2O_2, reacts with water. Peroxides may be formed by heating an oxide (or a metal) in air or oxygen:
$2BaO + O_2 \rightarrow 2BaO_2$
Here barium oxide is converted into barium peroxide. Peroxides are usually strong oxidising agents and may be considered to be salts of hydrogen peroxide, which is a very weak acid.

Perrin tube A cathode-ray tube used to demonstrate that a beam of cathode rays has a negative charge. A beam of rays is deflected, using Helmholtz coils, into a Faraday cylinder connected to a gold leaf electroscope. If the Faraday cylinder is connected to a negatively charged electroscope, the leaf stays up. If the electroscope is positively charged before it is connected to the Faraday cylinder, its leaf falls and then rises again. See Fig. 165.

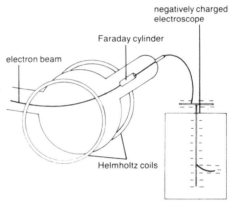

Fig. 165   Perrin tube

persistence of vision An effect which is made use of in the cine-projector and television. It is an effect of the retina of the eye and the brain combined, in which an image of an object is retained by the brain for a short time. A cinema film is made up of a sequence of still pictures, each slightly different, and is usually projected onto a screen at a rate of 24 pictures per second. The film is seen as a moving picture as a result of overlapping images on the retina.

Perspex® (poly(methyl 2-methylpropenoate); poly(methyl methacrylate), plexiglass) A thermoplastic, light, strong, transparent type of plastic formed by polymerisation of methyl 2-methylpropenoate (methyl methacrylate). It is used in aeroplane windows, lenses, packaging, etc., and is available in a wide range of colours.

perturbation A disturbance (deviation) in the motion of a heavenly body, e.g. a planet, from its true orbit. Perturbations are caused by the gravitational pull of other heavenly bodies.

pertussis See whooping cough.

perversion See lateral inversion.

pest An animal, usually an insect, which destroys crops.

pesticide A chemical used to destroy pests. It is usually an insecticide.

pestle *See* mortar.

petal One of the divisions of a flower which together make up the corolla. Petals are usually brightly-coloured and may also be scented. Both these characteristics attract insects which are necessary to bring about pollination. Wind-pollinated flowers very often have small petals or none at all. *See* Fig. 75.

petiole (leaf stalk) The stalk attaching a leaf blade to a stem. *See* Fig. 110 (a).

petri dish A shallow, circular, covered dish made of plastic or glass. It is used in bacteriology for growing cultures of micro-organisms on a jelly such as agar. Plastic petri dishes are purchased ready sterilised and cannot be resterilised in an autoclave after use as they melt.

petrifaction The process in which silica, calcite and various iron compounds are deposited, by water in which they are carried, in the pores of organic structures, e.g. wood, bones, etc. This makes them harder and denser, giving them a stony structure. Fossils, e.g. fossil wood, may be formed in this manner. Petrifaction rarely affects their appearance.

petrochemical A chemical which is obtained from petroleum (crude oil) or natural gas.

petrol A mixture of alkanes which have from five to ten carbon atoms, i.e. from pentanes to decanes. It is obtained during fractional distillation of petroleum (crude oil) and boils in the range 50–150°C. Petrol is used as a fuel in internal combustion engines. In the USA it is called gasoline. *See also* octane number *and* cracking.

petroleum *See* crude oil.

petroleum ether A mixture of pentanes and hexanes (alkanes) boiling in the range 30–70°C. The name is misleading as the substance is not an ether.

petroleum jelly *See* Vaseline.

petrology The study of rocks, including their origin and structure.

pewter An alloy of tin, antimony and copper used for decorative purposes. Low-grade pewter contains lead in place of antimony and copper.

Pfeffer's pot A porous pot used in the measurement of osmotic pressures. The walls of the pot contain copper(II) hexacyano-ferrate(II) which acts as a membrane. If the pot is filled with a solution, e.g. of sugar, and placed in distilled water at constant temperature, the water will enter the pot and the osmotic pressure of the sugar solution can be measured by a mercury manometer. This manometer also contains nitrogen and is connected to the interior of the pot. The mercury is forced up the manometer tube, compressing the nitrogen. The reverse pressure of the nitrogen will eventually equal the osmotic pressure of the solution; an equilibrium is reached and the osmotic pressure can be read on the manometer scale.

pH *See* hydrogen ion concentration.

phage *See* bacteriophage.

phagocyte A type of white blood corpuscle (cell) which is able to engulf foreign particles such as pathogens. Examples of phagocytes in mammals are monocytes, macrophages and polymorphs. Some phagocytes are fixed, e.g. those lining the lymph nodes, the spleen and the liver. Other phagocytes may find their way out of the blood vessels through the capillary walls into the surrounding tissue where they are attracted by invading bacteria (chemo-taxis). Phagocytes therefore play an important part in an organism's defence system against infection. *See also* granulocyte.

phagocytosis ('cell-eating') The process in which white blood corpuscles (phagocytes) engulf bacteria. Amoeba feeds in this manner. *Compare* pinocytosis.

phalanges *See* phalanx.

phalanx (pl. phalanges) One of the bones of the fingers and toes of tetrapods. In humans there are fourteen phalanges in each hand and foot. *See* Fig. 162.

phantom pain In people who have had a limb amputated, e.g. a leg, there may be a feeling of pain experienced as coming from the missing leg. This is called phantom pain. It is caused by severed nerves which, when healing, send impulses to the brain. *See also* referred pain.

pharmacology The study of drugs and their action on the body.

pharmacy The preparation and dispensing of drugs. The term is also used for a shop selling drugs.

pharynx In vertebrates, a muscular tube forming the part of the alimentary canal between the mouth and the oesophagus (gullet). During contraction of the pharynx food is forced into the oesophagus, and at the same time the epiglottis closes the glottis at the entrance to the trachea. The Eustachian tubes also open into the pharynx.

phase 1. Any of the apparent changes in shape of the Moon according to the amount of sunlight illuminating it. The phases of the Moon are caused by the relative positions of the Earth, Moon and Sun. **2.** A distinct, homogeneous part of a heterogeneous system. *Examples:* petrol added to water forms a two-phase liquid system; iron filings and powdered sulphur form a two-phase solid system. An aqueous solution of sugar is a single-phase system. **3.** A particular stage reached in a cycle or other periodic system, e.g. an alternating current. Two alternating currents which have the same frequency and whose maximum and minimum values coincide are said to be 'in phase'; otherwise they are 'out of phase'.

phase contrast A technique especially used in microscopy in the biology laboratory for viewing objects which have very little contrast, e.g. living cells which one does not want to kill or damage by staining. A microscope equipped with a special objective, diaphragm and condenser makes it possible to utilise the fact that different parts of a transparent object and its surroundings have different refractive indices. Light rays passing the object will therefore move with different velocity resulting in differences in phase of the rays. These differences in phase will produce lighter and darker areas in the object viewed through the microscope.

phellem See cork.

phelloderm (secondary cortex) The inner layer of the periderm, situated beneath the cork cambium.

phellogen See cork cambium.

phenate See phenolate.

phenol 1. C_6H_5OH An aromatic, organic compound which has an enol structure. It is a white, soluble crystalline solid. Phenol is not an alcohol but a very weak acid, whose salts are called phenolates. An aqueous solution of phenol is called carbolic acid and is used as a disinfectant. Originally it was used as an antiseptic in medicine, being introduced by Joseph Lister in 1867. However, less toxic and more effective antiseptics have since been discovered. Phenol can be obtained from coal-tar or by fusion of sodium hydroxide, NaOH, with the sodium salt of benzenesulphonic acid, $C_6H_5SO_2ONa$:

$C_6H_5SO_2ONa + 2NaOH \rightarrow$
$\qquad C_6H_5ONa + Na_2SO_3 + H_2O$

The sodium phenolate, C_6H_5ONa, is then treated with a strong, dilute acid:
$C_6H_5ONa + H^+ \rightarrow C_6H_5OH + Na^+$
Commercially this method has recently been superseded by a process in which benzene is brought to react with an alkene (propene), air (oxygen) and sulphuric acid. Phenol is used in organic synthesis, e.g. in the manufacture of certain plastics. **2.** The general name for a group of aromatic, organic compounds which contain one or more hydroxyl groups, —OH, connected directly to one or more carbon atoms in the ring. Phenol itself (*see* **1.**) is a member of this group but is less water-soluble and less easy to oxidise than the other members. Phenols are used in the manufacture of dyes and certain plastics.

phenolate (phenate) A salt of the acid phenol, C_6H_5OH.

phenolphthalein An acid–base indicator often used in titrations of a weak acid with a strong alkali, giving a weak alkaline end-point. The indicator is soluble in ethanol and is colourless in solutions which have a pH below 8,2 (8.2) or above 10,0 (10.0). In the pH range 8,2–10,0 (8.2–10.0), it is pink. Phenolphthalein is also used as a laxative. It can be prepared from phenol and benzene-1,2-dicarboxylic anhydride (phthalic anhydride).

phenotype The outward appearance of an organism, determined by the interaction of its genotype (genetic constitution) and the environment.

phenyl C_6H_5— An organic radical present in thousands of aromatic compounds. *See also* aryl.

phenylamine $C_6H_5NH_2$ (aniline, aminobenzene) The simplest of the aromatic amines. It can be prepared by the reduction of nitrobenzene, $C_6H_5NO_2$, using tin and hydrochloric acid:
$C_6H_5NO_2 + 3H_2 \rightarrow C_6H_5NH_2 + 2H_2O$
Phenylamine (aniline) is a colourless, oily liquid used in organic synthesis, e.g. for making dyes. With water it gives a weak alkaline solution.

phenylethene (styrene; ethenylbenzene, vinylbenzene) A colourless, viscous liquid with the formula $C_6H_5CH\!=\!CH_2$, i.e. it is an aromatic hydrocarbon. Phenylethene polymerises easily to form poly(phenylethene) (polystyrene). It can be prepared by dehydrogenation of ethylbenzene, C_6H_5—CH_2CH_3:
C_6H_5—$CH_2CH_3 \rightarrow C_6H_5CH\!=\!CH_2 + H_2$

phenylmethanol $C_6H_5CH_2OH$ (benzyl alcohol) An aromatic alcohol which is a good solvent for many organic compounds.

pheromone A substance produced by animals which influences the development or behaviour of other animals, especially those of the same species. Pheromones may be secreted in order to attract a mate, to mark out a territory or, in the case of queen bees, to prevent the development of other queens. Pheromones are secreted by glands or are present in the urine.

philosopher's stone See alchemy.

pH indicator See indicator.

phloem Vascular tissue which serves as a transport system for food, manufactured mainly in the leaves, to all parts of the plant. In flowering plants it consists of sieve-tubes, companion cells, parenchyma and fibres.

phlogiston theory

There are two types of phloem: primary phloem, which develops from procambium, and secondary phloem, which is produced by the activity of cambium. *See* Fig. 110; *compare* xylem.

phlogiston theory This was the first scientific theory ever proposed in chemistry. It was a theory concerning combustion stated by Johann Becher in 1669, refined by Georg Stahl and eventually disproved by Antoine Lavoisier in 1774. The theory was based on the idea that combustible materials contained a substance, phlogiston, which was liberated when the material burned. This phlogiston could also be taken up by other substances. *Example:* charcoal was thought to be rich in phlogiston; when it was heated with a metal ore (poor in phlogiston), the phlogiston was transferred from the burning charcoal to the metal ore, converting it to the metal. Lavoisier pointed out that combustion was a chemical process involving a reaction between a burning substance and oxygen in the air. Oxygen had recently been discovered by Joseph Priestley.

pH meter A potentiometer recording the e.m.f. of a glass electrode placed together with a reference electrode, e.g. a calomel electrode, in a solution whose hydrogen ion concentration (pH) is to be measured.

phosgene *See* carbonyl chloride.

phosphagen *See* phosphocreatine.

phosphatase An enzyme present in nearly all living organisms which is capable of hydrolysing organic phosphate esters to release phosphate groups. Two main types of phosphatase exist, namely acid and alkaline phosphatase, distinguished by their pH optimums. ATPase is a phosphatase.

phosphate *See* phosphate(V).

phosphate(V) (orthophosphate) A salt of phosphoric(V) acid (orthophosphoric acid), H_3PO_4.

phosphate test Dilute nitric acid, HNO_3 is added to a small amount of the sample to be tested. This mixture is poured into a hot solution of ammonium molybdate(VI), $(NH_4)_2MoO_4$. A yellow precipitate with a complex composition is formed either immediately or after some time, indicating that phosphate is present in the sample.

phosphine PH_3 A colourless, flammable, poisonous gas which is slightly soluble in water and has an unpleasant smell. PH_3 is the phosphorus analogue to ammonia, NH_3. When impure, PH_3 is spontaneously flammable. It can be prepared by heating a strong solution of sodium hydroxide, NaOH, with white phosphorus, P_4:
$P_4 + 3NaOH + 3H_2O \rightarrow PH_3 + 3NaH_2PO_2$

PH_3 is a much stronger reducing agent than ammonia, NH_3.

phosphite *See* phosphonate.

phosphocreatine (phosphagen) A substance found in the resting muscles of vertebrates. It is formed from creatine, phosphoric acid and energy, the latter being provided by carbohydrate breakdown (glycolysis).

phosphonate (phosphite) A salt of phosphonic acid (phosphorous acid), H_3PO_3.

phosphonic acid H_3PO_3 (phosphorous acid, orthophosphorous acid) A moderately strong, colourless, solid, deliquescent acid whose salts are called phosphonates (phosphites). It can be prepared by treating phosphorus trichloride, PCl_3, with water:
$PCl_3 + 3H_2O \rightarrow H_3PO_3 + 3HCl$
When heated, H_3PO_3 decomposes into phosphine, PH_3, and phosphoric(V) acid, H_3PO_4:
$4H_3PO_3 \rightarrow PH_3 + 3H_3PO_4$
In H_3PO_3, only two of the hydrogen atoms are acidic; the third has a direct bond to the phosphorus atom. Both the acid and its salts are strong reducing agents.

phosphor Any substance capable of fluorescence or phophorescence. *See also* scintillator.

phosphor bronze A hard alloy of copper, tin and very small amounts of phosphorus. It is used for making springs and bearings.

phosphorescence A kind of luminescence whereby light is emitted from certain substances and persists for some time after these are irradiated by light or certain other radiations. The absorbed light is usually emitted at a greater wavelength than the incident light. Examples of phosphorescent substances are white phosphorus, zinc sulphide and calcium sulphide. *See also* fluorescence *and* bioluminescence.

phosphoric acid The name phosphoric acid is usually applied to H_3PO_4; however, several phosphoric acids exist, e.g. phosphoric(V) acid (orthophosphoric acid), H_3PO_4, metaphosphoric acid, HPO_3, and pyrophosphoric acid, $H_4P_2O_7$.

phosphoric(V) acid H_3PO_4 (orthophosphoric acid) A colourless solid which is very soluble in water, forming a viscous solution. The acid is rather strong. It may be prepared from white phosphorus, P_4, treated with concentrated nitric acid, HNO_3:
$P_4 + 10HNO_3 + H_2O \rightarrow 4H_3PO_4 + 5NO + 5NO_2$
or by reacting phosphorus(V) oxide, P_4O_{10}, with water:
$P_4O_{10} + 6H_2O \rightarrow 4H_3PO_4$.
Salts of H_3PO_4 are called phosphate(V)s (orthophosphates). They are used in water treatment and as fertilisers. H_3PO_4 forms three series of salts containing the following anions

respectively: alkaline phosphate(V)s, PO_4^{3-}, acidic dihydrogenphosphate(V)s, $H_2PO_4^-$, and neutral hydrogenphosphate(V)s, HPO_4^{2-}.

phosphorus An element with the symbol P; atomic number 15; relative atomic mass 30,97 (30.97); state, solid. Phosphorus is a nonmetallic element occurring in several minerals, e.g. apatite from which it is extracted. Calcium phosphate(V), $Ca_3(PO_4)_2$, is heated with sand, SiO_2, and coke, C, in an electric furnace and is then reduced according to the following reaction:
$$2Ca_3(PO_4)_2 + 6SiO_2 + 10C \rightarrow$$
$$P_4 + 6CaSiO_3 + 10CO$$
The two main allotropes of phosphorus are white (yellow) and red phosphorus. Several other less important allotropes are known, e.g. black and violet phosphorus. Yellow phosphorus is white phosphorus which contains small amounts of red phosphorus. White phosphorus and red phosphorus have quite different properties: white phosphorus is poisonous, soluble in carbon disulphide, CS_2, and spontaneously flammable at room temperature, whereas red phosphorus is none of these. When white phosphorus is heated in the absence of air (oxygen), red phosphorus is formed. When the red or the white allotrope is heated in air, phosphorus(V) oxide, P_4O_{10}, is formed:
$$P_4 + 5O_2 \rightarrow P_4O_{10}$$
Molecules of both white and red allotropes are represented by the formula P_4. Phosphorus is used in making phosphoric(V) acid, H_3PO_4, and in making insecticides, matches, phosphor bronze, etc.

phosphorus chlorides Two important chlorides of phosphorus are phosphorus trichloride, PCl_3 and phosphorus pentachloride, PCl_5.
Phosphorus trichloride, PCl_3, is a colourless liquid which can be prepared by passing chlorine, Cl_2, over white phosphorus, P_4:
$$P_4 + 6Cl_2 \rightarrow 4PCl_3$$
PCl_3 hydrolyses to give phosphonic acid (phosphorous acid), H_3PO_3, and hydrogen chloride, HCl:
$$PCl_3 + 3H_2O \rightarrow H_3PO_3 + 3HCl$$
Phosphorus trichloride is used in organic synthesis. It reacts with many compounds containing the hydroxyl group, —OH, which it replaces with a chlorine atom.
Phosphorus pentachloride, PCl_5, is a white solid which can be prepared by passing chlorine, Cl_2, over phosphorus trichloride, PCl_3:
$$PCl_3 + Cl_2 \rightarrow PCl_5$$
PCl_5 hydrolyses to give phosphorus trichloride oxide (phosphorus oxychloride), $POCl_3$, and hydrogen chloride, HCl:
$$PCl_5 + H_2O \rightarrow POCl_3 + 2HCl$$

$POCl_3$ reacts with excess water to give phosphoric(V) acid, H_3PO_4, and hydrogen chloride, HCl:
$$POCl_3 + 3H_2O \rightarrow H_3PO_4 + 3HCl$$
Phosphorus pentachloride has almost the same uses as phosphorus trichloride.

phosphorus(V) oxide P_4O_{10} (phosphorus pentoxide) A white solid which can be prepared by burning phosphorus in oxygen:
$$P_4 + 5O_2 \rightarrow P_4O_{10}$$
It readily combines with water to form phosphoric(V) acid, H_3PO_4:
$$P_4O_{10} + 6H_2O \rightarrow 4H_3PO_4$$
and is therefore commonly used as a drying agent. However, it should be noted that phosphoric(V) acid cannot be dehydrated to form P_4O_{10}. The name phosphorus pentoxide originates from its empirical formula, P_2O_5.

phosphorus pentachloride *See* phosphorus chlorides.

phosphorus pentoxide *See* phosphorus(V) oxide.

phosphorus trichloride *See* phosphorus chlorides.

phosphorylase An enzyme capable of catalysing phosphorylation.

phosphorylation An enzymatic process in which an organic substance, e.g. a sugar, combines with one or more phosphate groups. The necessary phosphate and the energy for this process are derived from the splitting of ATP. Phosphorylation therefore results in an increase in the energy level of the sugar molecules so that useful energy can be liberated from them later.

photo- Prefix meaning light.

photoautotroph(e) An organism which uses inorganic material, e.g. carbon dioxide, as its main source of carbon, and light as its energy source. Photoautotrophs include all green plants and some bacteria. *See also* chemoautotroph(e).

photocathode A cathode emitting electrons as a result of the photoelectric effect.

photocell *See* photoelectric cell.

photochemical reaction A chemical reaction initiated by light of a particular wavelength. *Examples:* in photography, light-sensitive silver bromide is activated by incident light; a mixture of hydrogen, H_2, and chlorine, Cl_2, produces hydrogen chloride, HCl, when exposed to light:
$$H_2 + Cl_2 \rightarrow 2HCl$$
See also photosynthesis.

photochromic substance (phototrophic substance) A substance, e.g. a dye, which darkens when exposed to light. This change may be reversible. Photochromic glass contains extremely small crystals of silver chloride, silver bromide or silver iodide

photoconductivity

dispersed throughout a special type of glass. Silver chloride is especially sensitive to ultraviolet radiation; when it is struck by this radiation the silver chloride splits into the elements silver and chlorine. The silver causes the darkening. The chlorine does not react with the glass and cannot penetrate it, i.e. it stays in the reaction zone. When the ultraviolet radiation ceases, the silver and chlorine recombine forming silver chloride again and the glass clears. Photochromic glass is used in sunglasses, in windows, in bottles for light-sensitive drugs and in special optical systems for information storage and display.

photoconductivity The property possessed by certain materials, usually semiconductors, by which their electrical conductivity increases when there is an increase in the intensity of electromagnetic radiation incident on them. This is caused by an increased absorption of photons passing on more energy to electrons and promoting them to the conduction band.

photoelectric cell (photocell) A device for the detection and measurement of light. Its action is based on photoconductivity, the photoelectric effect or the photovoltaic effect. In a photoelectric cell whose action depends on the photoelectric effect, the photocathode is placed inside an evacuated glass bulb which may contain a noble gas at very low pressure. The light-sensitive layer on the photocathode may be made of the metal caesium. A potential is applied to the cell. When light of a suitable wavelength passes through the window, electrons are emitted from the metal and are attracted by the anode, thus giving rise to a current which can be detected by a galvanometer. The intensity of the incident light is proportional to the strength of the current. Photoelectric cells have a wide range of uses, e.g. as exposure meters, in photometers, in burglar alarms and in solar cells. See Fig. 166; see also selenium cell.

photoelectric effect The liberation of electrons from the surface of a conductor when it is struck by electromagnetic radiation of a certain wavelength. Usually the radiation is visible light or radiation of a shorter wavelength. The number of electrons released depends on the intensity of the incident radiation and not on its frequency. For a given material, the wavelength of the incident light must be shorter than a critical value (for emission of electrons) whose corresponding frequency f_0 is called the threshold frequency. The kinetic energy of a liberated electron (photoelectron) of mass m and velocity v is $\tfrac{1}{2}mv^2$. The work done when the electron is released is Φ (phi). Einstein found that
$$\tfrac{1}{2}mv^2 = hf - \Phi$$

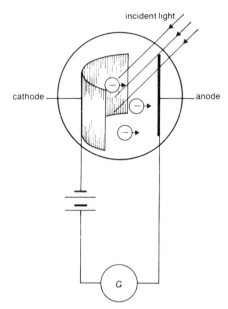

Fig. 166 Photoelectric cell

where h is Planck's constant and f is the frequency of the electron-emitting radiation. For $hf < \Phi$, there is no emission; if $hf = \Phi$, then $f = f_0$.

photoelectron An electron emitted from the surface of a material as a result of the photoelectric effect or by photoionisation.

photography The reproduction of an object on a film or other suitable material by means of a camera. The film is coated with a light-sensitive emulsion of a silver salt, e.g. silver bromide. Light passing through the camera lens and striking the film causes an activation of grains of silver bromide depending on the intensity of the incident light. During 'developing', when the film is placed in a solution of a reducing agent, these activated grains are more easily reduced to black, metallic silver than the non-activated grains, so black areas appear on the film where the incident light was strongest. Since silver bromide is sensitive to light, the non-activated silver bromide must be removed, otherwise the whole film would turn black eventually. This is done using a strong solution of sodium thiosulphate which dissolves and deactivates silver bromide. This is called 'fixing' and results in the formation of a negative. From the negative, a positive (photograph) can be made by placing the negative above a sheet of light-sensitive paper and exposing it to light. The positive is developed and fixed in the same manner as the negative.

photoheterotroph(e) An organism using organic material as its main source of carbon and light as its energy source. Only a few organisms are photoheterotroph(e)s, e.g. the non-sulphur purple bacteria. *See also* chemoheterotroph(e).

photoionisation The ionisation of an atom or molecule as a result of electromagnetic radiation. Each photon will have the energy hf, where h is Planck's constant and f is the frequency of the radiation. Photons which have energies greater than the ionisation energy (potential) of the irradiated atom or molecule may cause ionisation.

photolysis The decomposition or dissociation of a substance when struck by electromagnetic radiation. *Example:* radiation with a wavelength less than 480 nanometres (nm) may break the chemical bond between the two atoms in a chlorine molecule, Cl_2:

$$Cl_2 \rightarrow 2Cl$$

i.e. the photon energy exceeds the strength of this bond.

photometer An instrument for measuring luminous intensity. It is widely used in quantitative analysis.

photometry The measurement of luminous intensity using a photometer.

photomicrograph *See* micrograph.

photomultiplier An electronic device (a photoelectric cell) used to detect very small quantites of light or other electromagnetic radiation with a greater frequency than light. It consists of a row of electrodes, called dynodes, arranged between an anode and a cathode contained in an evacuated glass tube. Each dynode has a special coating. The cathode emits electrons when radiation falls on it (primary emission). These electrons strike the first dynode, which ejects more electrons (secondary emission). These electrons are accelerated to the second dynode and even more electrons are emitted. This process is repeated at successive dynodes of which there may be ten or more. The secondary emission of electrons from the last dynode produces a current which is large enough to be detected and measured. Photomultipliers are used to record scintillations produced by high-energy particles, as sensitive photometers and in television cameras.

photon A single quantum of electromagnetic radiation. Its energy is given by the product hf, where h is Planck's constant and f the frequency. Sometimes it may be useful to consider the photon as a particle.

photonasty A plant movement in response to a change in light intensity. *Example:* some flowers open during daylight and close at night.

photoperiodism The influence of the relative length of day and night on the activities of plants. The length of day is known as the photoperiod. Flowering is an example of an activity which may be controlled by photoperiodism. However, the photoperiod also controls other plant responses. Long-day plants, e.g. wheat and radish, only flower when the days are longer than a certain critical length. Short-day plants, e.g. chrysanthemum and soybean, only flower when the days are shorter than a certain critical length. Day-neutral plants, e.g. tomato and tobacco, flower whatever the day-length, i.e. they are virtually unaffected by the photoperiod.

photosphere The visible surface of the Sun. It is a layer of dense gas with a thickness of about 300 km and a temperature ranging from approximately 4300°C (at its extremity) to 9000°C. Almost all the Sun's radiation is emitted from the photosphere.

photosynthesis A photochemical reaction in which green plants manufacture organic compounds using light, chlorophyll, carbon dioxide, CO_2, and water. The process is complex, but may be represented by the equation:

$6CO_2 + 12H_2O \rightarrow C_6H_{12}O_6 + 6H_2O + 6O_2$

showing that water and oxygen, O_2, are produced together with carbohydrate during the reaction. Some bacteria (photosynthetic bacteria) are able to carry out photosynthesis. These sometimes use compounds other than water as their source of hydrogen, e.g. hydrogen sulphide, H_2S, producing waste products other than oxygen, e.g. sulphur.

photosynthetic bacteria *See* photosynthesis.

phototaxis Locomotory movement of an organism towards or away from a source of light.

phototrophic substance *See* photochromic substance.

phototrophism A type of nutrition performed by photoautotrophs and photoheterotrophs.

phototropism (heliotropism) Plant growth in response to the stimulus of light.

photovoltaic effect The production of an e.m.f. across the junction of two dissimilar materials, e.g. copper(I) oxide and copper, or selenium and gold, when struck by electromagnetic radiation such as light. This effect is used in photoelectric cells such as exposure meters.

pH scale *See* hydrogen ion concentration.

phthalate *See* benzene-1,2-dicarboxylate.

phthalic acid *See* benzene-1,2-dicarboxylic acid.

phthalic anhydride Benzene-1,2-dicarboxylic anhydride. *See* benzene-1,2-dicarboxylic acid.

Phycomycetes A class of fungi including many well-known parasites and moulds, e.g. *Mucor*. They reproduce sexually by conjugation and may reproduce asexually by zoospores.

phyllotaxis The arrangement of leaves on a stem. *Example:* leaves may be alternate.

phylum (pl. phyla) One of the major groups used in the classification of living organisms. A phylum consists of similar classes. In plant classification, the term division is often used instead of phylum.

physical change A change in which no new substances are formed and which is usually easily reversible. Often a physical change only involves a change of state. *Compare* chemical change.

physics The study of the properties of matter and energy without any reference to chemical changes.

physiological saline A salt solution which is isotonic with a body fluid of an animal and is used to keep cells alive *in vitro* for some time. The simplest physiological saline is an 0,9% (0.9%) aqueous solution of sodium chloride. However, most cells require more than one salt to survive, so most solutions contain several. *Example:* Ringer's solution for insects contains certain concentrations of sodium chloride, potassium chloride, calcium chloride and magnesium chloride, plus sodium hydrogencarbonate and sodium dihydrogenphosphate(V), which act as buffers to keep the correct pH. Ringer's solutions for amphibians, mammals, etc., have other compositions. In order to prolong the survival of cells in physiological saline, nutrients may also be added. *See also* tissue culture.

physiology The study of the functions and activities of organisms.

physiotherapy The treatment of a disease, injury or disability by physical means such as massage, exercise and heat treatment.

phyto- Prefix meaning plant.

phytokinin *See* cytokinin.

phytoplankton *See* plankton.

pi (π) The sixteenth letter of the Greek alphabet.

pia mater The thin inner membrane enveloping the brain and spinal cord. *See also* meninges.

pico- Symbol: p. A prefix meaning 10^{-12}.

piezo-electric effect The production of a small potential difference when certain crystals, e.g. quartz and Rochelle salt, are distorted by pressure. Conversely, if a potential difference is applied to such crystals, they distort slightly. This effect is used in gas lighters, microphones, crystal clocks, etc.

pig-iron *See* cast iron.

pigment A colouring-matter occurring in animals and plants, e.g. melanin and chlorophyll. The term is also applied to other colouring-matters used in dyes and paints.

piles *See* haemorrhoids.

pileus The umbrella-shaped cap of a mature mushroom.

piliferous layer The part of the epidermis of a root bearing root hairs.

pill, the An oral contraceptive for women. The pill contains female sex hormones, e.g. progesterone or oestrogen, either singly or in combination. When taken on a regular schedule (the packets of pills carry a precise instruction), ovulation is avoided. This method of contraception is regarded 100% safe, and since it was first introduced in the beginning of 1960 only a few serious side-effects have been recorded. These include an increased risk of blood clotting in the blood vessels of the brain or in the deep veins of the legs and pelvis. This risk can be reduced by using so-called minipills or micropills with a very low concentration of hormones, but this may increase the risk of pregnancy a little.

pineal body (pineal gland) In vertebrates, a gland which is believed to have an endocrine function as it has not yet definitely been proved to produce any hormones. In humans the pineal body develops as an outgrowth from the third ventricle of the brain and lies under the posterior part of the corpus callosum. It is a small cone-shaped structure, the glandular tissue of which is gradually replaced by connective tissue after puberty.

pineal gland *See* pineal body.

pinene *See* turpentine.

pinhole camera *See* camera.

pink-eye *See* conjunctivitis.

pinking *See* knocking.

pinna (pl. pinnae) A leaflet on a pinnate leaf. *See* Fig. 68; *see also* auricle.

pinnate Describes a compound leaf which has leaflets arranged on each side of a common petiole.

pinocytosis ('cell-drinking') The process by which cells ingest the liquid (solution) surrounding them through the cell membrane into the cytoplasm. *Compare* phagocytosis.

pipeclay triangle A triangular piece of iron wire covered with pipeclay. It is used in the laboratory as a crucible support when placed on a tripod. *See* Fig. 167.

pipette A glass tube for measuring and transferring volumes of liquids. Two common types are shown in Fig. 168(a) and (b). The first two pipettes are graduated to measure several volumes in the range 0–10 ml. The third pipette bears only one graduation and can therefore measure one volume (20 ml) only. A range of pipettes is manufactured, measuring volumes varying from less than 1 ml to 100 ml.

pituitary gland (body)

Fig. 167 Pipeclay triangle

pistil An alternative name for the gynoecium or each separate carpel.

piston A moving part of pumps and engines. It consists of a round plate or cylinder which is attached to a piston rod and fits closely inside another cylinder in which it moves. In the internal combustion engine pistons move under fluid pressure to drive the engine; in the force and lift pumps they are used to pump liquids.

pit A thin region (depression) in a cell wall of plant tissue where a secondary cell wall has not been laid down, i.e. only the primary wall remains. A pit in one cell wall is usually exactly adjacent to a pit in a neighbouring cell wall, and the thinness of these regions permits the passage of water in and out of the cell. In xylem tissue, the pits are bordered by a rim of lignin where the secondary wall has partially overgrown the pits. These pits are known as bordered pits. In certain plants they contain a thickened centre rather like a plug. This plug is known as a torus. It may function to control the movement of water between cells.

pitch 1. The black residue from the destructive distillation of wood or coal-tar. It is hard when cold and a viscous liquid when hot. Pitch is insoluble in water and is used to seal joints in roofs, in road-paving, etc. **2.** (of a note) The quality of a sound which depends on the frequency of its source. A high-pitched note has a high frequency, a low-pitched note has a low frequency. **3.** (of a screw) The distance between consecutive threads. *See* Fig. 169.

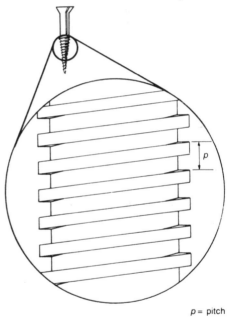

p = pitch

Fig. 169 Pitch of a screw

pitchblende A mineral used as a source (ore) of uranium and radium. It can be represented by the formula U_3O_8. In 1902, Marie and Pierre Curie extracted one-tenth of a gram of radium from several tons of pitchblende.

pith The central, spongy region (medulla) of a plant stem, consisting of parenchyma and surrounded by vascular tissue. Pith may also be found in roots. It functions as a food store.

pituitary gland (body) (hypophysis) A small organ attached to the floor of the diencephalon of the brain by a funnel-shaped stalk called the infundibulum. The pituitary gland is an

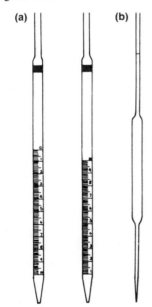

Fig. 168 Pipettes

endocrine gland producing several hormones such as ACTH and prolactin from its anterior lobe. Several other hormones such as oxytocin and vasopressin are released from its posterior lobe, but are produced elsewhere. This gland also regulates the action of other endocrine glands. *See* Fig. 16.

pivot *See* fulcrum.

placebo A substance that has no medical effect towards a certain disease. Placebos may be used in the testing of a new drug. Volunteer patients taking part in a controlled therapeutic trial are randomly divided into two groups. One group of patients will receive placebos and the other group the new drug. However, neither group knows who gets the drug. Attempts are made to give the placebo side effects similar to those produced by the active drug. Sometimes the idea that he has been given a powerful drug (medicine) makes a patient who is given a placebo feel better. In double blind trials, neither the patients nor the doctors know who has received the placebo.

placenta 1. In plants, the part of the ovary wall to which the ovules are attached. 2. In pregnant mammals, the spongy, vascular organ which is situated and develops where the fertilised egg cell first implants itself in the wall of the uterus. It allows exchange of nutritive material and respiratory substances between the mother and the foetus. The placenta is attached to the foetus by the umbilical cord, but there is no direct connection between the blood of the mother and the foetus. Hormones are produced by the placenta, some of which prevent further ovulation and menstruation during pregnancy. After the birth of the offspring, the placenta is discharged together with the foetal membranes.

placental Pertaining to mammals that develop a placenta. *See also* Metatheria *and* Monotremata.

Placentalia (Eutheria) A sub-class of Mammalia comprising the majority of mammals, including human beings. Characteristics include a placenta attaching the embryo to the uterus, full development of the embryo in the uterus and a well-developed brain.

plague *See* bubonic plague.

planarians *See* Turbellaria.

Planck's constant Symbol: h. A fundamental constant relating the frequency, f, of an electromagnetic radiation to its quantum energy, E; i.e.
$$E = hf$$
This expression is known as Planck's law of radiation. The value of the constant is $6,626 \times 10^{-34}$ joule second (6.626×10^{-34} J s).

Planck's law of radiation *See* Planck's constant.

plane-polarised light *See* polarisation.

planet A relatively small heavenly body which orbits a star and does not emit its own light as a star does. In our Solar System there are nine planets orbiting the Sun and reflecting its light. In order of proximity to the Sun, they are Mercury, Venus, Earth, Mars, Jupiter, Saturn, Uranus, Neptune and Pluto.

planetoid *See* asteroid.

plankton Minute animals and plants floating freely near the surface in seas, lakes, rivers, etc. Plant plankton is called phytoplankton and serves as food for animal plankton, called zooplankton. Plankton form the basis of many foodchains and are of great ecological importance. *See also* diatom.

plant cell *See* cell.

plaque A film of mucus on teeth formed from certain bacteria in the mouth when they feed on the residue of carbohydrates. If the plaque is not removed, harmful acids produced by the bacteria attack the tooth enamel. *See also* caries.

plasma *See* blood plasma.

plasma membrane *See* cell membrane.

plasmodesmata In plant cells, cytoplasmic bridges consisting of very fine threads connecting adjacent cells by passing through the cell walls. *See* Fig. 205.

plasmolysis The contraction of cytoplasm away from the wall of a plant cell due to the withdrawal of water by osmosis.

plaster of Paris $CaSO_4 \cdot \frac{1}{2}H_2O$ Partly-dehydrated gypsum, $CaSO_4 \cdot 2H_2O$. When mixed with water, it expands slightly to form gypsum and sets as a rather hard mass. It is used for making casts and moulds and in bandages to prevent movement of broken limbs.

plastic 1. A general term for various organic polymers, usually synthetic, which at some stage in their manufacture become plastic under the influence of heat and pressure. Plastics are either thermoplastic or thermosetting. Thermoplastics become soft when heated and can be moulded and remoulded; thermosetting plastics become hard when heated and cannot be softened. Thermoplastics have a molecular chain structure, whereas thermosetting plastics have a structure of interlinked chains. *See* Fig. 170 (a) *and* (b). 2. *See* plasticity.

plasticity The ability of a substance to retain its shape after being distorted by a force. Substances with this ability are said to be plastic. *Compare* elasticity.

plastic sulphur An unstable form of sulphur obtained as a soft, brown solid when a thin stream of boiling sulphur is poured into cold water. Because of the rapid cooling, no S_8

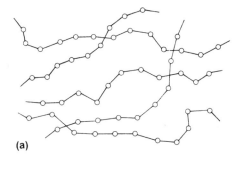

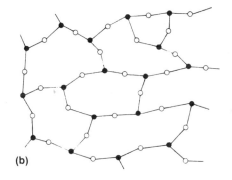

Fig. 170 Arrangement of atoms in plastics
(a) Thermoplastic
(b) Thermosetting plastic

molecules have time to form and open chains of sulphur atoms are produced. However, after a short time S_8 molecules begin to form and the plastic sulphur changes gradually to rhombic sulphur.

plastic surgery The reconstruction and replacement of damaged, lost or deformed parts of the body, mainly by the transfer of tissue, e.g. a skin graft.

plastid An organelle in the cytoplasm of plant cells which contains pigments or nutritive material. Examples of plastids are leucoplasts, chloroplasts and chromoplasts.

plastron See Chelonia.

platelet See blood platelet.

plating See electroplating.

platinum An element with the symbol Pt; atomic number 78; relative atomic mass 195,09 (195.09); state, solid. Platinum is a heavy transition metal, occurring naturally either free or in various minerals, e.g. ores of nickel-copper sulphide. The most important source from which it is extracted is sperrylite, $PtAs_2$. It is often found associated with other platinum metals. The metal is rather inert: it is not attacked by air, water or any strong acids except aqua regia. However, fused alkali oxides and, even more, alkali peroxides attack it, and so do fluorine and chlorine at red heat. Platinum is capable of absorbing large volumes of molecular hydrogen. It is used for electrical contacts, for printed circuitry, for plating, as a catalyst, for jewellery, as an electrode material, etc. As platinum has almost the same coefficient of expansion as glass, platinum wire can be inserted through soft glass without the glass cracking when it is cooled. Platinum is also used for coating missile nose cones and for jet engine fuel nozzles because it is unchanged for long periods of time at high temperatures.

platinum black A black form of platinum obtained when certain salts of the metal are reduced. It is widely used as a catalyst and as a gas absorber.

platinum metals Six transition metals (elements) with related chemical properties. These metals are: ruthenium, Ru, rhodium, Rh, palladium, Pd, osmium, Os, iridium, Ir, and platinum, Pt. Iridium is the most corrosion-resistant metal known.

platinum resistance thermometer See resistance thermometer.

Platyhelminthes A phylum of triploblastic Metazoa, commonly called flatworms, including Cestoda (tapeworms), Trematoda (flukes) and Turbellaria (planarians). Characteristics include bilateral symmetry, no coelom, no anus, no blood system, a complex hermaphrodite reproductive system, and a gut, if present, with only one opening.

Pleistocene A relatively recent geological period of approximately 1½ million years which ended approximately 10 000 years ago. Four ice ages occurred during this period.

pleura In mammals, a double, serous membrane covering the lungs and lining the thorax. The pleura secretes a fluid which acts as a lubricant, allowing the lungs to move without friction. See Fig. 118.

pleural cavity The potential space between the pleura covering the lungs and the pleura lining the thorax. In health, there is no actual space: however, in disease, fluid may collect in the cavity. See also pleurisy.

pleurisy Inflammation of the pleura. Symptoms include fever, coughing, chest pains and shallow breathing. The inflammation may be caused by the production of little or no fluid by the pleura (dry pleurisy) or may be accompanied by the production of excess fluid (wet pleurisy). Pleurisy caused by bacteria can be treated with antibiotics.

plexiglass See Perspex.

plexus See solar plexus.

pliable Describes a material which bends easily.

Plimsoll line (mark) A safety marking painted on both sides of a ship to indicate the deepest level to which it can be safely loaded in different climates and seas. In Fig. 171, freshwater marks are on the left and salt-water marks are on the right of the Plimsoll line. The letters LR represent Lloyd's Register of Shipping, TF tropical fresh water, F fresh water, T tropics, S summer, W winter and WNA winter in the North Atlantic.

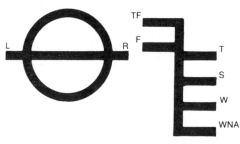

Fig. 171 Plimsoll line

Pliocene A geological period of approximately 5½ million years which ended approximately 1½ million years ago. It is a sub-division of the Tertiary period.

plumbago *See* graphite.

plumbic Term formerly used in the names of lead compounds to indicate that the lead is tetravalent; e.g. plumbic acetate, $(CH_3COO)_4Pb$, now called lead(IV) ethanoate. *Compare* plumbous.

plumbous Term formerly used in the names of lead compounds to indicate that the lead is divalent; e.g. plumbous oxide, PbO, now called lead(II) oxide. *Compare* plumbic.

plumbum The Latin name for lead.

plumule 1. The terminal bud in the embryo of a mature seed. It develops into the shoot. **2.** A down feather.

Pluto One of the nine planets in our Solar System. It is the outermost planet, orbiting the Sun once in 247,7 years (247.7 years). Pluto has one satellite (moon).

plutonium A transuranic element (metal) with the symbol Pu; atomic number 94; relative atomic mass, 242; state, solid. It belongs to the actinoids and is made by neutron bombardment of uranium. Plutonium is used in nuclear reactors and nuclear weapons.

pneumatic Operated by air pressure.

pneumonia An acute inflammation of the lungs caused by many different bacteria and virus. Symptoms include fever, pains in the chest and coughing. Bacterial pneumonia is treated with antibiotics; viral pneumonia seldom responds to this treatment and is therefore dangerous, requiring intensive hospital care.

pneumonic plague *See* bubonic plague.

***p-n-p* transistor** *See* transistor.

pod *See* legume.

podagra *See* uric acid.

poikilothermic Cold-blooded.

Poiseuille's formula A formula expressing the volume (flow), v, of a liquid running out of a capillary tube in a certain time. E.g. in a time of 1 second,

$$v = \frac{\pi p r^4}{8l\eta}$$

in which p is the difference of pressure at the ends of the tube, r its radius, l its length and η the coefficient of viscosity (dynamic viscosity). *See also* Stokes' law.

poison 1. A substance which destroys the activity of a catalyst. **2.** Any substance causing damage or death to a living organism when introduced into it.

polar bond *See* ionic bond.

polarimeter An optical instrument used in qualitative and quantitative analyses for the measurement of the amount of rotation of the plane of polarised, monochromatic light. *See also* optical activity.

polarisation 1. In an electric cell, the reduction of the e.m.f. mainly caused by the formation of small bubbles of gas on the electrodes. This changes the nature of the electrodes, e.g. reduces their surface area, and produces a back e.m.f. Polarisation can be prevented by adding a depolariser to the electrolyte. **2.** (of light) A restriction of the direction of vibration of the transverse waves that constitute light. Ordinary light is made up of transverse waves with vibrations in all possible directions in a plane perpendicular to the direction of the waves. If the vibrations are restricted to one direction only, then the light is called plane-polarised light. This can be done by passing a ray of unpolarised light through a polariser such as a Nicol prism. However, plane-polarised light may also be produced by reflection. Plane-polarised light has many uses, e.g. in chemical analysis and to measure the distribution of stress in materials. *See also* Polaroid.

polariser *See* Nicol prism, Polaroid *and* polarisation.

polarity 1. The distinction between positive and negative charges, terminals, potentials, etc. **2.** The distinction between the north and south magnetic poles of a magnet.

polar molecule A molecule with a positive and a negative terminal. *Example:* in water, H_2O, the oxygen atom is more electronegative than the hydrogen atoms. It therefore pulls the electrons of the covalent bonds towards it, giving rise to a dipole. Polar molecules are important because they affect chemical properties such as solvent action. A polar molecule should not be confused with a dipolar ion (*see* zwitterion). *Compare* non-polar molecule; *see also* polar solvent.

Polaroid® A thin transparent film made of sheets of plastic containing many minute double refracting crystals aligned with their axes parallel. Polaroid polarises ordinary light and produces plane-polarised light. It is used in sunglasses and in camera filters to reduce glare.

polar solvent A solvent such as water, H_2O, which has highly-polar molecules, i.e. molecules which have a high dipole moment. It is a good solvent for ionic and polar solutes, e.g. most salts and sugars, but generally a poor solvent for covalent solutes which are non-polar, e.g. paraffin wax and petroleum jelly (vaseline). *Compare* non-polar solvent.

pole 1. In animal or plant cells, one end of the spindle that forms between the centrioles during mitosis and meiosis. **2.** One of two regions in a magnet where the magnetic force is concentrated. Lines of force of the magnetic field surrounding the magnet either converge to or diverge from a pole. A freely-suspended bar magnet comes to rest with its north pole pointing towards the Earth's magnetic North Pole and its south pole pointing towards the magnetic South Pole. Unlike poles attract and like poles repel each other. *See also* magnetism, terrestrial. **3.** (of a mirror) *See* mirror. **4.** One of the terminals (positive or negative) of a battery, circuit or other electrical apparatus.

poliomyelitis (polio) An acute, infectious disease caused by a group of viruses affecting the central nervous system and sometimes leading to paralysis. The disease may attack people of any age and epidemic outbreaks may occur in late summer and autumn. Symptoms include fever, headache, vomiting, stiff neck, sore throat and sometimes diarrhoea. There is no specific treatment for the disease, but most cases recover in a short time. The disease can be prevented by immunisation.

pollen The powdery substance produced by the anthers of flowers. It is usually yellow in colour and consists of pollen grains containing male gametes. Pollen is carried by insects, wind or (less frequently) water during pollination. Insect-pollinated flowers produce sticky pollen, whereas wind-pollinated flowers produce smooth, light pollen.

pollen flower A flower which produces no nectar, but attracts insects by producing large amounts of pollen.

pollen grain *See* pollen.

pollen sac A cavity in an anther in which pollen is produced.

pollen tube A fine tube growing from a pollen grain when it germinates on a stigma. The pollen tube grows through the style and carries the male gametes to the ovary where it enters an ovule via the micropyle and releases its gametes.

pollex (pl. pollices) In tetrapod vertebrates, the innermost digit of the fore-limb. It is the thumb in humans and consists of two phalanges.

pollination The transfer of pollen from an anther to a stigma prior to fertilisation. *See also* cross-pollination *and* self-pollination.

pollution Contamination of the environment by Man. Pollution interferes with food-chains, health, the growth of species, etc. Its causes include the discharge of industrial by-products into air, water or soil, and unsatisfactory disposal of waste or sewage.

poly- Prefix meaning many.

polyamide A polymer characterised by the linkage

—C—N—
‖ |
O H

found between the monomers from which it is formed. These monomers are diamines and dibasic acids. Nylon is the best known polyamide.

polybasic Describes an acid with many hydrogen atoms which can be replaced by a metal or a metal-like group such as the ammonium group, NH_4^+, thus forming salts. *Example:* citric acid (2-hydroxypropane-1,2,3-tricarboxylic acid),

HOOC—CH_2—COH—CH_2—COOH
|
COOH

is a polybasic acid.

poly(chloroethene) *See* PVC.

polychromatic radiation Light or any other electromagnetic radiation consisting of a mixture of colours or wavelengths. *Compare* monochromatic radiation.

polyester A polymer obtained from the condensation polymerisation of polybasic carboxylic acids and polyhydric alcohols. Terylene is an example of a polyester.

poly(ethene) *See* polythene.

polyethylene *See* polythene.

polygamous 1. Bearing male, female and hermaphrodite flowers on the same or different plants. **2.** Having more than one mate.

polyhydric Describes a molecule containing several hydroxyl groups, —OH. *Example:* propane-1,2,3-triol (glycerol), $C_3H_5(OH)_3$, is a polyhydric alcohol.

polymer A substance with very large molecules which are built up from a series of small basic units (monomers). *Example:* a polysaccharide is built up from many units of monosaccharide. The term is most often applied to a plastic. Owing to variation in the lengths of the chains of molecules forming polymers, polymers do not have fixed relative molecular masses.

polymerisation The formation of a polymer from monomers. Polymerisation is usually divided into two types: addition polymerisation and condensation polymerisation. In addition polymerisation, the polymer is built up from the monomers without any elimination of other molecules. This means that the empirical formulae of both monomer and polymer are the same. Poly(ethene) (polythene) is an example of a polymer produced in this manner. In condensation polymerisation, a small molecule is eliminated, usually water. Terylene is an example of a polymer produced in this manner.

poly(methyl methacrylate) See Perspex.

poly(methyl 2-methylpropenoate) See Perspex.

polymorph See granulocyte.

polymorphism 1. The existence of different types of individuals within the same species. 2. The occurrence of different forms, or different forms of organs, during the life cycle of an individual. 3. The ability of chemical elements and compounds to crystallise in more than one form. Allotropy is a special type of polymorphism as it is not restricted to crystalline form. Each polymorphic form of a substance is thermodynamically stable within a certain range of temperature and pressure. The transformation of one polymorphic form to another takes places at any given pressure at a fixed temperature called the transition temperature or transition point. *See also* monoclinic sulphur.

polyoestrous See oestrus.

polyp 1. A small tumour of the mucous membrane lining the nose, bladder, intestine or womb. Polyps may cause obstructions. They are usually benign; however, some may develop into malignant tumours if not removed. 2. An individual member of a colony of sedentary coelenterates. Polyps have a cylindrical body with a mouth and tentacles at one end. They can reproduce both sexually and asexually (by budding). For some coelenterates, the polyp is a stage in their life cycle. *See also* medusa.

polypeptide See peptide *and* protein.

poly(phenylethene) See polystyrene.

polyphyodont Having many successive dentitions; describes an animal whose teeth may be replaced throughout its life. This is characteristic of frogs and lizards.

polyploid Describes a nucleus of a cell which has more than two sets of chromosomes. This is common in plants, rare in animals.

polyribosome (polysome) *See* ribosome *and* microtrabecular lattice.

polysaccharide The general name for a group of carbohydrates which may be represented by the formula $(C_6H_{10}O_5)_n$. They are built up from many units of monosaccharide; when they are broken down (by hydrolysis) molecules of smaller polysaccharides, disaccharides or monosaccharides may be formed. In the living organism this breakdown is assisted by enzymes (carbohydrases). Examples of polysaccharides are cellulose, glycogen, starch and inulin. Polysaccharides have no sweet taste as the sugars have.

polysome (polyribosome) *See* ribosome *and* microtrabecular lattice.

polystyrene (poly(phenylethene)) A highly-transparent plastic made by polymerisation of phenylethene (styrene). It can be used as a substitute for glass, and when expanded with air or carbon dioxide it forms an opaque, solid foam which is used for insulation and as a packing material.

poly(tetrafluoroethene) (PTFE) *See* Teflon.

polythene (poly(ethene); polyethylene) A polymer made from ethene, C_2H_4, ($CH_2\!\!=\!\!CH_2$). It is very resistant to chemical attack and is used in film or sheet form for packaging, coating electric wires and cables, making containers, etc.

polyunsaturate A term especially used for fats containing double bonds between carbon atoms. Such fats contain units of unsaturated fatty acids such as oleic acid (octadecenoic acid), $C_{17}H_{33}COOH$, and linoleic acid, (octadecadienoic acid) $C_{17}H_{31}COOH$. Polyunsaturates are found in vegetable oils and are considered to be of high value in the diet.

polyurethane A polymer made from polyhydric alcohols and organic isocyanates. It is used in making paints, adhesives and rigid or flexible foams used in carpets, upholstery, insulation, etc.

polyvinyl chloride *See* PVC.

pome A fleshy, false fruit (pseudocarp) mainly developed from the receptacle (the fleshy part) and not the ovary. The core, containing seeds, is the true fruit which has developed from the ovary. A tough pericarp surrounds the seeds. Examples of pomes are apples and pears.

pons Part of the hindbrain. It is a bridge of nerve-fibres joining the two hemispheres of the cerebellum. *See* Fig. 16.

population The total number of animals or plants in a given region. The term is often restricted to the number of a given species.

porcelain A translucent, non-porous ceramic material made by firing a mixture of kaolin, quartz and feldspar which is then dipped into a glaze and fired once more at a higher temperature. Porcelain is, to a large extent, unaffected by temperature variations. It is used in the manufacture of crockery and laboratory ware, as an electric insulator, etc.

pore A minute opening through which fluids may pass, e.g. sweat pore and stoma.

Porifera *See* Parazoa.

porous pot An unglazed earthenware pot.

portal vein *See* hepatic portal vein *and* renal portal system.

position isomerism *See* structural isomerism.

positive *See* photography.

positive catalyst *See* catalyst.

positive charge *See* charge.

positive electrode *See* anode.

positive ion *See* cation.

positive lens *See* lens.

positive pole *See* pole.

positive taxis *See* tactic movement.

positive tropism *See* tropic movement.

positron Symbol: $_{+1}^{0}e$. An elementary particle with the mass of an electron and a charge of the same magnitude but positive. In the presence of any matter, a positron rapidly combines with an electron producing gamma radiation. Positrons may be produced during radioactive disintegration of certain isotopes. *See also* antiparticle.

posterior Situated at or near the hind end of an animal. In human anatomy, the posterior side is equivalent to the dorsal side of other mammals.

post-mortem (autopsy) The examination of a dead body to discover the cause of death.

Post Office box A type of Wheatstone bridge in which the resistors consist of resistance coils contained in a special box. The resistance coils can be brought into circuit by removing brass plugs from the brass blocks connecting the coils. It can also be used as a potentiometer.

post precipitation The contamination of a precipitate standing in contact for some time with the mother liquor. The foreign compound precipitates after the desired precipitate has formed. *Example:* zinc sulphide may post precipitate onto sulphides of cadmium, copper or mercury. In cases where post precipitation may occur, the desired precipitate should not be allowed to stand too long before being filtered.

pot *See* cannabis.

potash alum *See* alum.

potassium An element with the symbol K; atomic number 19; relative atomic mass 39,10 (39.10); state, solid. It is a very electropositive, soft, light element belonging to the alkali metals. Potassium is found naturally in several minerals such as carnallite and sylvine, and in saltpetre (nitre). It is extracted by electrolysis of fused potassium chloride. Like all the alkali metals, potassium reacts with water to form a hydroxide and hydrogen:
$2K + 2H_2O \rightarrow 2KOH + H_2$
Potassium is sometimes used in the laboratory as a reducing agent.

potassium carbonate K_2CO_3 A white, solid, deliquescent, soluble salt giving an alkaline aqueous solution. Potassium carbonate cannot be prepared by the Solvay process (as is sodium carbonate) because potassium hydrogencarbonate, $KHCO_3$, is too soluble in water. K_2CO_3 is prepared by the Precht process. It is used as a drying agent, in soap-making and in the manufacture of hard glass.

potassium chlorate(V) $KClO_3$ A white, solid, soluble salt which can be prepared by electrolysis of a concentrated aqueous solution of potassium chloride, KCl, at a temperature of about 70°C. The products of this electrolysis are potassium hydroxide (solution), KOH, and chlorine, Cl_2, which are allowed to mix:
$6KOH + 3Cl_2 \rightarrow KClO_3 + 5KCl + 3H_2O$
The potassium chlorate(V) is then separated from the potassium chloride, KCl, by fractional crystallisation. When heated, potassium chlorate(V) decomposes to give potassium chloride and oxygen, O_2:
$2KClO_3 \rightarrow 2KCl + 3O_2$
This process may be catalysed by manganese(IV) oxide, MnO_2. $KClO_3$ is a strong oxidising agent. It is used in matches, fireworks, explosives, weed-killers, etc.

potassium chloride KCl A white, solid, soluble salt which can be prepared by neutralising a solution of potassium hydroxide, KOH, with hydrochloric acid, HCl:
$KOH + HCl \rightarrow KCl + H_2O$
Potassium chloride occurs naturally in several minerals such as carnallite and sylvine. It is used as a fertiliser and in the manufacture of potassium and potassium hydroxide.

potassium chromate(VI) K_2CrO_4 A yellow, solid, soluble salt which can be prepared by the oxidation of a chromium(III) salt by sodium peroxide, Na_2O_2, in aqueous solution:
$2Cr^{3+} + 3O_2^{2-} + 4OH^- \rightarrow 2CrO_4^{2-} + 2H_2O$
The Na_2O_2 provides the necessary hydroxyl ions, OH^-:
$Na_2O_2 + 2H_2O \rightarrow H_2O_2 + 2NaOH$
If an acid is added to an aqueous solution of chromate(VI) ions, CrO_4^{2-}, the orange-red dichromate(VI) ions, $Cr_2O_7^{2-}$, are formed:
$2H^+ + 2CrO_4^{2-} \rightarrow Cr_2O_7^{2-} + H_2O$
Potassium chromate(VI) is used as an indicator in titrations of halogens with silver nitrate, as a mordant and as a pigment.

potassium cyanate *See* Wöhler synthesis.

potassium cyanide KCN A white, solid, extremely poisonous, soluble salt which has a smell of bitter almonds. It can be prepared by

heating potassium carbonate, K_2CO_3, carbon, C, and nitrogen, N_2, to a high temperature:
$K_2CO_3 + 4C + N_2 \rightarrow 2KCN + 3CO$
In aqueous solution, KCN hydrolyses to some extent to give hydrocyanic acid, HCN:
$KCN + H_2O \rightleftharpoons HCN + K^+ + OH^-$
With acids, HCN is formed readily:
$KCN + H^+ \rightarrow HCN + K^+$
Potassium cyanide is used in the extraction of gold and silver and in various electrolytes. *See also* hydrogen cyanide.

potassium dichromate(VI) $K_2Cr_2O_7$ An orange-red, solid, soluble salt which can be prepared by acidifying an aqueous solution of potassium chromate, K_2CrO_4:
$2HCl + 2K_2CrO_4 \rightarrow K_2Cr_2O_7 + 2KCl + H_2O$
$K_2Cr_2O_7$ is a strong oxidising agent, used in tanning and in the manufacture of paints and dyes.

potassium hexacyanoferrate(II) *See* salt.

potassium hexacyanoferrate(III) *See* salt.

potassium hydrogentartrate *See* cream of tartar.

potassium hydroxide KOH A white, solid, deliquescent, soluble alkaline substance. An aqueous solution of KOH is a strong alkali. KOH is more soluble in water and ethanol than is sodium hydroxide, NaOH. Potassium hydroxide is produced in the Castner–Kellner process or in the diaphragm cell. It is used in soap-making, in certain electric cells and for absorbing acidic gases such as carbon dioxide, CO_2, and sulphur dioxide, SO_2.

potassium iodide KI A white, solid, soluble salt which can be prepared by adding iodine, I_2, to a hot, concentrated aqueous solution of potassium hydroxide, KOH:
$3I_2 + 6KOH \rightarrow 5KI + KIO_3 + 3H_2O$
The potassium iodate(V), KIO_3, which is produced is then reduced with charcoal, C:
$2KIO_3 + 3C \rightarrow 2KI + 3CO_2$
i.e. both processes yield potassium iodide. Iodine dissolves in an aqueous solution of potassium iodide because of the formation of tri-iodide ions, I_3^-:
$I_2 + KI \rightarrow KI_3 \rightleftharpoons K^+ + I_3^-$
Potassium iodide is used in medicine.

potassium manganate(VII) $KMnO_4$ (potassium permanganate). A purple, solid, soluble salt which can be prepared by oxidising potassium manganate(VI), K_2MnO_4, with chlorine, Cl_2:
$2K_2MnO_4 + Cl_2 \rightarrow 2KMnO_4 + 2KCl$
When heated, $KMnO_4$ decomposes to give oxygen, O_2, potassium manganate(VI) and manganese(IV) oxide, MnO_2:
$2KMnO_4 \rightarrow O_2 + K_2MnO_4 + MnO_2$
Potassium manganate(VII) is a powerful oxidising agent, used as a disinfectant and as a volumetric reagent in acidic solutions. It acts as its own indicator, as the manganate(VII) ion, MnO_4^-, is pink even in extremely dilute solutions, and the manganese(II) ion, Mn^{2+}, which is formed in acidic solution, is almost colourless. As pH increases, $KMnO_4$ becomes less strong as an oxidising agent. In a neutral or weak alkaline solution, the MnO_4^- ion is reduced, to MnO_2; in a stronger alkaline solution it is reduced to the green manganate(VI) ion, MnO_4^{2-}.

potassium nitrate KNO_3 (nitre, saltpetre) A white, solid, soluble, non-deliquescent salt which can be prepared by boiling and mixing saturated solutions of sodium nitrate, $NaNO_3$, and potassium chloride, KCl:
$NaNO_3 + KCl \rightarrow KNO_3 + NaCl$
This process is a double decomposition; it only takes place because of different solubilities of the salts. At the boiling temperature, sodium chloride, NaCl, is the least soluble of the four salts and therefore crystallises out. On cooling to room temperature, potassium nitrate crystallises out as it is the least soluble salt at this temperature. Potassium nitrate can also be prepared by neutralising an aqueous solution of potassium hydroxide, KOH, with nitric acid, HNO_3:
$KOH + HNO_3 \rightarrow KNO_3 + H_2O$
Potassium nitrate occurs naturally as saltpetre (nitre). When heated, KNO_3 decomposes to give potassium nitrite, KNO_2, and oxygen, O_2:
$2KNO_3 \rightarrow 2KNO_2 + O_2$
However, if it is heated very strongly, oxides and peroxides may also be formed. Potassium nitrate is a strong oxidising agent, used in making gunpowder and fireworks and as a fertiliser.

potassium permanganate *See* potassium manganate(VII).

potassium sodium tartrate KOOC—CHOH—CHOH—COONa·$4H_2O$ (Rochelle salt, Seignette salt) A soluble, crystalline double salt of tartaric acid, HOOC—CHOH—CHOH—COOH. It is used as an antacid and in Fehling's solution. *See also* piezo-electric effect.

potassium sulphate K_2SO_4 A white, solid, soluble salt which can be prepared by neutralising an aqueous solution of potassium hydroxide, KOH with sulphuric acid, H_2SO_4:
$2KOH + H_2SO_4 \rightarrow K_2SO_4 + 2H_2O$
K_2SO_4 is found in the Stassfurt deposits, from which it is extracted and used as a fertiliser.

potential (electric) Symbol: V. The electrical condition, e.g. the electromotive force of a cell, which can cause electric charges to move. A potential is a relative condition; however, by convention the Earth's potential is said to be zero. The SI unit of electric potential is the volt (V). *See also* potential difference.

potential difference Symbol: U or V. The difference in electric potential between two points in an electric field or circuit. The potential difference (p.d.) between two points in an electric circuit is the work done in moving unit charge from one point to another. The SI unit of potential difference is the volt (V). *See also* electromotive force.

potential divider *See* voltage divider.

potential energy The energy possessed by a body due to its position. The potential energy of a body of mass m kg at a height h metres above the ground is given by the expression

$$E = mgh$$

where g is the acceleration of free fall.

potentiometer An instrument (a type of voltage divider) for the accurate measurement of electromotive force or potential difference (p.d.). In its simplest form it consists of a length (l) of uniform resistance wire stretched over a scale XY and carrying a constant current. The source of p.d. under test is connected to X in series with a galvanometer, and a sliding contact S is moved until the galvanometer shows no deflection, i.e. the potentials of the accumulator and the source under test are balanced and no current flows. The p.d. can be found from the expression

$$E_2 = \frac{E_1 l_1}{l}$$

where E_1 and E_2 are the potential differences and $l_1 = $ XS (the balance length). Sometimes a standard cell (with a known p.d. E_3) is substituted for the source of p.d. under test and a new balance length l_2 is found. Then

$$\frac{E_2}{E_3} = \frac{l_1}{l_2} \quad \text{or} \quad E_2 = \frac{E_3 l_1}{l_2}$$

See Fig. 172.

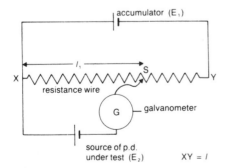

Fig. 172 Potentiometer

potentiometric titration A type of titration in which the hydrogen ion concentration or pH of a sample is determined during an acid–base titration. The two electrodes from a pH meter are immersed in the sample and the pH or the change in e.m.f. between the two electrodes is recorded after each addition of titrant. In this way a pH curve (titration curve) can be obtained. Near the end-point there are very marked changes in e.m.f. for small additions of titrant. This type of titration is particularly useful when the sample is coloured, making an indicator colour change difficult or impossible to observe. *See* Fig. 173.

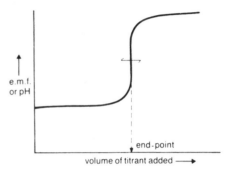

Fig. 173 Graph of a potentiometric titration (acid titrated with alkali)

potometer An apparatus for measuring the rate of absorption of water by a cut leafy shoot. It is normally used to compare the rates of transpiration of the same shoot under different conditions by timing the movement of a water meniscus or air bubble over a given distance on the scale. *See* Fig. 174.

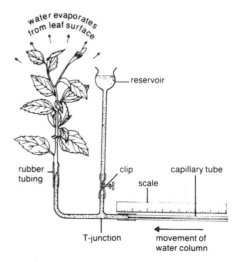

Fig. 174 Potometer

pouch In marsupials and monotremes, a bag-like structure in which the offspring are carried.

power 1. (electrical) *See* electric power. **2.** (in optics) *See* dioptre. **3.** (mechanical) Symbol: P; unit: watt (W). The rate at which energy is transferred or the rate of doing work.

power-station A building where electric power is produced. A conventional power-station uses fossil fuels as a source of heat for converting water into steam which drives turbines connected to generators. *See also* hydroelectric power-station *and* nuclear reactor.

power transmission (distribution) *See* grid.

pre- Prefix meaning before.

Precambrian The period of time (era) preceding the Palaeozoic era. It is estimated to have commenced about 4500 million years ago. *See* Appendix.

Precht process A process for the industrial manufacture of potassium carbonate, K_2CO_3. Solid hydrated magnesium carbonate, $MgCO_3 \cdot 3H_2O$, is added to a concentrated aqueous solution of potassium chloride, KCl, and carbon dioxide, CO_2, is then passed through the mixture. This causes the formation of a hydrated precipitate consisting of magnesium carbonate and potassium hydrogencarbonate, $KHCO_3$:

$2KCl + 3MgCO_3 \cdot 3H_2O + CO_2 \rightarrow$
$2MgCO_3 \cdot KHCO_3 \cdot 4H_2O + MgCl_2$

The precipitate is removed by filtration and treated with an aqueous suspension of magnesium oxide, MgO, at below 20°C. This causes the formation of a precipitate of hydrated magnesium carbonate (which can be re-used) and potassium carbonate can be crystallised from the solution:

$2MgCO_3 \cdot KHCO_3 \cdot 4H_2O + MgO \rightarrow$
$K_2CO_3 + 3MgCO_3 \cdot 3H_2O$

precipitate *See* precipitation.

precipitation A process in which an insoluble substance (the precipitate) is formed as a result of a chemical reaction. *Example:* when a solution of silver nitrate, $AgNO_3$, is added to a solution containing chloride ions, Cl^-, a white precipitate of silver chloride, AgCl, is formed: $AgNO_3 + Cl^- \rightarrow AgCl + NO_3^-$

predator An animal which attacks, kills and eats other animals (the prey). *See also* scavenger.

pregnancy The state of a female animal in whose womb an embryo is developing.

prehensile Describes a limb or tail which is capable of grasping.

pre-ignition *See* knocking.

premaxilla In most vertebrates, a paired bone forming the front part of the upper jaw. In mammals it bears the incisors, and in birds it forms most of the upper half of the beak.

premolar (bicuspid) A tooth with two points (cusps) in its crown, suited for crushing and grinding. In humans, premolars are not present in the milk dentition; in the permanent dentition there are four in each jaw. Premolars usually have a single root, except for the first premolar in the upper jaw which normally has a double root.

prepuce (foreskin) In mammals, the fold of skin covering the glans penis when not erected. In circumcision the prepuce is cut off.

preputial glands In mammals, rudimentary glands situated in the prepuce and the glans penis. The glands secrete a greasy fluid.

presbyopia A defect of the eye in elderly people in which the lens becomes less elastic and the ciliary muscles cannot produce enough accommodation. Distant objects can be seen clearly, but near objects can not be brought to a focus on the retina. The defect is corrected by convex spectacle lenses. Note that presbyopia should not be confused with hypermetropia.

pressure Symbol: p; unit: pascal (Pa). Pressure is the force acting per unit area. In a liquid at rest, pressure increases with depth and, at any point, acts equally in all directions. The pressure at a depth h in a liquid of density ρ (rho) is given by the expression

$p = h\rho g$,

where g is the acceleration of free fall. Other units of pressure are the atmosphere, the bar and millimetres of mercury (mmHg).

pressure-cooker A thick-walled container (usually made of aluminium) with a tightly-fitting lid. In the lid there is a valve, which can be adjusted to blow at preset pressures in excess of atmospheric pressure, and a safety valve. The pressure-cooker is used for fast, high-temperature cooking because increased pressure on the surface of the liquid in the cooker causes its boiling-point to be raised. *See* Fig. 175; *see also* autoclave.

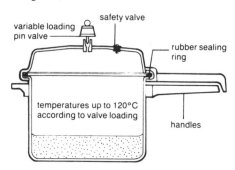

Fig. 175 Pressure-cooker

prey *See* predator.

prickly heat A skin rash consisting of minute red pimples, usually accompanied by itching and a burning sensation. It is caused by obstruction of sweat ducts, preventing sweat from reaching the surface of the skin; the sweat builds up, causing pimples. This disorder is usually seen among children and overweight people. Exposure to hot, moist air (as in the tropics) promotes the condition. Cooling of the body, baths without use of soap and the wearing of lightweight, loose-fitting clothing (especially at night) often make the rash disappear.

primary *See* primary feather.

primary cell *See* cell.

primary coil *See* induction coil *and* transformer.

primary colour *See* colour mixing.

primary consumer *See* consumer.

primary emission The emission of electrons from a surface when it is struck by radiation or has an electric field applied to it. *See also* photomultiplier.

primary feather (primary) One of the larger flight feathers on a bird's wing.

primary growth (primary thickening) The growth in length of roots and shoots by the action of meristems. *Compare* secondary growth.

primary leaf *See* cotyledon.

primary medullary ray *See* medullary ray.

primary phloem *See* phloem.

primary thickening *See* primary growth.

primary winding *See* transformer.

primary xylem *See* xylem.

Primates An order of placental mammals including humans, apes and monkeys. Characteristics include limbs with five digits (pentadactyl), a mainly arboreal life, complete dentition, a large and complicated brain, forward-facing eyes for binocular vision, and a long period of parental care after birth.

principal axis *See* lens *and* mirror.

principal focus *See* lens *and* mirror.

principle of moments *See* moments, principle of.

printed circuit An electric circuit in which the conductor (thin strips of copper) and certain components are coated onto an insulating board. Other circuit components are inserted through holes in the board and soldered to the strips of copper. To begin with, the insulating board is completely covered with a thin layer of copper. The required conductor pattern for the circuit is produced by covering (masking) the copper with a protective film in the shape of the required pattern and then dissolving away the unprotected copper with an acid. The conductor pattern remains on the insulating board when the protective film is removed. Printed circuits may be formed on both sides of the insulating board, and some are produced consisting of many layers of conductor patterns insulated from each other.

prismatic Shaped like a prism or containing prisms.

prismatic binoculars An optical instrument involving the use of both eyes, used for viewing distant objects. It consists of two telescopes, each containing two right-angled prisms arranged so that an upright, magnified image is obtained at the eye. The prism arrangement effectively shortens the length of the instrument and the binocular arrangement gives distance perception. *See* Fig. 176.

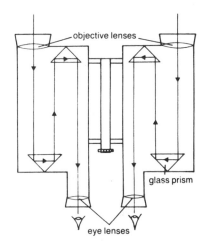

Fig. 176 Prismatic binoculars

prismatic sulphur *See* monoclinic sulphur.

prism, optical A block of transparent material (usually glass or plastic), often with a triangular section, used in optical instruments and experiments. Prisms are used for the dispersion of light, for erecting (or inverting) images and for total internal reflection. *See also* erecting prism or lens, prismatic binoculars *and* periscope.

prism periscope *See* periscope.

pro- Prefix meaning before.

Proboscidea An order of placental mammals including the elephants and extinct mammoths. Characteristics include large size, pillar-like legs, very long incisors (tusks), large grinding molars and a proboscis (trunk) used for putting food in the mouth and for sucking up and expelling water. The trunk has movable flaps at the opening.

proboscis A tube-like projection from the head of certain animals such as butterflies, mos-

239

quitoes, elephants and some marine worms. In insects, it is used for sucking food.

procambium (provascular tissue) Tissue in roots and shoots which gives rise to the first vascular tissue.

Procaryota One of the main groups into which all living organisms are divided. This group includes bacteria and blue-green algae. Their cells are characterised by having no true nuclei, i.e. the genetic material consists of a single filament of DNA not separated from the cytoplasm by a nuclear membrane. During cell division, the DNA divides without the formation of a spindle as in mitosis. Organelles such as mitochondria and chloroplasts are lacking in the cytoplasm. Procaryotic cells are much smaller than eucaryotic ones. *See also* Eucaryota *and* Protista.

processor *See* microprocessor *and* computer.

producer A living organism, usually a green plant, which manufactures complex organic substances from simple inorganic substances. Producers form the first link in a food chain and are eaten by primary consumers.

producer gas A mixture, mainly consisting of carbon monoxide, CO, and nitrogen, N_2, which can be made by reacting coke (carbon) with air in a special furnace (the 'producer') at a high temperature:
$2C + O_2 \rightarrow 2CO$
The process is exothermic. The nitrogen present in air does not react with the coke. In practice, producer gas is 34% CO, 64% N_2 and 2% carbon dioxide, CO_2, by volume, and is therefore a rather poor fuel. Producer gas is often prepared with water gas. It is used as a fuel in industrial heating.

product A substance produced during a chemical process. *Example:* when hydrochloric acid, HCl, reacts with iron(II) sulphide, FeS, two products are formed, namely hydrogen sulphide, H_2S, and iron(II) chloride, $FeCl_2$:
$FeS + 2HCl \rightarrow H_2S + FeCl_2$
Compare reactant.

proenzyme *See* zymogen.

progesterone A female sex hormone (steroid) produced in the ovaries by the corpus luteum after ovulation. Progesterone initiates changes in the womb preparing it for pregnancy. *Example:* it prevents further ovulation: the next oestrous cycle cannot take place until after birth. It also initiates the development of the placenta and mammary glands. *See also* oestrogen *and* prolactin.

proglottis (pl. proglottides) One of the many strings of segments of which tapeworms consist. The adult worm has at least one set of reproductive organs on each proglottis.

progressive wave *See* travelling wave.

projector, cinema or slide An optical device for producing a magnified real image of a film or slide on a screen. The arrangement of lenses is shown in Fig. 177.

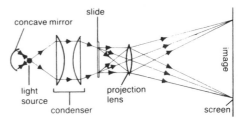

Fig. 177 Projector

prolactin (lactogenic hormone) A hormone produced and secreted by the anterior lobe of the pituitary gland in vertebrates. In female mammals, prolactin stimulates and controls the production of milk from the developed mammary glands and stimulates the production of progesterone by the corpus luteum.

proleg In most larvae (caterpillars) of Lepidoptera, an unjointed, fleshy projection on the abdomen. Prolegs are used in locomotion.

promoter A substance which increases the activity of a catalyst, but has no catalytic action itself. In the manufacture of a catalyst, the promoter is often incorporated in it.

Prontosil® *See* sulphonamide.

proof-plane A small metal plate on an insulated handle. It is used for transferring charges from conductors in electrostatic experiments.

propagule A general term for any part or form of a plant which is capable of developing into a new individual, e.g. a seed, spore or cutting.

propane C_3H_8 A saturated hydrocarbon belonging to the alkanes. It is a flammable gas occurring in natural gas and petroleum from which it is extracted. It is also prepared by cracking higher alkanes. It is used as a fuel and in organic synthesis. M.p. $-189°C$, b.p. $-42°C$.

propane-1,2,3-triol *See* glycerol.

propane-1,2,3-triyl trinitrate *See* nitroglycerine.

propanone CH_3COCH_3 (acetone). A colourless, volatile, flammable liquid. Propanone is extensively used as a solvent for several organic substances and as a raw material in the manufacture of many organic compounds. It is miscible with water. M.p. $-95,4°C$ ($-95.4°C$), b.p. $56,2°C$ ($56.2°C$). Propanone is the simplest of the aliphatic ketones.

prophase The first stage of meiosis and mitosis.

prophylactic A medicine or procedure which helps to prevent disease. *Example:* a vaccine is a prophylactic.

prophylaxis (pl. prophylaxes) Any treatment for the prevention of disease.

prop roots Adventitious roots growing from and supporting the branches of certain trees, e.g. the banyan tree. In maize, prop roots grow from the stem close to the ground.

prosencephalon *See* forebrain.

prosenchyma Tissue made up of elongated cells with pointed ends. It forms supporting and conducting tissues in plants.

prostate gland A gland accessory to the male reproductive system of mammals. In humans it is situated below the bladder, completely surrounding the first portion of the urethra. It secretes a thin, slightly alkaline, milky liquid which is added to the semen. This liquid is believed to increase the activity of the sperm.

prosthetic group A non-protein substance (molecule) attached to a protein. *Example:* in haemoglobin, haem is a prosthetic group. A coenzyme is also a prosthetic group. Without the prosthetic group, the protein is inactive.

protandrous Describes a hermaphrodite animal or plant in which the male gametes are produced and released before the female gametes. *Compare* protogynous.

protein A large, complex polypeptide. Proteins are important substances in the nutrition, structure and function of living organisms. A simple protein is made up of amino acids and contains the elements carbon, oxygen, hydrogen, nitrogen and occasionally sulphur and phosphorus. A conjugated protein has a non-protein substance (prosthetic group) attached to its chain of amino acids. Proteins may be divided into fibrous proteins and globular proteins. Fibrous proteins are found in skin, hair and muscle and they are insoluble in water. Globular proteins are soluble in water; enzymes and protein hormones such as insulin are examples of globular proteins. Physical and chemical properties of a protein are determined by its structure, i.e. by the nature of the amino acids present, the sequence of these acids and the shape of the entire protein molecule. The types of chemical bonds in a protein determine its shape. *Example:* hydrogen bonds between N—H groups and C=O groups:

$$\text{>N—H---O=C<}$$

may lead to coiling of the protein chain, forming a helix. This is seen in fibrous proteins. *See* Fig. 178; *see also* denaturation.

proteinase A general term for an enzyme capable of splitting the peptide bonds in proteins, e.g. erepsin, pepsin and trypsin.

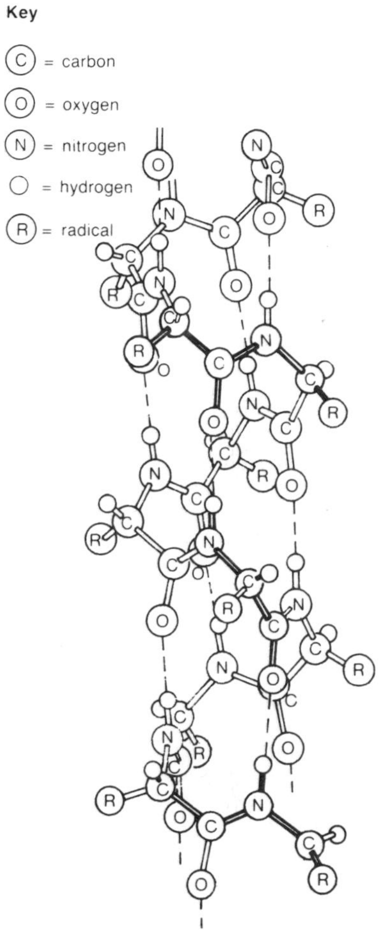

Key

C = carbon
O = oxygen
N = nitrogen
○ = hydrogen
R = radical

Fig. 178 Arrangement of atoms in a protein

proteolysis The process by which proteinases hydrolyse proteins to smaller peptides or amino acids.

prothallus The gametophyte of pteridophytes and gymnosperms, bearing antheridia or archegonia or both.

prothrombin A zymogen which is a protein found in blood plasma and produced in the liver in the presence of vitamin K. At a wound, it is converted to thrombin by the action of thrombokinase in the presence of calcium ions. *See also* blood clotting.

Protista A kingdom of simple organisms, many of which possess both animal and plant characteristics and are therefore difficult to classify. It includes bacteria, algae, protozoa and fungi. *See also* Eucaryota *and* Procaryota.

protium Symbol: H or $^1_1$H. The ordinary (lightest) isotope of hydrogen.

proto- Prefix meaning first or original.

protogynous Describes a hermaphrodite animal or plant in which the female gametes mature before the male gametes. Compare protandrous.

protolysis The process in which protons are transferred. *Example:* in an acid–base reaction, a proton (hydrogen ion) is transferred from the acid to the base: when hydrochloric acid, HCl, reacts with sodium hydroxide, NaOH, the reaction can be simplified to:
$H^+ + OH^- \rightleftharpoons H_2O$

proton (hydrogen ion) One of the three basic particles in an atom. It is situated in the nucleus of the atom and its mass is $1,673 \times 10^{-27}$ kg (1.673×10^{-27} kg), i.e. about the same mass as the neutron and about 1840 times that of the electron. The proton carries a positive electric charge of $1,602 \times 10^{-19}$ coulombs (1.602×10^{-19} C). It is found in all atomic nuclei. In a neutral atom, the number of protons equals the number of electrons.

protonema (pl. protonemata) *See* Musci.

proton number *See* atomic number.

protoplasm *See* cytoplasm.

Prototheria *See* Monotremata.

Protozoa A subkingdom and phylum of mainly microscopic, unicellular animals. There are more than 30 000 recognised species, all living in damp environments on land or in fresh or sea water. They are classified according to their methods of locomotion. Examples of protozoa are *Amoeba*, *Paramecium* and Plasmodium. *See also* Metazoa *and* Parazoa.

Proust's law *See* chemical combination (laws of).

provascular tissue *See* procambium.

proventriculus 1. In birds, the glandular part of the stomach leading to the gizzard. It secretes digestive enzymes. **2.** In crustaceans and insects, an alternative name for the gizzard.

Proxima Centauri The nearest star to the Earth (apart from the Sun) found in the triple star system Alpha Centauri. It is 0,1 light-year (0.1 light-year) nearer to Earth than the other two stars of the system. Proxima Centauri is approximately 4,3 light-years (4.3 light-years) away from Earth.

proximal Situated towards the place of attachment or origin. *Example:* the thigh is the proximal part of the leg. Compare distal.

pruning The cutting away of dead or overgrown branches of a tree in order to promote growth.

Prussian blue *See* cyanoferrates.

prussic acid *See* hydrogen cyanide.

psalterium (manyplies, omasum) The third stomach in ruminants.

pseudo- Prefix meaning false.

pseudocarp *See* false fruit.

pseudopodium (pl. pseudopodia) A temporary protrusion of the cytoplasm of phagocytes and certain protozoa, e.g. *Amoeba*. Pseudopodia are concerned with locomotion and feeding.

psoriasis A benign skin disease which may begin at any age and usually remains for life. However, the disease may disappear completely for many months or years. Symptoms include plaques; these may be found anywhere on the body, but the scalp, elbows and knees are favoured sites. Sometimes deformity of the joints is seen, similar to that caused by rheumatoid arthritis. The cause of the disease is unknown. Psoriasis is treated by applying various ointments to the affected skin. These ointments may contain steroids or coal-tar. This treatment may be combined with exposure of the skin to ultraviolet radiation.

psychiatry The branch of medicine dealing with mental disorders and diseases.

psychology The science dealing with the study of the mind, including human behaviour.

psychotherapy The treatment of psychic (mental) disorders, largely through verbal means. Psychotherapy was developed by Sigmund Freud who worked out a system of psychoanalysis. However, several modifications of psychoanalysis have been introduced since.

psychrometry The measurement of atmospheric humidity.

psychrophile A micro-organism which grows best at a low temperature, e.g. is capable of growing at 0°C. The optimum temperature of growth is usually in the range 20–30°C; however, some psychrophiles isolated from the depths of oceans and lakes have their optimum at 15°C and die at 20°C. Psychrophiles are said to be psychrophilic. *See also* mesophile *and* thermophile.

Pt Chemical symbol for platinum.

Pteridophyta A major division of the plant kingdom, comprising ferns, club-mosses and horse-tails. They are mainly terrestrial, non-flowering plants whose characteristics include a vascular system, no cambium, true stems, roots and leaves, sporangia borne on leaves, and well-marked alternation of generations.

PTFE *See* Teflon.

ptyalin An amylase present in the saliva of humans and some other mammals. It catalyses the hydrolysis of starch to maltose.

p-type semiconductor *See* semiconductor.

Pu Chemical symbol for plutonium.

puberty In animals, the beginning of sexual maturity, i.e. the onset of adolescence.

pubic bone *See* pubis.

pubic symphysis The joint between the two pubes (pubic bones) in most mammals and some reptiles.

pubis (pl. pubes) (pubic bone) The anterior, ventral bone of the pelvic girdle in tetrapod vertebrates. *See also* pubic symphysis.

pulley A simple machine consisting of a grooved wheel over which a rope or chain passes. Several pulleys may be arranged together to form a block and tackle. Pulleys are used for raising loads, e.g. in cranes. The mechanical advantage (force ratio) and velocity ratio (distance ratio) of a system of pulleys depend on the arrangement of the pulleys and ropes. In Fig. 12, the lower moveable block (the tackle) is supported by four parts of the rope with a total upward force on the load (L) equal to four times the effort (E). Neglecting friction and the weight of the tackle, the mechanical advantage of the block and tackle is therefore equal to

$$\frac{4E}{E} = 4$$

If the load is raised a distance of one metre, each part of the rope supporting the tackle must be shortened by one metre. Therefore, the effort moves a distance of four metres, i.e. the velocity ratio is 4.

pulmonary Pertaining to the lungs.

pulmonary artery In mammals, a paired artery carrying deoxygenated blood from the right ventricle of the heart to the lungs. The pulmonary artery is the only artery in the body carrying deoxygenated blood.

pulmonary tuberculosis *See* tuberculosis.

pulmonary vein In mammals, a paired vein carrying oxygenated blood from the lungs to the left atrium (auricle) of the heart. The pulmonary vein is the only vein in the body carrying oxygenated blood.

pulp cavity The central part of a tooth, surrounded by dentine and communicating with the exterior of the tooth via canals in the roots. The cavity contains connective tissue, blood vessels and nerves.

pulse A wave of pressure passing down the arteries, caused by the action of the heart. This pressure causes the arteries to expand and contract periodically. The pulse can be felt where an artery is close to the surface of the skin, e.g. at the wrist, where an artery runs over the bones below the base of the thumb. The pulse moves much faster than the actual blood flow.

pulse rate The number of times the pulse can be felt in a given period, e.g. per minute. The pulse rate of a healthy adult at rest is 70–80 beats per minute. In children, the pulse rate is higher.

pumice Lava containing many holes formed when it solidified whilst gases were still bubbling through it. Pumice floats in water because of the many sealed air-spaces it contains. It is used as an abrasive.

pump *See* force pump, lift pump *and* vacuum pump.

pupa (pl. pupae) The resting stage which follows the larval stage in the life cycle of insects with complete metamorphosis. During this stage locomotion and feeding stop, the insect's body is usually covered with a case (cocoon) and bodily changes take place before the insect emerges as the adult (imago). *See also* chrysalis.

pupil The opening in the centre of the iris of the eye. Its size is controlled by muscles in the iris, thus regulating the amount of light entering the eye. *See* Fig. 66.

purine 1. A heterocyclic, aromatic amine consisting of a six-membered ring linked to a five-membered ring:

2. The general term for a group of derivatives of purine (*see* 1.). Examples of purines include adenine, guanine and caffeine. Adenine and guanine are constituents of the nucleic acids.

pus A creamy yellow secretion from inflamed tissue. Pus contains decaying leucocytes (phagocytes), dead or living bacteria and damaged tissue. It is a result of the defensive reaction of the body against bacterial infection.

putrefaction The process in which protein-containing material is decomposed by micro-organisms (usually anaerobic) to produce bad-smelling products such as putrescine, a di-amine with the formula NH_2—$(CH_2)_4$—NH_2. Badly spoiled meat smells of putrescine.

putrescine *See* putrefaction.

PVC (poly(chloroethene); polyvinyl chloride) A polymer made from chloroethene (vinyl chloride), C_2H_3Cl (CH_2=$CHCl$). It is a very tough material used in the manufacture of artificial leathers, drainage pipes and containers, as a floor covering, etc.

pyloric orifice The opening at the lower end of the stomach, leading to the duodenum. The sphincter around this orifice is well-developed. *See also* cardiac orifice.

pyloric sphincter A circular muscle situated around the pyloric orifice. It is able to open and close the pyloric orifice, thus controlling

pyramid

the entry of food from the stomach to the duodenum. *See also* cardiac sphincter.

pyramid A conical-shaped process in the medulla of the kidney projecting into the pelvis. A collective duct, formed from the joining of many collecting tubules, passes through the apex of each pyramid.

pyranose *See* sugar.

pyrenoid In some algae, e.g. green algae, a spherical particle made of protein, present in the chloroplast either singly or in numbers. Starch produced as a result of photosynthesis is stored around the pyrenoids.

pyrethrin *See* pyrethrum.

pyrethrum A plant belonging to the family Compositae. A powerful insecticide (pyrethrin) is extracted from the dried flowers of the plant. Pyrethrum is grown in Africa.

pyretic Pertaining to fever.

Pyrex® Heat-resisting glass containing increased amounts of boron trioxide in place of some of the silica, and containing low amounts of alkalis. Pyrex glass has a low expansion coefficient and so is not likely to crack on heating or cooling. It softens at a higher temperature than ordinary glass. Pyrex is widely used in the manufacture of laboratory heat-resistant glassware and domestic glassware. *See also* Vitreosil.

pyrexia *See* fever.

pyridine C_5H_5N An aromatic, heterocyclic, very stable, colourless liquid which is a very weak base and has an unpleasant smell. It can be extracted from coal-tar. It is used as a solvent, as a denaturant and in organic synthesis of other compounds.

pyridoxin(e) A vitamin of the B-complex group. It is present in yeast, whole grains, liver and milk. Deficiency of pyridoxin(e) can lead to dangerous convulsions in babies and possibly to anaemia in adults.

pyrimidine 1. A heterocyclic, aromatic amine, consisting of a six-membered ring:

2. The general term for a group of derivatives of pyrimidine (*see* 1.). Examples of pyrimidines are thymine, cytosine and uracil. They are all constituents of the nucleic acids.

pyrite *See* iron(II) disulphide (pyrites).

pyrites Naturally occurring metal sulphides, e.g. iron pyrites, FeS_2.

pyrogallic acid *See* benzene-1,2,3-triol.

pyrogallol *See* benzene-1,2,3-triol.

pyrolusite *See* manganese.

pyrolysis The decomposition of a substance by heat. The term is sometimes used as a synonym for cracking.

pyrometer An instrument for measuring high temperatures, e.g. temperatures inside a furnace. There are several types of pyrometers, of which the optical pyrometer is the most common. It consists of a telescope in which a red filter and a small electric bulb are mounted. The bulb is connected to a battery, a rheostat and an ammeter which is calibrated, from previous experiments at known temperatures, to read temperature directly. When the pyrometer is directed towards an opening in a furnace, an observer looking through the telescope will see the dark filament of the bulb against the bright background of the furnace. By adjusting the rheostat, the brightness of the filament can be matched with the brightness of the background, so that the filament cannot be seen against the bright background. The temperature can now be read.

pyrometry The measurement of temperatures using pyrometers.

pyrophoric 1. Describes a material which gives sparks when struck. *Example:* mischmetall is a pyrophoric alloy used in cigarette-lighter flints. **2.** Describes a substance which ignites spontaneously in air, e.g. white phosphorus and phosphine. Several metals, e.g. zinc and iron, are pyrophoric when produced as fine powders.

pyrophosphate A salt of pyrophosphoric acid, $H_4P_2O_7$.

pyrophosphoric acid $H_4P_2O_7$ A white, crystalline solid which can be prepared by heating phosphoric(V) acid, H_3PO_4:
$2H_3PO_4 \rightarrow H_4P_2O_7 + H_2O$
The acid is soluble in water. Salts of $H_4P_2O_7$ are called pyrophosphates. They are used in water-softening and electroplating.

pyrosulphuric acid *See* oleum.

pyruvate (2-oxopropanoate) A salt of pyruvic acid, $CH_3—CO—COOH$.

pyruvic acid $CH_3—CO—COOH$ (2-oxopropanoic acid). A keto-acid which can be prepared by treating lactic acid, $CH_3—CHOH—COOH$, with a mild oxidising agent such as silver oxide, Ag_2O:
$CH_3—CHOH—COOH + Ag_2O \rightarrow$
$CH_3—CO—COOH + 2Ag + H_2O$
Salts of pyruvic acid are called pyruvates (2-oxopropanoates). Pyruvic acid plays an important part in metabolism. *Example:* it is formed during glycolysis by the reduction of carbohydrates. *See also* Krebs cycle.

Q

quadrat A square area (usually one metre square) marked out at random on the ground for the purpose of studying the vegetation in a certain region. The area may be marked out with string or with a square frame. *See also* transect.

quadriceps The muscle in front of the thigh. It has four origins and extends the lower leg.

quadrivalent (tetravalent) Having a valency of four.

quadruped *See* tetrapod.

qualitative analysis The analysis of a sample of unknown composition in order to find out what element or compound it is, or what chemicals it contains. This analysis can be carried out using chemical or physical tests. *Compare* quantitative analysis.

quality of sound (timbre) This is determined by the number of overtones accompanying the fundamental. A note played on a piano can easily be distinguished from the same note played on a trumpet because of the difference in the quality of sound produced by each instrument.

quantitative analysis The analysis of a sample of known composition in order to find out the concentration of one or more of its constituents. This form of analysis can be carried out using simple volumetric and gravimetric methods, though more sophisticated methods are also used. *Compare* qualitative analysis.

quantity of electricity Symbol: Q; unit: coulomb (C). The amount of electricity in coulombs passing any point in a circuit is given by the expression

$Q = It$

where I is the current in amperes and t is the time in seconds for which the current flows. Quantity of electricity also refers to the amount of static charge.

quantum (pl. quanta) The smallest definite (discrete) amount of energy which can be gained by a system or lost from it. The photon is a single quantum of electromagnetic radiation.

quantum energy *See* Planck's constant.

quantum mechanics A theory developed from the quantum theory, explaining the behaviour of atoms, molecules and elementary particles such as the electron, neutron and proton. The branch of quantum mechanics which states that particles may have wavelike properties is called wave mechanics.

quantum theory A theory first put forward by Max Planck, stating that energy can be gained by or lost from a system only in certain definite (discrete) amounts. These amounts (quanta) can only exist in whole numbers. *See also* Planck's constant.

quarantine The isolation of an individual who is suffering from or has been exposed to a communicable disease, in order to protect others against infection. The incubation period of the disease may determine the length of quarantine. The term also applies to imported animals.

quartz A widely distributed mineral composed of crystalline silicon(IV) oxide (silica), SiO_2, found in many varieties such as rock crystal (colourless), amethyst (purple), rose quartz (pink) and smoky quartz (brown to nearly black). Agate and flint are also varieties of quartz. Sand consists mainly of quartz; it is used in glass-making. Glass containing a high percentage of quartz transmits ultraviolet light.

Quaternary The geological period extending from the beginning of the Pleistocene (approximately $1\frac{1}{2}$ million years ago) to the present day.

quaternary ammonium ion A cation which resembles an ammonium ion in which the hydrogen atoms have been replaced by alkyl groups. The ammonium ion is NH_4^+; the quaternary ammonium ion is $N(CH_3)_4^+$, containing four methyl groups. The quaternary ammonium ion forms salts which are ionic and water-soluble. The ion can be prepared by treating ammonia, NH_3, with an alkyl halide, e.g.

$NH_3 + 4CH_3Cl \rightarrow N(CH_3)_4^+Cl^- + 3HCl$

Quaternary ammonium salts are used in certain ion exchangers and as disinfectants.

queen The fertile female of the order Hymenoptera.

quenching The hardening of a metal (usually steel) by rapidly cooling it in water or oil, or with a blast of air.

quicklime *See* calcium oxide.

quicksilver *See* mercury.

quill 1. The hollow part of a feather. *See* Fig. 67. 2. One of the hollow spines of a porcupine or hedgehog.

quill-feather *See* contour feather.

quinine A white or colourless crystalline substance (an alkaloid) obtained from the bark of the cinchona tree. It has a very bitter taste. Originally it was used in the treatment of malaria, but other less toxic and more effective drugs have now replaced it. Quinine has an antipyretic effect.

R

Ra Chemical symbol for radium.

rabies (hydrophobia) An acute viral disease affecting the central nervous system, transmitted to humans by the saliva of an infected animal, e.g. a dog. The saliva must enter the body, e.g. through a wound. Symptoms include restlessness and muscle spasms, including painful spasms of the larynx making it difficult for the infected person to drink. The disease must be treated before symptoms appear. The treatment is a series of injections containing weakened rabies virus. A vaccine is available to protect against rabies.

racemate *See* racemic mixture.

raceme A type of inflorescence in which the separate flowers are attached to a main, undivided peduncle by pedicels of equal length at equal distances. *See* Fig. 179.

racemic mixture (racemate) An equimolar mixture of the dextrorotatory and laevorotatory isomers of the same compound. Such a mixture shows no rotation of the plane of polarised light, i.e. it is optically inactive (a ± form). The separation of a racemic mixture is described as resolution. It is often a difficult process to carry out in the laboratory. However, it is sometimes possible to use a chemical method in which each of the two isomers is brought to react with another compound which is optically active. The products formed in this process (called diastereoisomers) are not mirror images and do not have identical properties, e.g. one may be less soluble in a particular solvent than the other. This means that the racemic mixture can be separated by fractional crystallisation. After the separation of the diastereoisomers they are converted into the original reactants. In practice the type of compound formation that takes place most readily is salt formation using optically active acids or bases; that is, a racemic acid can be separated by an active base, and a racemic base by an active acid. *See also* optical isomerism.

racemose Describes inflorescences of the following types: capitulum, corymb, raceme, spike and umbel. *Compare* cymose. *See* Fig. 179.

rachis 1. The main axis of an inflorescence. 2. In a compound leaf, the axis to which leaflets are attached. 3. The shaft of a feather.

radar A word derived from the phrase Radio Detection and Ranging. Radar is a system used to detect and locate objects by reflected radio waves. The radio waves (microwaves) may be transmitted from an aerial as beams of continuous waves or as pulses. If these waves strike an object they are reflected back to a receiver which is usually the same aerial as the transmitter. Radio waves travel with the speed of light; so if the time taken for the transmitted and reflected beam is measured, together with its direction, the location of the object is found. The receiver may be a cathode-ray tube, on the screen of which detected objects appear as bright spots. Radar has many applications such as in air traffic control, in navigation at sea and in meteorology.

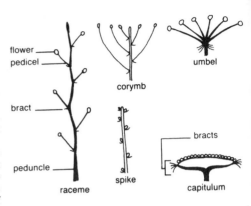

Fig. 179 Types of racemose inflorescence

radially symmetrical Capable of being cut along any diameter to produce two (almost) identical halves. Coelenterates are radially symmetrical. The term also applies to plants. *See also* bilaterally symmetrical.

radian Symbol: rad. The SI unit of angle. One radian is the angle subtended at the centre of a circle by an arc which is equal in length to the radius of the circle.
2π radians = 360° and
1 radian = 57,296° (57.296°)
See Fig. 180.

radiant energy Energy transmitted in the form of radiation, especially electromagnetic radiation. Heat is a form of radiant energy consisting of invisible electromagnetic waves which are partly absorbed and partly reflected by surfaces on which they fall. The amount of absorption and emission of heat radiation depends on the nature of the surface. Matt black surfaces are both good absorbers and good emitters of radiation. Polished surfaces are good reflectors of radiation. *See also* Leslie's cube.

radiant heat *See* infra-red radiation.

radiation 1. A method of transfer of heat energy which does not require a material medium. *Example:* heat energy from the Sun

radius of curvature

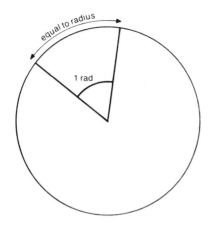

Fig. 180 Radian

to the Earth passes through a vacuum (space). **2.** Any form of energy emitted as waves or particles, e.g. light, sound, alpha particles and beta particles.
radiator *See* heat exchanger.
radical A group of atoms which constitutes a part of many inorganic and organic molecules. *Examples:* sulphate, SO_4, nitrate, NO_3, ammonium, NH_4, and hydroxyl, OH, are inorganic radicals; methyl, CH_3, ethyl, C_2H_5, and phenyl, C_6H_5, are organic radicals. Radicals have an incomplete electron structure and are therefore usually very reactive. During chemical reactions free radicals may have a brief existence.
radicle The embryonic root, i.e. the first structure to emerge from a germinating seed.
radio The transmission and reception of communications by means of radio waves, i.e. without connecting wires. *See also* modulation.
radioactive Describes isotopes which exhibit radioactivity.
radioactivity The property of spontaneous disintegration of certain isotopes, accompanied by the emission of radiation. *See also* alpha particle, beta particle *and* gamma rays.
radio-biology The branch of science concerned with the effects of radiation and radioactivity on living organisms.
radio-carbon dating A method of determining the age of ancient organic material. Natural carbon contains carbon-12, $^{12}_{6}C$, and a small constant proportion of the radioisotope carbon-14, $^{14}_{6}C$, with a half-life of 5730 years. This radioactive carbon is produced constantly in the atmosphere as a result of cosmic radiation. All living organisms contain carbon derived from atmospheric carbon dioxide; after death they cease to absorb carbon and the ratio of carbon-12 to carbon-14 increases because of the decay of carbon-14. By measuring the carbon-14 concentration in a specimen of unknown age, it is possible to estimate how long the specimen has been dead.
radio-chemistry The branch of science concerned with the use of radioisotopes in chemistry.
radio-frequency (RF) Electromagnetic radiation in the frequency range 3 kilohertz (kHz)–300 gigahertz (GHz).
radiography The production of an image of the internal structure of a body on fluorescent screens or photographic plates by passing X-rays or gamma rays through it.
radioisotope An isotope possessing radioactivity. The isotope may be natural or artificial.
radiology The study of X-rays, gamma rays and other penetrating radiations.
radiometer *See* Crookes' radiometer.
radio telescope An instrument used by astronomers for the detection of radio waves emitted by heavenly bodies. It may consist of a large parabolic reflector (dish) which collects and focuses radio waves onto an aerial connected to a radio receiver. Many heavenly bodies not visible in optical telescopes have been discovered using radio telescopes.
radio-therapy The treatment of various diseases such as cancer using X-rays, gamma rays or other sources of short wave radiation. This may be carried out by exposing the patient to external radiation or to internal radiation from radioisotopes introduced into the body.
radio waves A form of electromagnetic radiation with wavelengths ranging from less than one millimetre to many kilometres. Radio waves are produced as a result of the oscillations of electrons in conductors. They travel at the speed of light.
radium An element with the symbol Ra; atomic number 88; relative atomic mass 226; state, solid. Radium belongs to the alkaline earth metals. It is radioactive; the most stable isotope is $^{226}_{88}Ra$ with a half-life of 1602 years. Radium is found in several ores such as pitchblende, from which it was first extracted. It is used as a radioactive source in research and medicine.
radius **1.** One of the two distal bones of the forelimb in tetrapods. In humans, it is the shorter and smaller bone located between the elbow and the thumb side of the wrist. *See* Fig. 162. **2.** In insects, a wing vein.
radius of curvature The radius of the sphere of which the surface of a lens or mirror is a part. For a spherical mirror, the radius of curvature (r) is equal to twice its focal length (f).

247

radon An element with the symbol Rn; atomic number 86; relative atomic mass 222; state, gas. Radon is a very inert, radioactive element belonging to the noble gases. It is the heaviest of all gases, is colourless and is found in very small amounts in air. The most stable isotope is $^{222}_{86}$Rn, with a half-life of 3,824 days (3.824 days). Radon is used in radio-therapy.

radula (pl. radulae) In the mouth of most molluscs, a horny, movable strip bearing rows of teeth. It is used as a feeding mechanism and is continually replaced as it wears away.

raffinose A carbohydrate consisting of three molecules (units) of monosaccharide, $C_6H_{12}O_6$: i.e. raffinose is a trisaccharide (oligosaccharide) with the formula $C_{18}H_{32}O_{16}$. Raffinose is found in sugar-beet, cereals and some fungi. When completely hydrolysed it is broken down to the monosaccharides glucose, fructose and galactose.

rainbow An arc of the spectrum colours produced as a result of refraction and total internal reflection of rays of sunlight passing through raindrops. A bright, primary bow with violet on the inside of the arc is produced as a result of one total internal reflection. A dimmer, secondary bow with violet on the outside of the arc may also be seen. This is caused by two internal reflections within the raindrops. A rainbow can only be seen when the Sun is low in the sky and behind the observer. *See* Fig. 181 (a) *and* (b).

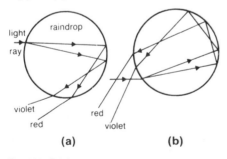

Fig. 181 Rainbow
(a) Primary bow (b) Secondary bow

Ranvier's node *See* node of Ranvier.

Raoult's law A law stating that equimolecular, unionised electrolytes of different solutes in the same amount of the same solvent at the same temperature elevate the solvent's boiling-point, lower its freezing-point and reduce its vapour pressure equally. The law only holds for dilute solutions.

rare earths *See* lanthanoids.

rarefaction (of sound) Part of a sound wave (longitudinal wave) in which molecules of a gas, liquid or solid are spread out, making the pressure at that place a little below normal at that instant. *See* Fig. 238; *compare* compression.

rare gas *See* noble gas.

rate constant *See* mass action (law of).

ratemeter A type of counter for detecting ionising particles or radiation. *See also* Geiger–Müller counter.

rate of reaction In a chemical reaction, a measure of either the amount of reactants used or the amount of products formed in unit time. The rate of reaction is dependent on several factors such as concentration, temperature, and sometimes catalyst and pressure. *See also* mass action (law of) *and* order of reaction.

ray **1.** *See* medullary ray. **2.** A narrow beam of radiation used to represent the path of radiations such as light. **3.** *See* fin-ray.

ray diagram A diagram (drawing) made to show the paths of rays of light through optical systems or to measure image positions and magnifications, etc., produced by lenses and mirrors.

rayon Artificial fibre derived from cellulose. There are two types of rayon, acetate rayon and viscose rayon. The former is made from cellulose ethanoate (cellulose acetate) which is dissolved in an organic solvent and then forced through fine holes into hot air. The solvent evaporates and fibres of acetate rayon form and harden. Viscose rayon is made by treating cellulose with a concentrated solution of sodium hydroxide and carbon disulphide. This solution is then forced through fine holes into a bath of dilute sulphuric acid and cellulose is re-formed as fibres. Rayon is mainly used in the textile industry.

reactance Symbol: X; unit: ohm (Ω). A component of the impedance of a circuit to the flow of an alternating current. It is caused by the presence of capacitance or inductance.

reactant A substance which takes part in a chemical reaction and which is there at the beginning. *Example:* in the reaction in which hydrochloric acid, HCl, reacts with iron(II) sulphide, FeS, to form hydrogen sulphide, H_2S, and iron(II) chloride, $FeCl_2$:

FeS + 2HCl → H_2S + $FeCl_2$

there are two reactants, hydrochloric acid and iron(II) sulphide. *Compare* product.

reaction **1.** *See* chemical reaction. **2.** *See* Newton's laws of motion.

reaction time (latent period) The interval between the time when a stimulus is applied to an organism and the time when a response to it is detected.

reactivity series of metals *See* activity series of metals.

reactor *See* nuclear reactor.

reagent Any substance used in qualitative analysis, quantitative analysis and synthesis, i.e. any substance which causes a chemical reaction.

real depth *See* apparent depth.

real focus *See* focus.

real image *See* image.

'real-is-positive' convention *See* lens formula *and* mirror formula.

Réaumur scale A scale of temperature in which the freezing-point (ice-point) and boiling-point (steam-point) of water are 0°R and 80°R respectively. The scale is rarely used for scientific purposes.

receiver 1. A container in which a distillate is collected. 2. The part of a communications system, e.g. radio, telephone and television, that receives transmissions and converts them into the desired form. *See also* transmitter.

receptacle (thalamus, torus) In flowering plants, the apex of the flower stalk, bearing the floral organs. *See* Fig. 75.

receptor (sense organ) A cell, tissue or organ receiving and responding to internal and external stimuli. *Examples:* the ears, eyes and nose are receptors.

recessive *See* dominant.

reciprocal ohm *See* mho *and* siemens.

recovery period 1. In a Geiger–Müller counter, the very short time (10^{-4} seconds) taken for the instrument to return to a state of readiness to detect another pulse of current. 2. *See* impulse.

rectification 1. Purification of a liquid by distillation. The term is most often used in the distilled liquor industry to denote the process in which alcoholic products are manufactured by distillation. 2. Conversion of an alternating current to a direct current. *See also* half-wave rectification *and* full-wave rectification.

rectified spirit A mixture of 96% ethanol and 4% water (by volume). It is a constant boiling (azeotropic) mixture, boiling at 78°C.

rectifier An electrical component or circuit which converts alternating current to direct current. A common type of rectifier is the diode.

rectilinear propagation of light Light travelling in straight lines through a homogeneous medium. This effect is seen in the formation of shadows.

rectum The posterior part of the large intestine. It stores and also expels faeces through the anus or cloaca. *See* Fig. 49.

red blood cell *See* blood corpuscle.

red blood corpuscle *See* blood corpuscle.

red lead *See* lead oxides.

red marrow *See* bone marrow.

redox potential A measure of the reducing and oxidising powers of redox systems. This can be expressed quantitatively as a series of numbers in volts. The higher the reducing power of a metal, the lower the redox potential; the higher the oxidising power of a metal ion, the higher the redox potential. As a single electrode potential cannot be measured, the redox potential may be measured against a standard hydrogen electrode at 25°C with the metal in contact with a 1 molar solution of its ions. The redox potential obtained may then be called the standard redox potential or the standard electrode potential. A series of metals arranged according to increased standard redox potentials is called the electrochemical series of these metals. *Examples:*

$Na^+ + 1$ electron $\rightleftharpoons Na$ $(-2,71\,V;\ -2.71\,V)$
$Ag^+ + 1$ electron $\rightleftharpoons Ag$ $(+0,80\,V;\ +0.80\,V)$

The two standard redox potentials, $-2,71\,V$ $(-2.71\,V)$ and $+0,80\,V$ $(+0.80\,V)$, indicate that sodium, Na, is a stronger reducing agent than silver, Ag, or that the silver ion, Ag^+, is a stronger oxidising agent than the sodium ion, Na^+. Redox potentials can also be obtained for non-metallic redox systems. The well-known oxidising ability of the halogens is reflected in their standard redox potentials:

$F_2 + 2$ electrons $\rightleftharpoons 2F^-$ $(+2,87\,V;\ +2.87\,V)$
$Cl_2 + 2$ electrons $\rightleftharpoons 2Cl^-$ $(+1,36\,V;\ +1.36\,V)$
$Br_2 + 2$ electrons $\rightleftharpoons 2Br^-$ $(+1,08\,V;\ +1.08\,V)$
$I_2 + 2$ electrons $\rightleftharpoons 2I^-$ $(+0,53\,V;\ +0.53\,V)$

From this it is seen that the ability of the halogens to act as oxidising agents decreases from fluorine, F_2, to iodine, I_2. *Example:* bromine, Br_2, will displace iodine from an aqueous solution of an iodide, I^-:

$Br_2 + 2I^- \rightarrow I_2 + 2Br^-$

whereas the opposite process is not possible. *See also* Appendix.

redox reaction A chemical process in which one substance is reduced and another is oxidised at the same time. *Example:* iron, Fe, reacts with chlorine, Cl_2, to form iron(III) chloride, $FeCl_3$:

$2Fe + 3Cl_2 \rightarrow 2FeCl_3$

In this reaction the iron is oxidised, its oxidation number rising from 0 to $+3$, and the chlorine is reduced, its oxidation number falling from 0 to -1. Redox reactions involve the transfer of electrons from the reducing agent, i.e. iron in the above example, to the oxidising agent, i.e. chlorine in the above example. In this example 6 electrons have been transferred. This transfer of electrons means that reduction is always accompanied by oxidation and vice versa. *See also* disproportionation.

reducing agent (reductant) A substance which can reduce another and, at the same time, be oxidised itself. *Example:* in the

reducing sugar

reaction between iron, Fe, and chlorine, Cl_2, iron(III) chloride, $FeCl_3$, is formed:
$2Fe + 3Cl_2 \rightarrow 2FeCl_3$ (Fe^{3+}, $3Cl^-$)
Here the iron is the reducing agent, reducing Cl to Cl^- and being oxidised itself to Fe^{3+}. A reducing agent is therefore a donor of electrons. *Compare* oxidising agent; *see also* redox reaction.

reducing sugar A sugar which is able to reduce Benedict's solution or Fehling's solution, i.e. a sugar containing an atomic group which is easily oxidised (aldehyde and ketone groups). All the monosaccharides, e.g. glucose and fructose, and some of the disaccharides, e.g. lactose and maltose, are reducing sugars.

reductant *See* reducing agent.

reduction Originally, reduction was considered to be a process in which oxygen was removed from a substance. *Example:* when copper(II) oxide, CuO, reacts with carbon, C, to form copper, Cu, and carbon dioxide, CO_2:
$2CuO + C \rightarrow 2Cu + CO_2$
the copper(II) oxide is said to be reduced. The addition of hydrogen, H_2, to a substance was also called reduction. *Example:* when chlorine, Cl_2, reacts with hydrogen:
$Cl_2 + H_2 \rightarrow 2HCl$
the chlorine is said to be reduced. However, reduction has developed into a far broader concept and is now considered to be a process in which one or more electrons are taken up by an atom, ion or molecule, thus producing reduction of these particles. *Example:* when a zinc rod, Zn, is placed in a solution containing copper(II) ions, Cu^{2+}, free copper is deposited on the zinc rod:
$Cu^{2+} + 2e \rightarrow Cu$
i.e. two electrons are donated by the zinc, which is therefore oxidised. Reduction may also take place electrically at the cathode during electrolysis. *Compare* oxidation; *see also* redox reaction, reducing agent, oxidising agent *and* oxidation number.

reduction division *See* meiosis.

reduction-oxidation process *See* redox reaction.

reed *See* abomasum.

reference electrode *See* calomel electrode.

referred pain A pain which a person feels some distance away from its true origin. *Example:* a person with a certain heart trouble may feel pain in the left arm and shoulder because nerve-impulses from the heart and the left arm run to the same part of the spinal cord. *See also* phantom pain.

refine Purify.

reflecting telescope (reflector) *See* telescope.

reflection The process in which radiation incident on a surface 'bounces back' instead of penetrating the surface. *See also* diffuse reflection *and* regular reflection.

reflection, angle of *See* angle of reflection.

reflection of light, laws of Two laws, stating that (1) The incident ray, the reflected ray and the normal at the point of incidence all lie in the same plane. (2) The angle of incidence is equal to the angle of reflection. *See* Fig. 4.

reflection, total internal *See* total internal reflection.

reflector *See* reflecting telescope *and* parabolic reflector.

reflex An automatic, involuntary response to a stimulus. The stimulus and response follow a sequence of nerves connected to the spinal cord and, in the case of head reflexes, the brain. This sequence of nerves is called a reflex arc. It causes the response to be constant and immediate. A common reflex is the knee-jerk reflex. *See* conditioned reflex *and* unconditioned reflex.

reflex arc The path followed by nerve-impulses during a reflex. A reflex arc consists of a chain of neurones. A receptor receives the stimulus, causing impulses to be transmitted via a sensory neurone to a synapse in the central nervous system. From here they are relayed via a motor neurone to an effector, e.g. a muscle. *See* Fig. 182.

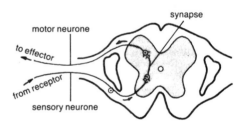

Fig. 182 Reflex arc

reflux A method of boiling a liquid for a long time while ensuring that almost none of it is lost as vapour. This is carried out by connecting a condenser vertically above a flask containing the liquid. Vapour from the boiling liquid condenses and runs back into the flask.

refracting telescope (refractor) *See* telescope.

refraction The change of direction (bending) of light or other electromagnetic radiation as it passes from one medium into another. This is because the radiation travels at different velocities in different media. *See* Fig. 183.

refraction, angle of *See* angle of refraction.

refraction of light, laws of Two laws, stating that (1) The incident ray, the refracted ray and the normal to the point of incidence all lie in the same plane. (2) The ratio of the sine of the

regelation of ice

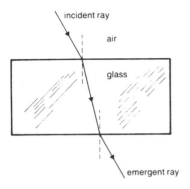

Fig. 183 Refraction of light passing through glass

angle of incidence to the sine of the angle of refraction is a constant for a given pair of media:

$$\frac{\sin i}{\sin r} = n$$

This is called Snell's law. *See also* refractive index.

refractive index Symbol: n. The constant which denotes the amount of refraction taking place when light passes from one medium to another. If the first medium is a vacuum, the refractive index is given by the ratio

$$n = \frac{c_0}{c}$$

where c_0 is the velocity of light in a vacuum and c is the velocity of light in the second medium. This is called the absolute refractive index. The relative refractive index of two other media (medium 1 and medium 2) may be found from the expression

$$_1n_2 = \frac{c_1}{c_2}$$

where c_1 and c_2 are the velocities of light in the respective media. The relative refractive index can also be found from the expression

$$n = \frac{\sin i}{\sin r}$$

where i is the angle of incidence and r is the angle of refraction. In general, the refractive index for any substance varies with the wavelength of the refracted light and the temperature. *See also* apparent depth, total internal reflection *and* refraction of light, laws of.

refractometer An optical apparatus used in qualitative and quantitative analyses for the measurement of the refractive index of a substance.

refractor Refracting telescope. *See* telescope.

refractory A substance with a very high melting-point, which is therefore able to withstand high temperatures. A refractory may be used as a lining material for furnaces.

refractory period *See* impulse.

refrigerant A volatile liquid or an easily liquefied gas used as the cooling medium (coolant) in a refrigerator. Examples of refrigerants are Freons, ammonia and sulphur dioxide.

refrigerator A device for keeping a freezing compartment at a low temperature. A volatile liquid may act as a refrigerant by evaporating, thus absorbing heat from the surroundings (the freezing compartment). When the vapour is compressed it condenses, giving off latent heat to external cooling fins. This cycle is then repeated. The temperature inside a refrigerator is regulated by a thermostat that switches the compressor motor on and off. There are several types of refrigerators, some of which operate without a compressor. *See* Fig. 184.

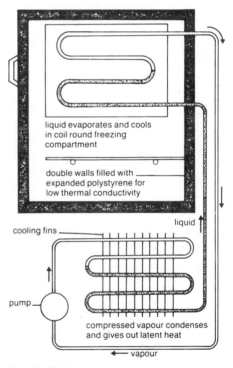

Fig. 184 Refrigerator

regelation of ice The effect of increased pressure on ice is to lower its melting-point. Once the pressure is removed, the melted ice refreezes. This process is known as regelation. Regelation will allow a copper wire with heavy weights attached at each end to pass through a block of ice leaving the block in one piece. *See* Fig. 185.

regeneration

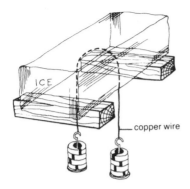

Fig. 185 Regelation

regeneration The renewal of a tissue or organ that has been damaged. It is most common in plants; in mammals it is restricted to the healing of wounds and the replacement of peripheral nerve-fibres. Planarian worms are often used in school laboratories in experiments to demonstrate the animals' ability to regenerate.

regular reflection A type of reflection from regular (smooth) reflecting surfaces in which the angle of incidence is equal to the angle of reflection. *See* Fig. 186; *compare* diffuse reflection.

Fig. 186 Regular reflection

relapsing fever A disease caused by certain spirochaetes and transmitted to humans by lice and ticks. The spirochaetes enter through cuts and bites on the skin, and their multiplication in the blood causes a fever which recurs at regular intervals. The disease is treated by maintaining a high blood concentration of penicillin for up to fourteen days.

relative atomic mass (atomic weight) Symbol: A_r. The relative atomic mass of an atom (element) is the ratio of the average mass of one atom of the naturally-occurring element to $1/12$ of the mass of a carbon-12 atom, $^{12}_{6}C$. *Example:*
A_r (Fe) = 55,85 (55.85)
a dimensionless number. *See also* Dulong and Petit's law.

relative density (specific gravity) Symbol: d. The relative density of a substance is defined as the density of the substance at 20°C divided by the density of water at 4°C, or as the ratio of the mass of any volume of the substance to the mass of an equal volume of water measured at the same temperature as the substance. Relative density for a gas is often given with respect to hydrogen. Relative density has no unit as it is a ratio.

relative density bottle *See* density bottle.

relative front *See* R_F-value.

relative humidity A measure of the amount of water vapour in air. It can be expressed in the following ways: (1) Relative humidity is the ratio of the mass of water vapour in a given sample of air to the mass of water vapour required to saturate the same volume of air at the same temperature. (2) Relative humidity is the ratio of the saturated vapour pressure of water at dew point to the saturated vapour pressure of water at the original air temperature. Note that the higher the air temperature the more water vapour the air can hold. *See also* absolute humidity *and* hygrometer.

relative molecular mass (molecular weight) Symbol: M_r. The relative molecular mass of an element or chemical compound is the ratio of the average mass of one molecule of the naturally-occurring element or compound to $1/12$ of the mass of a carbon-12 atom, $^{12}_{6}C$. The relative molecular mass of a molecule is the sum of the relative atomic masses of all atoms present in that molecule. *Example:*
M_r(CaO) = 56,08 (56.08)
a dimensionless number.

relative refractive index *See* refractive index.

relative vapour density *See* vapour density

relaxin A hormone produced by the corpus luteum to reduce uterine contractions. During parturition, relaxin stimulates dilation of the cervix and relaxation of the pubic symphysis to enable the pelvic girdle to widen.

relay An electrically-operated switch in which a current passing through one circuit opens or closes a second separate circuit. In Fig. 187, a current passing through a solenoid causes an armature to be attracted to it, thus pushing together two contacts to complete another circuit. In this way a small current can control a large supply of current.

renal Pertaining to the kidneys.

renal artery In vertebrates, one of a pair of arteries which convey blood to the kidneys. In humans the left renal artery is shorter than the right renal artery.

renal corpuscle *See* Malpighian corpuscle.

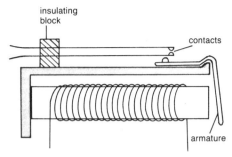

Fig. 187 Relay

renal portal system In fish, amphibians and some reptiles, the part of the circulatory system which conveys blood from the posterior end of the body to the kidneys. The blood is carried to the kidneys in a pair of renal portal veins and returns to the heart via renal veins.
renal portal vein See renal portal system.
renal vein In vertebrates, one of a pair of veins which convey blood from the kidneys to the heart.
reniform Shaped like a kidney.
rennet See rennin.
rennin (rennet) A proteinase which is a coagulative enzyme found in the gastric juice of mammals. It catalyses the precipitation of casein of milk. Rennin is used in cheese-making.
replication The process by which exact copies (replicas) of a complex molecule such as DNA or an organelle are made. This is necessary for growth and reproduction.
reproduction See asexual reproduction and sexual reproduction.
Reptilia A class of cold-blooded vertebrates including alligators, crocodiles, lizards, snakes, tortoises and turtles. Characteristics include skin covered with scales or horny plates, paired limbs which, when present, are short, breathing by lungs, and an embryo protected by an amnion and allantois in an eggshell.
reserve cell An electric cell to which the electrolyte is first added at the time of use. Such a cell can be stored for a very long time before it is activated. Reserve cells are used for emergency lighting, communications equipment, etc.
residual air The volume of air remaining in the lungs after forced expiration.
residue The solid substance which remains in a container after evaporation of a liquid, or on a filter paper after filtration.
resin A complex, organic, acidic, insoluble substance, usually obtained from the sap of certain plants. Resins are soluble in many organic solvents and may produce important substances such as turpentine and balsams. Resins are used in the manufacture of certain varnishes, adhesives, etc.
resistance Symbol: R. A measure of the ability of a material to oppose the passage of an electric current. It is equal to the ratio of the potential difference across the material to the current flowing through it. The resistance depends on the nature of the material, its dimensions, its temperature and sometimes on the amount of radiation falling on it. The SI unit of resistance is the ohm (Ω). See also impedance and Ohm's law.
resistance box A device containing resistance coils of fixed resistance, connected in a specially constructed box. The resistance coils in one type of resistance box may be brought into circuit by removing brass plugs from the brass blocks connecting the coils. The resistance box is used to provide resistances of known values. See Fig. 188.

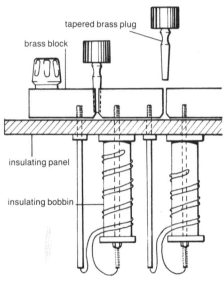

Fig. 188 Constructon of a resistance box

resistance thermometer A type of thermometer whose action depends on the changes in electrical resistance of a wire with changes in temperature. The wire is usually a coil of thin platinum wire through which a current is passed. It is connected to a meter which indicates the temperature of the body in which the coil is immersed.
resistivity Symbol: ρ (rho). A measure of the resistance of a material in terms of its dimensions. It is equal to the resistance of a 1 metre length of a conductor made of the material of

cross-sectional area 1 square metre. It can be given by the equation

$$\rho = \frac{RA}{l}$$

where R is the resistance of a uniform conductor of length l and cross-sectional area A. The SI unit of resistivity is ohm metre ($\Omega\,m$). The resistivity is the reciprocal of the conductivity.

resistor A component in an electric circuit used to introduce resistance. Resistors may be made of carbon or of resistance wire such as Nichrome.

resolution 1. *See* racemic mixture. 2. *See* resolving power.

resolving power (resolution) The ability of an optical instrument such as a microscope or telescope to produce distinguishable images. The resolving power depends on the wavelength of the radiation and on the aperture. *See also* electron microscope *and* oil immersion.

resonance 1. For many organic and inorganic compounds, one single structural formula is not enough to account for all their known properties. For example, the benzene molecule may be written

However, the formula

is also correct. The two structures for benzene may be written together in this way:

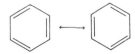

The double-headed arrow indicates that there is resonance between the two structures, i.e. that the actual structure of benzene lies between the two representations. The bonding electrons are delocalised in the molecule; each carbon–carbon bond is neither a simple single bond nor a simple double bond, but a bond intermediate between these. *See also* tautomerism. 2. A condition in which a particular body or system capable of oscillation responds by vibrating at its natural frequency as a result of impulses received from another vibrating system oscillating at the same frequency. This condition may set up vibrations of large amplitude.

respiration The process in living organisms in which chemical energy is liberated by means of a series of chemical reactions (redox reactions), the waste products of which may be carbon dioxide, CO_2, and water, H_2O. Respiration may be divided into two consecutive processes, aerobic respiration and anaerobic respiration. In aerobic respiration, oxygen is consumed. If glucose, $C_6H_{12}O_6$, is being used as a basis, the reaction is

$C_6H_{12}O_6 + 6O_2 \rightarrow 6CO_2 + 6H_2O + \text{energy}$

This process takes place partly in the cytoplasm by means of respiratory enzymes and partly in the mitochondria, and is called tissue respiration or internal respiration. It provides energy for all the chemical activities of the cell. Anaerobic respiration takes place in the absence of oxygen. However, the end-products are not CO_2 and H_2O as in aerobic respiration, but CO_2 and ethanol, C_2H_5OH (in plants), or lactic acid, CH_3—CHOH—COOH (in animals):

$C_6H_{12}O_6 \rightarrow 2CO_2 + 2C_2H_5OH + \text{energy}$

or

$C_6H_{12}O_6 \rightarrow 2CH_3$—CHOH—COOH + energy

Anaerobic respiration yields much less energy than aerobic respiration because the breakdown of glucose is less complete in anaerobic respiration. The energy released in respiration is stored in ATP. Many aerobic organisms can respire anaerobically for some time if their oxygen supply is cut off (oxygen debt). A few organisms respire anaerobically, including some yeasts, other fungi and bacteria. Some bacteria are strictly anaerobic: they may be poisoned by oxygen even in small concentrations. The term external respiration is used as an alternative name for breathing. *See also* fermentation, glycolysis *and* Krebs cycle.

respiratory pigment A pigment frequently found in blood. It combines readily and reversibly with oxygen at a certain high partial pressure of the gas and it releases oxygen as readily at a low partial pressure. In vertebrates and many invertebrates, haemoglobin is a respiratory pigment located in the red blood corpuscles. Some invertebrates do not have haemoglobin but have other respiratory pigments such as haemocyanin, which is often dissolved in the plasma instead of being carried in corpuscles. Such respiratory pigments have a lower capacity of carrying oxygen than haemoglobin. *See also* cytochrome.

response The activity in a living organism when exposed to a stimulus.

rest mass The mass of a particle when it is at rest or moves with a velocity which is low compared with the velocity of light. The rest mass of the electron is $9,109 \times 10^{-31}$ kg (9.109×10^{-31} kg); however, when the electron is moving at very high velocities its mass increases. The relation between rest mass and velocity can be seen from the equation:

$$m = \frac{m_0}{\sqrt{1-(v/c)^2}}$$

in which m_0 is the rest mass of the electron, m the mass of it when it moves with velocity v, and c the velocity of light. The equation also applies to any other body of rest mass m_0 moving with an extremely high velocity.

resultant A single vector which is equivalent to the sum of a set of vectors, e.g. velocities, forces. The resultant is equal and opposite to the equilibrant. See Fig. 159.

retardation See acceleration.

reticulum (pl. reticula) The second stomach in ruminants.

retina In vertebrates, the innermost, light-sensitive layer of the eye. It consists of rods and cones. See Fig. 66.

retinol See vitamin.

retort 1. A glass container consisting of a bulb with a long, bent neck. It is used in the laboratory for distillation. See Fig. 189. 2. In the chemical industry, a retort may be any container in which a chemical reaction is performed.

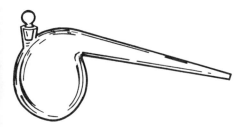

Fig. 189 Retort

reverberatory furnace A furnace in which the fuel e.g. coal, gas or oil, is not mixed with the charge. *Example:* the heat from the burning fuel may be directed down onto the charge from a curved roof.

reversible reaction A chemical reaction which proceeds in either direction, e.g. the production of ammonia, NH_3, from its elements:

$$N_2 + 3H_2 \rightleftharpoons 2NH_3$$

A reversible reaction will reach a state of equilibrium which is not static but dynamic, i.e. the rate of the forward reaction is just equalled by the rate of the backward reaction. The equilibrium may be displaced to the right or the left by changing conditions such as temperature, pressure and concentrations of reactants or products. A catalyst will have no effect on the displacement of the equilibrium as it alters the rate of reaction to the same extent for both processes. *Compare* irreversible reaction.

RF See radio-frequency.

R_F-value (relative front) A value used in chromatography such as paper chromatography. It represents the ratio of the distance travelled by the sample or by a reference substance to the distance travelled by the solvent (mobile phase). As the mobile phase always travels at least as far as the sample, the R_F-value is ≤ 1. By comparing R_F-values of unknown substances with R_F-values of known substances (reference substances), a sample may be identified.

Rh See Rhesus factor.

rheostat A variable resistor which may be constructed in various ways. Fig. 190 illustrates a type commonly used in school laboratories.

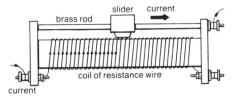

Fig. 190 Rheostat

Rhesus factor (Rh-factor) An inherited substance (antigen) found in the red blood corpuscles of most human beings and Rhesus monkeys. People whose blood contains this substance are called Rh-positive; those who lack it are called Rh-negative. In blood transfusions it is essential that a Rh-negative person does not receive blood accidentally from a Rh-positive donor as antibodies (anti-Rh factor) may form in the blood of the recipient. However, this is not dangerous unless a second, similar transfusion is given. If this happens, the antibodies cause agglutination in the recipient's blood which may prove fatal. Blood tests are always performed before a transfusion is given to guard against this. Agglutination may occur during the pregnancy of a Rh-negative mother bearing a Rh-positive child. Red blood cells from the foetus may pass across the placenta into the mother's bloodstream which responds by producing Rhesus antibodies. These return to the bloodstream of the foetus, destroying its red blood cells. Usually the first child is not seriously affected as the antibodies are not formed sufficiently quickly. However, subsequent children suffer the most serious form of the disease, and unless their blood is replaced by Rh-negative blood the disease may

rheumatic fever

prove fatal. Recently it has been found that the formation of antibodies can be prevented by injecting the mother with anti-Rh globulin.

rheumatic fever An acute, infectious disease often affecting children and young adults. It is caused by a toxin produced by streptococci and usually follows an untreated infection such as strep throat. Symptoms include fever, swollen joints and inflammation of the pericardium and heart valves. The disease is treated with penicillin.

rheumatism A general term for many conditions involving pain in the joints, bones and muscles. *See also* arthritis.

rhizoid A unicellular or multicellular hair-like structure found on the base of moss stems, and on liverworts, fern prothalli, some algae and fungi. Rhizoids function as roots. *See* Fig. 191.

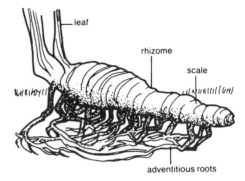

Fig. 192 Iris rhizome

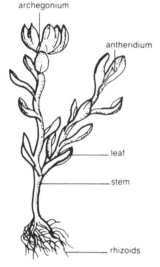

Fig. 191 Rhizoids on a moss stem

rhizome A horizontal stem (usually underground) serving as a means of perennation and vegetative reproduction. It bears both roots and shoots. *See* Fig. 192.

rhodochrosite *See* manganese.

rhodopsin *See* visual purple.

rhombencephalon *See* hindbrain.

rhombic sulphur *See* monoclinic sulphur.

rib In most vertebrates, a curved strip of bone forming part of the rib cage. The ribs are attached to the vertebral column. Intercostal muscles are situated between the ribs. *See* Fig. 193; *see also* false rib, floating rib *and* true rib.

rib cage In most vertebrates, a cage formed by the ribs, sternum and vertebral column. It protects the heart, major blood vessels and lungs. Its expansion and contraction produce volume changes of the thorax during breathing. *See* Fig. 193.

riboflavin (lactoflavin) Vitamin B_2. Riboflavin is widely distributed, occurring in milk, yeast, liver, green vegetables, eggs, etc. In humans, a deficiency leads to skin and digestive disorders and to retarded growth in children.

ribonucleic acid (RNA) A substance mainly found in the ribosomes of a cell and concerned with the synthesis of proteins from amino acids. RNA consists of alternating ribose units and phosphate groups with the nitrogenous bases adenine, cytosine, guanine and uracil attached to each ribose unit. RNA is manufactured on DNA in the nucleus of the cell and consists of a single-stranded helix. Several different types of RNA exist, each with a different structure and function. *See also* messenger RNA *and* transfer RNA.

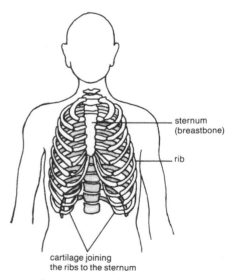

Fig. 193 Rib cage

ribose A carbohydrate (monosaccharide) with the formula $C_5H_{10}O_5$. Ribose is an aldopentose with a furanose structure. It is found as a constituent of RNA, AMP, ADP and ATP.

ribosome A structure in the cytoplasm of a cell, often lining the endoplasmic reticulum. A ribosome consists chiefly of RNA and protein and is concerned with the synthesis of proteins from amino acids. Polysomes (polyribosomes) are clusters of ribosomes linked together by messenger RNA (mRNA). *See also* nucleolus.

Richter scale The scale most commonly used in determining the magnitude of earthquakes. It is a measure of the amplitude of seismic waves, expressed in whole numbers and decimals. Each whole number interval on the scale represents an increase of 10 times in the amplitude of the seismic waves and an increase of 31 times in the energy released by the earthquake. The scale has no lower or upper limits; the largest recorded magnitude is approximately 9 and the smallest −3. A scale value of 2 represents the smallest earthquakes that human beings can feel; those with values of 7 or more are said to be major earthquakes. *See also* seismograph.

rickets A childhood disease caused by a deficiency of vitamin D and resulting in the softening of the bones which may lead to deformities. *Compare* osteomalacia.

rickettsia (pl. rickettsiae) A parasitic micro-organism, intermediate between a bacterium and a virus, living in certain arthropods such as fleas, ticks, lice and mites. Rickettsiae are harmless to the invertebrate host, but cause several disorders such as typhus in a mammal bitten by an infected animal.

right-hand rule *See* Fleming's rules.

Rigil Kent *See* Alpha Centauri.

rigor mortis The temporary stiffening of all the muscles which occurs from one to seven hours after death. This is a result of coagulation of glycogen and decomposition of ATP, resulting in the formation of lactic acid and phosphoric acid. After rigor mortis has set in, the body remains rigid for between one and four days.

ring *See* cyclic.

ring closure *See* cyclisation.

ringed worm *See* Annelida.

Ringer's solution *See* physiological saline.

ring main circuit A domestic electric circuit consisting of a loop of cable connected to the mains supply. From this ring a number of individual power outlets are connected in parallel. *See* Fig. 194.

ringworm A contagious itchy infection of the skin which occurs especially at the roots of hairs and sometimes of nails. It is caused by a fungus and the infection spreads in a circular fashion. The disease is treated with various ointments containing anti-fungal drugs, or with antibiotic tablets.

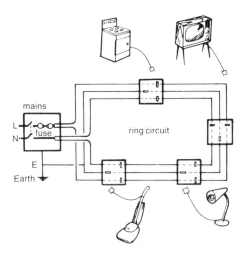

Fig. 194 Ring main circuit

ripple tank A rectangular tank in which the properties of water waves are demonstrated. The apparatus can be used to demonstrate wave properties such as reflection, refraction, diffraction and interference.

river blindness (onchocerciasis) An infection occurring in the tropical regions of Africa and America. It is caused by filarial worms which produce swellings on any part of the body and sometimes damage to the eyes. The infection occurs in people living near to rivers and streams where the intermediate host breeds. Treatment is by means of drugs such as diethyl-carbamazine, antihistamines and cortisone, and by surgical removal of the swellings.

r.m.s. value *See* root-mean-square value.

Rn Chemical symbol for radon.

RNA *See* ribonucleic acid.

roasting A process in which sulphide ores are heated in air to oxidise them. *Example:* the roasting of iron pyrites, FeS_2, yields iron (III) oxide, Fe_2O_3, and sulphur dioxide, SO_2:
$$4FeS_2 + 11O_2 \rightarrow 2Fe_2O_3 + 8SO_2$$
Iron is then obtained by reducing iron(III) oxide.

Rochelle salt *See* potassium sodium tartrate.

rock Any mass of mineral matter forming part of the Earth's crust. Rocks may consist of one or more minerals. *See also* metamorphic rock.

rock crystal *See* quartz.

rock salt (halite) A naturally-occurring, crystalline, mineral form of sodium chloride, NaCl. It is often found as an underground deposit, and is usually mined by pumping water into and out of the ground to dissolve the salt. This produces brine which is boiled and evaporated to obtain the salt. Rock salt may also be formed by evaporation of salt water in lagoons, seas and inland lakes.

rock wool *See* mineral wool.

rod 1. A specialised, light-sensitive cell in the retina of the eye of most vertebrates. Rods are concerned in non-colour vision and vision in dim light. They are rod-shaped and are connected to the brain by nerve-fibres in the optic nerve. *Compare* cone; *see also* visual purple. **2.** *See* bacterium.

rodent *See* Rodentia.

Rodentia The largest and most widespread order of placental mammals, including rats, mice, porcupines, squirrels, guinea-pigs and beavers. Characteristics include a pair of chisel-shaped incisors in each jaw (these teeth grow throughout the animal's life and have enamel only on the front surface), no canines, molar teeth for grinding, rapid breathing, a long intestine and highly-developed caecum. Rodents are either herbivorous or omnivorous, their young are generally born blind and hairless and they are usually nocturnal. Many rodents are pests and may carry fleas which transmit diseases such as bubonic plague.

Roentgen rays *See* X-rays.

Röntgen rays *See* X-rays.

root 1. A plant structure that anchors the plant to the ground and is normally below the surface. Roots absorb water containing dissolved mineral salts from the soil. They are distinguished from underground stems in that they bear neither buds nor leaves. A true root develops from the radicle and may have root hairs and a root cap. *See* Fig. 195 (a) *and* (b). **2.** The embedded portion of an organ or structure, e.g. the root of a hair, tooth, tongue or nail.

root cap A protective cap at the apex of a root. It is constantly replaced during growth. *See* Fig. 195 (a).

root hair A tubular, unicellular outgrowth from an epidermal cell of a young root. Root hairs increase the surface area of the root, thus increasing the absorption of water and mineral salts from the soil. *See* Fig. 195 (a) *and* (b).

root-mean-square value (r.m.s. value, effective value) The square root of the average of the squares of a group of values. *Example:* the r.m.s. value of the values 1, 2, 3, 4 is

$$\sqrt{\frac{(1^2+2^2+3^2+4^2)}{4}}$$

root-mean-square value of an alternating current or voltage This is defined as the square root of the average of the squares of the current (I^2) or voltage (V^2) taken

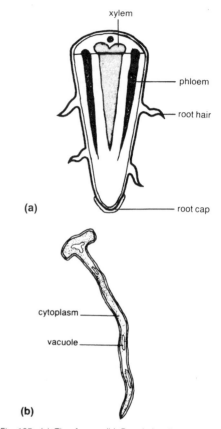

Fig. 195 (a) Tip of root (b) Root hair cell

over a complete cycle. The r.m.s. value of an alternating current may also be defined as the value of the direct current that would produce the same heating effect per second as the alternating current. The r.m.s. value is always less than the peak value and may be given by the expression

$$\text{r.m.s. value} = \frac{\text{peak value}}{\sqrt{2}}$$

root nodule A small swelling on the root of a leguminous plant. Root nodules contain extremely important nitrogen-fixing bacteria. *See* Fig. 196; *see also* Leguminosae.

root pressure A pressure that pushes water up the stem of a plant from its roots.

root tuber *See* tuber.

Rosenmund reaction A reaction in which an acyl halide (acid halide) is reduced to an aldehyde by passing hydrogen gas into a solution of the acyl halide in the presence of a palladium catalyst precipitated on barium sulphate. To prevent further reduction, a

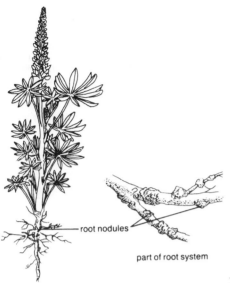

Fig. 196 Root nodules

sulphur-containing substance is added which poisons the catalyst. *Example:*
$CH_3CO—Cl + H_2 \rightarrow CH_3—CHO + HCl$
Here ethanoyl chloride (acetyl chloride), $CH_3CO—Cl$, is reduced to ethanal (acetaldehyde), $CH_3—CHO$.

rose quartz *See* quartz.

Rose's metal An alloy of 2 parts bismuth, 1 part lead and 1 part tin. It has a low melting-point (94°C) and is used in fire-alarms.

rostrum The beak of a bird or any beak-like structure.

rotor The rotating part of an electric motor, generator or turbine.

roughage Indigestible material which stimulates peristalsis and is therefore essential in maintaining good health. Roughage may be obtained by eating plenty of vegetables, fruit and brown bread.

round window *See* fenestra rotunda.

roundworm *See* Nematoda.

royal jelly A secretion produced by glands of young honey-bee workers. Larvae that are to develop into queen bees are fed only on royal jelly.

rubber *See* latex.

rubella (German measles) An acute, infectious, virus disease. Symptoms include fever, red rash and swollen glands in the neck. One attack of rubella generally gives life-long immunity. If a pregnant woman contracts the disease during early stages of her pregnancy, the baby is likely to be born with serious defects. A vaccine against rubella is available.

rubeola *See* measles.

ruby A precious stone composed of alumina, Al_2O_3. Rubies may have a variety of colours but are commonly red.

rumen (paunch) The first stomach in ruminants.

ruminant An animal which chews the cud, i.e. regurgitates partially digested food for further chewing. *See also* Artiodactyla.

runner A specialised stolon which produces adventitious roots at its nodes or at its tip, developing into new (daughter) plants. This is a type of vegetative reproduction. *See* Fig. 197.

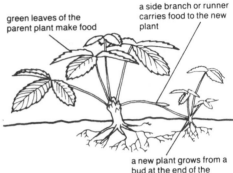

Fig. 197 Strawberry runner

rust 1. A plant disease caused by parasitic fungi. Brownish-yellow (rust-coloured) patches are produced on the host plant. **2.** *See* iron *and* rusting of iron.

rusting of iron The process in which iron (or steel) is attacked by both air (oxygen) and water, forming rust on its surface. Rust which has already formed on iron promotes further rusting. *See also* iron.

rutherfordium *See* transuranic elements.

rutile *See* titanium.

S Chemical symbol for sulphur.

saccharase *See* invertase.

saccharide *See* carbohydrate.

saccharin A solid, white, organic substance used in its sodium form (sodium salt) as an artificial sweetener; when pure, it is about 500 times sweeter than cane-sugar *and* beet-sugar.

Saccharomyces A genus of fungi including those used in the production of bread and alcohol. Yeasts belong to this genus.

saccharose See cane-sugar *and* beet-sugar.

saccule (sacculus) A small, membranous sac in the inner ear, from which the cochlea arises. The saccule is filled with endolymph and its base is covered with a sensory area called a macula. *See* Fig. 55.

sacculus *See* saccule.

sacral vertebrae In humans, the five fused vertebrae forming the sacrum, situated between the coccyx and lumbar vertebrae. *See* Fig. 242.

sacrificial protection A method used to prevent corrosion, especially of iron and steel. This is done by attaching a block of a more electro-positive metal to the metal to be protected. *Example:* steel pipelines and the hulls of ships are protected in this manner by attaching blocks of zinc at intervals along their lengths. The zinc blocks are corroded instead of the steel. *See also* electrolytic corrosion.

sacrum *See* sacral vertebrae.

safety lamp *See* Davy lamp.

safranin A red organic dyestuff which is soluble in water and in alcohol (ethanol). It is used as a biological stain.

sal ammoniac *See* ammonium chloride.

salicylate *See* 2-hydroxybenzoate.

salicylic acid *See* 2-hydroxybenzoic acid.

saline Denotes a solution containing one or more salts, especially those of the alkali metals and magnesium.

salinity The measure of the concentration of salt (sodium chloride), NaCl in a solution or a soil.

saliva The watery secretion of the salivary glands. In humans and most mammals, saliva prepares food for swallowing and rinses the mouth and teeth. The enzyme ptyalin present in it aids digestion of starch. Saliva also surrounds hard and edged particles of food so that damage to the pharynx and oesophagus is avoided.

salivary glands Glands secreting saliva into the mouth. In humans there are three pairs, the parotid, sublingual and submandibular glands.

Salmonella A genus of rod-shaped bacteria (bacilli). All species of *Salmonella* are pathogenic to animals and cause food poisoning in humans.

salt A substance that can be considered to arise from an acid in which all or part of the acidic hydrogen atoms are replaced by a metal or a metal-like group, i.e. a salt consists of anions and cations. Salts may be divided into various groups such as acid salts, basic salts, normal salts, double salts and complex salts.

Acid salts In these salts not all of the acidic hydrogen atoms of an acid have been replaced by a metal ion or other cation. *Examples:* $NaHSO_4$ is sodium hydrogensulphate; $NaHCO_3$ is sodium hydrogencarbonate. Some acid salts, such as $NaHSO_4$, give an acidic solution when dissolved in water. Others, such as $NaHCO_3$, give an alkaline solution when dissolved in water:

$NaHSO_4 \rightarrow Na^+ + HSO_4^-$ and
$HSO_4^- \rightleftharpoons H^+ + SO_4^{2-}$
$NaHCO_3 \rightarrow Na^+ + HCO_3^-$ and
$HCO_3^- + H^+ \rightleftharpoons CO_2 + H_2O$

Acid salts are typical for acids containing more than one acidic hydrogen atom.

Basic salts Salts which are partly salts and partly hydroxides or oxides. *Examples:* $Zn(OH)Cl$ and $Mg(OH)Cl$ are basic zinc chloride and basic magnesium chloride respectively. Such salts are often insoluble in water.

Normal salts Salts in which all the acidic hydrogen atoms in an acid have been replaced by metal ions or other cations. *Examples:* NaCl is sodium chloride, K_2CO_3 is potassium carbonate and NH_4Cl is ammonium chloride. Normal salts may give acidic, alkaline or neutral solutions when dissolved in water. NaCl gives a neutral solution, K_2CO_3 gives an alkaline solution and NH_4Cl gives an acidic solution:

$K_2CO_3 \rightarrow 2K^+ + CO_3^{2-}$ and
$CO_3^{2-} + H^+ \rightleftharpoons HCO_3^-$
$NH_4Cl \rightarrow NH_4^+ + Cl^-$ and
$NH_4^+ \rightleftharpoons H^+ + NH_3$

Double salts Salts containing different cations, which may be prepared by crystallising two salts in equivalent proportions. Alums are double salts. *Example:* $AlK(SO_4)_2 \cdot 12H_2O$ is aluminium potassium sulphate-12-water (potash alum). When dissolved in water, a double salt behaves as a mixture of the salts of which it consists. This distinguishes it from a complex salt.

Complex salts Salts which give complex ions when dissolved in water. *Example:* $K_4[Fe(CN)_6]$ is potassium hexacyanoferrate(II) (a yellow solid) and $K_3[Fe(CN)_6]$ is potassium hexacyanoferrate(III) (an orange solid). When dissolved in water the following ions are present in solution.

$K_4[Fe(CN)_6] \rightarrow 4K^+ + [Fe(CN)_6]^{4-}$ and
$K_3[Fe(CN)_6] \rightarrow 3K^+ + [Fe(CN)_6]^{3-}$

This distinguishes them from double salts.

salt bridge A tube containing potassium chloride in a gel used to connect two half-cells so that their electrolytes do not mix.

salt-cake *See* sodium sulphate.

salt, common *See* sodium chloride.

salting out A process in which a substance is made to separate from a solution by adding salt. The term is especially used in soap-making, in which a saturated solution of sodium chloride causes the soap to separate out as a white solid at the end of the saponification process.

saltpetre *See* potassium nitrate.

Salvarsan® (arsphenamine) The first effective drug against syphilis. Salvarsan was made by Ehrlich in 1909–10; with it he can be said to have founded chemotherapy. Salvarsan is an organic compound containing arsenic which is highly toxic towards the spirochaete *Treponema pallidum* which causes syphilis. Today syphilis is treated with penicillin. *See also* Wasserman test.

samara A dry, one-seeded, winged, indehiscent fruit, e.g. fruits of elm, ash and maple.

sand A hard powder which usually consists of granules of impure silicon(IV) oxide (silica), SiO_2. It is formed from rocks by weathering. Soils containing large amounts of sand retain water poorly.

sandfly A small insect which transmits sandfly fever, a viral disease similar to dengue, and leishmaniasis.

sandstone A sedimentary rock composed of sand-grains cemented together with clay, calcium carbonate, iron oxide, etc. Sandstone is used as a building material.

sap An aqueous solution containing nutrients, found in the vascular system of plants.

saponification The process in which an ester is treated with a solution of a strong alkali, e.g. sodium hydroxide, NaOH. The process yields a salt and an alcohol. *Example:* saponification of the ester ethyl ethanoate (ethyl acetate), $CH_3COOC_2H_5$, with sodium hydroxide produces sodium ethanoate (sodium acetate), CH_3COONa, and ethanol, C_2H_5OH:

$CH_3COOC_2H_5 + NaOH \rightarrow$
$CH_3COONa + C_2H_5OH$

In soap-making, fats and oils which are esters of glycerol and long-chain fatty acids are treated with a solution of a strong alkali to give soap (an alkali salt of a long-chain fatty acid) and the alcohol glycerol. At the end of this process, the soap is salted out by adding a saturated solution of sodium chloride.

sapphire A blue, very hard, precious stone composed of alumina, Al_2O_3.

saprophyte A micro-organism (bacterium or fungus) which lives on dead organic material and obtains nutrients from it. *Compare* parasite.

sapwood (alburnum) Xylem tissue found in the outer layer of a tree trunk or a branch. It consists of living cells and conducts water. Sapwood is softer than heartwood and is not coloured.

sarcoma (pl. sarcomata) A malignant tumour arising from connective tissue such as bones and muscles. *See also* carcinoma.

satellite 1. Any small body orbiting a larger body, i.e. the Moon is the Earth's satellite and the Earth is the Sun's satellite. **2.** Since 1957, artificial satellites have been put into orbit around the Earth and other planets. Some of these are communications satellites and weather satellites. Others are information satellites which transmit data to Earth about space or heavenly bodies.

saturated compound A chemical compound which contains only single bonds in its structure. *Example:* ethane, $CH_3\text{—}CH_3$, is a saturated hydrocarbon. Saturated compounds cannot take part in addition reactions, but may take part in substitution reactions. *Compare* unsaturated compound.

saturated solution A solution which can hold no more solute at a given temperature. In a saturated solution, there is an equilibrium between the solution and its solute. *Compare* unsaturated solution; *see also* supercooling.

saturated vapour Vapour which is in a dynamic equilibrium with the liquid from which it has evaporated at a given temperature, i.e. the rate at which atoms or molecules evaporate is equal to the rate at which they condense.

saturated vapour pressure (SVP) The pressure exerted by any saturated vapour at a given temperature. Saturated vapour pressure depends only on temperature and rises as the temperature rises.

Saturn One of the nine planets in our Solar System orbiting the Sun between Jupiter and Uranus once in 29,46 years (29.46 years). A unique feature of Saturn is its ring system, first observed by Galileo. In 1980, the spacecraft Voyager I flew close to Saturn and its television cameras revealed more details of the ring system. Scientists now believe that the planet has several hundred rings. Some of these are circular, some eccentric and others are inter-twined. Saturn has ten satellites (moons). The largest of these, Titan, is the only moon in the Solar System known to have an atmosphere.

savanna(h) Tropical or subtropical grassland with scattered trees and a vegetation which is capable of withstanding a dry, hot climate.

scabies A highly contagious disease caused by the female scabies mite which burrows into the skin. Symptoms include severe itching. The parts of the skin most often attacked are the sides and backs of the fingers, the wrists, the palms of the hands, the soles of the feet and the buttocks. When the skin is scratched, boils and abscesses occur. Scabies is treated with sulphur ointments, Balsam of Peru or benzene hexachloride (BHC).

scalar A quantity which has magnitude but no direction, e.g. mass, length and time. *Compare* vector.

scale *See* fur.

scale leaf A membranous, tough leaf, usually reduced in size, which protects a bud. *See* Fig. 18.

scalpel A small, sharp knife made of hard, stainless steel. Scalpels are used by surgeons and in the biology laboratory for dissection.

scapula The bone forming the posterior part of the shoulder girdle in many vertebrates. In humans the scapula is a flat, triangular bone commonly called the shoulder-blade. It is connected to the head of the humerus and the lateral end of the clavicle.

scattering The deflection of any radiation by matter (solid, liquid or gas) in its path. Light from the Sun is scattered as it passes through the atmosphere. This occurs when the light strikes dust particles and air molecules. Blue light, with a high frequency (short wavelength), is scattered more than any other colour in the spectrum of white light. The appearance of a blue sky is caused by this effect. *See also* Tyndall effect.

scavenger A carnivore, e.g. the vulture, which eats the dead remains of an animal which has been killed by a predator. However, sometimes a scavenger kills an animal for food, and a predator may sometimes feed on a dead animal.

scheelite *See* tungsten.

schistosomiasis *See* bilharziasis.

Schultze's reagent A solution containing zinc chloride, potassium iodide and iodine. It is used as a test for cellulose, with which it gives a blue colour.

Schwann cells Cells forming the myelin sheath of medullated nerve-fibres. A single Schwann cell forms the part of the sheath between adjacent nodes of Ranvier.

Schweizer's solution *See* cellulose.

sciatic Pertaining to the hip region.

scintillation The production of small flashes of light in a certain substance called a scintillator. This is caused by the impacts between high-energy radiations (such as alpha and beta particles and gamma rays) and the scintillator.

scintillation counter An apparatus consisting of a scintillator combined with a photomultiplier enabling even very weak scintillation to be detected and measured. *See also* spinthariscope.

scintillator A substance such as zinc sulphide which is able to produce a small flash of light when struck by a high-energy radiation.

scion A bud, branch or shoot which is cut and inserted into the growing stem of a tree (the stock). *See also* graft.

sclera (sclerotic) The tough, fibrous, opaque, outer coat of the vertebrate eye. It is continuous with the cornea and the eye muscles are attached to it. *See* Fig. 66.

sclerenchyma 1. Tissue made up of thick-walled cells containing lignin. Sclerenchyma supports and protects the underlying softer tissue. **2.** The hard tissue of corals.

scleroprotein A fibrous protein which is insoluble in water and salt solutions. Scleroproteins are found in animals as a surface covering, e.g. keratin, or as fibres that bind cells together, e.g. collagen and elastin.

sclerosis (pl. scleroses) Hardening of tissue. *See also* arteriosclerosis *and* multiple sclerosis.

sclerotic *See* sclera.

scolex The part of a tapeworm which is often called the 'head'. In the adult worm, the scolex is equipped with hooks and suckers with which it clings firmly to the intestine of the host. The scolex has no mouth.

screening *See* Faraday's cage *and* magnetic shielding (screening).

screw A cylinder or cone of metal with a spiral groove cut in it forming the thread. One revolution of the head of the screw causes the point to move forward against a resistance for a distance equal to the pitch. *See* Fig. 169.

scrotum In most mammals, a sac-like structure of skin containing the two testes. These are kept at a slightly lower temperature than the rest of the body as the higher body temperature inhibits sperm formation. The epididymis and the beginning of the vas deferens of each testis are also contained in the scrotum. In humans the scrotum is situated behind the penis and a short distance in front of the anus. It is divided by a septum into two compartments each containing a testis. *See also* labia majora.

scrub An area of land characterised by a dense cover of brushwood, shrubs and small forest trees.

scum *See* detergent *and* hard water.

scurvy A disease caused by severe lack of vitamin C (ascorbic acid) in the diet. Symptoms include swollen and bleeding gums, loose teeth and blue marks on the skin. The patient is short of breath and feels weak.

Se Chemical symbol for selenium.

sea-water Sea-water, apart from that of inland seas, consists of about 96% water. The remaining 4% is made up of the following salts in varying proportions: sodium chloride, magnesium chloride, magnesium sulphate, calcium sulphate and potassium chloride. Sea-water contains a mixture of the ions of these salts.

sebaceous glands Glands secreting an oily substance called sebum into the hair follicles. The sebum lubricates the shaft of the hair and the skin, also making them waterproof. *See* Fig. 208.

sebum *See* sebaceous glands.

second Symbol: s. The SI unit of time, defined as the time taken by 9 192 631 770 cycles of a certain energy change in the caesium-133 atom, $^{133}_{55}$Cs. *See also* clock.

secondary *See* secondary feather.

secondary cell *See* cell.

secondary coil *See* inducton coil *and* transformer.

secondary colour *See* colour mixing.

secondary consumer *See* consumer.

secondary cortex *See* phelloderm.

secondary emission The emission of electrons from a surface when it is bombarded by electrons from another source. *See also* photomultiplier.

secondary feather (secondary) One of the smaller flight feathers on a bird's wing.

secondary growth (secondary thickening) The growth in diameter of roots and shoots by the activity of cambium. *Compare* primary growth.

secondary medullary ray *See* medullary ray.

secondary phloem *See* phloem.

secondary sexual characteristic A characteristic which develops at the onset of sexual maturity. Secondary sexual characteristics develop as a result of the secretion of androgens in the male and oestrogens in the female. In humans, these characteristics include growth of hair on the pubis and in the armpits, deepening of the voice and growth of beard in boys, and development of the breasts and hip-girdle in girls.

secondary thickening *See* secondary growth.

secondary winding *See* transformer.

secondary xylem *See* xylem.

secretin A polypeptide hormone secreted by the duodenum and jejunum. It enters the blood and is carried to the pancreas and liver, stimulating them to produce pancreatic juice and bile respectively.

secretion 1. The process in which a liquid or solid substance is produced by a cell or gland and passes out of it. **2.** Any substance, e.g. sweat, enzymes and saliva, produced by a cell or gland.

sedative A drug acting on the nervous system reducing irritability, excitement and activity.

sediment 1. Material which falls (settles) to the bottom of a liquid. **2.** Particles of rock carried by wind, water or ice which settle and may be compacted and cemented to form sedimentary rocks.

sedimentary rock Rock formed from fragments of already existing rock or organic material which have been compacted and cemented together. Examples include sandstone, coal and limestone.

sedimentation The process in which particles in a suspension settle to the bottom of a container, either because of gravity or when centrifuged. The rate of sedimentation may provide useful information about the size of the particles.

Seebeck effect When the junctions between two different metals forming a circuit are kept at different temperatures, an electric current flows round the circuit. For a small temperature difference the current is proportional to this difference. This effect is used in thermocouples. *Compare* Peltier effect.

seed 1. The reproductive structure of flowering plants, formed from a fertilised ovule. A seed consists of the embryo and a food store, protected by a tough, outer coat called the testa. After germination, the seed develops into a new plant. **2.** A small crystal added to a saturated or supersaturated solution to bring about crystallisation.

seed-coat *See* testa.

seed-leaf *See* cotyledon.

seedling A plant which has recently emerged from seed.

seed plant *See* Spermatophyta.

Seger cone A small cone made of a ceramic material which can be made to soften and bend at different temperature ranges. Seger cones are used to estimate the temperature in kilns and furnaces.

segmentation *See* cleavage.

segmented worm *See* Annelida.

segregation, law of *See* Mendel's laws.

Seignette salt *See* potassium sodium tartrate.

seismic waves Shock waves produced naturally by earthquakes or artificially by exploration geologists using explosives, etc.

seismogram The record of seismic waves, usually produced in the form of a graph by a seismograph.

seismograph An instrument used to record earthquakes. There are various types of seismographs, but all are vibrating systems designed to measure and record seismic waves. In one type, a heavy pendulum which can move horizontally is suspended from a rigid frame which is firmly bolted to the ground. Another type consists of an inverted pendulum supported on springs to measure vertical vibrations. Movements of the pendulum caused by earthquakes are recorded by an inked pen on paper, or on heat-sensitive or light-sensitive paper on a revolving drum by mechanical, electromagnetic or optical devices. When the seismograph is not vibrating, the record on the paper (the seismogram) is a smooth line. However, as soon as an earthquake causes vibrations in the Earth, the line changes to a series of zigzags on the paper. Recently, more sophisticated seismographs have been developed using laser beams. They may be connected to computers for rapid analysis of the data obtained. *See also* Richter scale.

seismology The study of the structure of the Earth by means of seismic waves produced by earthquakes.

seismonasty A plant movement in response to the stimulus of shock or vibration. *Example:* the sensitive plant *Mimosa pudica* rapidly folds its leaflets and lowers its leaves when shaken.

selectively permeable membrane *See* semi-permeable membrane.

selenium An element with the chemical symbol Se; atomic number 34; relative atomic mass 78,96 (78.96); state, solid. Selenium occurs naturally in several metal sulphide ores. There are three allotropes of selenium, of which two are non-metallic and one has certain metallic properties: it is a semiconductor whose electrical conductivity improves in the presence of light and also with an increase in temperature. Selenium is used in photoelectric cells.

selenium cell A type of photoelectric cell making use of the photovoltaic effect. It consists of a disc of selenium resting on a metal plate and covered with a very thin layer of gold through which light can pass. When light falls on the gold a potential difference is set up between it and the selenium layer. This will cause a deflection on a galvanometer connected to the cell, the deflection being proportional to the brightness of the light. The selenium cell is used in the exposure meter. *See* Fig. 198.

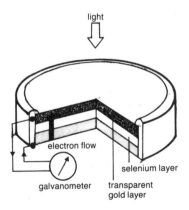

Fig. 198 Selenium cell

self-fertilisation (autogamy) The process in which male and female gametes fuse, the two gametes coming from the same individual. *Compare* cross-fertilisation.

self inductance Symbol: L; unit: henry (H). A coil through which a changing current passes has an e.m.f. induced in it. This property of the coil is called its self inductance. The self inductance of the coil is given by the expression

$E = L \times$ (rate of change of current)

where E is the induced e.m.f. *See also* mutual inductance.

self induction The generation of a back e.m.f. in a circuit (coil) as a result of variations in the current flowing through it. *See also* electromagnetic induction, mutual induction *and* Lenz's law.

self-pollination The transfer of pollen from the anther of a flower to the stigma of the same flower or one on the same plant. *See* Fig. 199; *compare* cross-pollination.

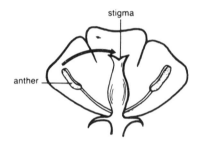

Fig. 199 Self-pollination

Seliwanoff test A test for a ketose sugar such as fructose. The test is based on the action of resorcinol (1,3-dihydroxybenzene), $C_6H_4(OH)_2$ and hydrochloric acid, HCl, on the sugar. A red colour develops almost immediately.

semen The fluid produced by the reproductive organs of male animals. It is composed of spermatozoa produced by the testes suspended in secretions from the prostate gland and the seminal vesicles.

semicircular canal One of three semicircular tubes situated in the bony labyrinth of the vertebrate ear. They contain a fluid endolymph and are concerned with balance and not hearing. *See* Fig. 55.

semiconductor A material with an electrical resistance (resistivity) between those of insulators and conductors. The most commonly used semiconductors are covalent crystals of the elements silicon, selenium and germanium, all of which are used in the electronics industry in the manufacture of transistors, resistors, diodes, thermistors, etc.

Metal conductors have free electrons as the carriers of current (*see* conduction); in electrolytes current is carried by ions. However, in semiconductors the carriers of current are free electrons carrying negative charges and 'holes' carrying positive charges. A hole is a vacancy left behind in an atom of the semiconductor as a result of the movement of a 'free' electron from it. This positively-charged hole then attracts an electron from a neighbouring atom, leaving it with a positive charge or hole. In this manner a positive charge is transferred from atom to atom and is equivalent to an electron moving in the opposite direction. A pure semiconductor has an equal number of electrons and holes when a current flows through it. However, the addition of certain impurities (often in controlled amounts) to the semiconductor can change the relative numbers of electrons and holes. If an impurity increases the number of free electrons, the semiconductor is known as an *n*-type semiconductor. An impure semiconductor in which the holes are in the majority is known as a *p*-type semiconductor. The '*n*' and '*p*' stand for negative and positive respectively. The electrical resistance of semiconductors decreases with increases in temperature because more electrons then gain enough energy to escape from their positions in the covalent crystal and become free electrons. *See also* thermistor.

semi-heavy water *See* heavy water.

semi-lunar valves Half-moon shaped valves found between the right ventricle and the pulmonary artery, and between the left ventricle and the aorta. They prevent blood from flowing back into the heart. Semi-lunar valves are also found in veins.

semi-metal *See* metalloid.

seminal vesicles In most mammals, two club-shaped, lobed, glandular structures situated immediately above the base of the prostate gland in males, between the posterior surface of the bladder and the anterior surface of the rectum. Each vesicle opens into a vas deferens. The seminal vesicles secrete a viscous, yellowish, sticky, alkaline secretion which forms the major constituent of semen, transporting and providing nutrients for spermatozoa. In lower vertebrates and some invertebrates, the seminal vesicles store sperm.

seminiferous tubules Long, coiled tubules found in the testes of vertebrates and surrounded by connective tissue in which occur numerous interstitial cells. The seminiferous tubules are responsible for the production of spermatozoa (spermatogenesis). From the tubules the spermatozoa pass to the epididymis and from here to the vas deferens. *See also* Sertoli cell *and* vas efferens.

semi-permeable membrane A membrane which allows certain small particles, e.g. small atoms, molecules or ions, to pass through it, but not other bigger ones. A semi-permeable membrane may allow the passage of the solvent of a solution through it, but not the solute. Since semi-permeable membranes only allow certain particles to pass through them, they may be called selectively permeable membranes. The outer layer of protoplasm of living cells acts as a semi-permeable membrane. *See also* dialysis *and* osmosis.

senility A physical or mental deterioration due to old age.

sense organ *See* receptor.

sensor *See* transducer.

sensory neurone (afferent neurone) A nerve-cell which conducts impulses from a receptor towards the central nervous system. *See* Fig. 150; *compare* motor neurone.

sepal One of the leaf-like structures forming the calyx of a flower. Sepals are situated just below the petals and are usually green. They enclose and protect the flower bud. *See* Fig. 75.

separating funnel (dropping funnel, tap funnel) A glass funnel fitted with a tap, used for the separation of immiscible liquids or the addition of a liquid drop by drop, e.g. to a chemical reaction. *See* Fig. 200.

Fig. 200 Separating funnel

sepiolite *See* meerschaum.

sepsis (pl. sepses) The condition resulting from infection of blood or other tissue by pathogens. Tissue affected in this way is said to be septic.

septic *See* sepsis.
septicaemia *See* blood-poisoning.
septum (pl. septa) A partition or wall separating two cavities or tissue masses. Septa are found in the nose, in the heart and in fruits.
Sequestrol® *See* edta.
sera *See* serum.
series Components of an electric circuit are 'in series' when they are connected in such a way that the current flows through each in turn. The effective (total) resistance R of three resistors in series, as in Fig. 201, is given by the expression:
$R = R_1 + R_2 + R_3$
Compare parallel.

Fig. 201 Resistors connected in series

serology The study of sera.
serosa *See* serous membrane.
serous Watery; pertaining to serum or any other watery liquid found in living animals.
serous membrane (serosa) One of the delicate membranes lining the internal body cavities of vertebrates. The membranes consist of connective tissue.
serrate With margin or edge notched like a saw. *See* Fig. 111.
Sertoli cell A nutritive cell found in the testis, to which developing spermatids are attached.
serum (pl. sera) *See* blood serum.
sessile 1. Describes a flower, leaf, eye, etc. directly connected to a base without a stalk, pedicel or peduncle. **2.** Describes an animal which remains in one place, either attached to a substrate or without means of locomotion, e.g. a sponge.
sewage The waste liquid expelled from a house, factory or town. Sewage contains drainage waters, faeces, urine, etc.
sex The structural and functional characteristics that distinguish male and female organisms.
sex cell *See* gamete.
sex chromosomes The chromosomes that determine the sex of an individual animal. In humans, female cells have a pair of homologous sex chromosomes called X-chromosomes. Male cells have two different sex chromosomes: one is an X-chromosome, the other a smaller Y-chromosome. *Compare* autosome; *see also* sex determination *and* sex linkage.
sex determination In higher animals, including human beings, sex determination is genetic, i.e. the sex of the offspring depends on which type of sperm (X or Y) fertilises the ovum. From Fig. 202 it can be seen that the chances of the offspring being a male (XY) or a female (XX) are equal. The mechanism of sex determination is absent in hermaphrodite animals and in plants.

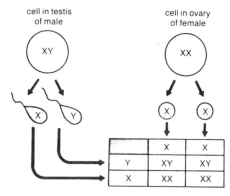

Fig. 202 Sex determination

sex hormone Any of several hormones involved in the regulation of sexual development and reproduction. The pituitary gland produces certain master hormones which stimulate the testes and the ovaries (the gonads) to produce spermatozoa and androgens in males, and ova, oestrogens and progesterone in females.
sex-linkage The transmission of characteristics controlled by genes and carried on sex chromosomes. In human beings, the male receives his X chromosome and sex-linked genes from his mother. The characteristics controlled by these genes can only be transmitted to his daughters, i.e. his sex-linked characteristics may only be transferred to his grandchildren via his daughters. Examples of sex-linked characteristics in humans are red-green colour blindness and haemophilia.
sextant A navigational instrument used for measuring the angle between two objects, especially between a star or the Sun and the horizon. The observer sights the horizon by looking through a telescope T and through the unsilvered part of the half-silvered mirror M. He then rotates an adjustable mirror A mounted on a movable arm until light from a star is reflected from it to the lower silvered half of mirror M and from there to the telescope. The image of the star and the horizon then coincide and the angle between them can be read on the graduated scale. *See* Fig. 203.
sexual reproduction Reproduction involving the fusion of two gametes to form a zygote.
shadow A dark patch formed by an obstacle in the path of rays of light. Fig. 204 (a) shows the

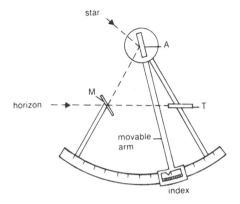

Fig. 203 Sextant

shadow formed by an obstacle in the path of light from a point source. In this case the shadow is total and is called an umbra. Fig. 204 (b) shows the shadow formed when an extended source is used. Here the shadow consists of an umbra and a partial shadow or penumbra.

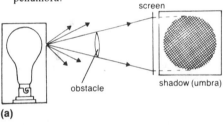

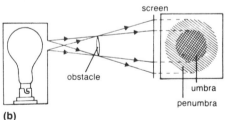

Fig. 204 Shadows: umbra and penumbra

shaft *See* Fig. 67; *see also* rachis.
shell *See* energy level.
shellac varnish A varnish prepared from shellac and methylated spirits. It dries by evaporation of the methylated spirits. Shellac is a resin excreted by the lac insect. Shellac varnish is used on furniture, on floors and for several other purposes.
shielding *See* Faraday's cage *and* magnetic shielding (screening).

shock An emergency condition in which a person is in a state of collapse which, if not controlled, can lead to death. Causes of shock may include excessive bleeding, severe burns, injuries, heart attack or extensive loss of body fluid. Symptoms include a cold, sweaty, pallid skin, faintness, blurring of vision and rapid and shallow breathing. The patient's blood pressure is low. Shock is treated by stopping the loss of fluid and then replacing it. The patient should lie down with the head lower than the rest of the body, should be kept warm and be encouraged to drink any fluid (except alcohol). A doctor may decide to inject adrenalin(e) to raise the blood pressure. There are several types of shock: circulatory shock (described above) should not be confused with shocks such as electric shock and diabetic shock.
shoot A stem bearing buds, leaves and sometimes flowers. A shoot develops from the plumule.
shooting star *See* meteor.
short circuit A connection, usually made accidentally, in which two points in an electric circuit are connected through a low resistance, i.e. the current bypasses other components in the circuit.
short-day plant *See* photoperiodism.
short sight *See* myopia.
shoulder-blade *See* scapula.
shoulder girdle *See* pectoral girdle.
shrub A woody multi-stemmed plant which is generally smaller than a tree, but taller than a herb: e.g. rose, hibiscus.
shunt A resistor with a known and usually low resistance connected in parallel with an instrument, e.g. an ammeter or galvanometer. The shunt diverts most of the current away from the instrument and therefore increases its sensitivity. A galvanometer may be converted for use as an ammeter by connecting a suitable shunt to it.
Si Chemical symbol for silicon.
sial The upper, less dense layer of the Earth's crust, composed mainly of silicon and aluminium compounds. *See also* sima.
sibling One of a number of sisters or brothers or both having the same male parent and female parent, but not born at the same birth.
sickle-cell anaemia A hereditary blood disorder caused by deformed red blood cells which have assumed irregular, pointed shapes resembling sickles. Such blood cells are less efficient at carrying oxygen than normal cells and are unable to pass easily through blood capillaries, with the result that blood vessels become clogged and circulation is impaired. Sickle-cell anaemia is found among American Negroes, Africans, West Indians and the

peoples of Mediterranean countries. Symptoms include severe pain in the joints and abdomen after strenuous exercise or during certain illnesses. The disorder is not very common but this type of anaemia is generally fatal.

Sickle-cell anaemia occurs in people who are homozygous for the sickle-cell gene. People who are heterozygous for the gene suffer from the much milder sickle-cell trait, and are also resistant to malaria. This means that the sickle-cell gene is relatively common in the populations in which it is found.

side chain A group of two or more atoms attached at some point(s) to a chain or ring of atoms in a molecule. The term is commonly used in organic chemistry. *Example:* in the hydrocarbon 3-methyl-4-ethylheptane,

$$\overset{1}{C}H_3-\overset{2}{C}H_2-\overset{3}{C}H-\overset{4}{C}H-\overset{5}{C}H_2-\overset{6}{C}H_2-\overset{7}{C}H_3$$
$$\underset{}{|}\underset{}{|}$$
$$CH_3CH_2$$
$$|$$
$$CH_3$$

the methyl group, CH_3-, and the ethyl group, CH_3-CH_2-, attached to carbon atoms number 3 and number 4 respectively, are side chains. In methylbenzene (toluene),

the methyl group, CH_3-, is the side chain.

siderite *See* iron.

siemens Symbol: S. The SI unit of electrical conductance. This unit replaces the reciprocal ohm or mho. It is defined as the conductance of a circuit with a resistance of one ohm (Ω). A circuit with a resistance of 2 ohms has a conductance of 0,5 siemens (0.5 S), and so on.

Siemens—Martin process *See* Open-Hearth process.

sieve-plate *See* sieve-tube.

sieve-tube In plants, a column of living cells arranged end to end in long rows. The end walls of the cells are perforated to form sieve-plates which allow passage of manufactured food. *See* Fig. 205; *see also* phloem.

silage Fodder prepared by placing grass or any green vegetable matter in a pit (silo) sealed up with earth. Certain bacteria act anaerobically upon this material breaking it down to a usable fodder. Instead of a pit, a tower or tank may be used as a silo.

silane *See* silicide.

silica *See* silicon(IV) oxide.

silica gel A substance with a large surface area, used as a drying agent and as an inert supporting material for certain finely-divided catalysts. Silica gel is a gel of polymeric silicic acid molecules, formed when an aqueous solution of a silicate is acidified. When the gel is dried it loses most of its water. However, it takes up water again when exposed to moisture. By adding a water indicator such as cobalt(II) chloride, which is blue when dry and pink when moist, it is possible to check whether or not the silica gel is still capable of taking up more water.

silicate A salt of one of the silicic acids.

silicic acid Any of several very weak acids obtained by acidifying an aqueous solution of a silicate. Examples include H_2SiO_3 (metasilicic acid) and H_4SiO_4 (orthosilicic acid). H_2SiO_3

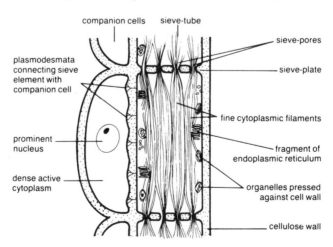

Fig. 205 Sieve-tube

can be obtained from H_4SiO_4 when this loses water:
$H_4SiO_4 \rightarrow H_2SiO_3 + H_2O$
Both these acids have a marked tendency to polymerise, and it is doubtful whether any discrete silicic acid exists at all. Salts of silicic acids are called silicates. *Examples:* Na_2SiO_3 is sodium metasilicate; Na_4SiO_4 is sodium orthosilicate. *See also* water glass.

silicide A compound consisting of silicon and a more electropositive element, e.g. Mg_2Si is magnesium silicide. Silicides may be prepared by strongly heating silicon and a metal. With acids, silicides produce silanes, which are compounds of silicon and hydrogen:
$Mg_2Si + 4H^+ \rightarrow SiH_4 + 2Mg^{2+}$
SiH_4 is monosilane. Silanes are strong reducing agents.

silicon An element with the symbol Si; atomic number 14; relative atomic mass 28,09 (28.09); state, solid. Silicon is the second most abundant element in the Earth's crust. It is found as silicon(IV) oxide (silica) and silicates in sand, rocks and clay. It can be extracted from silicon(IV) oxide, SiO_2, by reduction with carbon at a high temperature:
$SiO_2 + 2C \rightarrow Si + 2CO$
Silicon is a hard, brittle semi-metal and is chemically rather inert. It is used in the manufacture of transistors.

silicon carbide SiC (carborundum) An abrasive which is almost as hard as diamond. It is made by heating silicon(IV) oxide (silica), SiO_2, with carbon in an electric furnace.

silicon dioxide *See* silicon(IV) oxide.

silicone A chemical compound consisting of chains or a network of alternating atoms of silicon and oxygen with organic groups such as the methyl group, CH_3-, bonded to the silicon atoms:

$$\begin{array}{ccc} CH_3 & CH_3 & CH_3 \\ | & | & | \\ \diagdown O \diagup Si \diagdown O \diagup Si \diagdown O \diagup Si \diagdown \\ | & | & | \\ CH_3 & CH_3 & CH_3 \end{array}$$

Silicones are very unreactive. They are used as lubricants, water-repellents, electrical insulators and in waxes, polishes, varnishes, etc.

silicon(IV) oxide SiO_2 (silicon dioxide, silica) A hard, crystalline substance found in nature in different pure and impure varieties such as quartz, sand and flint. In a crystal of SiO_2, silicon atoms are strongly bonded tetrahedrally to four oxygen atoms by means of single bonds, i.e. there are no Si—Si bonds. Silicon(IV) oxide (silica) is used in the manufacture of glass.

silk *See* cocoon.

silt A type of soil with particle size and properties intermediate between sand and clay. Water applied to silty soil does not pass quickly through it because of air trapped in the pores between the silt particles. This means that the greater the amount of silt in a soil, the greater the amount of water available to plants.

Silurian A geological period of approximately 35 million years, occurring between the Ordovician and the Devonian, and which ended approximately 395 million years ago.

silver An element with the symbol Ag; atomic number, 47; relative atomic mass, 107,87 (107.87); state, solid. Silver is a transition metal, occurring naturally as the free element, as silver sulphide (silver glance or argentite), Ag_2S, as silver chloride (horn silver), AgCl, and together with certain sulphide ores such as lead sulphide. After extraction, the silver is usually purified by electrolysis. Silver is the best electrical conductor known and is resistant to attack by air. However, the presence of hydrogen sulphide, H_2S, will stain the silver black as a result of the formation of silver sulphide. Silver is used in coins, tableware and jewellery. Certain silver compounds are used in the photographic industry.

silver bromide AgBr An insoluble light-yellow salt which can be formed when silver ions, Ag^+, are added to an aqueous solution containing bromide ions, Br^-:
$Ag^+ + Br^- \rightarrow AgBr$
Silver bromide is less water-soluble than silver chloride, AgCl, but it dissolves in a concentrated ammonia solution, NH_3, to form the complex ion $[Ag(NH_3)_2]^+$:
$AgBr + 2NH_3 \rightarrow [Ag(NH_3)_2]^+ + Br^-$
Silver bromide is sensitive to light, slowly decomposing into silver and bromine, Br_2:
$2AgBr \rightarrow 2Ag + Br_2$
It is used in the photographic industry.

silver chloride AgCl An insoluble white salt which can be formed when silver ions, Ag^+, are added to an aqueous solution containing chloride ions, Cl^-:
$Ag^+ + Cl^- \rightarrow AgCl$
Silver chloride is more water-soluble than silver bromide, AgBr. In a dilute ammonia solution, NH_3, it dissolves to form the complex ion $[Ag(NH_3)_2]^+$:
$AgCl + 2NH_3 \rightarrow [Ag(NH_3)_2]^+ + Cl^-$
Silver chloride is sensitive to light, slowly decomposing into silver and chlorine, Cl_2:
$2AgCl \rightarrow 2Ag + Cl_2$
It is used in the photographic industry. AgCl occurs naturally as horn silver.

silver glance *See* silver.

silver iodide AgI An insoluble yellow salt which can be formed when silver ions, Ag^+, are added to an aqueous solution containing iodide ions, I^-:
$Ag^+ + I^- \rightarrow AgI$

Silver iodide is even less water-soluble than silver bromide, AgBr, and it does not dissolve in ammonia solution of any concentration. It is sensitive to light, slowly decomposing into silver and iodine, I_2:
$2AgI \rightarrow 2Ag + I_2$
Silver iodide is used in the photographic industry.

silver mirror test *See* Tollens's reagent.

silver nitrate $AgNO_3$ A white, very soluble salt which can be prepared by the addition of nitric acid, HNO_3, to silver, Ag:
$Ag + 2HNO_3 \rightarrow AgNO_3 + NO_2 + H_2O$
When heated, it decomposes into silver, Ag, oxygen, O_2, and nitrogen dioxide, NO_2:
$2AgNO_3 \rightarrow 2Ag + O_2 + 2NO_2$
Silver nitrate is used in photography, in medicine and in the chemistry laboratory in qualitative and quantitative analysis. When testing for chloride ions, Cl^-, a little nitric acid, HNO_3, and silver nitrate solution are added to the liquid sample. A positive reaction is a white precipitate of silver chloride, AgCl:
$Cl^- + Ag^+ \rightarrow AgCl$
After a few minutes the precipitate turns violet due to the formation of free silver, Ag. The nitric acid present prevents the formation of a light-yellow precipitate of silver carbonate, Ag_2CO_3, or silver phosphate, Ag_3PO_4, if carbonate ions, CO_3^{2-}, or phosphate ions, PO_4^{3-}, should also be present in the sample.

silver nitride *See* Tollens's reagent.

silver oxide Ag_2O A brown, slightly soluble solid which can be prepared by adding a solution of a strong alkali to a solution containing silver ions, Ag^+. Silver hydroxide, AgOH, may be formed as an intermediate compound in this reaction:
$2Ag^+ + 2OH^- \rightarrow 2AgOH \rightarrow Ag_2O + H_2O$
Silver oxide turns red litmus blue, i.e. it produces some hydroxyl ions, OH^-, in the presence of water. When heated, Ag_2O decomposes to give silver and oxygen:
$2Ag_2O \rightarrow 4Ag + O_2$
Silver oxide dissolves readily in an ammonia solution, NH_3, forming the complex ion $[Ag(NH_3)_2]^+$:
$Ag_2O + 4NH_3 + H_2O \rightarrow 2[Ag(NH_3)_2]^+ + 2OH^-$
It is used in organic chemistry.

silver sulphide Ag_2S A black, very insoluble salt which precipitates when hydrogen sulphide, H_2S, is bubbled through a solution containing silver ions, Ag^+:
$2Ag^+ + H_2S \rightarrow Ag_2S + 2H^+$
The familiar black stain on the surface of silver articles is Ag_2S. Silver sulphide occurs naturally as argentite (silver glance).

sima The lower, denser layer of the Earth's crust, composed mainly of silicon and magnesium compounds. *See also* sial.

simple cell *See* voltaic cell.

simple eye (ocellus) In many insects and larvae, an eye which has only one lens. *Compare* compound eye.

simple leaf A leaf with a single blade. *Compare* compound leaf.

simple microscope *See* magnifying glass.

simple pendulum *See* pendulum.

simple protein *See* protein.

single bond *See* covalent bond.

sinuate With a wave-like margin or edge. *See* Fig. 111.

sinus A cavity, chamber, dilation or tube, e.g. the nasal sinuses. *See* Fig. 16.

siphon A bent tube to transfer liquids from a high level to a lower level. The siphon is filled with the liquid to be transferred. When its shorter limb dips into the container, liquid flows out of the longer limb (Fig. 206). The pressure at A and D in the siphon is equal to atmospheric pressure P and the pressure at E is therefore P plus the pressure $h\rho g$ (g is the acceleration of free fall). This excess pressure causes a flow of liquid from E.

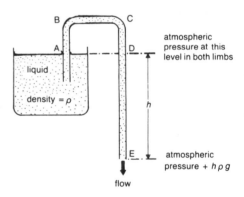

Fig. 206 Siphon

SI units (Système International d'Unités). An international system of units which is called SI in all languages. It is derived from the MKS system and is composed of seven basic units, two supplementary units and several derived units. The basic units are: ampere, candela, kelvin, kilogram, metre, mole and second. The supplementary units are the radian and the steradian. *See also* Appendix.

Six's thermometer *See* maximum and minimum thermometer.

skeletal muscles Muscles attached to the bones and found throughout the animal body, located between the skin and the bones of the skeleton. These muscles are under the control of the will (voluntary muscles). They function by moving various parts of the body by contraction of the striated (striped) muscle cells (fibres) of which they consist. The muscle fibres are held together by connective tissue. *Compare* involuntary muscle.

skeleton A hard framework composed of bones, cartilage, etc., which supports the animal, protects internal organs and serves as an anchorage for muscles. The skeleton may be internal (an endoskeleton) or external (an exoskeleton). *See* Fig. 207.

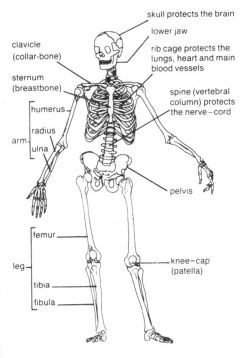

Fig. 207 Skeleton

skin The most widespread organ of the animal body, forming its outer covering. In vertebrates the skin consists of two distinct layers, the outer epidermis and the inner dermis. The skin is usually covered by varying quantities of hair. It acts as a barrier between the deeper structures of the body and the external environment. Its dead, outer, waterproof cells prevent excessive loss or uptake of fluid and prevent entry of micro-organisms. Its pigments protect the body from the damaging effect of sunlight. The skin is also involved in the regulation of body temperature in warm-blooded animals through the activity of the sweat glands. It acts as an excretory organ for water and other waste products and is an important sense organ with receptors for pain, touch and temperature. *See* Fig. 208.

skull (cranium) In vertebrates, the part of the skeleton enclosing the brain and surrounding the sense organs of the head.

slag Waste material formed on the surface of a molten metal during its extraction or refining. It consists mainly of oxides, phosphates, silicates and sulphides produced when impurities in the metal ore combine with the flux.

slag wool *See* mineral wool.

slaked lime *See* calcium hydroxide.

slaking *See* calcium hydroxide.

slate Fine-grained, metamorphic rock which occurs in a variety of colours, the most common being grey and black. Slate can be split into smooth, flat plates which are used as a roofing material.

sleeping sickness (trypanosomiasis) A tropical disease caused by a protozoon which is transmitted to humans by the bite of the tsetse-fly. Symptoms include recurrent fever, headache, drowsiness and swelling of the body. Drugs are available to treat the disease if it is diagnosed early. However, the disease is fatal if untreated. Sleeping sickness occurs only in Africa.

sliding friction *See* friction.

slime layer A layer of polysaccharide or protein material surrounding capsule-forming bacteria.

slip rings Two copper rings mounted on the spindle of a generator. A carbon brush presses against each slip ring in order to draw off the current.

slough 1. The skin cast off (shed) periodically by a snake. **2.** To cast off or shed the skin.

small intestine The part of the alimentary canal between the stomach and the large intestine. It consists of the duodenum, jejunum and ileum.

smallpox (variola) A highly contagious, viral disease. Early symptoms include fever, headaches and aching limbs. These are followed by the appearance of a rash which gradually develops into scabs containing the virus. Once widespread, the disease has now been eliminated throughout the world by vaccination.

smaltite *See* cobalt.

smear *See* blood smear.

smelting The extraction of a metal from its ores. This is usually done by heating them in a furnace together with a flux and a reducing agent such as carbon or carbon monoxide.

smithsonite *See* zinc.

smog Fog containing industrial pollutants such as smoke, sulphur dioxide, etc.

smoke A suspension of fine, solid particles in a gas.

smoke cell

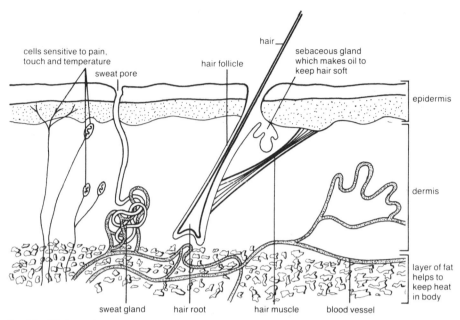

Fig. 208 Skin in section

smoke cell A small, transparent vessel fitted with a coverslip and illuminated from the side. Smoke, e.g. collected by a syringe, is transferred to the vessel and a microscope is focused on the cell. The fine, solid particles making up smoke can now be seen moving in an irregular way, i.e. they exhibit Brownian movements. See Fig. 209.

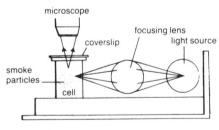

Fig. 209 Smoke cell

smoky quartz See quartz.
smooth muscle See involuntary muscle.
Sn Chemical symbol for tin.
snail See Gastropoda.
snake See Squamata.
Snell's law See refraction of light, laws of.
soap Any of several sodium or potassium salts of long-chain fatty acids such as palmitic acid, $C_{15}H_{31}COOH$, and stearic acid, $C_{17}H_{35}COOH$. Soap is manufactured from vegetable oils or animal fats by treating them with a hot, strong alkaline solution containing sodium hydroxide or potassium hydroxide. This produces the above-mentioned salts and glycerol. The soap is then obtained by a process known as salting out. See also saponification and detergent.
soapless detergent See detergent.
soapstone (steatite) Any of various soft rocks which are greasy to the touch and are usually composed of talc. Soapstone is used in sculpture.
soda Commonly used sodium compounds such as caustic soda, NaOH, bicarbonate of soda, $NaHCO_3$, and washing soda, $Na_2CO_3 \cdot 10H_2O$, all giving an alkaline aqueous solution.
soda-acid fire extinguisher See fire extinguisher.
soda-ash See sodium carbonate.
soda lake See trona.
soda-lime A greyish-white, solid mixture of sodium hydroxide and calcium hydroxide. It is made by slaking calcium oxide with a solution of sodium hydroxide; it is used as an absorber for carbon dioxide and as a drying agent.
soda-water See carbonation.
sodium An element with the symbol Na; atomic number 11; relative atomic mass 22,99 (22.99); state, solid. It is a very electropositive, soft, light element belonging to the alkali metals. Sodium is found naturally in several minerals such as sodium chloride (rock salt) and Chile saltpetre, and in sea-water. It is extracted by electrolysis of fused sodium chloride. Like all

the alkali metals, sodium reacts with water to form a hydroxide and hydrogen:
$2Na + 2H_2O \rightarrow 2NaOH + H_2$
Sodium is sometimes used as a reducing agent, as a coolant in some nuclear reactors and in organic synthesis. Almost all sodium compounds are soluble in water.

sodium aluminium fluoride *See* cryolite.

sodium bicarbonate *See* sodium hydrogencarbonate.

sodium carbonate Na_2CO_3 (soda-ash) A white, solid, soluble salt forming an alkaline aqueous solution. Most sodium carbonate is prepared by the Solvay process. When it is crystallised from an aqueous solution, the hydrate $Na_2CO_3 \cdot 10H_2O$ (washing soda) is formed. This is efflorescent, forming another hydrate, $Na_2CO_3 \cdot H_2O$. Sodium carbonate is used in the manufacture of glass and sodium hydroxide, as a water-softener, in detergents, etc. *See also* trona.

sodium chlorate(V) $NaClO_3$ A white, solid, soluble salt. It can be prepared by electrolysis of a concentrated aqueous solution of sodium chloride, NaCl, at a temperature of about 70°C; the products formed in this electrolysis are sodium hydroxide, NaOH, (in solution) and chlorine, Cl_2. They are allowed to mix:
$6NaOH + 3Cl_2 \rightarrow NaClO_3 + 5NaCl + 3H_2O$
The sodium chlorate(V) is then separated from the sodium chloride by fractional crystallisation. When heated, sodium chlorate(V) decomposes to give sodium chloride and oxygen, O_2:
$2NaClO_3 \rightarrow 2NaCl + 3O_2$
This process may be catalysed by manganese(IV) oxide, MnO_2. $NaClO_3$ is a strong oxidising agent, used in matches, fireworks, explosives, weed-killers, etc.

sodium chloride NaCl (common salt) Sodium chloride occurs naturally as rock salt (halite) and in sea-water. In the crystal structure of NaCl, sodium ions, Na^+, and chloride ions, Cl^-, are arranged in a lattice in which each chloride ion is surrounded by six sodium ions and vice versa. Distinct ions of each element are present and therefore there are no separate NaCl molecules in a crystal of NaCl. The ions (Na^+ and Cl^-) attract each other strongly, forming a rather hard solid with a high melting-point (801°C) and a high boiling-point (1413°C).
Sodium chloride is used in the preservation of food and is an essential part of the diet. In the chemical industry it is used in the manufacture of sodium, sodium hydroxide, sodium carbonate, chlorine, etc. It is also used in the regeneration of certain ion exchangers and in the salting out of soap. *See* Fig. 210.

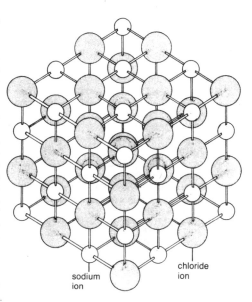

Fig. 210 Arrangement of atoms in a crystal of sodium chloride

sodium dihydrogen phosphate(V) *See* sodium phosphates.

sodium hydrogencarbonate $NaHCO_3$ (sodium bicarbonate, bicarbonate of soda) A white, solid, soluble acid salt which can be prepared by passing a large amount of carbon dioxide, CO_2, through a strong aqueous solution of sodium carbonate, Na_2CO_3:
$Na_2CO_3 + CO_2 + H_2O \rightarrow 2NaHCO_3$
Although termed an acid salt, it forms an alkaline solution when dissolved in water because the hydrogencarbonate ion, HCO_3^-, is a stronger base than it is an acid. This gives the reaction:
$HCO_3^- + H^+ \rightleftharpoons CO_2 + H_2O$
When heated, $NaHCO_3$ decomposes to give sodium carbonate, carbon dioxide and water:
$2NaHCO_3 \rightarrow Na_2CO_3 + CO_2 + H_2O$
$NaHCO_3$ is used in baking-powder, antacids, effervescent powders, fire extinguishers, etc. *See also* trona.

sodium hydroxide NaOH (caustic soda) A white, solid, deliquescent, soluble, alkaline substance. An aqueous solution of NaOH is a strong alkali. NaOH is less soluble in water and in ethanol than potassium hydroxide, KOH. Sodium hydroxide is produced in the Castner–Kellner process or in the diaphragm cell and is used widely in the chemistry laboratory, in soap-making, in certain electric cells and for absorbing acidic gases such as carbon dioxide, CO_2, and sulphur dioxide, SO_2.

sodium monoxide *See* sodium oxides.

sodium nitrate $NaNO_3$ (Chile saltpetre, caliche) A white, solid, soluble, deliquescent salt which can be prepared by neutralising an aqueous solution of sodium hydroxide, NaOH, with nitric acid, HNO_3:
$NaOH + HNO_3 \rightarrow NaNO_3 + H_2O$
$NaNO_3$ is found in large quantities as Chile saltpetre. When heated, it decomposes into sodium nitrite, $NaNO_2$, and oxygen, O_2:
$2NaNO_3 \rightarrow 2NaNO_2 + O_2$
However, if it is heated very strongly oxides and peroxides may also be formed. $NaNO_3$ is a strong oxidising agent, used as a fertiliser and in the small-scale manufacture of nitric acid.

sodium oxides Sodium oxide (sodium monoxide), Na_2O, sodium peroxide, Na_2O_2, and sodium superoxide, NaO_2.
Sodium oxide, Na_2O, is a powerful oxidising agent. It is a white, soluble solid which can be prepared by heating sodium, Na, in a poor supply of oxygen:
$4Na + O_2 \rightarrow 2Na_2O$
Sodium oxide reacts violently with water to form sodium hydroxide, NaOH:
$Na_2O + H_2O \rightarrow 2NaOH$
With acids, it forms salts.
Sodium peroxide, Na_2O_2, is a white (when pure), soluble solid which can be prepared by heating sodium in excess air (oxygen):
$2Na + O_2 \rightarrow Na_2O_2$
With water, sodium peroxide forms sodium hydroxide, NaOH, and hydrogen peroxide, H_2O_2:
$Na_2O_2 + 2H_2O \rightarrow 2NaOH + H_2O_2$
The H_2O_2 rapidly decomposes to give oxygen and water:
$2H_2O_2 \rightarrow O_2 + 2H_2O$
With acids, Na_2O_2 produces hydrogen peroxide:
$Na_2O_2 + 2H^+ \rightarrow H_2O_2 + 2Na^+$
Na_2O_2 absorbs carbon dioxide, CO_2, liberating oxygen:
$2Na_2O_2 + 2CO_2 \rightarrow 2Na_2CO_3 + O_2$
and it can therefore be used in submarines to regenerate expired air. Na_2O_2 is a strong oxidising agent, used as a bleaching agent in the textile and paper industries.
Sodium superoxide, NaO_2, is a powerful oxidising agent. It is a white (when pure), soluble solid which can be prepared from sodium peroxide, Na_2O_2, when this is treated with oxygen at a certain high pressure and temperature:
$Na_2O_2 + O_2 \rightarrow 2NaO_2$
With water, sodium superoxide forms a mixture of sodium hydroxide, NaOH, oxygen, O_2, and hydrogen peroxide, H_2O_2:
$2NaO_2 + 2H_2O \rightarrow 2NaOH + O_2 + H_2O_2$

sodium peroxide *See* sodium oxides.

sodium phosphates Trisodium phosphate(V), Na_3PO_4, disodium hydrogenphosphate(V), Na_2HPO_4, and sodium dihydrogenphosphate(V), NaH_2PO_4. Salts derived from phosphoric(V) acid (orthophosphoric acid), H_3PO_4. All three salts are white, soluble solids.
Trisodium phosphate(V), Na_3PO_4, dissolves in water to give an alkaline solution:
$Na_3PO_4 \rightarrow 3Na^+ + PO_4^{3-}$ and
$PO_4^{3-} + H^+ \rightleftharpoons HPO_4^{2-}$
Disodium hydrogenphosphate(V), Na_2HPO_4, dissolves in water to give an alkaline solution:
$Na_2HPO_4 \rightarrow 2Na^+ + HPO_4^{2-}$ and
$HPO_4^{2-} + H^+ \rightleftharpoons H_2PO_4^-$
Sodium dihydrogenphosphate(V), NaH_2PO_4, dissolves in water to give an acidic solution:
$NaH_2PO_4 \rightarrow Na^+ + H_2PO_4^-$ and
$H_2PO_4^- \rightleftharpoons H^+ + HPO_4^{2-}$
The salts can be made by treating phosphoric(V) acid, H_3PO_4, with sodium carbonate, Na_2CO_3, or sodium hydroxide, NaOH. They are used in making detergents and as water-softeners.

sodium silicate *See* silicic acid *and* water glass.

sodium sulphate Na_2SO_4 A white, solid, soluble salt which can be prepared by heating sodium chloride, NaCl, and concentrated sulphuric acid, H_2SO_4, at a high temperature:
$2NaCl + H_2SO_4 \rightarrow Na_2SO_4 + 2HCl$
Na_2SO_4 produced in this way is called saltcake. The hydrogen chloride, HCl, is dissolved in water to form hydrochloric acid as a by-product.
Sodium sulphate exists in two hydrated forms: $Na_2SO_4 \cdot 7H_2O$ and $Na_2SO_4 \cdot 10H_2O$. $Na_2SO_4 \cdot 10H_2O$ is known as Glauber's salt. It is an efflorescent substance, forming anhydrous sodium sulphate. Sodium sulphate is used in medicine and in the manufacture of glass.

sodium sulphide Na_2S A white, solid, soluble, deliquescent salt which can be prepared from an aqueous solution of sodium hydroxide, NaOH, through which hydrogen sulphide, H_2S, is bubbled:
$2NaOH + H_2S \rightarrow Na_2S + 2H_2O$
Na_2S liberates H_2S when treated with acids:
$Na_2S + 2H^+ \rightarrow H_2S + 2Na^+$
When dissolved in water, Na_2S forms an alkaline solution.

sodium sulphite Na_2SO_3 A white, solid, soluble salt which can be prepared from an aqueous solution of sodium hydroxide, NaOH, through which sulphur dioxide, SO_2, is bubbled:
$2NaOH + SO_2 \rightarrow Na_2SO_3 + H_2O$
Na_2SO_3 liberates SO_2 when treated with acids:
$Na_2SO_3 + 2H^+ \rightarrow SO_2 + 2Na^+ + H_2O$
In air, Na_2SO_3 is readily oxidised to sodium

sulphate, Na_2SO_4. When heated strongly, Na_2SO_3 decomposes into sodium sulphate, Na_2SO_4, and sodium sulphide, Na_2S:
$4Na_2SO_3 \rightarrow 3Na_2SO_4 + Na_2S$
Na_2SO_3 is used in the paper industry to remove chlorine after bleaching, and in the manufacture of sodium thiosulphate.

sodium superoxide *See* sodium oxides.

sodium tetraborate *See* borax.

sodium thiosulphate $Na_2S_2O_3$ ('hypo') A white, solid, soluble salt which can be prepared by boiling an aqueous solution of sodium sulphite, Na_2SO_3, with sulphur, S:
$Na_2SO_3 + S \rightarrow Na_2S_2O_3$
When crystallising from an aqueous solution, sodium thiosulphate forms crystals of $Na_2S_2O_3 \cdot 5H_2O$. When acidified, an aqueous solution of $Na_2S_2O_3$ forms free sulphur and sulphur dioxide, SO_2:
$Na_2S_2O_3 + 2H^+ \rightarrow S + SO_2 + 2Na^+ + H_2O$
Sodium thiosulphate is used in fixing in photography, as antichlor in the textile industry and in volumetric analysis.

soft iron Iron containing very little carbon. *See also* magnetic material.

soft solder *See* solder.

software The programs etc. used in a computer. *Compare* hardware.

soft water *See* hard water.

softwood *See* conifer.

soil The upper layer of the Earth's crust, consisting mainly of sand, clay and humus. *See also* subsoil *and* topsoil.

sol A liquid solution or suspension of a colloid. *See also* gel.

solar Pertaining to the Sun.

solar cell A type of photoelectric cell used to generate electricity in artificial satellites, spacecraft, etc.

solar constant The amount of solar energy striking unit area of the Earth per second. The atmosphere absorbs some of this energy; if it were absent, the solar constant would amount to approximately 1400 joules per second per square metre.

solar eclipse *See* eclipse.

solar energy *See* fusion.

solar plexus In higher mammals, the largest concentration of cell bodies and nerve-fibres from which nerves radiate to the stomach, adrenal glands and intestines. In humans it is situated below the diaphragm and behind the stomach.

Solar System The system consisting of the Sun and the planets, also comprising asteroids, comets and meteors.

solar time Time measured with respect to the Sun. *See also* mean solar time.

solar wind All stars are surrounded by a 'wind' which actually consists of a fast-moving, continuous stream of particles (mostly protons and electrons) emitted from the surface of the star. This emission of particles is called stellar wind or solar wind (if talking of the Sun). The solar wind makes the tail of a comet always point away from the Sun.

solder A general term for an alloy used to join metals. Soft solder contains tin and lead and may include a little antimony when greater strength is required. This type of solder melts at about 200°C. Hard solder contains copper and zinc and melts at about 850°C. Joints made with hard solder are considerably stronger than soft-soldered joints. Hard solder is also called brazing solder and the process of applying it is known as brazing. Before two metals can be joined using solder, the metal surfaces should be clean and free from any oxide layer. This is done by using a flux.

solenoid A coil of insulated wire, usually cylindrical in shape and with a length greater than its diameter. An electric current passing through the solenoid produces a magnetic field similar to that of a bar magnet. Solenoids are used in electromagnets, relays, etc.

solid A state of matter in which the particles of a substance are not free to move, but have fixed positions about which they vibrate. Solids may be either amorphous or crystalline.

solid angle Symbol: ω or Ω (omega); unit: steradian (sr). A three-dimensional angle, e.g. the angle at the corner of a cube. It is measured by the ratio of the area (A) of the surface of that part of a sphere that the angle intercepts to the square of the radius (r) of the sphere, i.e.

$$\omega = \frac{A}{r^2}$$

The solid angle completely surrounding a point is 4π steradians. *See* Fig. 211.

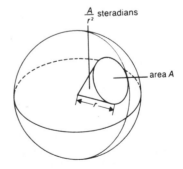

Fig. 211 Solid angle

solid-state Describes an electronic device which uses transistors or other semiconductors in place of themionic valves.

solid-state physics The branch of physics concerned with the structure and properties of solids, particularly the properties of solid-state devices.

solubility A measure of the amount of solute that dissolves in a certain amount of solvent to form a saturated solution under particular conditions of temperature and (sometimes) pressure. The solubility of a solute at a particular temperature is defined as the number of grams of the solute necessary to saturate 100 grams of the solvent at that temperature.

solubility curve A graph showing the solubility of a solute in a solvent (usually water) at different temperatures. *See* Fig. 212.

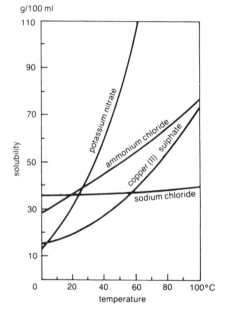

Fig. 212 Solubility curves for various substances

solubility product The ionisation equation of an electrolyte A_mB_n is:
$$A_mB_n \rightleftharpoons mA + nB$$
The solubility product of A_mB_n is then expressed as:
$$[A]^m \times [B]^n = K_{sp}$$
where the brackets indicate molar concentrations and K_{sp} is the solubility product of the electrolyte. The expression
$$[A]^m \times [B]^n = K_{sp}$$
applies to all electrolytes in their saturated solutions but is only used for sparingly soluble compounds, though it forms a useful guide even in solutions of high salt concentration. K_{sp} is a constant at a particular temperature and in a particular solvent, and it is a property of the electrolyte concerned. If, therefore, the ionic product in a solution is smaller than K_{sp}, the solution is unsaturated. If the ionic product is greater than K_{sp}, the solution is supersaturated: i.e. precipitation is likely to occur, thus reducing the concentrations of the ions to the point at which their product equals the solubility product constant. If the value of K_{sp} is known, the concentration of the ions in a saturated solution can be found by using the solubility product expression; K_{sp} can be found when the molar concentration of the ions is known, i.e. when the solubility of the electrolyte is known. *Example:* calcium fluoride, CaF_2, ionises in aqueous solution according to the equation:
$$CaF_2 \rightleftharpoons Ca^{2+} + 2F^-$$
The solubility product expression is therefore:
$$[Ca^{2+}] \times [F^-]^2 = K_{sp}$$
If the solubility of CaF_2 is 0,0167 g l^{-1} (0.0167 g l^{-1}) at 26°C, it corresponds to
$$\frac{0,0167}{78} = 2,14 \times 10^{-4} M$$
$$\left(\frac{0.0167}{78} = 2.14 \times 10^{-4} M\right)$$
where 78 is the relative molecular mass of CaF_2. $[Ca^{2+}]$ is then $2,14 \times 10^{-4}$ (2.14×10^{-4}) and $[F^-]$ is
$$2 \times 2,14 \times 10^{-4} = 4,28 \times 10^{-4}$$
$$(2 \times 2.14 \times 10^{-4} = 4.28 \times 10^{-4})$$
By substituting into the solubility product expression one gets:
$$[2,14 \times 10^{-4}] \times [4,28 \times 10^{-4}]^2 = K_{sp}$$
$$([2.14 \times 10^{-4}] \times [4.28 \times 10^{-4}]^2 = K_{sp})$$
or
$$K_{sp} = 3,92 \times 10^{-11} \ (3.92 \times 10^{-11})$$
at 26°C. *See also* Appendix.

soluble A relative term describing a substance which dissolves in a certain solvent at a given temperature.

solute A substance which dissolves in another substance (the solvent) to form a solution.

solution A homogeneous mixture of one or more solutes in a solvent. The term is usually applied to solids dissolved in liquids. However, other types of solutions exist such as gases in liquids, gases in solids, etc.

solvate A compound formed as a result of solvation.

solvation The process in which ions of a solute become attached to molecules of the solvent. In an aqueous solution there is always a sheath of water molecules surrounding the ions and some of these molecules are bound to the ions by coordinate bonds or by Van der Waals forces. *See also* hydration.

Solvay process (ammonia-soda process) A process for the industrial manufacture of sodium carbonate, Na_2CO_3, in which the raw materials are calcium carbonate, $CaCO_3$, and sodium chloride, NaCl. The chemical processes involved are as follows:
(1) $CaCO_3 \rightarrow CaO + CO_2$
(2) $CaO + H_2O \rightarrow Ca(OH)_2$
(3) $2NH_4Cl + Ca(OH)_2 \rightarrow$
$CaCl_2 + 2NH_3 + 2H_2O$
(4) $2NaHCO_3 \rightarrow Na_2CO_3 + CO_2 + H_2O$
(5) $2NaCl + 2NH_3 + 2CO_2 + 2H_2O \rightarrow$
$2NaHCO_3 + 2NH_4Cl$
In process (1), calcium carbonate, $CaCO_3$, is heated to produce calcium oxide, CaO, and carbon dioxide, CO_2. The carbon dioxide produced is used in process (5), where it is bubbled through a solution of sodium chloride (brine), NaCl, saturated with ammonia, NH_3. The calcium oxide, CaO, is used in process (2) to produce calcium hydroxide (slaked lime), $Ca(OH)_2$, which reacts with ammonium chloride, NH_4Cl, in process (3) to produce ammonia, NH_3, which is used in process (5). During process (5) sodium hydrogencarbonate, $NaHCO_3$, is precipitated. When heated (process (4)) this produces sodium carbonate, Na_2CO_3, and carbon dioxide, CO_2, the latter being used in process (5). It can be seen that calcium chloride, $CaCl_2$, is the only by-product which is not re-used in the Solvay process.
Adding the five processes gives the following reaction:
$CaCO_3 + 2NaCl \rightarrow Na_2CO_3 + CaCl_2$
However, this process does not take place directly because calcium carbonate, $CaCO_3$, is insoluble in water. Potassium carbonate, K_2CO_3, cannot be prepared in a similar process because potassium hydrogencarbonate, $KHCO_3$, is more soluble than sodium hydrogencarbonate.

solvent A substance (usually a liquid) which dissolves another substance (the solute) to form a solution.

solvent extraction The separation of a mixture of solids by treating it with immiscible solvents, e.g. water and an organic solvent, in which the solids of the mixture have different solubilities. The method has proved especially useful in separating mixtures of solid inorganic and organic substances.

somatic cell A cell which is not a reproductive cell (germ-cell).

somatotrophic hormone *See* growth hormone.

somatotrophin *See* growth hormone.

sonar A word derived from the phrase sound navigation ranging. *See* echo-sounder.

sonic Pertaining to sound.

sonic boom The loud noise heard as a result of shock waves projected outwards and backwards from an aircraft flying at supersonic speed.

sonometer (monochord) An apparatus used to study the vibrations of a string (wire). It consists of a string stretched over two movable bridges on a hollow wooden box. The string may be stretched by attaching weights to one end of it. The vibrating length of the string and its tension may be altered by moving the bridges and changing the weights. *See* Fig. 213.

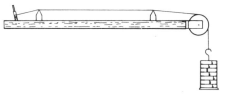

Fig. 213 Sonometer

sonorous Describes a material which produces a metallic, ringing sound when struck.

soot *See* carbon black.

sorbitol $CH_2OH—(CHOH)_4—CH_2OH$ A white, crystalline, soluble, polyhydric alcohol with a sweet taste, found in many fruits. It can be made by reduction of glucose, $C_6H_{12}O_6$:
$C_6H_{12}O_6 + H_2 \rightarrow C_6H_{14}O_6$
Mannitol, an isomer of sorbitol, has a different spatial arrangement of hydroxyl groups, —OH. It also occurs in many fruits. Both sorbitol and mannitol are used as sweeteners and in the manufacture of certain synthetic resins.

sorosis (pl. soroses) A pseudocarp formed from a whole inflorescence, e.g. the pineapple and the mulberry.

sorption A term used to mean either absorption or adsorption.

sorus (pl. sori) A group of sporangia found on the under-surface of a fern leaf. The sorus is protected by a scale-like cover called an indusium. *See* Fig. 214.

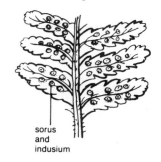

Fig. 214 Fern leaf showing sori

SOS *See* Morse code.

sound Energy transmitted through a medium (solid, liquid or gas) by longitudinal waves as a series of alternate compressions and rarefactions. Sound is produced by vibrating objects and cannot travel through a vacuum. It is characterised by its frequency (pitch), intensity (loudness) and quality. The speed of sound in a medium depends upon the nature of the medium. In dry air at 0°C it is 331,46 metres per second (331.46 m s^{-1}), increasing with temperature. The speed of sound in air is also increased by the addition of water vapour, e.g. an increase of 0,05 m s^{-1} (0.05 m s^{-1}) for an additional 0,10% (0.10%) water vapour by volume. In distilled water at 25°C it is 1497 metres per second, and in aluminium at 20°C it is 6374 metres per second. There is the following relation between wavelength, frequency and speed of sound:

$$\text{wavelength} = \frac{\text{speed of sound}}{\text{frequency}}$$

south pole *See* pole.

SP Suction pressure. *See* diffusion pressure deficit.

space The vacuum beyond the Earth's atmosphere in which are found all the heavenly bodies of the universe. Space is not a total vacuum, but contains scattered particles of gases and solids.

spark *See* electric spark.

spark counter An apparatus for detecting and counting alpha particles. It consists of a pair of parallel electrodes across which a variable high potential difference is maintained at a level just below that needed to produce a spark between the electrodes. The passage of alpha particles causes ionisation of the air between the electrodes, producing sparks which may be counted photographically or electronically.

sparking-plug A device which produces an electric spark to ignite the petrol-air mixture in the cylinder of an internal combustion engine. *See* Fig. 215; *see also* induction coil.

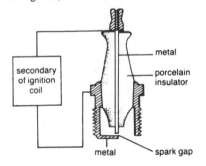

Fig. 215 Sparking-plug

spawn 1. The mycelium of certain fungi, e.g. mushroom. 2. A collection of eggs produced by frogs, fish, molluscs, etc.

specialisation 1. The adaptation of a cell or organ to perform a particular function. 2. During the evolution of an organism, its adaptation to a particular habitat or way of life.

species A group of animals or plants with certain common characteristics. The species is the basic unit of classification. Members of the same species breed to produce fertile offspring; if members of different species interbreed to produce hybrid offspring, the offspring will be infertile. A group of similar species forms a genus; related genera are further grouped into a family, etc.

specific gravity *See* relative density.

specific heat Obsolete term for specific heat capacity.

specific heat capacity Symbol: c. The specific heat capacity of a substance (solid, liquid, gas) is the amount of heat required to raise the temperature of unit mass (1 kg) of it through 1 kelvin. For gases there are two specific heat capacities, one measured at constant pressure (c_p), the other at constant volume (c_v). The SI unit of specific heat capacity is the joule per kilogram kelvin (J kg^{-1} K^{-1}).

specific latent heat Symbol: l; unit: joule per kilogram (J kg^{-1}). 1. The specific latent heat of fusion of a substance is the heat required to convert unit mass (1 kg) of the substance from solid to liquid without change in temperature. 2. The specific latent heat of vaporisation of a substance is the heat required to convert unit mass (1 kg) of the substance from liquid to gas (vapour) without change in temperature.

specific resistance An obsolete term for resistivity.

spectral line *See* line spectrum.

spectral series *See* Balmer series.

spectrogram The photographic record produced by a spectrograph.

spectrograph An instrument producing a photographic record (spectrogram) of a spectrum. Spectrographs are used in both qualitative and quantitative analysis.

spectrometer (spectroscope) An instrument for producing a pure spectrum for observation and measurement. There are various designs of spectrometers depending upon their uses: e.g. a spectrometer may be used to measure refractive indices, prism angles, etc.

spectroscope *See* spectrometer.

spectrum (pl. spectra) The pattern produced when electromagnetic radiations are separated into their constituent wavelengths (frequencies). A well-known example is the spectrum of white light consisting of the colours red, orange, yellow, green, blue, indigo and violet. Spectra may be classified as either absorption or emission spectra. An absorption spectrum is produced when radiation with a range of wavelengths is passed through a sample and one or more wavelengths are absorbed by the sample. An emission spectrum is that produced by matter emitting electromagnetic radiation. *See also* band spectrum, continuous spectrum, line spectrum *and* electromagnetic spectrum.

spectrum colour *See* spectrum, rainbow *and* colour.

speed The distance moved in unit time by an object or particle. Speed is a scalar quantity. Speed in a specified direction is called velocity, and is a vector.

speed of light *See* light energy.

speed of sound *See* sound.

sperm *See* spermatozoon.

spermatid A reproductive cell which is formed from a spermatocyte and which changes into a mature spermatozoon. This takes place in the seminiferous tubules in the testis.

spermatocyte A reproductive cell formed from a spermatogonium when it grows. This takes place in the seminiferous tubules in the testis. Spermatocytes give rise to spermatids.

spermatogenesis The process in which spermatozoa are produced in the seminiferous tubules in the testis.

spermatogonium (pl. spermatogonia) A reproductive cell formed when a gamete divides by mitosis. This takes place in the seminiferous tubules in the testis. Spermatogonia give rise to spermatocytes.

Spermatophyta (seed plants) A major division of the plant kingdom, comprising all seed-bearing plants. Spermatophytes are divided into gymnosperms and angiosperms. All flowering plants, grasses, trees and shrubs are spermatophytes with root, leaf and stem systems. They are chiefly terrestrial; they have well-defined sexual reproduction, well-developed vascular systems and well-marked alternation of generations; and they produce seeds in which the embryo is protected by a seed-coat.

spermatozoon (pl. spermatozoa) (sperm) A mature male gamete formed from a spermatid. It usually consists of a head with nucleus, a middle portion (the neck) containing mitochondria and a tail (flagellum) for locomotion. Spermatozoa are produced in the seminiferous tubules of the testis. *See* Fig. 216.

spermology The study of seeds.

sperrylite *See* platinum.

sphalerite *See* zinc blende.

spherical aberration *See* aberration.

spherical mirror *See* mirror.

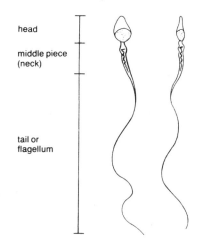

Fig. 216 Spermatozoon, front and side view (×1250)

sphincter A ring of muscle surrounding an opening and serving to narrow or close it by contraction, e.g. cardiac sphincter, pyloric sphincter.

sphygmomanometer An instrument for measuring blood pressure. It consists of an inflatable cuff connected to a manometer (aneroid or mercury). The cuff is wrapped tightly around the upper arm of a patient and then inflated with air from a rubber squeeze-bulb. This stops the flow of blood in a major artery. With a stethoscope placed just below the cuff, a doctor releases air from the cuff and listens for the first thudding sound which indicates the return of blood flow. The reading on the manometer at this moment indicates the systolic pressure resulting from the contraction of the heart. By releasing more air from the cuff, the doctor reaches a pressure where he can no longer hear the flow of blood. At this moment the manometer indicates the diastolic pressure resulting from the relaxation of the heart.

spider *See* Arachnida.

spike A type of inflorescence in which the flowers are sessile and are arranged on a long common axis. *See* Fig. 179.

spikelet One of the units making up a grass inflorescence. The spikelet encloses one or more flowers. *See* Fig. 217.

spinal canal *See* neural arch.

spinal column *See* vertebral column.

spinal cord *See* nerve-cord.

spinal nerve One of a number of pairs of nerves arising from the spinal cord at regular intervals. In humans there are thirty-one pairs of spinal nerves: eight cervical, twelve thoracic, five lumbar, five sacral and one coccygeal.

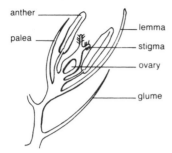

Fig. 217   Spikelet

spinal puncture See lumbar puncture.

spindle A spindle-shaped structure which forms between centrioles in a cell during metaphase in meiosis and mitosis.

spindle attachment See centromere.

spine See vertebral column.

spinneret A tube connected to silk-producing glands in spiders and in the larvae of butterflies and moths.

spinthariscope (Crookes' spinthariscope) An instrument for observing the scintillations produced when alpha particles strike a fluorescent substance (a scintillator). It consists of a short tube fitted at one end with a screen coated with zinc sulphide and at the other end with a magnifying glass (eye lens). A radioactive substance emitting alpha particles is placed just above the screen. The alpha particles cause minute flashes (scintillations) to appear on the screen. These can be observed through the eye lens. See Fig. 218.

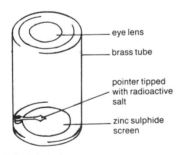

Fig. 218   Spinthariscope

spiracle 1. In insects and some arachnids, an external opening from the trachea.   2. In cartilaginous fish, a small opening situated in front of the first gill slit through which water is drawn in.   3. In frog tadpoles, an opening on the left of the body leading to the gill chambers.

spiral See bacterium.

spirillum A spiral-shaped bacterium.

spirits of salt See hydrochloric acid.

spirochaetes Slender, unicellular, spirally twisted bacteria. They are motile, whirling rapidly about the long axis of the cell and moving forward in a corkscrew or serpentine fashion. Spirochaetes are Gram-negative and may be free-living or parasitic. Some species are pathogenic to animals, e.g. *Treponema pallidum*, which causes syphilis. See also Salvarsan.

Spirogyra A genus of bright green, slimy algae found floating in stagnant fresh water. The alga consists of long strands (filaments) made up of rows of cylindrical cells with spiral chloroplasts. Reproduction is by conjugation.

spirometer An instrument for measuring lung capacity, respiratory and metabolic rates, etc.

spleen In vertebrates, an organ in which lymphocytes and antibodies are produced, blood is stored and red blood corpuscles and bacteria are destroyed. In humans it is situated in the upper part of the abdominal cavity, to the left of and a little behind the stomach.

spodumene See lithium.

sponge See Parazoa.

spongy mesophyll See mesophyll.

spontaneous combustion Combustion of a substance resulting from heat generated by slow oxidation within it.

sporangiophore In many fungi, e.g. *Mucor*, the stalk on which one or more sporangia are borne. The sporangiophore is separated from a sporangium by a columella. See Fig. 219.

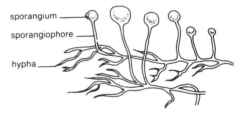

Fig. 219   *Mucor* showing sporangiophore

sporangium (pl. sporangia) A structure in which spores are produced asexually. See Fig. 220.

spore In plants, bacteria and Protozoa, a unicellular or multicellular reproductive body.

sporogenesis The formation of spores.

sporophyll A specialised leaf which bears or subtends one or more sporangia.

sporophyte The phase in the life cycle of a plant which produces spores. The sporophyte is diploid; the diploid nuclei produce haploid spores which develop into gametophytes.

spring balance An instrument used to measure the weight of an object. The object is placed in a pan or fastened to a hook attached

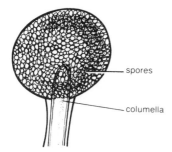

Fig. 220 Ripe sporangium of *Mucor*

to a vertical spring. The extension of the spring is proportional to the weight of the object; a pointer attached to the spring indicates the weight in newtons on a calibrated scale. *See also* force meter.

spring tide *See* tide.

sputum (pl. sputa) Matter, especially mucus, secreted in the mouth, nose, throat and lungs. Examination of sputum is used as an aid to the diagnosis of certain diseases such as tuberculosis.

Squamata An order of reptiles comprising lizards and snakes. The term 'lizard' includes geckos, iguanas, chameleons, slow-worms and monitors.

Sr Chemical symbol for strontium.

stabiliser Any substance which when added to another substance prevents or retards a chemical change in the latter for a considerable time. A stabiliser may be an inhibitor, an emulsifying agent, etc.

stable equilibrium The equilibrium of a body such that when it is displaced slightly it tends to return to its original position. *See also* neutral equilibrium *and* unstable equilibrium.

staining A technique used to highlight certain parts of biological material, e.g. in bacteria and cells, before they are examined under a microscope. Staining is usually carried out after the material is fixed, i.e. attached to a slide by slightly heating it. This is usually done by passing the slide horizontally two or three times through a bunsen flame. In staining, one or more stains may be used. When two stains are used (double staining), the second stain is called the counterstain. Stains are either acidic, basic or neutral dyes; a negative stain colours the background and not the material. *See also* Gram's stain *and* granulocyte.

stainless steel A general name for a number of steels with a high resistance to corrosion by air or by weak chemical reagents. The composition of these steels varies; however, they all contain a high percentage (up to 25%) of chromium, and sometimes other metals such as nickel and manganese.

stalactite A needle-shaped projection from the roof of a limestone cave. Stalactites form when hard water evaporates and drips from the roof, leaving a deposit of calcium carbonate. The calcium carbonate 'grows' down from the roof; after a long time it may join with a stalagmite 'growing' from the floor to form a continuous column. *See* Fig. 221.

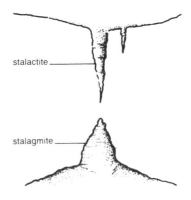

Fig. 221 Stalactite and stalagmite

stalagmite A squat projection from the floor of a limestone cave. Stalagmites form when hard water drips from the roof of the cave. When this water evaporates, it leaves a deposit of calcium carbonate. The calcium carbonate 'grows' up from the floor; after a long time it may join with a stalactite 'growing' from the roof to form a continuous column. *See* Fig. 221.

stamen The male reproductive organ of a flowering plant. It consists of a filament bearing the anther. *See* Fig. 75.

standard cell A cell, such as the Weston cell, which is capable of producing a constant reproducible e.m.f. Standard cells may be used for calibrating voltage measuring devices.

standard electrode potential *See* electrode potential *and* redox potential.

standard hydrogen electrode *See* hydrogen electrode.

standard near point distance *See* near point.

standard redox potential *See* redox potential.

standard solution Any solution of known concentration. A standard solution is usually prepared by the exact weighing of a solute which is then dissolved in a certain volume of solvent in a volumetric flask. However, if the solute is unstable, e.g. deliquescent or efflorescent, or if it cannot be obtained in a pure form, an approximate weighing is performed and the solution is then standardised by titrating it against another standard solution.

standard temperature and pressure (STP; normal temperature and pressure, NTP) A temperature of 0°C (273,16 K; 273.16 K) and a pressure of 101 325 Pa (760 mmHg) used as a standard in comparing the properties of gases.

standing crop The abundance of organisms in an ecosystem at any given time which are available to organisms at the next highest level in a food chain. Each feeding level within the ecosystem has its own standing crop. *Examples:* plants are the standing crop for herbivores; phytoplankton are the standing crop for zooplankton.

standing wave (stationary wave) The interference pattern produced when two waves of the same type, frequency, amplitude and speed travel at the same time in opposite directions through a medium. The wave form produced shows points of maximum displacement (antinodes) and minimum displacement (nodes). A vibrating string produces a standing wave.

stannic Term formerly used in the names of tin compounds to indicate that the tin is tetravalent; e.g. stannic chloride, $SnCl_4$, now called tin(IV) chloride.

stannous Term formerly used in the names of tin compounds to indicate that the tin is divalent; e.g. stannous chloride, $SnCl_2$, now called tin(II) chloride.

stannum The Latin name for tin.

stapes (stirrup) The third of the three small bones in the middle ear. It is attached to the surface of the fenestra ovalis (oval window) and transmits vibrations from the incus to the perilymph in the membranous labyrinth of the inner ear. *See* Fig. 55.

Staphylococcus A Gram-positive bacterium occurring singly, in pairs, in fours and in grapelike clusters. Staphylococci respire both aerobically and anaerobically and some types produce toxins causing food poisoning. They are frequently found on the skin and in the nasal membranes and are the cause of various diseases such as blood-poisoning and pus-forming infections.

star A large, self-luminous ball of hot gas appearing as a bright point of light in the night sky. Stars produce heat, light and other forms of electromagnetic radiation as a result of fusion processes in their interiors. They lie at different distances from the Earth and they all move relative to one another. However, because of their great distances from the Earth this motion is not easily seen and they appear to be 'fixed' in the sky. Groups of stars forming a recognisable pattern are called constellations; because of the motion of the stars, these patterns will change over the centuries. The nearest star to the Earth (apart from the Sun) is Proxima Centauri, which is approximately 4,3 light-years (4.3 light-years) away. The brightest star (apart from the Sun) is Sirius, which is approximately 8,6 light-years (8.6 light-years) away. Almost all stars emit X-rays. The study of these rays gives scientists much important information. The first satellite especially constructed for X-ray studies in space was launched by the USA in 1970. *See also* galaxies *and* Galaxy, the.

starch A polysaccharide, $(C_6H_{10}O_5)_n$, built up of many units (molecules) of glucose, $C_6H_{12}O_6$. It consists of a mixture of soluble amylose, a linear polymer of glucose, and insoluble amylopectin, a glucose polymer with interlinked chains which give the molecule a highly-branched structure. Amylopectin is a much bigger molecule than amylose. Starch is found in all plants and serves as a storage material. When extracted from plants it is a white powder forming a sol with hot water which, on cooling, sets into a gel. Starch is an important food for animals as it is enzymatically hydrolysed to glucose during digestion. Digestion of starch begins in the mouth, in which starch is converted into maltose with the aid of amylase found in the saliva. This process is continued and finished in the small intestine, in which the enzyme maltase aids in the hydrolysis of maltose to glucose. Iodine solution turns blue-black on contact with starch; this reaction is used as a test for either starch or iodine.

Stassfurt deposits Natural deposits of various inorganic salts (minerals) such as carnallite, kainite and rock salt. These deposits are found in Germany and are important sources of potassium, magnesium and sodium compounds.

states of matter *See* solid, liquid *and* gas.

static A general term for all radio interference caused by electrical disturbances in the atmosphere (atmospherics).

static electricity *See* electricity *and* frictional electricity.

static friction *See* friction.

statics The branch of mechanics concerned with forces acting upon bodies in equilibrium, i.e. systems in which there is no motion.

stationary phase *See* chromatography.

stationary wave *See* standing wave.

stator The stationary part of an electric motor or generator.

steam Water, H_2O, in the gaseous state, formed when water boils. It is an invisible gas and should not be confused with the white clouds associated with it. These are often referred to as steam, but are in fact tiny water drops formed by condensation of steam.

steam distillation A type of distillation which is useful for separating a volatile liquid, immiscible with water, from non-volatile substances by passing steam through the mixture. This technique makes it possible to perform the distillation below the normal boiling-point of the volatile liquid because the boiling-point of a mixture of two immiscible liquids (in this case condensed steam and the volatile liquid) is always lower than the separate boiling-points of the two liquids. Steam distillation can also be used in determining relative molecular masses. *See* Fig. 222.

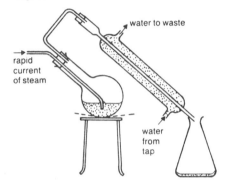

Fig. 222 Steam distillation

steam-point *See* fixed point.
stearate (octadecanoate) A salt of stearic acid (octadecanoic acid), $C_{17}H_{35}COOH$.
stearic acid $C_{17}H_{35}COOH$ (octadecanoic acid) A solid, saturated fatty acid. It is found naturally as a part of many fats and oils. The salts of stearic acid are called stearates (octadecanoates). It is used in the manufacture of soaps, candles and lubricants.
steatite *See* soapstone.
steel An alloy of iron and carbon (usually containing less than 2% carbon). *See also* carbon steel, mild steel *and* Bessemer process.
steel wool A mass of very thin steel threads, used as an abrasive for cleaning purposes.
Stefan—Boltzmann constant *See* black body.
Stefan—Boltzmann's law *See* black body.
stele In roots and stems, the central vascular cylinder consisting of xylem, phloem and the surrounding pericycle. There are various arrangements of the tissue making up steles. Some may also contain pith and medullary rays.
stellar Pertaining to the stars.
stellar wind *See* solar wind.

stem The part of a plant which bears leaves, flowers etc. It is generally erect and growing above ground. Water and food are transported by the vascular bundles in the stem, whose cells may also carry out photosynthesis. It serves to hold the leaves in the light for photosynthesis, and to hold flowers and fruits above ground for maximum pollination and dispersal. There are other types of stem which grow underground. Examples of these are bulbs, corms and rhizomes.
stem tuber *See* tuber.
step-down transformer *See* transformer.
step-up transformer *See* transformer.
steradian Symbol: sr. The SI unit of solid angle. One steradian is equal to the angle subtended at the centre of a sphere by an area of its surface numerically equal to the square of the radius.
stereochemistry The branch of chemistry concerned with the spatial arrangement of atoms or groups of atoms in a molecule and how this arrangement affects its chemical properties.
stereoisomerism A type of isomerism in which a substance exists with several different spatial arrangements of its atoms or groups of atoms. Stereoisomerism is divided into geometrical isomerism and optical isomerism.
sterile 1. Describes an organism which is unable to reproduce asexually or sexually. **2.** Free from any form of living organism. *See also* sterilisation.
sterilisation 1. The act of making an organism sterile, e.g. by removing the womb, ovaries or Fallopian tubes from a woman, or by cutting or tying off the vasa deferentia in a man. **2.** The process in which all living organisms are removed or destroyed, including bacterial endospores. This may be done using heat, radiation or chemicals.
sternum (breastbone) In vertebrates, a wedge-shaped or rod-shaped bone situated in the middle of the ventral side of the thorax. In humans it articulates with the clavicles and the first seven ribs. *See* Fig. 193.
steroid The general name for a large group of organic substances with four carbon rings of which three are six-membered and one is five-membered. Steroids are found in animals and plants. Examples include certain hormones such as cortisone and progesterone. Steroids are often called sterols.
sterol 1. The general name for a large group of organic substances derived from steroids but containing one or more aliphatic side chains and hydroxyl groups. Sterols are alcohols and are found in animals and plants. An important sterol is cholesterol. **2.** *See* steroid.
stethoscope An instrument used to listen to the sounds made by the heart, lungs, major

stigma

blood vessels and other internal organs. It consists of two ear-pieces attached by flexible rubber tubes to a disc or cone. A stethoscope is used in the diagnosis of a disease or abnormality.

stigma The part of the carpel that receives pollen. *See* Fig. 75.

still Any apparatus used for distillation.

stilt roots Adventitious roots growing from the main stem and supporting it in muddy soil whose level may change with the seasons.

stimulated emission *See* laser.

stimulus (pl. stimuli) A change in the surroundings (internal or external) of an organism or part of it producing a reaction (response). The stimulus itself does not provide any energy for the response.

stipe The stalk bearing the pileus of a mushroom.

stipule One of a pair of leaf-like outgrowths from the leaf bases of many plants, e.g. the rose. The stipule protects the axillary bud and often takes part in photosynthesis.

stirrup *See* stapes.

stock *See* scion.

Stock notation A notation recommended by IUPAC and now widely used in naming certain chemical compounds. It gives the valency (oxidation number) of an element in a compound in Roman numerals in parentheses. *Example:* copper has the valencies 1 (monovalent) and 2 (divalent). Monovalent copper was formerly called cuprous in compounds, and divalent copper was called cupric. However, using the Stock notation they are written copper(I) and copper(II) respectively. Other examples are: iron(III) oxide, Fe_2O_3; lead(IV) oxide, PbO_2.

stoichiometry A branch of chemistry dealing with the gravimetric or volumetric composition of substances, i.e. calculations regarding the proportions in which elements or compounds (molecules) react with one another. *See also* chemical combination (laws of).

Stokes' law A law stating that when a spherical particle falls through a liquid, the velocity of sedimentation, v, depends on the particle size, the difference between the density of the particle, d, and the density of the liquid, d_1, and on the viscosity of the liquid, η. This can be expressed mathematically as:

$$v = \frac{2r^2(d - d_1)g}{9\eta}$$

where r is the radius of the particle and g the acceleration of free fall. The law can also be used to determine particle size or viscosity when the value of v is known. *See also* Poiseuille's formula.

stolon A horizontal stem which may lie on the surface of the soil or grow underground. Adventitious roots develop from the nodes, giving rise to new plants. This is a type of vegetative reproduction. A stolon which grows above ground is sometimes called a runner.

stolzite *See* tungsten.

stoma (pl. stomata) A small opening in the epidermis of plants, especially on the lower epidermis of leaves. Stomata allow the gaseous exchange of carbon dioxide and oxygen to take place between the plant and the atmosphere, and allow the escape of water vapour during transpiration. The opening and closing of a stoma is controlled by the turgidity of a pair of guard cells. *See* Fig. 223.

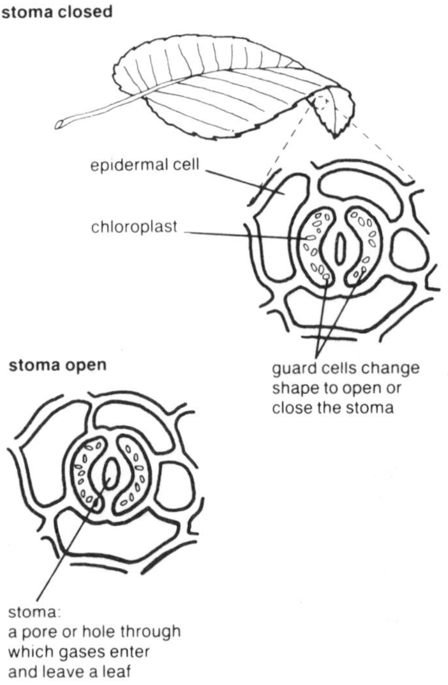

Fig. 223 Stoma

stomach In vertebrates, the part of the alimentary canal between the oesophagus and the small intestine. The stomach secretes enzymes and hydrochloric acid (gastric juice); these mix and react with the food which is churned by the muscular contractions of the stomach walls. When the food is converted to chyme, it is allowed to pass into the duodenum. In humans the stomach is a sac-like structure with a sphincter muscle at its entrance and exit. Practically no absorption of food takes place in the stomach. *See* Fig. 49.

stomata *See* stoma.

STP *See* standard temperature and pressure.

strain The temporary or permanent distortion of a body when a stress is applied to it. Strain is measured by the ratio of the change in length, area or volume to the original dimensions of the body.

stratification The separation of vegetation into layers according to its height, e.g. the herb layer, shrub layer and tree layer in a forest.

stratosphere The region of the Earth's atmosphere situated between the troposphere and the mesosphere.

strep throat A common name for a throat infection caused by streptococci. It is always accompanied by fever and it should be promptly treated with penicillin, otherwise it may lead to diseases such as nephritis and rheumatic fever.

Streptococcus A Gram-positive bacterium occurring in short or long chains. Streptococci are found in the mouth and intestines of many animals, including humans, and also in many types of food. Some streptococci are pathogenic, causing rheumatic fever and strep throat, whereas others are harmless.

streptomycin An antibiotic used in the treatment of various diseases such as pneumonia, tuberculosis, intestinal infections and infections of the gall-bladder and urinary tract. *See also* glucosamine.

stress A force or system of forces producing a distortion (strain) in a body. Stress is measured as a force per unit area, i.e. in newtons per square metre (Nm^{-2}).

striated muscles *See* skeletal muscles.

stridulation In some insects, such as grasshoppers and crickets, the production of sound by rubbing together parts of the body. It is usually carried out by males during courtship.

striped muscles *See* skeletal muscles.

strobilus (pl. strobili) *See* cone.

stroboscope An instrument used to study objects which are vibrating or rotating. A simple stroboscope can be made from a circular piece of card with a number of equidistant radial slits cut in it. The card is pivoted at its centre and spun so that vibrating strings, water waves, etc. may be observed through the slits. At certain speeds of rotation, the object being viewed appears to move slowly or to be stationary. An accurate measuring stroboscope is one consisting of a flashing light whose frequency can be controlled. A rotating disc, e.g. the turntable of a record player, will appear to be stationary when illuminated by a stroboscope flashing at the same frequency as the number of revolutions per second of the turntable. Stroboscopes may be used to inspect rapidly moving parts of machinery, and to measure frequencies of vibrations and the speeds of rotating objects.

stroke A condition caused by bleeding from a ruptured artery in the brain or blockage of such an artery. Strokes occur most often in people with high blood pressure (hypertension) or arteriosclerosis, and in elderly people. Symptoms may include headache, vomiting, confusion, temporary or permanent paralysis, loss of consciousness, etc.

'stroking' *See* magnetise.

stroma *See* granum.

strontianite *See* strontium.

strontium An element with the symbol Sr; atomic number 38; relative atomic mass 87,62 (87.62); state, solid. Strontium is one of the alkaline earth metals. It occurs naturally in the minerals celestine (strontium sulphate, $SrSO_4$) and strontianite (strontium carbonate, $SrCO_3$). Strontium can be extracted by electrolysis of its fused chloride. It is used in certain alloys. Strontium-90, $^{90}_{38}Sr$, is a radioactive isotope with a half-life of 28 years. It is produced in nuclear explosions and is a serious hazard because it is metabolised like calcium, i.e. it may be included in the bones of animals.

strontium carbonate $SrCO_3$ A white, insoluble, crystalline salt occurring naturally as strontianite. It can be prepared by adding a solution of strontium nitrate, $Sr(NO_3)_2$, or strontium chloride, $SrCl_2$, to a solution of sodium carbonate, Na_2CO_3:

$Sr^{2+} + CO_3^{2-} \rightarrow SrCO_3$

It is used in fireworks, where it produces a red flame.

strontium hydroxide $Sr(OH)_2$ A white, soluble, solid hydroxide which forms a strong alkali when dissolved in water. It may be prepared by dissolving strontium oxide, SrO, in water:

$SrO + H_2O \rightarrow Sr(OH)_2$

$Sr(OH)_2$ is used in sugar-refining.

structural formula *See* formula.

structural isomerism A type of isomerism in which different isomers have different structural formulae. Different forms of structural isomerism exist: (1) In position isomerism, the position of a functional group in a molecule differs from one isomer to another. *Example:* there are two structural isomers of propanol, $CH_3-CH_2-CH_2OH$ (propan-1-ol) and $CH_3-CHOH-CH_3$ (propan-2-ol). (2) In functional group isomerism, the isomers are different chemical compounds because the functional groups are different. Examples are propanone (acetone), $CH_3-CO-CH_3$, and propanal (propionaldehyde), CH_3-CH_2-CHO, which both have the formula C_3H_6O. (3) In chain isomerism there is a change in the arrangement of carbon atoms in a molecule.

Example: $CH_3-CH_2-CH_2-CH_3$ is butane whereas

$$CH_3-CH-CH_3$$
$$|$$
$$CH_3$$

is 2-methylpropane. Structural isomers usually have different chemical and physical properties. *See also* tautomerism.

strychnine A colourless, crystalline, slightly soluble, highly poisonous alkaloid extracted from the seeds of certain plants. It was formerly used as a stimulant.

style The part of the carpel connecting the ovary and the stigma. After pollination, pollen tubes grow through the style carrying gametes to the ovary. *See* Fig. 75; *see also* micropyle.

styrene *See* phenylethene.

subarachnoid space The space, containing cerebrospinal fluid, between the arachnoid and the pia mater.

subatomic particle A particle smaller than an atom. Subatomic particles include electrons, neutrons and protons.

subcutaneous Under the skin or dermis.

subdural space The space between the arachnoid and the dura mater.

suberin A wax-like substance, rather like cutin, which develops in thickened cell walls. It is found in cork tissue and in the roots and stems of many plants. It is impermeable to water and gases; it also serves to strengthen and protect the organism.

sublimate 1. *See* sublime. 2. The solid substance formed during sublimation.

sublimation A process in which there is a direct change from a solid to a vapour and vice versa. The solid does not melt before it becomes a gas:

$$\text{solid} \underset{}{\overset{\text{heat}}{\rightleftharpoons}} \text{gas}$$

Sublimation takes place when the vapour pressure of a solid reaches atmospheric pressure before it melts. Only a few substances can sublime. They include iodine, ammonium chloride, solid carbon dioxide (dry ice) and naphthalene. Substances which sublime at atmospheric pressure may be turned into liquids at higher pressures. Solids which do not sublime at atmospheric pressure may do so at lower pressures. Sublimation can be used as a method of refining (purifying).

sublimation point The temperature at which the vapour pressure of a solid becomes equal to the external (atmospheric) pressure.

sublime (sublimate) *See* sublimation.

sublingual glands In humans and other mammals, the salivary glands situated under the tongue and beneath the floor of the mouth.

submandibular glands (submaxillary glands) In humans and other mammals, the salivary glands situated beneath the angle of the lower jaw.

submaxillary glands *See* submandibular glands.

sub-shell *See* electron configuration *and* orbital.

subsoil The layer of soil beneath the topsoil. Subsoil is a light-coloured layer consisting of stones, gravel, clay, etc. In this layer the deeper roots of plants are also found. Subsoil contains little or no humus. Underneath the subsoil is rock from which all soil is formed.

subsonic Describes a speed below Mach 1, the speed of sound.

substituent An atom or group of atoms which substitutes for another atom or group of atoms in a chemical compound. *Example:* in the reaction:

$$C_6H_6 + Cl_2 \rightarrow C_6H_5Cl + HCl$$

one chlorine atom from the Cl_2 molecule substitutes for one hydrogen atom in benzene, C_6H_6, forming chlorobenzene, C_6H_5Cl, and hydrogen chloride, HCl.

substitution reaction A reaction in which one or more atoms or groups of atoms are replaced by other atoms or groups of atoms. This type of reaction is typical for saturated organic compounds. *Example:* in methane, CH_4, one or more of the hydrogen atoms may be replaced by one or more halogen atoms such as chlorine, Cl:

$$CH_4 + Cl_2 \rightarrow CH_3Cl + HCl$$

In this reaction chloromethane (methyl chloride), CH_3Cl, and hydrogen chloride, HCl, are formed. When benzene, C_6H_6, reacts with chloromethane (methyl chloride), CH_3Cl, and anhydrous aluminium chloride, $AlCl_3$, is used as a catalyst, methylbenzene (toluene), $C_6H_5CH_3$, and hydrogen chloride, HCl, are formed:

$$C_6H_6 + CH_3Cl \rightarrow C_6H_5CH_3 + HCl$$

The reactions of aromatic compounds such as benzene and derivatives of benzene with alkyl halides such as chloromethane, and with acyl halides, are known as Friedel–Crafts reactions, after the names of their two discoverers. *Compare* addition reaction.

substrate 1. A substance on which enzymes may act. 2. A substance on which microorganisms, etc., are cultured.

succession The chain of events in which changes in composition in a community of an ecosystem take place over a number of years, starting from the initial colonisation of a piece of bare soil or water.

succinite *See* amber.

succulent Describes plants or parts of plants which are thick and full of juice (sap). Examples of succulent plants are cacti and aloes.

succus entericus *See* intestinal juice.
sucker A shoot which grows from an underground stem or root to form a new plant. Suckers are produced by plants such as banana, sisal hemp and pineapple as a means of vegetative reproduction.
sucrase *See* invertase.
sucrose *See* cane-sugar *and* beet-sugar.
suction pressure (SP) *See* diffusion pressure deficit.
Sudan III An azo compound which is insoluble in water, but soluble in oils. It is used in microscopy in the staining of fat cells.
suffocation *See* asphyxia.
sugar The general term for members of two large groups of sweet-tasting very soluble carbohydrates, the monosaccharides and the disaccharides. Among the monosaccharides, only those containing five or six carbon atoms are of importance, i.e. those with the formula $C_5H_{10}O_5$ or $C_6H_{12}O_6$. Sugars with the formula $C_5H_{10}O_5$ are called pentoses, those with the formula $C_6H_{12}O_6$ are called hexoses. Monosaccharides are either aldoses or ketoses, i.e. they contain either an aldehyde group, —CHO, or a ketone group, —CO|

Pentoses and hexoses which contain an aldehyde group are called aldopentoses and aldohexoses. If instead they contain a ketone group they are called ketopentoses and ketohexoses. Disaccharides are built up of two molecules of monosaccharide. Apart from the aldehyde and ketone groups, sugars also contain hydroxyl groups, —OH, attached to a chain or ring of carbon atoms. When monosaccharides form cyclic molecules the aldehyde group or ketone group reacts with one of the other carbon atoms in the chain. This may give rise to a six-membered ring (a pyranose) or a five-membered ring (a furanose). *Examples:* glucose is a monosaccharide which forms a pyranose ring, fructose is a monosaccharide which forms a furanose ring. Glucose is an aldohexose, fructose is a ketohexose:

```
¹CHO           ¹CHOH
 |              |
²CHOH          ²CHOH
 |              |        
³CHOH          ³CHOH     O
 |              |
⁴CHOH          ⁴CHOH
 |              |
⁵CHOH         H—⁵C
 |              |
⁶CH₂OH         ⁶CH₂OH
glucose (chain)  glucose (ring) –
                       pyranose
```

Here carbon atoms number 1 and 5 are connected in a six-membered ring (pyranose).

```
¹CH₂OH          ¹CH₂OH
 |               |
²CO           HO—²C
 |               |
³CHOH           ³CHOH
 |               |        O
⁴CHOH           ⁴CHOH
 |               |
⁵CHOH          H—⁵C
 |               |
⁶CH₂OH          ⁶CH₂OH
fructose (chain)  fructose (ring) –
                       furanose
```

Here carbon atoms number 2 and 5 are connected in a five-membered ring (a furanose). Saccharose (sucrose) is a disaccharide which gives glucose and fructose on hydrolysis.
sugar of lead *See* lead(II) ethanoate.
sulpha drug *See* sulphonamide.
sulphanilamide $H_2N—C_6H_4—SO_2NH_2$ (4-aminobenzenesulphonamide; *p*-aminobenzenesulphonamide) An aromatic compound which is valuable in the treatment of certain infectious diseases. Some of its derivatives have proved to be even more effective than sulphanilamide itself. *See also* sulphonamide.
sulphate A salt of sulphuric acid, H_2SO_4.
sulphide A salt of hydrogen sulphide, H_2S.
sulphite A salt of sulphurous acid, H_2SO_3.
sulphonamide (sulpha drug) The general term for a group of drugs which inhibit the growth of some pathogenic micro-organisms. They are not antibiotics. Sulphonamides are very useful in the treatment of urinary infections. The first sulpha drug to be used in medicine was examined by Gerhard Domagk in 1932 and it proved to have a powerful effect against infections caused by streptococci. In 1935 the drug was marketed under the name Prontosil. Prontosil is an azo compound also containing sulphur. Inside an organism, it is changed into sulphanilamide (4-aminobenzenesulphonamide), which is the active substance.
sulphonation The process in which the sulphonic acid group, —SO_2OH, is introduced into an organic molecule which is either aliphatic or cyclic, giving a direct carbon–sulphur link. When benzene, C_6H_6, is treated with hot, concentrated sulphuric acid, H_2SO_4, or with oleum (fuming sulphuric acid), benzenesulphonic acid, $C_6H_5SO_2OH$, is produced:
$C_6H_6 + H_2SO_4 \rightarrow C_6H_5SO_2OH + H_2O$
Sulphur compounds other than H_2SO_4 may be used, such as sulphur(VI) oxide, SO_3, which is the active substance when using sulphuric acid or oleum.
sulphonic acid The general term for organic compounds containing the group —SO_2OH,

sulphur

e.g. benzenesulphonic acid, $C_6H_5SO_2OH$. Sulphonic acids are strong acids which are colourless and are either solids or viscous liquids. They may be prepared by sulphonation of aliphatic or cyclic compounds. Only aromatic sulphonic acids are important. They are used as intermediate compounds in organic synthesis.

sulphur An element with the symbol S; atomic number 16; relative atomic mass 32,06 (32.06); state, solid. Sulphur is a yellow non-metal, occurring in the free state in volcanic areas and in underground deposits from which it is extracted by the Frasch process. It is also found combined with various metals as insoluble sulphides e.g. zinc, lead, copper and mercury sulphides; as anhydrite, $CaSO_4$; as gypsum, $CaSO_4 \cdot 2H_2O$; and in crude oil and natural gases. Sulphur reacts with most metals when heated. *Example:* iron and sulphur form iron(II) sulphide, FeS:
$Fe + S \rightarrow FeS$
However, some non-metals, e.g. oxygen, carbon and chlorine, also react with sulphur. *Example:* sulphur and oxygen form sulphur dioxide, SO_2, when heated:
$S + O_2 \rightarrow SO_2$
Sulphur exhibits allotropy; the two main allotropes are rhombic sulphur and monoclinic sulphur, both consisting of S_8 molecules. Sulphur is used in the manufacture of sulphuric acid, in vulcanising rubber, in gunpowder, fireworks, matches, etc. See Fig. 143; *see also* amorphous sulphur *and* plastic sulphur.

sulphur bacteria Bacteria which are capable of oxidising sulphides such as hydrogen sulphide, H_2S, to sulphur and sulphur to sulphate, or reducing sulphate to sulphide. Sulphur bacteria are common in soil.

sulphur dioxide *See* sulphur oxides.

sulphuretted hydrogen *See* hydrogen sulphide.

sulphuric acid H_2SO_4 A strong, viscous, colourless, odourless acid. It is manufactured industrially in the contact process. When concentrated (and hot), sulphuric acid is a strong oxidising agent, attacking metals such as copper, Cu, and silver, Ag:
$Cu + 2H_2SO_4 \rightarrow CuSO_4 + SO_2 + 2H_2O$ and
$2Ag + 2H_2SO_4 \rightarrow Ag_2SO_4 + SO_2 + 2H_2O$
Concentrated sulphuric acid is also a powerful drying agent. Sulphuric acid forms two series of salts: sulphates, e.g. sodium sulphate, Na_2SO_4; and hydrogensulphates, e.g. sodium hydrogensulphate, $NaHSO_4$. H_2SO_4 is widely used in the chemical laboratory and in the manufacture of fertilisers (e.g. superphosphate), explosives, dyes, drugs, etc. *See also* calcium phosphate(V).

sulphurous acid H_2SO_3 A weak acid which has never been isolated, but which is found in solution when sulphur dioxide, SO_2, dissolves in water:
$SO_2 + H_2O \rightleftharpoons H_2SO_3$
Sulphurous acid smells strongly of SO_2; when heated, it decomposes completely into SO_2 and H_2O. Sulphurous acid forms two series of salts: sulphites, e.g. sodium sulphite, Na_2SO_3; and hydrogensulphites, e.g. sodium hydrogensulphite, $NaHSO_3$.

sulphur oxides Sulphur dioxide, SO_2, and sulphur(VI) oxide (sulphur trioxide), SO_3.
Sulphur dioxide, SO_2, is a colourless, soluble gas with a pungent smell. With water, it forms an acidic solution containing sulphurous acid, H_2SO_3:
$SO_2 + H_2O \rightleftharpoons H_2SO_3$
SO_2 can be pepared by burning sulphur in air:
$S + O_2 \rightarrow SO_2$
by roasting iron(II) disulphide (pyrites), FeS_2:
$4FeS_2 + 11O_2 \rightarrow 2Fe_2O_3 + 8SO_2$
or by acidifying a sulphite:
$SO_3^{2-} + 2H^+ \rightarrow SO_2 + H_2O$
Sulphur dioxide is used in the manufacture of sulphuric acid, as a food preservative, as a bleaching agent, etc.
Sulphur(VI) oxide (sulphur trioxide), SO_3, is a white, soluble solid which fumes in moist air and which exists in several polymorphic forms. It can be prepared by reacting sulphur or sulphur dioxide, SO_2, with oxygen using a catalyst (platinum or vanadium(V) oxide):
$2SO_2 + O_2 \rightarrow 2SO_3$
Sulphur(VI) oxide reacts violently with water forming sulphuric acid, H_2SO_4:
$SO_3 + H_2O \rightarrow H_2SO_4$
SO_3 continues to dissolve in the sulphuric acid so formed, producing oleum (fuming sulphuric acid or pyrosulphuric acid), $H_2S_2O_7$:
$SO_3 + H_2SO_4 \rightarrow H_2S_2O_7$
When oleum is diluted with water, sulphuric acid is formed:
$H_2S_2O_7 + H_2O \rightarrow 2H_2SO_4$

sulphur trioxide *See* sulphur oxides.

Sun The central heavenly body in our Solar System, a star which is visible as a gaseous globe. The Sun is the nearest star to Earth, being $1{,}496 \times 10^8$ km (1.496×10^8 km) away. In our galaxy, the Milky Way, the Sun is one of the millions of stars located at a distance of about 30 000 light-years from its centre. The Sun rotates at its equator once every 24,7 days (24.7 days). Its surface temperature is about 6000°C, whereas its internal temperature is more than 10 000 000°C. This is a result of fusion processes. The mass of the Sun is so great ($1{,}99 \times 10^{27}$ tonnes; 1.99×10^{27} t) that it controls the motion of the planets by its gravitational force. *See also* tide.

sunspot A dark area on the surface of the Sun. Sunspots have a lower temperature than the surrounding photosphere and therefore they emit less light, appearing dark by contrast. Each spot consists of a dark, central area (the umbra) around which there is a lighter region (the penumbra). Little is known about the formation of sunspots; however, it is believed that they are associated with strong magnetic fields and may be the cause of magnetic storms.

superconductivity The resistivity of all metals and alloys approaches zero as the temperature approaches absolute zero ($-273,16°C$; $-273.16°C$). For some metals, such as lead, mercury and tin, the resistivity drops sharply to zero (or almost zero) at a transition temperature above absolute zero (for tin, at $7,2 K$ ($7.2 K$)). This is known as superconductivity. The metals become perfect (or almost perfect) conductors. A current once started in a perfect conductor (a superconductor) will flow indefinitely. Superconductivity can be destroyed by placing the metal in a sufficiently large magnetic field. Certain metals do not become superconducting; instead, when the temperature falls, their resistance falls to a minimum and then rises again. Examples of such metals are copper and platinum.

superconductor *See* superconductivity.

supercooled water *See* supercooling.

supercooling The process in which a system is slowly cooled below a temperature at which a phase change would normally take place. A supercooled system is in a metastable state. *Example:* when water is cooled slowly, it does not crystallise at the freezing-point but remains liquid at this temperature. This happens because when the water reaches freezing-point its molecules are not lined up in an arrangement corresponding to the crystalline shape of ice. The water may be cooled below this temperature before the molecules form into the correct pattern by chance and begin to crystallise. Once the correct crystalline pattern has been built up to a sufficient size, other water molecules rapidly crystallise on it. The heat from this process makes the temperature rise to the freezing-point of water. Supercooling can be avoided if the liquid contains a seed crystal on which crystallisation can occur. In a similar way a solution may be cooled to a temperature at which the solvent contains more solute than it normally would hold at that temperature; the solution is then said to be supersaturated. Vapours may be supercooled if they are free from solid impurities such as dust on which drops of liquid may be produced. A supercooled vapour is a supersaturated vapour, i.e. it exists at a temperature below its normal condensation temperature (dew point) and has a vapour pressure that is higher than a saturated vapour at the same temperature. *See also* cloud chamber.

super giant *See* giant star.

superheated steam Steam which has a temperature above that of water boiling at one atmosphere, i.e. above 100°C. Superheated steam can be produced by increasing the pressure above boiling water, while supplying enough heat to keep the water boiling. It is used in conventional power-stations to drive the turbines.

superior Describes a structure which is situated above another. *Compare* inferior.

superior vena cava *See* vena cava.

supernatant A clear liquid found above a precipitate or sediment.

supernova (pl. supernovae or supernovas) An exploding star which rapidly flares up to a brightness which may be 100 million times its original brightness. The energy released in such an explosion is much greater than that of a nova. Supernovae may be classified according to their brightness. After a supernova explosion, either the star is totally destroyed or it collapses into a small and extremely dense neutron star. In a neutron star the matter is so tightly packed that protons, $_1^1H$, and electrons, $_{-1}^0e$, are compressed to form neutrons, $_0^1n$:
$$_1^1H + _{-1}^0e \rightarrow _0^1n$$
The mass of an average neutron star is roughly equal to that of the Sun, but its diameter is only a few kilometres. This results in fantastic densities. *Example:* a pin-head amount of material from a neutron star would have a mass of about 1 million tonnes. *Compare* nova; *see also* black hole.

superoxide An oxide which contains the O_2^- ion and is therefore a powerful oxidising agent. *Example:* NaO_2 is sodium superoxide. *See also* sodium oxides.

superphosphate *See* calcium phosphate(V).

supersaturated solution *See* supercooling.

supersaturated vapour *See* supercooling.

supersonic Describes a speed in excess of Mach 1, the speed of sound.

supplemental air The volume of air which can be expelled from the lungs after normal expiration.

suprarenal gland *See* adrenal gland.

surface tension Symbol: γ (gamma). The property of a liquid which makes it behave as if it were enclosed by a membrane. The molecules of a liquid attract one another equally in all directions, apart from those at the surface. These have no force attracting them outwards, so the resultant forces tend to pull them towards the interior of the liquid. For this reason liquid surfaces tend to become as small as possible. A drop will have a spherical shape because this is the form that has the smallest surface area for a given volume. However, water molecules are attracted to some substances more than they are to each other. This causes water to rise up the sides of a narrow glass tube, forming a concave meniscus. Mercury molecules are attracted to each other more than they are attracted to glass, so the meniscus formed by mercury in a narrow glass tube is convex. The surface tension of water is reduced by adding soap. This makes it easier for water to penetrate clothing during washing. Surface tension enables certain insects to walk on the surface of water. The surface tension of a liquid decreases with increasing temperature. *See also* capillary action.

suspension A type of dispersion in which small, solid particles are dispersed in a liquid or gas.

suspensory ligament One of the fine fibres which hold the eye lens in position and attach it to the ciliary body. *See* Fig. 66.

suture 1. A line showing the junction of bones, e.g. in the skull, or the junction of the two halves of a carpel, e.g. in legumes. 2. The joining of two sides of a wound by means of stitches or clamps. The term is also used for the actual material used to do this. Sutures may be made of silk or nylon thread, catgut, or wire and metal clamps. Most sutures are removed after some time; others, especially those used internally, are gradually absorbed by the body.

SVP *See* saturated vapour pressure.

swamp A more or less permanently waterlogged area of land, usually in tropical regions.

sweat A thin, watery fluid secreted from sweat glands. It mainly consists of water with a small amount of dissolved salts, e.g. sodium chloride, and urea.

sweat gland In mammals, a coiled tubular gland situated in the dermis and producing sweat. The number of sweat glands varies between species. Sweat is produced when there is a rise in body temperature, usually as a result of fever, increased air temperature or exercise. Evaporation of sweat from the skin serves to control body temperature. *See* Fig. 208.

swim bladder In bony fish, an air-filled bladder situated below the spinal column. It serves to adjust buoyancy so that the fish can swim easily at different depths. In some fish it is connected to the gut and the air pressure in it can be adjusted by intake or release of air through the mouth. In others the swim bladder has no opening to the exterior and the pressure in it is regulated by the absorption or secretion of oxygen by surrounding blood vessels.

switch A device for opening and closing an electric circuit.

syconium (pl. syconia) (syconus) A pseudocarp formed from a fleshy, hollow receptacle, e.g. the fig.

syconus *See* syconium.

sylvine A mineral consisting of potassium chloride, KCl. It is similar to halite, with which it is commonly associated.

symbiont One of the partners associated in symbiosis (mutualism).

symbiosis (pl. symbioses) A term often used for an association between dissimilar organisms (symbionts) which benefit mutually. However, the correct term for such an association is mutualism. *See also* commensalism *and* parasitism.

sympathetic nervous system A division of the autonomic nervous system, consisting of the sympathetic nerve and certain other cranial and spinal nerves. Noradrenalin(e) is produced at the ends of the nerve-fibres of the sympathetic nervous system. This system works in opposition to the parasympathetic nervous system. It controls such involuntary actions as dilation of the iris and coronary arteries, increasing the heartbeat, raising the blood pressure, decreasing peristalsis, etc.

symphysis (pl. symphyses) The union of bones by fusion, ligament or cartilage allowing a slight movement, e.g. pubic symphysis. *See also* vertebra.

synapse *See* neurone.

synergism The united effort of structures, organisms or substances. *Examples:* the muscles closing the fingers can only produce a strong grip if other muscles bend the wrist backwards at the same time; two hormones may interact to produce a total effect which is greater than the sum of their individual effects. *Compare* antagonism.

syngamy Fusion of gametes in fertilisation.

synovia *See* synovial fluid.

synovial fluid (synovia) A fluid secreted from the synovial membrane which serves to lubricate a joint, thus preventing friction between articulating bones.

synovial membrane A membrane lining the capsule which surrounds a joint. It is made of connective tissue and secretes synovial fluid (synovia).

synthesis (pl. syntheses) The preparation of chemical compounds (molecules) by building them up from simpler compounds or directly from elements. *Compare* analysis.

synthesis gas A mixture of carbon monoxide, CO, and hydrogen, H_2 produced when methane, CH_4, reacts catalytically with steam: $CH_4 + H_2O \rightarrow CO + 3H_2$

It is used as a starting mixture in the synthesis of several organic compounds. *See also* methanol *and* water gas.

synthetic Describes a material that has been prepared artificially, in a laboratory or by an industrial process, rather than being found in nature. *Example:* nylon is a synthetic material, cotton is a natural material.

syphilis A venereal disease caused by a spiral-shaped bacterium (spirochaete). The disease is almost always contracted during sexual intercourse with an infected person; however, it may be passed from an infected mother to her unborn child. Syphilis is a more serious disease than gonorrhoea. Its symptoms include a single, painless sore, usually on the penis or at the entrance to the vagina (or inside it). After this sore has healed, fever, sore throat and skin rash occur, If untreated, syphilis may lead to serious disease of the heart or brain. If it is detected in its early stages, it may be easily cured by antibiotics. *See also* Salvarsan.

syringe An instrument used for measuring and transferring fluids. It may be fitted with a hollow needle at its base and consists of a graduated glass or plastic cylinder through which a piston (plunger) moves. *See* Fig. 224.

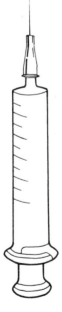

Fig. 224 Syringe

systematic name A name for a chemical substance which gives useful information about the structure of the substance. Examples of systematic names are: iron(III) oxide, Fe_2O_3; lead(IV) oxide, PbO_2; propanone, CH_3—CO—CH_3; ethanol, CH_3—CH_2OH. *See also* Stock notation *and* trivial name.

Système International d'Unités *See* SI units.

systole 1. In vertebrates, the phase of heartbeat in which the atria and ventricles are contracted, thus pumping blood into the aorta and pulmonary artery. *Compare* diastole. **2.** In a contractile vacuole, the phase in which it expels fluid (water). *Compare* diastole.

T

T Chemical symbol for tritium.

tactic movement (taxis) A movement of an organism or freely motile part of it, in response to an external directional stimulus. If the movement is towards the stimulus the taxis is positive, whereas a movement away from the stimulus is a negative taxis. *See also* chemotaxis *and* magnetotaxis.

tactile Pertaining to the sense of touch.

tadpole The aquatic, larval stage of a frog or toad.

talc A soft, greasy mineral containing magnesium silicate. It can be respresented by the formula $Mg_3Si_4O_{10}(OH)_2$, and has a hardness of 1 on the Mohs' scale. *See also* talcum powder *and* soapstone.

talcum powder Powdered talc, usually perfumed and used as a toiletry.

tallow A substance obtained by melting animal fat. It is used in the manufacture of candles, soaps and lubricants. *See also* lard.

tannic acid *See* tannin.

tannin (tannic acid) Any of several very complex, aromatic compounds which may contain saccharides combined with various phenols. Tannins are extracted from vegetable matter such as the bark of certain trees and shrubs. They are used in tanning and in making dyes and inks.

tanning The process in which skins or hides are converted into leather. This may be done by soaking the unhaired hides in solutions of tannins or chromium salts. During this process the proteins in the hides form insoluble compounds.

tapeworm *See* Cestoda.

tap funnel *See* separating funnel.

tap-root A main root growing vertically downwards with smaller, lateral roots growing from it. In some plants tap-roots serve as storage organs, e.g. beetroot and carrot.

tar The dark residue obtained by the destructive distillation of materials such as coal, wood or petroleum. *See also* coal-tar.

tarsal (tarsal bone) An ankle bone. In humans there are seven, one of which forms the heel. *See* Fig. 162.

tarsus 1. A medical name for the ankle. 2. The region of the hindlimb of vertebrates, containing the tarsal bones.

tartaric acid $HOOC-(CHOH)_2-COOH$ ((−)-2,3-dihydroxybutanedioic acid) A white, crystalline, soluble, organic acid whose normal salts are called tartrates and whose acid salts are called hydrogen-tartrates. It is found in many plants and fruits, especially grapes. Potassium sodium tartrate is a double salt called Rochelle salt or Seignette salt. It is used as an ingredient of Fehling's solution. Tartaric acid is used in dyeing and in effervescent powders.

tartrate A salt of tartaric acid, $HOOC-(CHOH)_2-COOH$.

taste-bud In vertebrates, a group of receptor cells which are modified epithelial cells and are sensitive to taste. In mammals, taste-buds are mainly found on the upper surface and sides of the tongue, on and between papillae. In humans there are four types of taste-buds, each type sensitive to one of the following tastes: sweet, sour, salt and bitter. They are distributed on different areas of the tongue.

tautomer *See* tautomerism.

tautomerism A type of structural isomerism in which the isomers, at room temperature, may so rapidly interconvert that they are in a dynamic equilibrium. Such isomers are called tautomers. Usually tautomers only differ from one another by the position of a hydrogen atom, e.g.

$$CH_3-\underset{\underset{O}{\|}}{C}-CH_2-COOC_2H_5 \rightleftharpoons$$

$$CH_3-\underset{\underset{OH}{|}}{C}=CH-COOC_2H_5$$

This equation shows that acetoacetic ester exists in two forms, normally in equilibrium. The hydroxy isomer (right) is called the enol form (ethyl 3-hydroxybut-2-enoate) and the keto isomer (left) is called the keto form (ethyl 3-oxobutanoate). Both tautomers can be isolated, but the presence of even minute amounts of acid or alkali establishes the above equilibrium. Acetoacetic ester has ketonic properties as well as properties which are characteristic of a double bond and a hydroxyl group. It is important to distinguish between tautomerism and resonance: tautomers are individual molecules, even if it is not always possible to isolate them.

taxis (pl. taxes) *See* tactic movement.

taxon (pl. taxa) (taxonomic unit) Any unit, e.g. genus, order, etc., used in taxonomy.

taxonomic unit *See* taxon.

taxonomy The study of the principles and practice of classification of plants and animals.

TB Originally an abbreviation for tubercle bacillus, but now used for tuberculosis.

tea 1. The dried, crushed leaves of the tea plant, which is an evergreen shrub grown in tropical and sub-tropical regions. 2. *See* cannabis.

tear *See* lachrymal gland.

tear-duct *See* lachrymal duct.

tear-gas *See* lachrymator.

tear-gland *See* lachrymal gland.

teat *See* nipple.

technical chemical *See* fine chemical.

Teflon® (Fluon; poly(tetrafluoroethene), PTFE) An organic substance (a fluorocarbon) made by polymerisation of tetrafluoroethene, $CF_2=CF_2$. Teflon is very resistant to chemicals, heat and wear. It is used to coat 'non-stick' cooking utensils, in bearings, as electrical insulation, etc.

tele- Prefix meaning at a distance.

telecommunications Communications by radio, telegraph, telephone, television, telex, etc.

telegraph An apparatus for transmitting and receiving messages sent in the form of electrical impulses along a wire. By pressing and releasing a tapping-key (switch) the operator makes and breaks an electric circuit, thus sending a message in the form of a code as a series of dots and dashes. The small currents reaching the receiver operate a relay which makes and breaks a circuit carrying a larger current. This circuit may contain a buzzer or a device for recording the dots and dashes on a paper strip. *See also* Morse code.

telencephalon In vertebrates, the anterior part of the forebrain, giving rise to the cerebral hemispheres, olfactory lobes and optic lobes.

Teleostei A group comprising more than 20 000 species, including most bony fish. Characteristics include a body covered with thin, bony scales, an ossified endoskeleton, a symmetrical tail and filamentous gills.

telephone An apparatus for transmitting and receiving speech. Sound waves are converted by a carbon microphone to variable electrical impulses which are transmitted along a wire or cable to the receiver (earpiece). The receiver consists of a U-shaped permanent magnet whose arms are wound with two solenoids. The electrical impulses cause changing magnetic fields which cause a magnetic diaphragm to vibrate, thus producing audible sound waves which are a copy of those which entered the microphone. *See* Figs. 24 *and* 225.

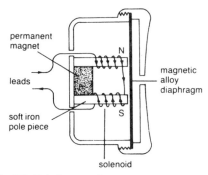

Fig. 225 Telephone receiver

telephoto lens A lens system consisting of a concave and convex lens which increases the effective focal length without changing the distance between the film and the lens. This type of lens produces a magnified image and is used in place of the normal camera lens when photographing distant objects. *See also* zoom lens.

telescope An optical instrument for viewing images of distant objects. Telescopes may be refracting (using lenses) or reflecting (using mirrors). The Galilean telescope was the first astronomical refracting telescope. It consists of a concave lens of short focal length as the eye lens and a converging lens of long focal length as the object lens, separated by a distance equal to the difference between the focal lengths of the two lenses. This telescope produces an erect final image. It is no longer used in astronomy. The Newtonian telescope was the earliest atronomical reflecting telescope. It consists of a large concave mirror which reflects light onto a small plane mirror. This type of telescope is often used today. *See* Fig. 226 (a). A simple atronomical refracting telescope (Kepler telescope) consists of two convex lenses. The object lens has a long focal length, the eye lens a short focal length, and they are separated by a distance equal to the sum of their focal lengths. This telescope produces a virtual, inverted, greatly magnified image. *See* Fig. 226 (b). Modern astronomical telescopes are of the reflecting type, with large concave mirrors several metres in diameter to collect as much light as possible. *See also* radio telescope.

television (TV) Television pictures begin with the formation of an optical image of a scene by a television camera. The television camera has a zoom lens at its front which focuses an image onto a light-sensitive target in the camera tube. The rear surface of this target is scanned by a beam of electrons emitted by an electron gun in the camera. The beam of electrons moves over

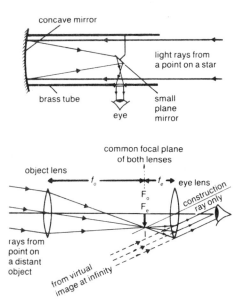

Fig. 226 (a) Astronomical reflecting telescope
(b) Astronomical refracting telescope

the target in the same way that the eye moves when reading a printed page. When the beam reaches the bottom right hand corner of the image it rapidly returns to the top left hand corner of the image and the process is repeated. The image is completely scanned 25 times per second. The target may consist of as many as 200 000 tiny parts (picture elements), each of which gains a positive charge whose size depends on the amount of light falling on it. As the electron beam moves over the image on the target, each picture element is discharged by the beam, producing an electrical impulse corresponding in strength to the amount of light falling on that particular element. Thus the image is copied in the form of electrical impulses which are amplified and pass to the transmitter, where they are converted to radio waves (or are transmitted to a receiver, the TV set, by cable). In the TV set, the transmitted radio waves are first converted to electrical impulses which are amplified and pass to an electron gun at the rear of the tube (a cathode-ray tube). The electron gun emits a beam of electrons which scans the fluorescent screen in the same way as the electron beam in the television camera. The strength of the beam varies according to the strength of the electrical impulses and a picture of the original scene is formed on the screen. As the camera scans the image on the target 25 times per second, a complete picture is received 25 times per second.

telophase

This system forms the basis of black-and-white television. Colour TV is basically the same system with additional electrical impulses transmitting information about colour. Colour television cameras may have three or four tubes which split the image into its red, green and blue component colours. Three pictures of the image are transmitted simultaneously to a colour TV receiver in which the three separate colour images are superimposed on the screen and the correct colours of the image are reproduced by addition of the primary colours red, green and blue. *See* colour mixing. The sound accompanying a TV picture is collected by a microphone and is transmitted as radio waves to the TV receiver.

telophase The stage of mitosis or meiosis following anaphase, when new nuclei are formed.

temperature The degree of hotness of a body on a chosen scale, e.g. the Celsius scale. Temperature is measured by means of a thermometer. *See also* thermodynamic temperature.

tempering of steel The reheating of steel after quenching to give it a definite hardness and remove any unwanted brittleness. The steel is allowed to cool in air.

temporary hardness *See* hard water.

temporary magnet A piece of magnetic material, e.g. soft iron, which possesses magnetism only when it is placed in a magnetic field. *Compare* permanent magnet; *see also* electromagnet *and* electromagnetic induction.

tendon A white cord or band of parallel collagen fibres attaching a muscle to a bone, cartilage or other muscles. Tendons have great strength and do not stretch.

tendril A specialised, slender structure which may be a modified stem, leaf, terminal bud, etc. It supports a climbing plant by sticking to or twining round the support. *See* Fig. 227.

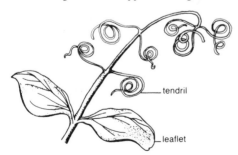

Fig. 227 Pea tendrils

tensile strength The stress required to break a material under tension.

tension A force tending to stretch a body which is part of a system in equilibrium or motion. The body is said to be under tension.

tensor A muscle which tightens or stretches a part of the body.

tentacle 1. A sticky structure of insectivorous plants such as the sundew. *See* Fig. 228 (a). **2.** In many invertebrates, a slender, unjointed, flexible appendage situated on the head. Tentacles are used for feeling, grasping, moving, etc. *See* Fig. 228 (b).

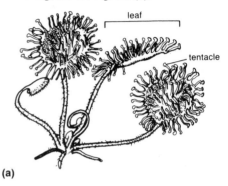

Fig. 228 Tentacles
(a) Sundew   (b) Snail

tera- Symbol: T. A prefix meaning 10^{12}.

terephthalic acid Benzene-1,4-dicarboxylic acid. *See* benzene-1,2-dicarboxylic acid.

terminal 1. Pertaining to or situated at the end of a structure. **2.** A point to which an electrical connection is made, e.g. a battery has a positive terminal and a negative terminal.

terminal bud A bud situated at the end of a stem. *See* Fig. 86.

terminal velocity The constant velocity reached by a body acted upon by a constant force and moving freely through a fluid. A body falling under the force of gravity through the atmosphere accelerates until air resistance is equal to the pull of gravity. It then continues to fall at constant velocity, which rarely exceeds 225 kilometres per hour.

termite *See* Isoptera.

terpene The general term for a group of volatile, organic substances found in the leaves, flowers or fruits of many plants such as conifers and citrus fruits, from which they are extracted. Terpenes are unsaturated, cyclic compounds with the empirical formula $(C_5H_8)_n$: i.e. they contain units of isoprene (methylbuta-1,3-diene). Terpenes are classified as monoterpenes, $C_{10}H_{16}$ (containing 2 isoprene units), sesquiterpenes, $C_{15}H_{24}$ (containing 3 isoprene units), diterpenes, $C_{20}H_{32}$ (containing 4 isoprene units), etc. Terpenes are found in turpentine and used in perfumes and cosmetics.

Terramycin® *See* tetracycline.

terrapin *See* Chelonia.

terrestrial Pertaining to the Earth.

terrestrial magnetism *See* magnetism, terrestrial.

terrestrial meridian *See* meridian, terrestrial.

territory An area defended by an animal or group of animals (especially vertebrates) against other animals of the same species. Territories are often only occupied during a breeding season; however, some are occupied at all times. A territory may be marked by scent or, in the case of birds, by singing.

Terylene® (Dacron) A linear polyester made from ethane-1,2-diol (ethylene glycol) and the aromatic acid benzene-1,4-dicarboxylic acid (terephthalic acid), $C_6H_4(COOH)_2$, by esterification and condensation polymerisation. Terylene is widely used in the manufacture of textile fibres.

tesla Symbol: T. The SI unit of magnetic flux density, equal to one weber per square metre.

testa (pl. testae) (seed-coat) The tough, protective, outer coat of a seed, formed from the integument.

testicle *See* testis.

testis (pl. testes) (testicle) In animals, the male reproductive organ which produces spermatozoa. In vertebrates, it also produces sex hormones. *See also* scrotum.

testosterone A male sex hormone (steroid) produced by certain cells of the testes. Testosterone is important in the normal development of the reproductive organs and the male secondary sexual characteristics. *See also* luteinising hormone.

test-tube A slender glass tube, closed at one end, widely used in science laboratories.

tetanus (lockjaw) A very serious infectious disease caused by *Clostridium tetani*, a bacterium which is commonly found in soil and in the intestines of many animals, including humans. The bacterium may invade the body through a wound. It produces a powerful toxin which causes prolonged muscle spasms. Immunisation has reduced the occurrence of this disease.

tetra- Prefix meaning four.

tetracarbonylnickel(0) [Ni(CO)$_4$] (nickel carbonyl) A colourless liquid, m.p. $-25°C$, b.p. $43°C$, prepared by passing carbon monoxide, CO, over nickel at a temperature of $60°C$. The gaseous [Ni[CO]$_4$] decomposes when heated to a temperature of about $200°C$. *See also* Mond process.

tetrachloromethane CCl_4 (carbon tetrachloride) A colourless, poisonous heavy liquid which is not flammable and has a sweetish smell. It is insoluble in water, but soluble in all organic solvents. Tetrachloromethane can penetrate the skin and its heavy vapour is very toxic. It is used as a solvent and in fire extinguishers. It is prepared by treating carbon disulphide, CS_2, with chlorine, Cl_2, or by chlorination of methane, CH_4.

tetracycline One of a group of antibiotics which consist of a basic structure of four rings (tetracyclic) and have the broadest spectrum of activity against bacteria, though some bacteria are resistant to them. Terramycin is a tetracycline.

tetraethyl-lead(IV) $(C_2H_5)_4Pb$ (lead tetraethyl) A substance (an inhibitor) added to petrol to prevent knocking, thus increasing its octane number. It is an insoluble, viscous, colourless liquid which dissolves in many organic solvents. It is prepared by heating a lead alloy containing sodium with chloroethane (ethyl chloride), C_2H_5Cl:
$4C_2H_5Cl + 4Na + Pb \rightarrow (C_2H_5)_4Pb + 4NaCl$

tetrafluoroethene *See* Teflon.

tetrahedral configuration of carbon The four valencies of a carbon atom radiate from the nucleus of the atom towards the points of a regular tetrahedron. The angle between any two of the forces that constitute the valency bonds is about $109.5°$ ($109.5°$): i.e. these forces are symmetrically distributed in space. Methane, CH_4, is tetrahedral, and this arrangement provides the maximum separation between the electrons. If the angle is distorted in a molecule it leads to decreased stability.

tetrapod (quadruped) An animal with four limbs.

tetravalent (quadrivalent) Having a valency of four.

Th Chemical symbol for thorium.

thalamus (pl. thalami) **1.** *See* receptacle. **2.** In higher mammals, one of two egg-shaped masses of grey matter on either side of the third ventricle in the forebrain. The thalami transmit nerve-impulses to the cerebral cortex.

Thallophyta A major division of the plant kingdom which includes algae, fungi and, in some classifications, bacteria and lichens. These 'plants' have plant bodies which are not differentiated into leaves, stems and roots.

thallus (pl. thalli) A simple plant body not differentiated into leaves, stems and roots. It is a characteristic of bryophytes and thallophytes.

theine *See* caffeine.

theory An idea or system of ideas put forward to explain certain facts, but which has yet to be verified by experiment.

thermal capacity A term formerly used for heat capacity.

thermal conductivity Symbol: λ (lambda). A measure of the ability of a material to conduct heat energy. It is measured as the rate of heat flow across a metre cube of the material whose opposite faces have a temperature difference of 1 kelvin. In SI units, thermal conductivity is measured in joules per second per metre per kelvin ($J\,s^{-1}\,m^{-1}\,K^{-1}$). *See also* Ingenhousz's apparatus.

thermal decomposition The process in which a chemical compound is broken down, by means of heat, into simpler substances which do not recombine on cooling to form the orginal compound. *Example:* calcium nitrate, $Ca(NO_3)_2$, decomposes into calcium oxide, CaO, nitrogen dioxide, NO_2, and oxygen, O_2:
$2Ca(NO_3)_2 \rightarrow 2CaO + 4NO_2 + O_2$

thermal dissociation The process in which a chemical compound is broken down, by means of heat, into simpler substances which recombine on cooling to form the original compound. *Examples:* ammonium chloride, NH_4Cl, dissociates thermally into ammonia, NH_3, and hydrogen chloride, HCl:
$NH_4Cl \rightleftharpoons NH_3 + HCl$
and dinitrogen tetraoxide, N_2O_4, dissociates thermally into nitrogen dioxide, NO_2, which may itself dissociate thermally into nitrogen monoxide, NO, and oxygen, O_2:
$N_2O_4 \rightleftharpoons 2NO_2 \rightleftharpoons 2NO + O_2$

thermionic emission The emission of electrons from a heated metal or metal oxide. This effect is made use of in the electron gun, X-ray tubes and thermionic valves.

thermionics A branch of electronics dealing with the study of the emission of electrons from hot substances and the behaviour and control of such electrons, especially in a vacuum.

thermionic valve An evacuated or gas-filled glass or metal tube containing two or more electrodes. The principal electrodes are a heated cathode which emits electrons and an anode, maintained at a positive potential, which attracts the electrons. There may be other electrodes (grids) placed between the cathode and anode to control the flow of electrons. A diode (i.e. with only two electrodes) is the simplest type of thermionic valve. Thermionic valves have been superseded by transistors in most of their applications.

thermistor A semiconductor whose electrical resistance decreases rapidly with increases in temperature. Thermistors are used in sensitive temperature-measuring instruments and in electronic equipment to compensate for changes in temperature.

thermit(e) A mixture of iron(III) oxide, Fe_2O_3, and aluminium powder, Al. When ignited, iron and aluminium oxide, Al_2O_3, are produced:
$Fe_2O_3 + 2Al \rightarrow 2Fe + Al_2O_3$
The process is strongly exothermic and melts the iron produced. This is used in welding and in certain types of explosives. *See also* Goldschmidt process.

thermocouple An instrument containing an electric circuit consisting of two wires of different metals joined at each end. When the two junctions are at different temperatures, a potential difference is produced. The resulting current is proportional to the difference in temperature between the two junctions (for small temperature differences). Thermocouples are used for measuring temperature over a wide range. Metals used in thermocouples may be copper/constantan, iron/constantan, etc. *See also* Seebeck effect *and* thermopile.

thermoduric A micro-organism which is heat-resistant, e.g. one which survives pasteurisation, but does not grow at high temperatures.

thermodynamics The branch of science concerned with the relationship between heat and other forms of energy.

thermodynamics, laws of Three laws which state: (1) Energy is neither created nor destroyed (conservation of energy), i.e. the total energy in a closed system is constant. (2) Heat can never flow spontaneously from a cold body to a hotter body, i.e. heat cannot flow against a temperature gradient. (3) The entropy of a substance approaches zero as its temperature approaches absolute zero, 0 K or $-273,16°C$ ($-273.16°C$). *See also* zeroth law of thermodynamics.

thermodynamic temperature (absolute temperature) Symbol: T. Temperature measured on the Kelvin scale.

thermoelectric effect *See* Seebeck effect.

thermograph A thermometer which produces a continuous record of temperature variations over a period of time. A simple type consists of a bimetallic spiral, one end of which is fixed and the other connected to a stylus (pen) which records temperature variations on slowly moving graph paper.

thermometer An instrument for measuring temperature. There are many types of thermometers, e.g. the clinical thermometer, gas thermometer and resistance thermometer.

thermonasty A plant movement in response to a change in temperature. *Example:* tulip flowers open when the temperature increases and close when it decreases.

thermonuclear reaction *See* fusion.

thermophile A micro-organism which grows best at 40°C or even higher temperatures. Such micro-organisms are said to be thermophilic. *See also* mesophile *and* psychrophile.

thermopile An instrument consisting of a number of thermocouples connected in series. The metals used are usually antimony and bismuth. The 'cold' junctions are shielded, whereas the 'hot' junctions are exposed to the source of heat radiation. A thermopile produces an easily detectable current and is used to measure radiant heat.

thermoplastic *See* plastic.

Thermos® *See* Dewar flask.

thermosetting *See* plastic.

thermostat A device for automatically maintaining a temperature between certain limits. It usually consists of a bimetallic strip, arranged so that when it is heated or cooled it opens or closes a circuit containing a heating element. *See* Fig. 229.

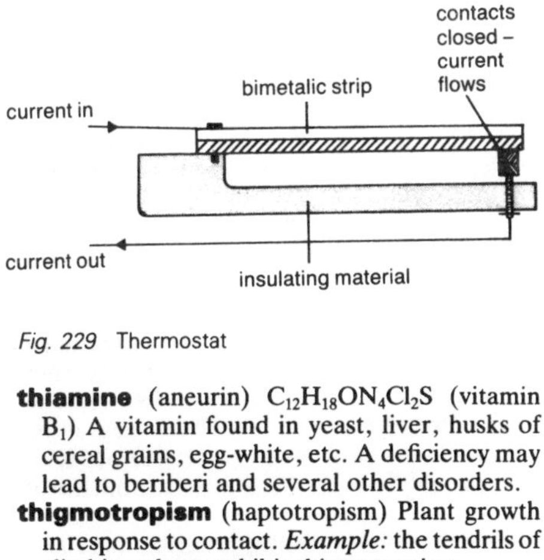

Fig. 229 Thermostat

thiamine (aneurin) $C_{12}H_{18}ON_4Cl_2S$ (vitamin B_1) A vitamin found in yeast, liver, husks of cereal grains, egg-white, etc. A deficiency may lead to beriberi and several other disorders.

thigmotropism (haptotropism) Plant growth in response to contact. *Example:* the tendrils of climbing plants exhibit thigmotropism.

thin-layer chromatography A type of chromatography in which the stationary phase is a thin layer of adsorbent material such as alumina or silica gel on a glass plate. After drying, the samples (pure substances or mixtures) and reference substances are applied to the thin layer and the plate is dipped upright in a mobile phase. This rises through the thin layer carrying samples and reference substances with it. After some time, before the mobile phase reaches the top of the plate, the plate is removed and dried. To locate the samples and reference substances the plate may be sprayed with a developer; the R_F-values are then obtained. This method was originally qualitative but it can be made quantitative.

thio- Prefix indicating that a chemical compound contains sulphur and used in naming certain compounds. *Example:* sodium thiosulphate, $Na_2S_2O_3$, which can be prepared by boiling an aqueous solution of sodium sulphite, Na_2SO_3, with sulphur powder:
$Na_2SO_3 + S \rightarrow Na_2S_2O_3$

thiol (mercaptan) The general term for a group of organic compounds derived from alcohols. In thiols an oxygen atom in the hydroxyl group, —OH, in the alcohol is replaced by a sulphur atom: i.e. thiols contain the thiol group, —SH. Thiols are usually colourless liquids or solids. They are more volatile and less soluble in water than the corresponding alcohols. Examples of thiols are CH_3SH (methanethiol or methyl mercaptan) and C_6H_5SH (thiophenol or phenyl mercaptan). CH_3SH (a gas) can be prepared by reacting methanol, CH_3OH, catalytically with hydrogen sulphide, H_2S:
$CH_3OH + H_2S \rightarrow CH_3SH + H_2O$
Thiols have a very offensive smell. *Example:* C_2H_5SH (ethanethiol) is a constituent of the fluid produced by the skunk. Thiols are used as intermediate compounds in organic synthesis and in the manufacture of certain sulphur-containing organic compounds. The name mercaptans is misleading as it indicates the presence of mercury: it was originally suggested because thiols form insoluble mercury compounds.

thiol group The group —SH, found in thiols (mercaptans). The thiol group is attached to an alkyl or aryl group.

thiosulphate A salt of thiosulphuric acid, $H_2S_2O_3$.

thiosulphuric acid $H_2S_2O_3$ An unstable acid which is derived from sulphuric acid, H_2SO_4, by replacing one oxygen atom with one sulphur atom. Thiosulphuric acid is only known in solution and chemically it has no importance. It is a weak acid whose salts are called thiosulphates.

third ventricle In vertebrates, a small cavity filled with cerebrospinal fluid and situated in the diencephalon. *See* Fig. 16.

thistle funnel A glass funnel consisting of a long, narrow tube with a thistle-shaped funnel at one end. *See* Fig. 230.

thixotropy A property of certain liquids by which changes in their viscosities are caused by changes in the force acting upon them. *Example:* the viscosities of some paints may be lowered by stirring.

thoracic Pertaining to the chest (thorax).

Fig. 230 Thistle funnel

thoracic duct In mammals, the main lymphatic duct. It receives lymph and chyle from most parts of the body and empties into a vein at the base of the neck.

thoracic vertebrae The vertebrae situated between the cervical and lumbar vertebrae. In humans there are twelve. *See* Fig. 242.

thorax In land vertebrates, the region of the body between the neck and the abdomen, containing the heart and lungs. In mammals it is separated from the abdomen by the diaphragm. In other animals, e.g. insects, it is the region of the body between the head and the abdomen, bearing the legs and wings (if present).

thorite *See* thorium.

thorium An element with the symbol Th; atomic number 90; relative atomic mass 232,04 (232.04); state, solid. Thorium is a radioactive, metallic element belonging to the actinoids. It is found naturally in minerals such as monazite, which may be represented by the formula $ThPO_4$, but which also contains elements such as cerium, lanthanum and ytterbium. Thorium is also found in thorite, $ThSiO_4$, from which it is extracted. The most stable isotope of thorium is thorium-232, $^{232}_{90}Th$, which has a half-life of $1,41 \times 10^{10}$ years (1.41×10^{10} years). Thorium is used in certain alloys and as a radioactive fuel.

threadworm *See* Nematoda.

threshold stimulus The minimum intensity of a stimulus which can produce a response in a given irritable tissue.

thrombin *See* fibrin *and* thrombokinase.

thrombocyte *See* blood platelet.

thrombokinase An enzyme which converts prothrombin to thrombin in the presence of calcium ions. It is liberated from blood platelets or injured tissue.

thrombosis (pl. thromboses) The blocking of a blood vessel by a blood clot. *See also* coronary thrombosis *and* stroke.

thunder A sound caused by pressure waves arising from temperature changes in the air. It is often accompanied by a flash of lightning.

thymine A nitrogenous, cyclic organic base (pyrimidine base). It is part of the genetic code in DNA, where it pairs with adenine.

thymus gland In mammals, a gland situated in the neck region and upper part of the thorax. It consists of two lobes, is irregular in shape and reaches a maximum size at puberty, after which it slowly degenerates. The thymus gland stimulates the formation of lymphocytes and controls immune responses of the body.

thyroid gland In vertebrates, the largest of the endocrine glands, situated near the larynx and trachea. It secretes thyroid hormones such as thyroxine, which contains iodine and regulates the rate of metabolism. *See also* goitre.

thyroxine *See* thyroid gland.

Ti Chemical symbol for titanium.

tibia The larger of the two distal bones of the hindlimb in tetrapods, the bone anterior to the fibula. In humans it is the larger and inner shin bone, located between the knee and the ankle. *See* Fig. 162.

tick *See* Acarina.

tidal air The volume of air which is inhaled and exhaled during normal breathing.

tide The twice-daily rise (high tide) and fall (low tide) of the sea level due to the gravitational attraction of the Moon and Sun. The tide-raising force of the Sun is less than half that of the Moon because of its greater distance from the Earth. A neap tide is a high tide occurring at the Moon's first or third quarter (i.e. at Half Moon) when the gravitational attractions of the Moon and Sun are working against each other. A spring tide is a high tide occurring at a New Moon or Full Moon when both gravitational attractions are working together. A spring tide therefore produces a maximum tide.

timbre *See* quality of sound.

time measurement *See* second.

tin An element with the symbol Sn; atomic number 50; relative atomic mass 118,69 (118.69); state, solid. Tin is a metal occurring naturally in the mineral cassiterite (tinstone), SnO_2, from which it is extracted by reduction with carbon:

$SnO_2 + 2C \rightarrow Sn + 2CO$

There are three allotropes of tin: grey, white and rhombic tin. Grey tin is stable below 13°C, white tin between 13°C and 161°C and rhombic tin above 161°C. However, at 232°C it melts. The change of tin from white to grey is also affected by impurities such as aluminium or zinc. It can be prevented or minimised by the addition of small amounts of antimony or bismuth. Tin is used in several alloys such as solder and pewter, in the manufacture of tin plate and foil and in glass manufacture.

tin chlorides Tin(II) chloride (stannous chloride), $SnCl_2$, and tin(IV) chloride (stannic chloride), $SnCl_4$. Salts of tin.
Tin(II) chloride, $SnCl_2$, is a white, crystalline solid which can be prepared by treating tin with concentrated hydrochloric acid, HCl:
$$Sn + 2HCl \rightarrow SnCl_2 + H_2$$
It is used as a reducing agent and as a mordant. With water, it forms a basic chloride.
Tin(IV) chloride, $SnCl_4$, is a colourless, volatile liquid. It is rapidly hydrolysed to form hydrated tin(IV) oxide and hydrogen chloride:
$$SnCl_4 + 4H_2O \rightarrow SnO_2 \cdot 2H_2O + 4HCl$$
It can be prepared by passing chlorine, Cl_2, over tin:
$$Sn + 2Cl_2 \rightarrow SnCl_4$$
Tin(IV) chloride is used as a mordant.

tincture A solution of a substance in alcohol (ethanol), e.g. tincture of iodine.

tin plaque Tin which has crumbled into a grey powder. This may happen at temperatures below 13°C.

tin plate Thin sheets of iron or steel coated with a layer of tin to prevent rusting. This may be carried out by dipping the sheets into a bath of molten tin or by an electrolytic method.

tinstone *See* tin.

tissue A mass of similar cells specialised for a particular function, e.g. connective tissue, nervous tissue. Blood and lymph are liquid animal tissues.

tissue culture (explantation) A technique in which living tissue removed from an animal or a plant is brought to live and undergo cell division under test-tube conditions. Such cultures are usually handled aseptically to prevent any contamination with microorganisms such as bacteria and viruses. A tissue culture is nourished with the appropriate amount of salts, oxygen, sugar, amino acids, etc. and given a correct temperature and pH so that life is maintained. Viruses are often cultured in this way. Some tissue, however, cannot be maintained living using this technique, e.g. nerve tissue. *See also* physiological saline.

tissue respiration *See* respiration.

Titan *See* Saturn.

titanium An element with the symbol Ti; atomic number 22; relative atomic mass 47,90 (47.90); state, solid. Titanium is a transition element (metal) occurring naturally in the minerals rutile, TiO_2, and ilmenite, $FeO \cdot TiO_2$. It is extracted by first converting the oxide to titanium(IV) chloride, $TiCl_4$, by mixing it with carbon and heating it in a stream of chlorine, Cl_2:
$$TiO_2 + C + 2Cl_2 \rightarrow TiCl_4 + CO_2$$
$TiCl_4$ is a volatile liquid which is purified by fractional distillation and then reduced to metallic titanium with magnesium or sodium at a temperature of about 1350°C:
$$TiCl_4 + 2Mg \rightarrow Ti + 2MgCl_2$$
This process is carried out in an inert atmosphere of argon because traces of air make titanium brittle. Titanium is very resistant to corrosion, has a low density and a high strength. It is used in the construction of supersonic aircraft, space vehicles and nuclear reactors.

titanium(IV) oxide TiO_2 (titanium dioxide) A brilliant white, very stable, crystalline solid occurring naturally in three different polymorphic forms. It is used as a pigment.

titrant A standard solution which is slowly added to a known volume of a solution of unknown concentration (the sample) during a titration.

titration A form of volumetric analysis in which a known volume of a sample of unknown concentration is mixed with a known volume of titrant from a burette until the reaction between them reaches the equivalence point (end-point). Titrations are often carried out using indicators so that the end-point is shown by a colour change. Knowing the volumes of the two liquids and the strength of the titrant, the strength of the sample can be calculated. *See also* potentiometric titration.

titration curve *See* potentiometric titration.

TNT *See* methyl-2,4,6-trinitrobenzene.

tobacco mosaic virus *See* mosaic disease.

tocopherol (Vitamin E) Four closely-related compounds: $C_{29}H_{50}O_2$ (alpha-tocopherol), $C_{28}H_{48}O_2$ (beta-tocopherol), $C_{28}H_{48}O_2$ (gamma-tocopherol) and $C_{27}H_{46}O_2$ (delta-tocopherol). Beta-tocopherol and gamma-tocopherol are isomers. Tocopherols are found in green plants such as lettuce and watercress and in some vegetable oils such as wheat germ oil. A deficiency of vitamin E may interfere with normal reproduction. *See also* ageing process.

Tollens's reagent A solution used in the test for aldehydes. It contains the complex ion $[Ag(NH_3)_2]^+$. This is prepared by mixing solutions of silver nitrate, $AgNO_3$, and sodium hydroxide, NaOH, to produce silver oxide, Ag_2O. The Ag_2O is then dissolved by treating it with a concentrated ammonia solution, NH_3:
$$2Ag^+ + 2OH^- \rightarrow Ag_2O + H_2O \quad \text{and}$$
$$Ag_2O + 4NH_3 + H_2O \rightarrow 2[Ag(NH_3)_2]^+ + 2OH^-$$
When mixed with an aldehyde, $[Ag(NH_3)_2]^+$ is reduced to metallic silver which forms a 'mirror' on the inside of the test-tube in which

the reaction is performed. Solutions of [Ag(NH$_3$)$_2$]$^+$ should never be stored as they slowly form silver nitride, Ag$_3$N, which may explode violently.

toluene *See* methylbenzene.

tone 1. A note with a single frequency, e.g. as produced by a tuning-fork. **2.** A tone may consist of the fundamental accompanied by partials.

tongue In vertebrates, a taste organ which is usually movable and situated in the mouth. In humans it is composed primarily of striated muscles, covered by a mucous membrane and attached to the floor of the mouth by its root. The upper surface is covered with taste-buds. The tongue is important in mastication of food, in swallowing and in speech. Some amphibians and reptiles have a sticky, forked tongue which can be flicked out to catch food, e.g. insects.

tonne Symbol: t. A metric unit of mass equal to 1000 kg.

tonoplast In plants, the membrane on the inside of a vacuole.

tonsillitis Inflammation of the tonsils. This may be caused by a number of infectious diseases such as diphtheria and strep throat. Symptoms include painful swelling of the tonsils, which become red and may have white specks (pus) on them; difficulty in opening the mouth; and fever. Tonsillitis is treated with antibiotics. If the disorder becomes chronic, the tonsils may be removed surgically.

tonsils In tetrapods, paired patches of tissue situated at the back of the throat. Lymph circulates through the tonsils, assisting in the removal of bacteria from the blood. Together with the adenoids, the tonsils help in guarding the body against micro-organisms entering through the mouth or nose.

tooth (pl. teeth) In most vertebrates, a hard, dense structure used for biting and chewing. Teeth are arranged in the jaws and vary in shape and size according to their function. *See* Fig. 231; *see also* dental formula.

topaz A mineral which may be represented by the formula Al$_2$SiO$_4$(F,OH)$_2$. Topaz is found in a variety of colours, of which brown and pink stones are valuable gemstones. Pink jewellery topaz is often formed from brown topaz as some brown topaz turn pink when heated.

topsoil The dark top layer of soil in which plants and other organisms live. Topsoil is formed by weathering of the soil surface and it usually contains decaying remains of dead organisms, i.e. humus. For this reason, topsoil provides nutrients for plants. *See also* subsoil.

torque *See* moment.

Torricellian vacuum *See* mercury barometer.

tortoise *See* Chelonia.

torus *See* receptacle *and* pit.

total internal reflection The total reflection of light travelling from one medium to another which has a lower refractive index, e.g. from glass to air. This occurs when the incident ray strikes the boundary between the two media at an angle of incidence greater than the critical angle. The relationship between critical angle and refractive index is given by:

$$_a n_g = \frac{1}{\sin c}$$

where $_a n_g$ is the refractive index for a ray passing from air to glass and c is the critical angle for an air-glass boundary. Total internal

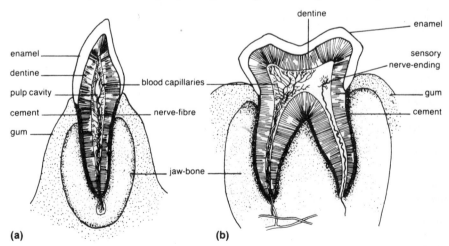

Fig. 231 (a) Incisor tooth in section
(b) Molar tooth in section (nerve shown on one side, blood vessels on the other)

reflection is used in erecting prisms, prismatic binoculars, periscopes, etc.

tourmaline A mineral which may be represented by the formula $NaMg_3Fe_3Al_6(OH)_4(BO_3)_3Si_6O_{18}$. Tourmaline is found in a variety of colours, of which green, blue and red are valuable gemstones. It exhibits the piezo-electric effect.

toxic Pertaining to poison.

toxicology The study of toxins, including their effects.

toxin Any poison of plant or animal origin.

TP See turgor pressure.

trabecula (pl. trabeculae) A small fibrous band of elongated cells. Trabeculae form frameworks across cavities in living tissue. See also microtrabecular lattice.

trace element (micro-element) An element which is only required in very small amounts by a living organism. Trace elements are often important constituents of enzymes, hormones and vitamins. Examples of trace elements are iodine, fluorine, zinc, copper, manganese and cobalt.

tracer A substance, usually a radioactive isotope, which is introduced into a system to determine the action or structure of the system. The tracer is followed with the aid of a detector, e.g. a Geiger-Müller counter. The method is used both in industry and in medicine.

trachea (pl. tracheae) 1. In insects and arachnids, one of several air-passages carrying air through the body and opening to the exterior through a spiracle. 2. (windpipe) In land vertebrates, a tube leading from the throat to the bronchi. In humans it is a tubular structure consisting of alternate rings of cartilage and dense, fibrous connective tissue. See Fig. 118.

tracheole In insects and arachnids, the part of a trachea nearest the cuticle (the terminal part). Gaseous diffusion occurs through the thin walls of the tracheoles.

tracheostomy See tracheotomy.

tracheotomy (tracheostomy) A surgical operation in which an opening is made in the lower part of the neck (below the voice-box) to the trachea (windpipe). Tracheotomy is an emergency measure used to bypass an obstruction in the air passage to the lungs. A small tube is inserted into the opening, allowing the free passage of air.

trachoma A contagious bacterial disease of the eyelids caused by the species *Chlamydia trachomatis*. The disease is widespread in the tropics and is endemic in Northern Africa, the Middle East and South East Asia. It leads to blindness if not treated. Symptoms include swollen and inflamed eyelids accompanied by a discharge. If the disease is untreated the conjunctiva becomes covered with scars and the edges of the eyelids turn inwards, causing the eyelashes to scratch the cornea and leading to further infection. The disease is treated with antibiotics such as tetracyclines.

trans-actinoids The elements following the last actinoid, lawrencium, Lr, in the Periodic Table of the Elements.

transcriptase See transcription.

transcription A process which takes place in the nucleus of a cell. The genetic code in chromosomal DNA is transferred to messenger RNA (mRNA) by the production of a molecule of mRNA from a length of DNA. The mRNA then leaves the nucleus and aims for the ribosomes, where it acts as a template for protein synthesis. The process of transcription is catalysed by the enzyme transcriptase.

transducer (sensor) A device which converts non-electrical energy into electrical energy and vice versa. *Example:* electrical energy is converted into sound energy by an electro-acoustic transducer in a loudspeaker. The electric motor is an example of an electro-mechanical transducer.

transect A line or strip across a region of vegetation chosen for study. See also quadrat.

transfer RNA (tRNA) A form of RNA concerned with the process of carrying amino acids to ribosomes in the cytoplasm of a cell and arranging them in the correct order for protein synthesis along a molecule of messenger RNA (mRNA). There are different molecules of tRNA for each of the fundamental amino acids. See also codon.

transformer A device for changing (transforming) an alternating current of one voltage to an alternating current of another voltage (with no change in the frequency), by electromagnetic induction (mutual induction). A transformer consists of two coils of wire wound on a laminated iron core. The coil to which the alternating voltage is applied is called the primary; the coil in which induced currents are produced is called the secondary. Fig. 232 (a) shows a step-up transformer: the number of turns in the primary coil (primary winding) is less than the number of turns in the secondary coil (secondary winding). Fig. 232 (b) shows a step-down transformer, for which the reverse is true. If E_p and E_s are the primary and secondary voltages and n_p and n_s the number of windings (turns) in the primary and secondary coils, then

$$\frac{E_p}{E_s} = \frac{n_p}{n_s}$$

transfusion See blood transfusion.

trans-isomer

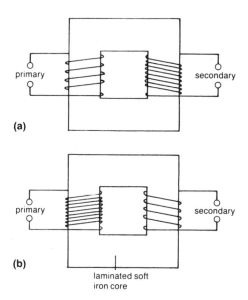

Fig. 232 Transformers
(a) Step-up (b) Step-down

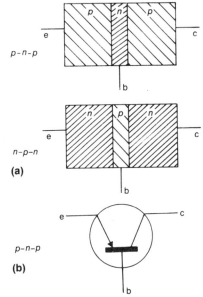

Fig. 233 Transistors
(a) Diagrams (b) Circuit symbol

trans-isomer *See* geometrical isomerism.

transistor A semiconductor device used to amplify electric currents flowing through it. The most common type is the junction transistor, consisting of three pieces of semiconductor with an electrode attached to each piece. The electrodes are called the emitter (e), base (b) and collector (c). There are two types of this transistor: a *p-n-p* transistor, consisting of a thin layer of *n*-semiconductor as base sandwiched between two outer layers of a *p*-semiconductor; and an *n-p-n* transistor, with a different arrangement of the semi-conductor layers. A flow of current between the emitter and collector can be controlled by a small, changing current flowing between the base and the emitter. In this way, the transistor can be used as a current amplifier. Transistors have, in most cases, replaced the thermionic valve as they are much smaller, less fragile and more efficient. *See* Fig. 233.

transition element A metallic element with an incomplete inner electron structure. There are three series of transition elements: the first starts with scandium and ends with zinc, the second starts with yttrium and ends with cadmium and the third starts with lanthanum and ends with mercury. The chemical properties of transition elements vary widely; they have several oxidation numbers (oxidation states), most of their compounds are coloured and they form a great number of complexes. Chemists disagree about the number of elements which should be called transition elements.

transition point *See* polymorphism.

transition temperature *See* polymorphism.

translation A process, taking place in the ribosomes, in which the genetic code carried in messenger RNA (mRNA) is decoded (translated), thus making protein synthesis possible. The process is carried out by transfer RNA (tRNA).

translocation 1. The transport of material such as food and mineral salts through a plant. This takes place in the phloem and xylem systems. 2. The transfer of a part of a chromosome to a different location on the same chromosome or to another chromosome. This gives rise to chromosome mutation.

translucent Describes a substance which partially transmits light.

transmittance Symbol: T. The transmittance of a beam of light encountering a substance is defined as

$$T = \frac{I}{I_o}$$

where I is the intensity of the emergent radiation and I_o is the intensity of the incident radiation. The logarithm (base 10) of the reciprocal of the transmittance is called the absorbance, A:

$$A = \log_{10}\left(\frac{1}{T}\right) \quad \text{or} \quad A = -\log_{10} T$$

In the expression
$A = abC$
called Beer's law, a is a constant called the absorptivity or extinction coefficient, b is the thickness of the absorbing sample and C is the concentration of the sample. The absorptivity depends upon the wavelength of the radiation and the nature of the absorbing material. From Beer's law it is seen that the absorbance is directly proportional to the concentration. This forms the basis for analysis in which light absorption is measured. In practice there are cases in which there is a deviation from Beer's law. However, this may be due to the concentration of the sample, as the law usually applies best when the absorbing sample is diluted.

transmitter The part of a communications system, e.g. radio, telephone and television, which transmits signals to the receiver.

transmutation The formation of one element from another, e.g. by alpha decay, beta decay or by bombardment of the nucleus of an element with subatomic particles such as neutrons. *Examples:* a radium isotope may change into a radon isotope by alpha decay:
$^{226}_{88}Ra \rightarrow ^{222}_{86}Rn + ^{4}_{2}He$
A radon isotope may change into a radium isotope by beta decay:
$^{222}_{86}Rn \rightarrow ^{222}_{88}Ra + 2_{-1}^{0}e$
A uranium isotope may change into a tellurium isotope and a zirconium isotope by bombardment with neutrons:
$^{235}_{92}U + ^{1}_{0}n \rightarrow ^{137}_{52}Te + ^{97}_{40}Zr + 2^{1}_{0}n$

transparent Describes a substance which transmits all incident light.

transpiration The loss of water vapour from a plant, mainly through the stomata. Transpiration results from the need for the stomata to be open for gaseous exchange to take place. Transpiration cools the leaves and the continuous movement of water through the plant (transpiration stream) also carries mineral salts from the roots to the leaves. Factors affecting transpiration are humidity, temperature, atmospheric pressure, air movements (wind), light and the plant's water supply. *See also* guttation *and* potometer.

transpiration stream The continuous flow of water through a plant from the roots, through the stem and out through the stomata of the leaves. During transpiration, most evaporation occurs from the walls of spongy mesophyll cells. Loss of water from these cells results in their cell sap becoming more concentrated than that of neighbouring cells. Thus the osmotic pressure of the surface cells is greater than that of cells deeper in the leaf and water is drawn into them. Cells which lose water to the surface cells will then take water from their neighbouring cells, and so on. Thus a continuous stream of water moves through the plant as long as evaporation continues to take place from the leaf surface. *See also* diffusion pressure deficit.

transplant The transfer of tissue or an organ from its normal position to another position on the same or another individual.

transuranic elements The elements following uranium (atomic number 92) in the Periodic Table of the Elements, including some of the actinoids. All transuranic elements are radioactive and are artificially produced by nuclear reactions. The first transuranic element is neptunium (atomic number 93). Some of the isotopes of the transuranic elements are very short-lived, but some theories have been put forward stating that nuclei of elements with very high atomic numbers will be relatively stable, e.g. elements 112–14, elements around 126 and 184 and elements with even higher atomic numbers. Consequently, properties of some of these super-heavy elements are already predicted. After lawrencium (atomic number 103), there has been some international disagreement about the names of the elements 104 and 105 and also about whether they were first prepared by American or Russian scientists. So far IUPAC has not decided upon specific names. However, Russian scientists suggest the name kurchatovium, Ku, for element 104 and nielsbohrium, Ns, for element 105 whereas American scientists suggest the name rutherfordium, Rf, for element 104 and hahnium, Ha, for element 105. To avoid future disagreements about names of elements, IUPAC has now suggested names for elements with atomic numbers greater than 103, e.g. element 104 would be called unnilquadium, Unq; element 105 would be called unnilpentium, Unp. This, however, gives symbols consisting of three letters instead of one or two as before.

transverse magnification *See* magnification.

transverse process A lateral projection from each side of a vertebra in tetrapods. Transverse processes either provide points of attachment for muscles or join one vertebra to the next. In humans, the transverse processes of the thoracic vertebrae articulate with the ends of ribs. *See* Fig. 242.

transverse wave A wave motion in which the vibration or displacement is perpendicular to the direction of propagation, e.g. electromagnetic radiation and water waves. *See also* longitudinal wave.

travelling wave (progressive wave) A wave motion in which the periodic displacement is

Trematoda

moving: i.e. energy is carried continuously away from the source. Longitudinal and transverse waves are travelling waves.

Trematoda A class of parasitic flatworms with a mouth leading to a gut, two suckers, and a body covered with protective cuticle to prevent digestion by the host. They are commonly called flukes. *See also* Platyhelminthes.

tri- Prefix meaning three.

triangle of forces If three forces acting at a point are in equilibrium, they can be represented by a triangle whose sides represent the magnitude and direction of the forces taken in order.

Triassic A geological period of approximately 32 million years which occurred at the beginning of the Mesozoic, and which ended approximately 193 million years ago.

triatomic A molecule consisting of three like or unlike atoms is said to be triatomic, e.g. ozone (trioxygen), O_3, and water, H_2O.

tribasic Describes an acid with three acidic hydrogen atoms which can be replaced by a metal or by a metal-like group such as ammonium, NH_4, thus forming salts. An example of a tribasic acid is phosphoric(V) acid (orthophosphoric acid), H_3PO_4. This acid gives rise to normal salts and to two acid salts, e.g. Na_3PO_4, Na_2HPO_4 and NaH_2PO_4. *See also* phosphoric(V) acid.

triboelectricity *See* frictional electricity.

tribology The study of friction, lubrication and wear of surfaces in relative motion.

tricarboxylic acid cycle *See* Krebs cycle.

triceps A muscle with three heads or origins. The muscle situated behind the humerus in the upper arm is a triceps muscle.

trichloroethanal CCl_3—CHO (chloral) An aldehyde which may be formed when chlorine, Cl_2 is brought to react with ethanol, CH_3—CH_2OH:
CH_3—$CH_2OH + 4Cl_2 \rightarrow CCl_3$—$CHO + 5HCl$
Chlorine both oxidises the ethanol molecule and undergoes a substitution reaction with it. It oxidises the primary alcohol group, —CH_2OH, to aldehyde, —CHO, and replaces the hydrogen atoms of the CH_3— group. In the above reaction, an intermediate compound is first formed and from this trichloroethanal is released by distillation with 85% sulphuric acid. Trichloroethanal is an oily liquid with a penetrating odour. It is soluble in most organic solvents and is mainly used in the manufacture of the insecticide DDT. With water, it forms a white crystalline solid, $CCl_3CH(OH)_2$, 2,2,2-trichloroethanediol (chloral hydrate) which has been used as a sleeping medicine.

trichloroethene $CHCl$=CCl_2 A colourless, stable, toxic, non-flammable liquid manufactured from ethyne, HC≡CH, and chlorine, Cl_2:
HC≡$CH + 2Cl_2 \rightarrow CHCl$=$CCl_2 + HCl$
It is used as a solvent, as a cleaning agent and as an anaesthetic.

trichloromethane $CHCl_3$ (chloroform) A heavy, colourless, volatile liquid. It is toxic, non-flammable and insoluble in water. Trichloromethane is used as an anaesthetic or as a solvent for many organic compounds. It can be made by the chlorination of methane, CH_4:
$CH_4 + 3Cl_2 \rightarrow CHCl_3 + 3HCl$

tricuspid Pertaining to the heart valve which separates the right atrium from the right ventricle. This valve has three flaps.

triglyceride *See* glyceride.

2,6,8-trihydroxypurine *See* uric acid.

tri-iodide *See* iodine.

tri-iron tetroxide *See* iron oxides.

tri-lead tetroxide *See* lead oxides.

trimer A compound obtained from the combination of three identical molecules called monomers. *See also* dimer *and* polymer.

trinitrotoluene *See* methyl-2,4,6-trinitrobenzene.

triode A type of thermionic valve with three electrodes: an anode, a cathode and a grid.

trioxygen *See* ozone.

triple bond *See* covalent bond.

triple point The point at which the three phases of a substance (solid, liquid and gas) are in equilibrium. The triple point of water is a temperature of 0,01°C (0.01°C) at 611 pascals (Pa). *See* Fig. 234.

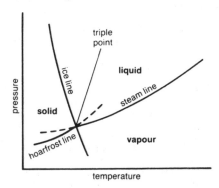

Fig. 234 Triple point

triploblastic Describes animals which have three cell layers, ectoderm, mesoderm and endoderm, making up their body wall. All Metazoa, except coelenterates, are triploblastic.

tripod An iron stand with three legs, widely used in science laboratories as a support on which apparatus is heated. *See* Fig. 235.

Fig. 235 Tripod

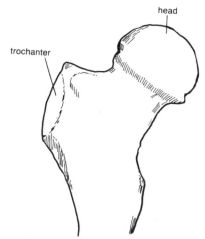

Fig. 236 Trochanter of femur

trip switch A type of circuit breaker which operates automatically and which may be reset. Two common types of trip switches are the magnetic type and the thermal type. In the magnetic type an electromagnet attracts an armature to break a circuit when an excessive current flows. In the thermal type the movement of a bimetallic strip breaks the circuit.

trisaccharide See raffinose.

trisodium phosphate(V) See sodium phosphates.

tritium Symbol: T or $^3_1$H. A radioactive isotope of hydrogen with a half-life of 12,33 years (12.33 years). The nucleus of the tritium isotope contains one proton and two neutrons, unlike hydrogen whose nucleus contains one proton only. It is very doubtful whether tritium exists naturally in ordinary hydrogen; however, it may be prepared by bombardment of a lithium isotope with neutrons:

$^6_3\text{Li} + ^1_0\text{n} \rightarrow ^3_1\text{H} + ^4_2\text{He}$

Tritium is used as a radioactive tracer.

triton The name given to the nucleus of a tritium atom (isotope).

trivalent Having a valency of three.

trivial name Usually an old name for a chemical substance. The trivial name often gives no information about the structure of the substance; sometimes it refers to a person. Examples of trivial names are Glauber's salt, limestone, saltpetre and acetone. See also systematic name.

tRNA See transfer RNA.

trochanter In vertebrates, one of several projections for the attachment of muscles on the femur. See Fig. 236.

trona Naturally occurring soda in the form of $Na_2CO_3 \cdot NaHCO_3 \cdot 2H_2O$, formed when water evaporates in soda lakes. Trona is an important natural source of sodium carbonate, Na_2CO_3.

tropic movement (tropism) A directional plant movement in response to a stimulus. Such movements are the result of growth curvatures. A movement towards the stimulus is a positive tropism; a movement away from the stimulus is a negative tropism. See also chemotropism, geotropism, hydrotropism, phototropism and thigmotropism.

tropism See tropic movement.

troposphere The region of the Earth's atmosphere situated beneath the stratosphere. The troposphere is the lowest of the atmospheric layers.

true rib A rib attached directly to the sternum by means of a short piece of cartilage. In humans the first seven ribs are true ribs.

true solution A clear solution, i.e. a solution in which the solute has dissolved completely in the solvent.

trunk See Proboscidea.

trypanosomiasis See sleeping sickness.

trypsin In mammals, an enzyme found in the pancreatic juice in the duodenum where it is formed by the activation of trypsinogen. Trypsin catalyses the breakdown of proteins and other peptides. The digestive juice of other animals contains trypsin-like enzymes.

trypsinogen A zymogen which is an inactive form of trypsin. It is activated and converted into trypsin by the enzyme (kinase) enterokinase.

tsetse fly A fly belonging to the genus Glossina. It is endemic to Africa south of the Sahara. Tsetse flies carry diseases which may affect cattle, horses and humans. In humans, tsetse flies are responsible for sleeping sickness (trypanosomiasis). The tsetse fly feeds by

tuber

piercing the skin of mammals, birds and reptiles with its mouthparts and sucking blood. Both male and female tsetse flies feed in this way. The tsetse fly is a little bigger than the house-fly, and it folds its wings, when at rest, in a scissor-like manner over its back. The female develops only one egg at a time and this hatches inside the female, where the larva develops and is nourished by secretions from certain glands. After birth, the larva pupates immediately in the soil.

tuber A specialised organ of vegetative reproduction, which may be either a swollen underground stem (stem tuber) or a root (root tuber). Tubers are storage organs. The potato is an example of a stem tuber and the dahlia of a root tuber. *See* Fig. 237.

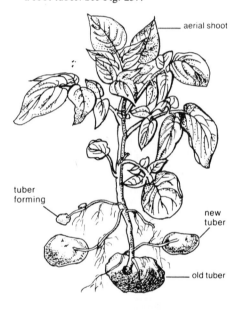

Fig. 237   Potato plant with tubers

tubercle (tuberculum) A small, rounded structure, e.g. a root swelling or a nodule; or a bony projection, e.g. an outgrowth from a rib which articulates with a transverse process of a vertebra.

tuberculin A substance, extracted from cultures of tubercle bacilli, which is used to test for the presence of tuberculosis. *See also* Mantoux test *and* patch test.

tuberculosis (TB) An infectious disease caused by the tubercle bacillus *Mycobacterium tuberculosis*, which lives within a protective membrane and so is difficult to kill. The disease may be contracted by breathing in the bacillus or by eating contaminated food. Any part of the body may be attacked by the bacillus, but the common site is the lungs (pulmonary tuberculosis) in which the bacilli break down the tissue. The disease is common in overcrowded societies with poor sanitation. People weakened by other diseases are more likely to get TB because their bodies' natural defences are low. Symptoms of pulmonary tuberculosis may be few or none at all in its early stages; however, an X-ray examination will detect it. Later symptoms are loss in weight, coughing, fever, blood in the sputum and general weakness. The disease is treated with antibiotics. A vaccine called BCG is available. This is usually given to children after a Mantoux test or a patch test.

tuberculum *See* tubercle.

tubule Any hollow structure with a small diameter.

tumour An abnormal swelling of tissue. Tumours may be benign or malignant.

tundra A treeless area of land in Arctic regions where the subsoil is permanently frozen.

tungsten (wolfram) An element with the symbol W; atomic number 74; relative atomic mass 183,85 (183.85); state, solid. Tungsten is a transition element (metal) occurring naturally in several minerals such as wolframite, $FeWO_4 \cdot MnWO_4$, scheelite, $CaWO_4$, and stolzite, $PbWO_4$, from which it is extracted by converting the ore into the oxide, WO_3, which is then reduced by hydrogen, H_2:

$$WO_3 + 3H_2 \rightarrow W + 3H_2O$$

Tungsten is used in certain alloys and in filaments in electric lamps.

tungsten carbides WC and W_2C. Greyish-white powders which can be prepared by direct combination of tungsten and carbon at a high temperature. Both carbides are extremely hard. They are used as abrasives and in toolmaking.

tuning-fork A metal fork which produces a pure tone of specified frequency when struck. *See* Fig. 238.

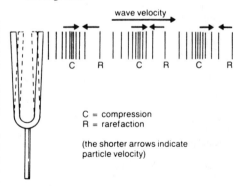

C = compression
R = rarefaction

(the shorter arrows indicate particle velocity)

Fig. 238   Sound waves from a tuning fork

Turbellaria A class of free-living, aquatic flatworms, e.g. planarians. Characteristics include a ciliated body, a simple nervous system, a simple or branched gut and a head often bearing tentacles and eyes. *See also* Platyhelminthes.

turbine A machine in which a rotor fitted with blades is rotated by the kinetic energy of a fluid, e.g. steam, air, water or gas.

turbo-generator A turbine coupled to a generator for the production of electricity. This arrangement is commonly found in power-stations.

turgid Describes a plant cell or tissue which is well supplied with water. The contents of turgid cells press against the cell walls and are the main means of support in herbaceous plants.

turgor The state of a plant cell which is turgid, i.e. well supplied with water and stretched by an increase in volume of vacuole and protoplasm, thus exerting a pressure (turgor pressure) on the cell wall.

turgor pressure (TP; wall pressure) In a plant cell, the pressure caused by water in the vacuole which pushes against the cytoplasm and the cell wall. The turgor pressure is balanced by the osmotic pressure when the cell is well supplied with water. When the cell loses water, the turgor pressure decreases and more water can enter the cell by osmosis. *See also* diffusion pressure deficit.

Turnbull's blue *See* cyanoferrates.

turpentine An oily liquid extracted from pine trees and mainly containing terpenes, e.g. pinene, $C_{10}H_{16}$. Turpentine is used as a solvent for lacquers, polishes and paints.

turquoise A mineral which may be represented by the formula $CuAl_6(PO_4)_4(OH)_8 \cdot 4H_2O$. The colour of turquoise is usually blue-green. It is used as a gemstone.

turtle *See* Chelonia.

tusk *See* Proboscidea.

tuyère One of a number of nozzles distributed evenly round the circumference of a blast-furnace about 2½ metres above its base. A blast of hot air is blown through the tuyères into the furnace.

TV *See* television.

twining The process by which a climbing plant twists around a support. *See* Fig. 239.

twins *See* dizygotic twins *and* monozygotic twins.

tympanic cavity *See* middle ear.

tympanic membrane *See* ear drum.

Tyndall effect The scattering of light from very small particles, as seen when a beam of sunlight passes through a dusty atmosphere. This gives a bluish light. True solutions show no Tyndall effect, whereas colloidal solutions do.

typhoid (typhoid fever) An infectious disease which attacks the intestines, caused by certain

Fig. 239 Bindweed twining

salmonella bacteria. The disease may be contracted from a carrier; its source is faeces and urine from infected persons. Flies can transmit the disease. Symptoms include headache, fever, diarrhoea, vomiting and red spots on chest and abdomen. The disease is treated with antibiotics. A vaccine is available against typhoid. Paratyphoid or paratyphoid fever is a disease which is also caused by certain salmonella bacteria, but they are different from those causing typhoid. The symptoms resemble those of typhoid, but the disease is much less severe. A vaccine is also available against paratyphoid.

typhoid fever *See* typhoid.

typhus A collective term for various infectious diseases caused by different *rickettsia* micro-organisms and spread by ticks, mites, fleas and lice.

U

U Chemical symbol for uranium.

ulcer An inflamed, open sore which may be external or internal, i.e. on the skin or on the mucous membrane lining a body cavity.

ulna One of the two distal bones of the forelimbs in tetrapods. In humans it is the longer bone located between the back of the elbow and the wrist. *See* Fig. 162; *see also* olecranon process.

ultracentrifuge *See* centrifuge.

ultramicroscopic Describes objects which are too small to be seen with a light microscope, but which can be seen with an electron microscope.

ultrasonic Describes sound waves with frequencies above the upper limit of human hearing, i.e. above 20 kHz.

ultrasonics The study of ultrasonic frequencies and their applications. Such applications include echolocation, detecting imperfections (flaws) in materials, and locating a brain tumour or a foetus.

ultra-violet radiation (UV radiation) The part of the electromagnetic spectrum which has a wavelength shorter than the violet end of the visible spectrum. Ultra-violet radiation is required by the human body for the manufacture of vitamin D. To prevent undue penetration of UV radiation, the production of the skin pigment melanin is increased, resulting in the production of a suntan. However, constant exposure to UV radiation may cause skin cancer. Ultra-violet radiation also converts atmospheric oxygen to ozone which effectively acts as a barrier, shielding the Earth from the harmful effects of the radiation.

umbel A type of inflorescence in which the pedicels of individual flowers arise from a common point on the stem, forming an umbrella-shaped cluster. *See* Fig. 179.

umbilical cord A cord through which two arteries and a vein pass, transporting blood to and from the embryo or foetus, i.e. it connects the embryo or foetus with the placenta. This cord is cut at birth, leaving a scar, the navel.

umbilicus *See* navel.

umbra *See* shadow.

unconditioned reflex An inborn reflex which does not become weaker with time, e.g. the release of saliva from salivary glands when food enters the mouth. *Compare* conditioned reflex.

ungulate A hoofed, grazing mammal. *See also* Artiodactyla *and* Perissodactyla.

unicellular Consisting of one cell only. *Compare* multicellular.

unisexual *See* dioecious.

unit A dimension or quantity chosen as a standard of measurement. *See also* SI units.

univalve An animal with a one-piece shell, e.g. the snail. *Compare* bivalve.

universal gas constant *See* molar gas constant.

universal indicators Mixtures of selected indicators giving a gradual colour change over a wide pH range.

Universe, the All of space and the matter (e.g. galaxies) contained in it. *See also* 'big bang' theory.

unnilpentium *See* transuranic elements.

unnilquadium *See* transuranic elements.

unsaturated compound A chemical compound containing one or more double or triple bonds in its structure, i.e. between carbon atoms. *Examples:* ethene, $CH_2\!=\!CH_2$ and ethyne, $CH\!\equiv\!CH$ are unsaturated hydrocarbons. Unsaturated compounds readily take part in addition reactions during which a double bond is converted into a single bond or a triple bond is converted into a double bond or single bond. *Examples:* Ethene may be converted into the saturated compound ethane, C_2H_6, by reaction with hydrogen:
$CH_2\!=\!CH_2 + H_2 \rightarrow CH_3\!-\!CH_3$
Ethyne may be converted into ethene:
$CH\!\equiv\!CH + H_2 \rightarrow CH_2\!=\!CH_2$
However, with more hydrogen, it may be converted to ethane:
$CH\!\equiv\!CH + 2H_2 \rightarrow CH_3\!-\!CH_3$
Compare saturated compound.

unsaturated solution A solution in which the solvent can dissolve more solute at a given temperature. *Compare* saturated solution.

unstable equilibrium The equilibrium of a body such that when it is displaced slightly, it tends to move away from its original position. *See also* neutral equilibrium *and* stable equilibrium.

upper partial *See* overtone.

upthrust An upward force, e.g. the upward force exerted on a body immersed in a fluid.

uracil A nitrogenous, cyclic organic base (pyrimidine base). It is part of the genetic code in RNA, where it pairs with adenine. In DNA, it is replaced by thymine.

uraninite *See* uranium.

uranium An element with the symbol U; atomic number 92; relative atomic mass 238,03 (238.03); state, solid. Uranium is a radioactive metal belonging to the actinoids and occurring naturally in several oxides, e.g. uraninite, one form of which is pitchblende, U_3O_8. Another uranium ore is carnotite, $K_2(UO_2)_2(VO_4)_2\cdot 3H_2O$ which is also an ore of vanadium. Naturally occurring uranium is found mainly as uranium-238, $^{238}_{92}U$ (99,27%; 99.27%) with a half-life of $4{,}50 \times 19^9$ years (4.50×10^9 years); and as uranium-235, $^{235}_{92}U$ (0,72%; 0.72%) with a half-life of $7{,}13 \times 10^8$ years (7.13×10^8 years). The $^{235}_{92}U$ isotope is used in nuclear fission reactors and in nuclear weapons.

Uranus One of the nine planets in our Solar System, orbiting the Sun between Saturn and Neptune once in 84,01 years (84.01 years). It is the seventh planet in order of distance from the

Sun and was the first planet to be discovered with the aid of a telescope (1781); the five planets (other than the Earth) which are closer to the Sun can be seen without the aid of a telescope. Uranus has five satellites (moons).

uranyl(VI) sulphate $UO_2SO_4 \cdot 3H_2O$. A yellow-green, crystalline, soluble salt which exhibits fluorescence.

urate A salt of uric acid, $C_5H_4N_4O_3$.

urea NH_2—CO—NH_2 (carbamide) Urea is the dicarbamide of carbonic acid, H_2CO_3, and is a white, crystalline, soluble solid. In ureotelic animals it is the main nitrogenous excretory product of protein catabolism, and is found in the urine. Urea is also found in some plants. When heated, urea decomposes into biuret, NH_2—CO—NH—CO—NH_2, and ammonia, NH_3:
$$2NH_2\text{—CO—}NH_2 \rightarrow$$
$$NH_2\text{—CO—NH—CO—}NH_2 + NH_3$$
See also Wöhler synthesis and ornithine cycle.

urea cycle See ornithine cycle.

urease An enzyme which catalyses the hydrolysis of urea into ammonia, NH_3, and carbon dioxide, CO_2:
$$NH_2\text{—CO—}NH_2 + H_2O \rightarrow 2NH_3 + CO_2$$
Urease is not common in animals but occurs in many bacteria and plants, e.g. soya beans.

ureotelic Describes animals which excrete nitrogen in the form of urea. Examples of such animals are adult amphibians and mammals. Compare uricotelic; see also ornithine cycle.

ureter In birds, reptiles and mammals, a duct conveying urine from the kidney to the cloaca or bladder.

urethra In mammals, a duct conveying urine from the bladder to the exterior. The urethra of a female mammal is shorter than that of a male, whose urethra also carries semen.

uric acid $C_5H_4N_4O_3$ (2,6,8-trihydroxypurine) A colourless, slightly soluble, crystalline organic acid belonging to the purines. It is the main nitrogenous excretory product of uricotelic animals, but it is also found in small amounts in the urine of mammals. Salts of uric acid are called urates. Sometimes crystals of uric acid and some of its salts are deposited in body tissues, particularly in cartilage in joints, giving rise to the disease called gout (podagra). This is caused by abnormal amounts of uric acid in the blood.

uricotelic Describes animals which excrete nitrogen in the form of uric acid. Examples of such animals are birds, terrestrial reptiles and some invertebrates. Compare ureotelic.

urinary Pertaining to urine.

urinary bladder See bladder.

urination (micturition) The act of discharging urine.

urine The liquid excreted from the kidneys and discharged through the urethra or cloaca in ureotelic animals; the semi-solid or solid excreted from the kidneys and discharged through the cloaca in uricotelic animals. In humans, urine mainly consists of water containing dissolved urea, salts and small amounts of other substances, e.g. uric acid.

uriniferous tubule The tubular portion of the nephron leading from a Bowman's capsule to a collecting tubule. It is subdivided into two convoluted regions and Henle's loop. See Fig. 149.

urinogenital Pertaining to the urinary and genital systems.

urotropine See hexamine.

uterine tube See Fallopian tube.

uterus (pl. uteri) (womb, matrix, metra) In female mammals, a hollow, thick-walled, muscular organ lying behind the bladder and in front of the rectum. Within the uterus, the fertilised egg is implanted, protected and nourished while undergoing growth and development before birth of the offspring.

utricle (utriculus) A small, membranous sac (larger than the saccule) in the inner ear, from which the semicircular canals arise. The utricle is filled with endolymph and its base is covered with a sensory area called a macula. See Fig. 55.

utriculus See utricle.

UV Abbreviation for ultraviolet. See ultraviolet radiation.

uvula (pl. uvulae) A small, conical process hanging from the soft palate in the roof of the mouth.

V Chemical symbol for vanadium.

vaccination Inoculation against smallpox; however, the term vaccination is often used as a synonym for inoculation.

vaccine A preparation of micro-organisms (including virus), which may be dead, weakened or still virulent, used in inoculation (vaccination) against infectious diseases. Vaccines are given by injection or orally. See also immunisation.

vacuole A fluid-filled cavity in the cytoplasm of a cell. A plant cell contains only one vacuole, which is larger than the sum of the several vacuoles that an animal cell may contain. *See also* food vacuole *and* contractile vacuole.

vacuum A region (space) containing no matter. A perfect vacuum is unobtainable: a vacuum is therefore considered to be a region at a very low pressure.

vacuum distillation A distillation carried out under reduced pressure which decreases the boiling-point of the liquid. This process enables unstable substances, which decompose at boiling temperatures corresponding to normal pressures, to be separated or purified.

vacuum evaporation A technique in which a thin layer of a solid, usually a metal or semiconductor, is deposited on another solid by heating the former in a vacuum. The atoms of the solid leave the hot surface and travel directly (without collisions) to the nearby cool solid which is to be coated.

vacuum flask *See* Dewar flask.

vacuum pump A device used to produce a low pressure in a vessel to which it is connected. The rotary oil pump can produce a pressure of about 10^{-3} millimetres of mercury (mmHg). *See also* air pump.

vagina In all female mammals except Monotremata, a short, thin-walled muscular tube connecting the uterus to the vestibule. It is situated behind the bladder and in front of the rectum.

vagus nerve The tenth cranial nerve, which is a motor and sensory nerve and the major nerve of the parasympathetic nervous system. It arises from the medulla oblongata, passes out of the skull and descends to the abdomen, giving off a number of branches to organs such as the larynx, pharynx, oesophagus, heart, lungs and stomach.

valence *See* valency.

valency (valence, combining power) The valency of an element (atom, ion or group) is the number of atoms of hydrogen (or its equivalent) which one atom of that element can combine with or can displace. *Examples:* carbon has a valency of 4 in methane, CH_4; and calcium has the valency of 2 in calcium oxide, CaO, in which one calcium atom has displaced two hydrogen atoms in water, H_2O. The valency of an element is determined by the number of electrons (valency electrons) in its outermost shell. An atom which easily loses or gains one electron is said to be monovalent (univalent), and so on. Transition elements (and certain others) can have more than one valency: e.g. in copper(I) oxide, Cu_2O, the valency of copper is 1, whereas it is 2 in copper(II) oxide, CuO. The valency of an ion is equal to the magnitude of its charge: e.g. Na^+ is monovalent, Ca^{2+} is divalent, SO_4^{2-} is divalent and Al^{3+} is trivalent. There is the following relation between valency, relative atomic mass and equivalent mass:

$$\text{valency} = \frac{\text{relative atomic mass}}{\text{equivalent mass}}$$

See also Dulong and Petit's law.

valency electron An electron which is situated in the outermost shell of an atom and which takes part in the formation of a chemical bond.

vanadinite *See* vanadium.

vanadium An element with the symbol V; atomic number 23; relative atomic mass 50,94 (50.94); state, solid. Vanadium is a transition element (metal) occurring naturally in several minerals such as carnotite, $K_2(UO_2)_2(VO_4)_2 \cdot 3H_2O$, and vanadinite, $Pb_5(VO_4)_3Cl$, from which it is extracted by first converting the ore into vanadium(V) oxide, V_2O_5, and then reducing this with aluminium, Al, in the presence of steel chippings:
$$3V_2O_5 + 10Al \rightarrow 6V + 5Al_2O_3$$
The vanadium forms an alloy with the steel; since the main use of vanadium is as an alloy ingredient in steel, pure vanadium is only rarely extracted. Steel alloys containing vanadium are very hard and have great strength; they are therefore used in toolmaking, in cutting-steels and in aircraft components.

vanadium(V) oxide V_2O_5 (vanadium pentoxide) One of several oxides of vanadium. It is an orange solid and is used as a catalyst in many chemical processes such as the contact process for the manufacture of sulphuric acid. It may be prepared by heating ammonium polytrioxovanadate(V) (ammonium metavanadate), NH_4VO_3:
$$2NH_4VO_3 \rightarrow V_2O_5 + 2NH_3 + H_2O$$

vanadium pentoxide *See* vanadium(V) oxide.

van de Graaff generator An electrostatic generator which can produce very high potential differences. A typical generator used for laboratory demonstrations consists of an endless fabric or rubber belt mounted vertically between two rollers. The upper roller is contained in a hollow metal sphere supported on an insulating stand mounted on a metal base which is usually earthed. The lower roller is driven by an electric motor or by hand. Each roller faces a series of metal points which are connected to the metal sphere and the metal base respectively. The rollers are covered with different materials chosen so that when the belt touches the upper roller it acquires a negative charge by friction with the roller, while on touching the lower roller it acquires a positive charge. As one side of the belt moves upwards it carries positive charges; as it passes the upper metal points the charges are transferred to the metal sphere. As the belt moves downwards it becomes negatively charged and carries negative charges (electrons) out of the sphere. These negative charges are removed from the belt as they pass the lower metal points, and flow to earth. Thus the upward movement of the belt carries positive charges to the sphere, and the downward movement taking negative charges out of the sphere also results in a build-up of positive charges on the sphere. *See* Fig. 240.

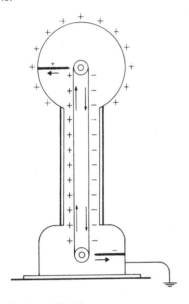

Fig. 240 Van de Graaff generator

Van der Waals forces Intermolecular or interatomic electrostatic forces which work between neutral molecules or atoms and are much weaker than the forces of a chemical bond. Between polar molecules there are marked Van der Waals forces, arising from a dipole–dipole attraction between the molecules when they are correctly orientated. However, there are also weak Van der Waals forces between non-polar molecules.

vane The flat part of a feather, consisting of barbs and barbules. *See* Fig. 67.

vaporisation The process in which a solid or a liquid changes into a vapour.

vapour A gas at a temperature below its critical temperature, which can therefore be liquefied by pressure alone.

vapour density (relative vapour density) The vapour density of a vapour (gas) is the ratio of the mass of a certain volume of the vapour to the mass of an equal volume of hydrogen (at the same temperature and pressure). The vapour density is numerically equal to half of the relative molecular mass of the vapour (gas).

vapour pressure The pressure exerted by a vapour given off by a solid or liquid. If the vapour is in equilibrium with the solid or liquid, then the vapour pressure is known as saturated vapour pressure (SVP).

variant A plant or animal which has characteristics different from those normal to the species.

variation 1. In animals and plants, a difference between individuals of the same species. It may be caused by environmental or genetic factors. **2.** *See* angle of declination.

varicella *See* chicken-pox.

variegated Irregular variation in colour of leaves, flowers or other plant organs. This may be caused by a disease or be an inherited character.

variola *See* smallpox.

vas (pl. vasa) Any small canal, vessel or duct.

vascular Pertaining to vessels or ducts for conducting fluids such as blood, water, etc.

vascular bundle A longitudinal strand of conducting tissue in plants consisting of xylem and phloem, which may be separated by cambium.

vascular cylinder *See* stele.

vascular plant A plant possessing vascular tissue.

vascular ray *See* medullary ray.

vascular system 1. The system mainly, consisting of xylem and phloem, for conducting water, mineral salts and food materials in plants. It also provides mechanical support. **2.** The system of vessels conducting fluids such as blood and lymph throughout the body.

vascular tissue Plant tissue consisting of xylem or phloem or both.

vas deferens (pl. vasa deferentia) A small duct conveying spermatozoa from the testis to the exterior. In male amniotes, it leads from the epididymis and opens into the urethra or cloaca.

vasectomy A small surgical operation in which the vasa deferentia are cut or tied, thus preventing spermatozoa passing from the testes to the seminal vesicles. Vasectomy is an effective method of contraception. It does not affect orgasm, ejaculation, virility or sexual desire. *See also* sterilisation.

vas efferens (pl. vasa efferentia) One of many straight ducts conveying spermatozoa from the seminiferous tubules to the epididymis in the testes of male amniotes.

Vaseline® A high-boiling residue obtained from the fractional distillation of petroleum. It is a colourless or pale-yellow semi-solid used in pharmacy. Vaseline is a trade name for petroleum jelly.

vasoconstriction The constriction (narrowing) of blood vessels, especially arterioles, as a result of the contraction of their muscular walls. This is mainly controlled by the sympathetic nervous system.

vasodilatation (vasodilation) The dilatation (widening) of blood vessels, especially arterioles, as a result of the relaxation of their muscular walls. This is mainly controlled by the parasympathetic nervous system.

vasomotor nerves Nerves of the autonomic nervous system which stimulate the smooth muscle in the walls of blood vessels to increase (dilate) or reduce (constrict) the diameter of the vessels.

vasopressin A polypeptide hormone secreted by the hypothalamus and released by the posterior lobe of the pituitary gland. In mammals, it stimulates the contraction of the smooth muscle in the walls of blood vessels, thus raising the blood pressure. Vasopressin is also an antidiuretic hormone (ADH), regulating the content of water in the body.

VD *See* venereal disease.

vector 1. A quantity which has both magnitude and direction, e.g. velocity, force and momentum. *Compare* scalar. **2.** A carrier of pathogens from one host to another. Most vectors are insects.

vegetable oil An oil obtained from plants, e.g. from fruits or seeds. Examples of vegetable oils are palm oil, coconut oil, maize oil, castor oil, olive oil, etc. *Compare* mineral oil.

vegetal pole (vegetative pole) **1.** The region of an egg cell containing the yolky cytoplasm. **2.** The side of a blastula where megameres (macromeres) collect. *Compare* animal pole.

vegetation Plants present in a certain area.

vegetative pole *See* vegetal pole.

vegetative propagation *See* vegetative reproduction.

vegetative reproduction (vegetative propagation) Asexual reproduction, i.e. reproduction that does not involve fusion of gametes. This is commonest in plants, but also occurs in some simple animals. In plants, vegetative reproduction may be by means of bulbs, corms, runners, tubers, stolons, etc. In animals, e.g. hydra, budding is a type of vegetative reproduction.

vein 1. A vascular bundle of a leaf. **2.** A thin-walled blood vessel which conveys blood from the body to the heart. Blood in veins is deoxygenated, apart from that carried by the pulmonary vein. In the veins, there are valves allowing the blood to flow in one direction only. The diameter of veins is greater than that of arteries. *Compare* artery. **3.** A hollow tube made of chitin which supports and strengthens an insect's wing.

velamen A water-absorbing, spongy tissue found on the outside of aerial roots in some plants such as orchids. The tissue consists of several layers of dead cells, which often have perforated and spirally thickened walls.

velocity The distance moved in a certain direction in unit time by an object or particle. Velocity is a vector quantity, not a scalar. Speed is a scalar quantity.

velocity constant *See* mass action (law of).

velocity ratio (VR; distance ratio) In a machine, the ratio of the distance moved by the effort to the distance moved by the load. *See also* efficiency.

vena cava (pl. venae cavae) In tetrapods, either of two large veins carrying deoxygenated blood from the body to the right atrium of the heart. In humans the inferior vena cava carries blood from the lower part of the body such as the legs and abdomen, and the superior vena cava carries blood from the head, neck and arms.

venation (nervation, neuration) **1.** The arrangement or distribution of veins in a leaf. There are two such arrangements: net venation (usually a characteristic of dicotyledons), in which the veins branch throughout the lamina from a main vein; and parallel venation (usually a characteristic of monocotyledons) in which all the veins are parallel on the lamina. **2.** The distribution of veins in an insect's wing.

venereal disease (VD) An infectious disease transmitted by sexual intercourse. The major venereal diseases are gonorrhoea and syphilis. Both are treated with penicillin. Since the micro-organisms causing these diseases only survive for a very short time outside the body, the chances of contracting VD by using an infected towel or lavatory seat are almost nil.

venom A poisonous fluid secreted by certain animals such as snakes and scorpions. The venom may be introduced into the victim by a bite or sting.

venous Pertaining to a vein.

ventral Pertaining to or situated on the lower surface or abdomen, e.g. ventral fin.

ventricle A body cavity, as in the brain and heart.

venule A very small vein.

Venus One of the nine planets in our Solar System, orbiting the Sun between Mercury and Earth once in 224,7 days (224.7 days). Venus is only slightly smaller than the Earth, and it is the most brilliant planet. Venus passes closer to the Earth than any other planet.

verdigris The green layer which forms on the surface of copper and copper alloys such as bronze which have been exposed to air for some time. If the air is not polluted by sulphur dioxide, SO_2, verdigris is composed of copper(II) ions, Cu^{2+}, carbonate ions, CO_3^{2-}, and hydroxyl ions, OH^-. These arise from carbon dioxide, oxygen and water vapour in the air. In air polluted with sulphur dioxide, the carbonate ions are replaced by sulphate ions, SO_4^{2-}. The term verdigris is also used for basic copper(II) ethanoate (acetate), which is used as a pigment.

vermiculite A brownish-yellow mineral which can be represented by the formula $Mg_3Al_4Si_4O_{10}(OH)_2 \cdot 4H_2O$
It belongs to the mica group. Vermiculite expands greatly when heated and in this form it is used as an insulating and packing material and as a soil-conditioner.

vernier A small, movable, graduated scale which moves over a main, fixed scale to obtain a more accurate value of measurement. Fig. 241 shows a fixed millimetre scale with a vernier that slides along it. The vernier is 9 mm long and divided into 10 divisions, i.e. the difference between a vernier division and a scale division is equal to 0,1 mm (0.1 mm) or 0,01 cm (0.01 cm). The scale is being used to measure the length of an object which is slightly more than 5,3 cm (5.3 cm) long. In order to obtain a second decimal place, the vernier is slid along the fixed scale until its zero touches the end of the object being measured. The second decimal place is the fraction of a scale division shown as *x*. On the vernier, the fourth division on it coincides with a mark on the fixed scale; by counting back to the left from this mark the distance *x* is shown to be 0,04 cm (0.04 cm). Therefore the length of the object is 5,34 cm (5.34 cm). Verniers are used in instruments such as the Fortin barometer and vernier callipers.

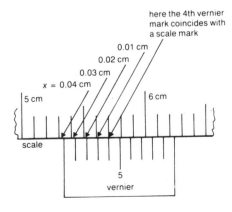

Fig. 241 Vernier scale

vernier callipers *See* callipers.
Versene® *See* edta.
vertebra (pl. vertebrae) One of the series of bones forming the vertebral column. Except for the first and second cervical vertebrae, the sacrum and coccyx, all other vertebrae bear processes for the attachment of muscles. All vertebrae bear processes for articulation with adjacent vertebrae and are separated from one another by a disc of cartilage (symphysis). *See* Fig. 242.

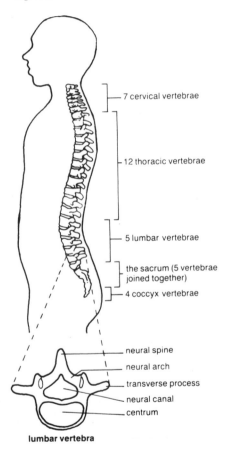

Fig. 242 Vertebral column and vertebra

vertebral column (backbone, spinal column, spine) In vertebrates, a flexible column of bones (vertebrae) running dorsally from the neck to the tail region and providing the chief support for the body. The vertebral column protects the spinal cord (nerve-cord). In humans, the vertebral column consists of 24 separate vertebrae, the sacrum and the coccyx. *See* Fig. 242.

Vertebrata (Craniata) A subphylum of Chordata comprising mammals, birds, fish, amphibians and reptiles. Characteristics include an endoskeleton of cartilage or bone, a well-developed brain protected by a bony or cartilaginous skull, and a vertebral column which replaces the notochord and encloses the spinal cord (nerve-cord).

vesicle A small bladder, sac, vacuole or hollow structure which may contain a fluid.

vessel 1. A duct, canal or tube for conducting a fluid, e.g. a blood vessel. 2. A hollow container for holding a fluid.

vestibule A cavity or channel communicating with another, as in the wall of the bony labyrinth between the middle and inner ear. Another example is the space between the labia minora, containing the opening of the urethra.

vestigial Describes a part or organ which is small and imperfectly developed, e.g. the appendix.

veterinary Pertaining to the diseases and injuries of animals.

vibration (oscillation) Any rapid to-and-fro motion of a fluid or elastic solid, e.g. a tuning-fork, pendulum or stretched string.

vibrios *See* bacterium.

video tape Magnetic tape on which recordings of television pictures and sound are made.

villus (pl. villi) 1. In vertebrates, one of numerous minute, finger-like projections on the lining of the small intestine. A villus is supplied with a network of blood capillaries and a lacteal, and it is covered by a layer of absorptive epithelium. It contains smooth muscle which moves it continually. The villi increase the surface area of the intestine, thereby increasing the rate of absorption of products of digestion. *See* Fig. 243.
2. In placental mammals, one of the projections on the chorion, increasing the surface area between the tissues of the embryo and the mother.

vinegar A liquid preservative and food-flavouring agent containing about 5% ethanoic acid (acetic acid). It is usually made by oxidation of ethanol formed by fermentation of carbohydrates.

vinyl *See* ethenyl.

vinylbenzene *See* phenylethene.

vinyl chloride *See* PVC.

viral Pertaining to virus.

virion A single mature virus.

viroid A disease-causing agent. Viroids are even smaller than any known virus and appear to consist only of free RNA. Their method of replication is not known. Viroids live symbiotically with their host, but can give rise to viruses by mutation. Viroids cause diseases

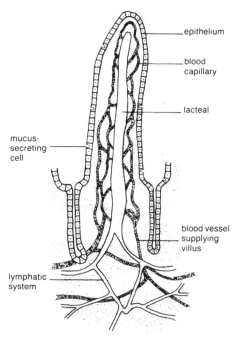

Fig. 243 Villus structure

of both plants and animals. They represent the smallest infectious agents known. Some plant diseases that previously were believed to be caused by plant viruses are now known to be caused by viroids.

virology The study of viruses.

virtual focus *See* focus.

virtual image *See* image.

virulence The ability of a micro-organism to induce disease.

virus An extremely small particle, consisting of nucleic acid around which there is a protective coat of protein. Viruses contain either RNA or DNA, but never both. *Examples:* poliomyelitis virus contains RNA, chicken-pox virus contains DNA. Viruses are much smaller than bacteria; they pass readily through a fine filter which will retain bacteria. Unlike bacteria, viruses are too small to be seen in a light microscope and are therefore studied using an electron microscope. As viruses cannot grow apart from living cells, they may be grown in tissue cultures when studied in the laboratory. Outside the living cell, viruses are inert.

Viruses are very simple. They possess no nucleus, cytoplasm or organelles, as found in animal and plant cells, and are therefore considered to be on the borderline between living and non-living material. Since viruses do not possess the capacity to construct proteins independently, they insert their nucleic acid into a functional cell for their replication. Viruses may grow in animal tissue or plant tissue, or in bacteria (bacteriophage). They are the cause of many diseases, e.g. mumps, measles, poliomyelitis, smallpox, chicken-pox, rabies, yellow fever, foot-and-mouth, influenza and the common cold.

When viruses enter the body they are attacked by antibodies, which are the body's main defence against them. Interferon is also produced in response to a viral attack. Antibiotics have little or no effect on most viruses. It is therefore the development of vaccines that controls the spread of viral diseases. There is convincing evidence that a number of specific cancers in fish and mammals are virus-induced, i.e. it is now known that viruses may cause cancer. However, it is still not known what proportion of cancers are virus-induced. See also viroid.

viscera (viscus) The internal organs in the cavities of the body, e.g. heart, lungs, liver, intestines, etc.

viscose rayon See rayon.

viscosity Symbol: η (eta). The property of fluids by which they offer resistance to flow, due to the internal friction which exists between layers of liquid or gas in motion. Viscosity, or dynamic viscosity (η), is measured in newton seconds per square metre (N s m^{-2}) or pascal seconds (Pa s). Kinematic viscosity, ν (nu), is the dynamic viscosity divided by the density of the fluid. It is measured in square metres per second (m^2 s^{-1}). The viscosity of a fluid usually decreases with an increase in temperature. η is also called the coefficient of viscosity. Compare fluidity; see also Poiseuille's formula.

viscous Having a high viscosity, e.g. treacle, glycerol and oils.

viscus See viscera.

visible spectrum The continuous spectrum of electromagnetic radiation which is visible to the human eye. See also spectrum.

Visking® tubing (dialysis tubing) A seamless, transparent, cellulose tube which is widely used in laboratories as a semi-permeable membrane.

visual purple (rhodopsin) In vertebrates, a light-sensitive protein with a prosthetic group containing a vitamin A derivative. It is found in the rods of the retina. When struck by light it changes chemically, providing energy for initiating a nerve-impulse. In darkness visual purple increases in amount, thus raising the sensitivity of the rods to faint light.

vital capacity The sum of the complemental air, the supplemental air and the tidal air.

vital force hypothesis See Wöhler synthesis.

vitamin A complex, organic substance which is essential in small amounts for metabolism. Vitamins have differing chemical structures and often act in conjunction with enzymes as catalysts. They are synthesised by plants and certain lower animals but are usually necessary in the diet of higher animals, where a deficiency may lead to disease. Vitamins are divided into two groups: those that are water-soluble (vitamins B and C) and those that are fat-soluble (vitamins A, D, E and K).

Vitamin A (axerophthol, retinol) A vitamin found in fish liver oils such as cod and halibut oil, dairy products, egg-yolk and several vegetables. A deficiency of vitamin A leads to abnormalities in the mucous membranes lining the respiratory and alimentary canals, softening or drying of the cornea and no regeneration of visual purple in the rods of the retina. See also night-blindness and carotene.

Vitamin B complex A large group of vitamins including the following: thiamine (B_1), riboflavin (B_2 or G), pantothenic acid (sometimes called B_3 or B_5), biotin (B_4 or H), pyridoxin(e) (B_6), nicotinic acid (B_7), folic acid (B_c) and cobalamine (B_{12}).

Vitamin C (ascorbic acid) A white, crystalline acid widely distributed in certain fruits and vegetables such as citrus fruits, tomatoes, cabbage and potatoes. Many species of animals are able to produce vitamin C. Humans cannot, but they can store it in the adrenal cortex. A deficiency of vitamin C may lead to scurvy.

Vitamin D A vitamin formed by the action of ultra-violet radiation upon certain sterols in the skin. It is found in fish liver oils and dairy products. A deficiency of this vitamin may lead to rickets and osteomalacia.

Vitamin E See tocopherol.

Vitamin G See vitamin B complex.

Vitamin H See vitamin B complex.

Vitamin K A vitamin formed by intestinal bacteria in humans and found in many green plants. It is concerned with blood clotting, and a deficiency (which rarely occurs) may lead to haemorrhage.

vitelline membrane A protective membrane covering the ovum after fertilisation, e.g. the membrane surrounding the yolk of an egg.

Vitreosil® Laboratory glassware made from almost pure silica, SiO_2. It has a very low coefficient of expansion and thus is able to withstand large, sudden temperature changes. See also Pyrex.

vitreous humour A transparent, jelly-like substance filling the space between the lens and the retina of the eye. See Fig. 66.

vivarium (pl. vivaria) An area prepared for keeping animals in a situation which closely resembles their natural surroundings.

viviparous 1. Describes plants which produce bulbs in place of flowers. 2. Describes seeds which germinate while still attached to the parent plant. 3. Describes animals which give birth to live, well-developed young. All placental animals are viviparous. *Compare* oviparous.

vivisection The dissection of living animals under anaesthesia for scientific purposes.

vocal cords Two folds of mucous membrane lining the larynx. Muscles alter the length and tension of the vocal cords, which vibrate when air is expelled from the lungs.

voice-box *See* larynx.

volatile Describes a substance which readily evaporates.

volcano An opening in the Earth's crust through which gas and solid or molten rock are forced at intervals. The material issuing from a volcano is believed to come from somewhere below the crust at a depth of approximately 40 km. There are thousands of volcanoes, of which only one-quarter are active. A volcano which has not erupted since ancient times is considered to be extinct and is not expected to erupt again. New volcanoes are still formed, many of which are under the sea. *See also* lava *and* magma.

volt Symbol: V. The SI unit of electric potential, potential difference or electromotive force, defined as the potential difference between two points in an electric circuit when one joule of energy is obtained as one coulomb moves between the points.

voltage Symbol: V. The potential difference or electromotive force expressed in volts.

voltage divider (potential divider) A resistor or series of resistors with a voltage applied across it and fitted with terminals so that a definite fraction of this voltage can be obtained. *See also* potentiometer.

voltaic cell (simple cell, galvanic cell) A primary cell consisting of a zinc plate (the negative electrode) and a copper plate (the positive electrode) dipping into dilute sulphuric acid. The cell produces an e.m.f. of about 1 volt; it suffers from local action and polarisation. *See* Fig. 244.

voltaic pile *See* Volta's pile.

voltameter *See* coulombmeter.

Volta's pile (voltaic pile) The first battery, constructed by Volta in 1800. It consisted of a pile of pairs of copper and zinc discs separated by layers of leather soaked in salt solution. Volta also suggested that silver could be used in place of copper and tin in place of zinc. *See* Fig. 245.

voltmeter An instrument for measuring voltage, having a high internal resistance. Two common types are the moving coil voltmeter and the moving iron voltmeter. *See also* moving coil instrument *and* moving iron instrument.

volume Symbol: V. The amount of space occupied by an object or substance.

volumetric analysis A quantitative method of analysis, including titration and measurement of gases, in which volume changes in a eudiometer are measured, or volume changes are measured before and after absorption of a gas in a suitable substance, e.g. pyrogallol (an absorber of oxygen).

volumetric flask (graduated flask) A glass flask with a long, thin neck on which there is a graduation. The flask is used to measure

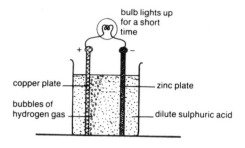

Fig. 244 Voltaic cell

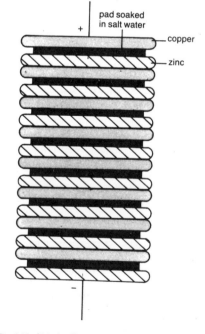

Fig. 245 Volta's pile

accurate volumes of liquid, e.g. in the preparation of standard solutions. *See* Fig. 246.

Fig. 246   Volumetric flask

voluntary action A bodily action which is under the control of the will and which occurs some time after a stimulus. Voluntary actions are controlled by the central nervous system. *Compare* involuntary action.

voluntary muscles *See* skeletal muscles.

vomit The product of vomiting.

vomiting (emesis) The forceful expulsion of the contents of the stomach and those of part of the small intestine through the mouth. Vomiting can arise in a variety of circumstances, e.g. it may follow the intake of certain drugs, be a complication of pregnancy, be caused by motion sickness or be a symptom of a disease. In vomiting, the glottis closes, the cardiac orifice dilates, the muscles of the stomach and oesophagus relax and those of the abdomen and diaphragm contract, forcing out the contents of the stomach.

vulcanisation The treatment of rubber by heating it with about 5% of sulphur at about 150°C. In this process, the rubber molecules become interlinked into a three-dimensional network with sulphur atoms between adjacent chains of rubber molecules. As a result, the rubber loses most of its plasticity and instead acquires a high degree of elasticity. This process was accidentally discovered by Charles Goodyear in 1839.

vulcanite *See* ebonite.

vulva The external female genitals, especially the opening of the vagina.

W

W Chemical symbol for tungsten.

Wacker process *See* ethanal.

wall pressure (WP) *See* turgor pressure.

Wankel engine A type of internal combustion engine constructed by F. Wankel in 1954. It consists of a triangular piston (rotor) which rotates inside an oval-shaped combustion chamber. The advantages of this type of engine are its smoothness of operation and the fact that it produces rotary motion directly.

warfarin A colourless, crystalline substance which acts as an anticoagulant and is used as a rodenticide (rat poison).

warm-blooded (homoiothermic) Pertaining to animals whose body temperature is maintained at a constant level and is independent of the surrounding temperature. Examples of warm-blooded animals are birds and mammals. *Compare* cold-blooded.

washing soda *See* sodium carbonate.

wasp *See* Hymenoptera.

Wasserman test A diagnostic test for syphilis developed in 1906 by Wasserman. The test may be carried out on blood or cerebrospinal fluid and proves positive after about six weeks of infection. The test shows whether the sample contains antibodies against the spirochaete *Treponema pallidum* which causes syphilis, and it also makes it possible to estimate the number of these antibodies. *See also* Salvarsan.

water H_2O A colourless, odourless, tasteless liquid widely distributed in nature, where it covers about 70% of the Earth's surface. Water is an oxide of hydrogen and is a covalent compound. Its three atoms do not lie in a straight line; the two hydrogen atoms form an angle of about 105° with the oxygen atom. Despite its covalent nature, water ionises to a very small extent into hydrogen ions, H^+ and hydroxyl ions, OH^-:

$$H_2O \rightleftharpoons H^+ + OH^-$$

On the pH scale, water has a pH value of 7 because it contains 10^{-7} moles of both H^+ and OH^- per litre, i.e.

$pH = -\log_{10} 10^{-7} = 7$

Water molecules are polar and are held together by hydrogen bonds. This polar nature makes water an excellent solvent, especially for ionic compounds such as many acids, bases (alkalis) and salts. The melting-point and boiling-point of water are 0°C and 100°C respectively. These temperatures are used as fixed points in the Celsius scale. *See also* maximum density of water.

water cultures Aqueous solutions containing different concentrations of various mineral salts. They are used in experiments to show the relative importance of individual elements to plant growth. *See also* hydroponics.

water cycle The circulation of water in nature. *See* Fig. 247.

water equivalent The mass of water which has the same heat capacity as a given body.

water gas A mixture of carbon monoxide, CO, and hydrogen, H_2, which can be made by reacting coke (carbon) with steam in a special furnace at a high temperature:
$$C + H_2O \rightarrow CO + H_2$$
The process is endothermic. Water gas is often prepared with producer gas and is used as a fuel in industrial heating. *See also* Fischer–Tropsch process *and* synthesis gas.

water glass A glassy, solid salt consisting of sodium or potassium metasilicate (Na_2SiO_3 or K_2SiO_3), which may be prepared by fusing anhydrous sodium carbonate, Na_2CO_3, together with sand, SiO_2:
$$Na_2CO_3 + SiO_2 \rightarrow Na_2SiO_3 + CO_2$$
Na_2SiO_3 is soluble in water, forming a colourless, viscous liquid. Water glass is used to preserve eggs, in fire-proofing of wood and textiles and in the manufacture of paper and cement.

water of constitution *See* water of crystallisation.

water of crystallisation The definite amount of water that a hydrated, crystalline compound contains. *Examples:* washing-soda, $Na_2CO_3 \cdot 10H_2O$, contains ten molecules of water of crystallisation; copper sulphate, $CuSO_4 \cdot 5H_2O$, contains five molecules of water of crystallisation. Some substances may exist with different amounts of water of crystallisation. *Example:* sodium carbonate forms $Na_2CO_3 \cdot H_2O$, $Na_2CO_3 \cdot 7H_2O$ and $Na_2CO_3 \cdot 10H_2O$. Water of crystallisation can be removed by heating. However, some hydrated compounds do not lose all the water at the same temperature. *Example:* $CuSO_4 \cdot 5H_2O$ loses four molecules of water at about 100°C; the fifth molecule is not removed until the temperature reaches 250°C. The strongly-bonded water is called water of constitution.

water potential *See* diffusion pressure deficit.

water-softening *See* hard water *and* demineralisation.

water table The level in the earth up to which the ground is saturated with water.

watt Symbol: W. The SI unit of power, defined as a rate of working of one joule per second ($J\,s^{-1}$). In an electric circuit, one watt is the product of one ampere and one volt.

wattage Symbol: W. Power measured in watts.

wattmeter An instrument for the direct measurement of electrical power in watts.

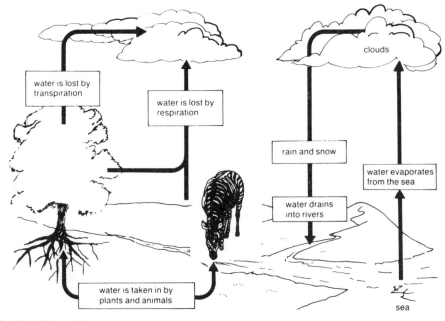

Fig. 247 Water cycle

wave A method of energy transfer through a medium or through space in which there is a periodic change involving the disturbance of particles (e.g. sound waves) or a periodic change in a physical quantity (e.g. electromagnetic waves). *See also* longitudinal wave *and* standing wave.

wave-form The graph showing the shape of a wave by plotting the values of the periodic quantity against time.

wave-front In a wave motion, the surface containing adjacent points that are in the same phase.

wavelength Symbol: λ (lambda). In a wave motion, the distance between adjacent points that are in the same phase. *See* Fig. 248; *see also* frequency.

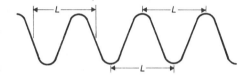

Fig. 248 Wavelength

wave mechanics *See* quantum mechanics.

wave motion The propagation of energy by the movement of waves as opposed to energy transfer by a stream of particles.

wave number Symbol: σ (sigma). The number of single waves in unit length, i.e. the reciprocal of the wavelength.

wax The general term for a group of insoluble, organic esters of fatty acids and a variety of higher monohydric alcohols. Waxes are produced in certain animals and plants, e.g. wax from the sperm whale, beeswax and carnauba wax from a Brazilian palm. Waxes are solids at room temperature. They form a waterproof layer on many stems, leaves, fruits, animal fur, etc. Waxes are used in the impregnation of paper and wood, in lacquers, polishes, paints, cosmetics, etc.

weathering The process in which rocks are broken down to smaller particles by the action of external agencies such as wind, rain, temperature changes, plants, etc. Weathering may take place when water expands on freezing in pores or cracks in rocks, when rocks expand when strongly heated by the Sun or when plant roots grow into cracks in rocks. Weathering may also take place when oxygen in the air oxidises the rock or when dissolved carbon dioxide in rainwater dissolves calcium carbonate present in rocks. *See also* erosion.

weber Symbol: Wb. The SI unit of magnetic flux, defined as the flux that, linking a circuit of one turn, produces an e.m.f. of one volt when reduced uniformly to zero in one second.

weed 1. A plant growing where it is not wanted. **2.** *See* cannabis.

weight Symbol: W. The force exerted on a mass (body) by the Earth's gravitational attraction. Weight depends on the body's distance from the Earth's surface and the local value of the acceleration of free fall (g). If the mass of a body is m kilograms, then its weight in newtons is given by
$W = mg$

weightlessness Someone who is accelerating at the same rate as his surroundings, e.g. in a box falling freely under gravity, has no sensation of weight, as there is no resultant force between him and his surroundings. These conditions are found in a spacecraft orbiting the Earth, as the centripetal force which keeps the spacecraft in orbit is identical to the Earth's gravitational force acting on an astronaut inside the spacecraft. There is no resultant force between the astronaut and the spacecraft, and so the astronaut experiences weightlessness.

welding The process in which metals (or plastics) are joined together by application of heat or pressure or both. Sometimes welding is carried out using a filler metal.

Weston cell (cadmium cell) A standard cell consisting of a mercury anode covered with a paste of mercury(I) sulphate, a cadmium amalgam cathode, and an electrolyte of a saturated solution of cadmium sulphate containing crystals of this salt. The Weston cell has a constant e.m.f. of 1.0186 V (1.0186 V) at 20°C.

wet and dry bulb hygrometer *See* hygrometer.

wet cell An electric cell in which the electrolyte is a liquid, e.g. a lead-acid cell. *Compare* dry cell.

wetting agent A substance which makes a liquid spread all over a solid material when in contact with it. This happens because the wetting agent lowers the surface tension of the liquid. *Example:* soap is a wetting agent for water.

Wheatstone bridge An apparatus for measuring electrical resistances. It consists of two parallel circuits made up of four resistances P, Q, R and S, with a battery connected across a pair of opposite junctions and a sensitive centre-zero galvanometer bridging the other pair of junctions. By adjusting the resistances, the current flowing in the galvanometer can be reduced to zero. Then the potential differences across P and R are the same as the potential differences across Q and S and
$$\frac{P}{Q} = \frac{R}{S}$$

wheel and axle

If the values of three of the resistances are known, then the value of the fourth can be calculated. *See* Fig. 249; *see also* metre bridge.

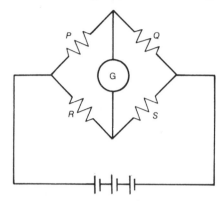

Fig. 249 Wheatstone bridge

wheel and axle A simple machine consisting of a wheel joined to an axle. The effort is applied to a rope attached to the rim of the wheel and the load is raised by a rope attached to the axle. The velocity ratio of the machine is equal to the ratio of the radius of the wheel to the radius of the axle, i.e.

$$\text{velocity ratio} = \frac{\text{load}}{\text{effort}} = \frac{R}{r}$$

See Fig. 250.

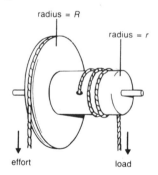

Fig. 250 Wheel and axle

white ant *See* Isoptera.
white blood cell *See* blood corpuscle.
white blood corpuscle *See* blood corpuscle.
white dwarf A very dense star with a small diameter. White dwarfs have a high surface temperature, thus appearing white, but have a low luminosity (brightness). They represent the final stage of stellar evolution; they are radiating their internal reservoirs of heat and light. In this way they slowly cool and eventually fade away as black globes. The Sun will eventually become a white dwarf.

white light Light containing all the spectrum colours and therefore appearing white, e.g. sunlight.
white matter Nervous tissue consisting of myelinated nerve-fibres with some supporting connective tissue. It is found in the central nervous system of vertebrates. White matter connects different parts of the central nervous system. *See also* grey matter.
white spirit A mixture, mainly consisting of alkanes, that boils in the range 150–200°C. It is obtained by fractional distillation of petroleum, and is used as a solvent and in the paint and varnish industries.
whooping cough (pertussis) An acute, contagious disease of childhood caused by a bacterium and affecting the bronchial tubes and the upper respiratory passages. The disease is very serious if it affects children under the age of one year.
Symptoms include a persistent cough which eventually reaches a stage in which the patient 'whoops' or 'barks', sometimes followed by vomiting. The cough may be accompanied by a light fever, or none at all. This stage usually lasts for 4–8 weeks. In its early stages the disease can be treated with tetracyclines, and a vaccine is available.
whorl A ring of leaves, flowers, etc., arising from the same level on a stem.
wild type The most common phenotype of an organism as found in nature.
Williamson synthesis A general method for the preparation of ethers, in which a metal alkoxide is brought to react with an alkyl halide. *Example:*
$CH_3—O—Na + CH_3I \rightarrow CH_3—O—CH_3 + NaI$
Here sodium methoxide, $CH_3—O—Na$, reacts with iodomethane (methyl iodide), CH_3I, to give the simple ether methoxymethane (dimethyl ether), $CH_3—O—CH_3$. Mixed ethers can also be prepared using this method.
Wilson cloud chamber *See* cloud chamber.
wilting The loss of turgidity in plant cells caused by inadequate water uptake.
wilting point *See* permanent wilting point.
Wimshurst machine An electrostatic generator consisting of two circular discs made of an insulating material such as glass or plastic and fitted with radial strips of metal foil. The discs rotate in opposite directions; the charge is produced by friction with tinfoil brushes and is collected from the discs by metal combs.
windpipe *See* trachea.
wing 1. In birds, a pentadactyl forelimb covered with feathers and modified for flying. *See* Fig. 251 (a). **2.** In bats, the wing is formed by a membrane joining the digits of the pentadactyl forelimb. **3.** In insects, a membranous outgrowth from the thorax. The wing is strengthened by veins. *See* Fig. 251 (b).

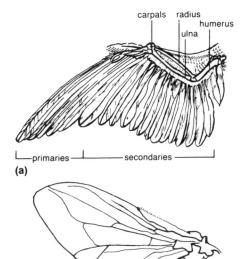

Fig. 251 (a) Bird's wing (b) Insect's wing

wisdom teeth In humans, the four rearmost teeth at each end of the upper and lower jaws. They may appear between the ages of 17 and 25 years but may also be absent from the permanent dentition. Sometimes they do not grow fully and are jammed by neighbouring teeth.

Wöhler synthesis The synthesis of urea, $CO(NH_2)_2$, from the inorganic compounds, potassium cyanate, KCNO, and ammonium sulphate, $(NH_4)_2SO_4$. When mixed, these two substances form ammonium cyanate, NH_4CNO, and potassium sulphate, K_2SO_4. By heating this mixture, urea is formed:
$2KCNO + (NH_4)_2SO_4 \rightarrow 2NH_4CNO + K_2SO_4$
and
$NH_4CNO \rightleftharpoons CO(NH_2)_2$
This synthesis was performed in 1828 by Wöhler and was the first synthesis of an organic substance from inorganic material. This made the first break with the 'vital force' hypothesis. Supporters of this hypothesis believed that organic compounds could only be formed by living organisms with the aid of a so-called 'vital force', the nature of which could not be explained. This hypothesis was widely held until Wöhler prepared urea in 1828, but was still believed by many scientists until 1845 when Kolbe prepared ethanoic acid (acetic acid), CH_3COOH, directly from its elements carbon, hydrogen and oxygen.

wolfram See tungsten.
wolframite See tungsten.
womb See uterus.
wood The hard, fibrous substance in tree stems, formed from secondary xylem.
wood alcohol See methanol.
wood naphtha See methanol.
Wood's metal An alloy consisting of bismuth (50%), lead (25%), tin (12,5%, 12.5%) and cadmium (12,5%, 12.5%). This alloy has a low melting-point, 70°C, and is used in devices for fire-protection.
wood sugar See xylose.
wool fat See lanolin.
work symbol: *w*. Work is done when a force overcomes a resistance and produces motion. Work is said to be done when a force moves its point of application. Its value is the product of the force and the distance moved in the direction of the force. The SI unit of work is the joule (J).
worker See honey-bee and Isoptera.
WP See wall pressure and turgor pressure.
wrist See carpus.
wrought iron Iron with a very small carbon content and small amounts of slag. It is very rust resistant and can easily be forged and welded. Wrought iron is used for making chains, hooks and bars.
wurtzite See zinc blende.
Wurtz reaction A reaction producing alkanes when an alkyl halide is treated with sodium in ethoxyethane, e.g.
$2CH_3I + 2Na \rightarrow CH_3-CH_3 + 2NaI$
Here iodomethane (methyl iodine), CH_3I, is converted into ethane, CH_3-CH_3. The reaction can also be carried out by the action of metallic sodium on two different alkyl halides, e.g.
$CH_3-CH_2I + CH_3I + 2Na \rightarrow$
$CH_3-CH_2-CH_3 + 2NaI$
Here sodium reacts with iodoethane (ethyl iodide), CH_3-CH_2I, and iodomethane (methyl iodide) CH_3I, to give propane, $CH_3-CH_2-CH_3$
However, the method is most useful in the synthesis of straight-chain alkanes with an even number of carbon atoms.

xanthophyll See carotene.
xanthoproteic test A test for proteins. The liquid sample is treated and warmed with concentrated nitric acid, producing a yellow precipitate if proteins are present.

X-chromosome *See* sex chromosomes.

Xe Chemical symbol for xenon.

xenon An element with the symbol Xe; atomic number 54; relative atomic mass 131,30 (131.30); state, gas. Xenon is a colourless, odourless, inert element found in the atmosphere in very small amounts. It belongs to the noble gases; it is used in certain lasers and in fluorescent tubes.

xenon hexafluoride *See* noble gas.

xenon trioxide *See* noble gas.

xerophyte A plant adapted to live in dry or drought conditions. Its adaptations may include an extensive root system, reduced leaf surface area (spines or scales), sunken or hair-covered stomata, waxy waterproofing of the leaves, fleshy stems for water storage and photosynthetic stems. Examples of xerophytes are cactus, heather and gorse. *Compare* hydrophyte *and* mesophyte.

X-ray crystallography The study of crystal structures by means of X-rays. The wavelengths of X-rays are about the same magnitude as the interatomic distances in a crystal, and the regular spacing of planes of atoms (ions or molecules) in a crystal can act as a diffraction grating for waves of X-rays. This gives information about the atomic arrangement in a crystal, and the technique may also be used to find the wavelengths of X-rays. A beam of monochromatic X-rays is passed through a crystal and the diffracted X-rays make a pattern on a photographic plate. This pattern consists of an arrangement of spots arising from diffraction from different crystal planes. By examining this pattern, the exact spacing of the atoms in a crystal can be determined (calculated). X-ray crystallography has been greatly refined. With the help of computer analysis it has been used to reveal the structure of proteins and other complex molecules.

X-ray diffraction *See* X-ray crystallography.

X-rays (Röntgen rays, Roentgen rays) Electromagnetic radiation with very short wavelengths in the range 10^{-12}–10^{-9} metres. X-rays are produced when a beam of high-energy electrons strikes a metal target (e.g. tungsten). The penetrating power of X-rays depends on the wavelength of the radiation and the nature of the material the rays are incident on. In gases through which they pass, they produce ionisation. X-rays are used in crystallography, radiography, medicine, etc.

X-ray tube Fig. 252 shows the design of an X-ray tube on which most modern tubes are based. It consists of an evacuated tube containing a heated tungsten filament cathode which emits electrons by thermionic emission. These electrons are accelerated by a high potential difference between the cathode and the copper anode and are focused on to a tungsten target which emits X-rays through a window in the tube. Only a small part of the energy carried by the electrons is converted into X-radiation. The remainder is converted into heat energy, causing the anode to become very hot. For this reason radiator fins are attached to it. Very large X-ray tubes are water-cooled.

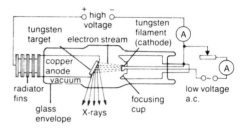

Fig. 252 X-ray tube

xylem Vascular tissue which conducts water and dissolved mineral salts from the roots to the leaves of a plant. It also gives mechnical support to a plant. There are two types of xylem: primary xylem, which develops from procambium, and secondary xylem, which is additional to primary xylem and is produced by the activity of cambium. *See* Fig. 110; *compare* phloem; *see also* heartwood *and* sapwood.

xylene *See* dimethylbenzene.

xylose $C_5H_{10}O_5$ (wood sugar) An aldopentose (monosaccharide) which is a white, crystalline, soluble solid obtained from wood or straw.

yaws A disease occurring in tropical areas of Africa, Asia and South America. It is caused by a spirochaete, *Treponema pertenue*. Symptoms include bone and joint pains, fever and skin rash. The disease is treated with penicillin.

Y-chromosome *See* sex chromosomes.

year A measure of time which may be defined in a number of ways, each of which gives slightly different values. The civil or calendar year is used for calendar purposes and is equal to 365,2425 (365.2425) mean solar days.

yeast One of a number of unicellular fungi which usually reproduce asexually by budding. However, yeasts can also reproduce sexually by conjugation. Yeasts can respire both aerobically and anaerobically. They are important substances in brewing and baking, where they produce ethanol (alcohol) and carbon dioxide by secreting certain enzymes. *See also* zymase *and Saccharomyces*.

yellow fever A viral disease of man transmitted by the mosquito *Aedes aegypti*. The first symptom is the development of fever, followed rapidly by the production of black vomit and diarrhoea. Liver damage occurs, which in Europeans results in the skin turning yellow, and kidney damage might also occur. Internal bleeding may eventually cause death. There is no specific drug for the treatment of yellow fever, but a vaccine is available.

yellow marrow *See* bone marrow.

yellow spot (macula lutea) A thickened, yellowish, oval area in the centre of the posterior part of the retina in many vertebrates. In the centre of the yellow spot is a small depression (fovea centralis) which contains only cones. This area is the area of most acute vision.

yield point The point at which a material under stress no longer obeys Hooke's law, i.e. the elastic limit is exceeded and the material is permanently deformed.

yolk The part of an egg (ovum) which provides nourishment for the developing embryo. Yolk contains proteins and granules of fat.

yolk-sac A membrane enclosing the yolk of an egg (ovum) and connected to the embryo by the yolk-stalk through which nutritive material passes into the alimentary canal of the embryo.

yolk-stalk *See* yolk-sac.

Young's double slits Two narrow parallel slits which are illuminated by monochromatic light from a single slit to demonstrate interference of light.

Z

zenith The point on the celestial sphere which is vertically above an observer on the Earth's surface.

zeolite A naturally occurring mineral or synthetic substance which has ion-exchange properties and is therefore used in the manufacture of soft water. Zeolites may consist of hydrated silicates of aluminium, sodium, potassium or calcium. They may be prepared in such a way as to have pores with uniform dimensions, thus trapping certain molecules from a mixture which is passed through them. Such zeolites are used in the selective separation of certain mixtures and they are then called molecular sieves. By heating the zeolites, the trapped molecules can be obtained. When used in ion exchange, zeolites are regenerated with concentrated solutions of sodium chloride, NaCl. *See also* Permutit *and* clathrate.

zero-point energy The vibrational energy possessed by atoms or molecules at absolute zero; it is equal to

$$\frac{hf}{2}$$

where h is the Planck constant and f the frequency of the vibration. At absolute zero, the thermal energy of a substance is zero.

zeroth law of thermodynamics A law stating that if two systems are separately in thermal equilibrium with a third then they are also in thermal equilibrium with one another. The law is called the zeroth law of thermodynamics because it is fundamental to the first, second and third laws of thermodynamics.

zinc An element with the symbol Zn; atomic number 30; relative atomic mass 65,37 (65.37); state, solid. Zinc occurs naturally in zinc blende (zinc sulphide), ZnS, and calamine (smithsonite) which is zinc carbonate, $ZnCO_3$, and from which zinc is extracted. Zinc blende usually contains galena (lead sulphide), PbS, and extracting zinc from this mixture is rather complicated. However, after roasting the ore, zinc oxide, ZnO, is obtained which is then reduced by carbon monoxide, CO, to zinc:
$$ZnO + CO \rightarrow Zn + CO_2$$
This takes place in a blast furnace in which zinc distils off from the top. Zinc is used in galvanising, in alloys such as brass, in dry cells, as a foil and as a reducing agent.

zinc blende ZnS A zinc sulphide mineral found in several varieties, e.g. sphalerite and wurtzite, which have different crystal structures. Zinc blende is an ore of zinc. Sphalerite, frequently found in association with galena (lead sulphide), PbS, is the commoner of the two minerals.

zinc carbonate $ZnCO_3$ A white, insoluble, crystalline salt which occurs naturally as calamine (smithsonite). In the laboratory, it can be prepared by adding an aqueous solution containing hydrogencarbonate ions, HCO_3^-, to an aqueous solution containing zinc ions, Zn^{2+}:
$$Zn^{2+} + 2HCO_3^- \rightarrow ZnCO_3 + CO_2 + H_2O$$

Using carbonate ions, CO_3^{2-}, instead of hydrogencarbonate ions gives a basic zinc carbonate, because CO_3^{2-} is more basic than HCO_3^-. When heated, zinc carbonate decomposes into zinc oxide, ZnO, and carbon dioxide, CO_2:
$ZnCO_3 \rightarrow ZnO + CO_2$
Zinc carbonate is used in the manufacture of zinc ointments.

zinc chloride $ZnCl_2$ A white, soluble salt which can be prepared in its anhydrous form by passing dry chlorine, Cl_2, or hydrogen chloride, HCl, over heated zinc, Zn:
$Zn + Cl_2 \rightarrow ZnCl_2$ or
$Zn + 2HCl \rightarrow ZnCl_2 + H_2$
Hydrated zinc chloride, $ZnCl_2 \cdot H_2O$, is prepared by treating the metal or the oxide, hydroxide or carbonate with dilute hydrochloric acid, HCl, e.g.
$ZnO + 2HCl \rightarrow ZnCl_2 \cdot H_2O$
Zinc chloride is used as a flux in soldering and as a preservative for timber.

zinc oxide ZnO An almost insoluble solid which is white when cold, but yellow when hot due to the loss of a small amount of oxygen on heating which is again absorbed on cooling. Zinc oxide is obtained by heating zinc in air, or by heating zinc carbonate, $ZnCO_3$, or zinc nitrate, $Zn(NO_3)_2$:
$ZnCO_3 \rightarrow ZnO + CO_2$ or
$2Zn(NO_3)_2 \rightarrow 2ZnO + 4NO_2 + O_2$
Zinc oxide is used as the white pigment zinc white, in glass-making, in the ceramic and rubber industries and in antiseptic ointments.

zinc sulphide ZnS (zinc blende) A white, insoluble salt which is phosphorescent when impure, especially if it contains small amounts of manganese. In the laboratory, it is prepared when hydrogen sulphide, H_2S, is passed through an aqueous solution containing zinc ions, Zn^{2+}:
$Zn^+ + H_2S \rightarrow ZnS + 2H^+$
A mixture of barium sulphate, $BaSO_4$, and zinc sulphide forms a white pigment called lithopone which is prepared when zinc sulphate, $ZnSO_4$, and barium sulphide, BaS, react:
$ZnSO_4 + BaS \rightarrow BaSO_4 + ZnS$

zinc white *See* zinc oxide.

zoo- Prefix meaning animal.

zoology The study of animals.

zoom lens A lens system consisting of concave and convex lenses which can be moved so that the focal length can be continually adjusted to produce changes in magnification without loss of focus. Zoom lenses are used in cine-cameras and television cameras.

zooplankton *See* plankton.

zoospore In Phycomycetes, Protozoa and some algae, an asexual, motile spore moving by means of cilia, by one or more flagella or by amoeboid movement.

zwitterion A dipolar ion such as
$H_3^+N-CH_2-COO^-$
formed from the amino acid glycine (H_2N-CH_2-COOH). Amino acids exist in aqueous solution as zwitterions, since their molecules have both an acidic terminal, the carboxyl group —COOH, and a basic terminal, the amino group $-NH_2$. Amino acids with an equal number of carboxyl and amino groups are neutral in aqueous solution. However, if acidified they become cations, and if basified, anions:
$H_3^+N-CH_2-COO^- + H^+ \rightarrow$
$\qquad\qquad H_3^+N-CH_2-COOH$
and
$H_3^+N-CH_2-COO^- + OH^- \rightarrow$
$\qquad\qquad H_2N-CH_2-COO^- + H_2O$
At a certain pH value (the isoelectric point), the degree of ionisation of the carboxyl groups and that of the amino groups are equal, i.e. the amino acid is neutral. Amino acids containing more carboxyl groups than amino groups form acidic aqueous solutions; conversely, amino acids containing more amino groups than carboxyl groups form basic aqueous solutions.

zygosis (pl. zygoses) *See* conjugation.

zygospore In some algae and fungi, a resting spore with thick walls. It may be formed by the conjugation of two similar gametes.

zygote The cell formed by fusion of two gametes.

zygotene *See* meiosis.

zymase A mixture of numerous enzymes (at least 14). Zymase was first extracted from yeast in 1903 and was then believed to be a single enzyme. Zymase catalyses the breakdown of glucose to ethanol and carbon dioxide in fermentation, requiring the presence of a number of co-enzymes and certain metal ions in order to act completely.

zymogen (proenzyme) An inert substance consisting of proteins. Zymogens are converted to active enzymes by the action of kinases. Pepsinogen, trypsinogen and prothrombin are examples of zymogens.

Appendices

Solubility products	325	The commonest elements in the Earth's crust	328
Density of water	326	Terms used in the classification of animals and plants	328
Molecular boiling-point elevations for some solvents	326	The Periodic Table of the elements	329
Molecular freezing-point depressions for some solvents	326	Plant groups	330
Values of some common standard electrode potentials at 25°C	326	Animal groups	330
International system of units	327	Melting-points and boiling-points of some elements	331
Other units	327	Some common acid-base indicators	332
The geological periods and time scale	328	Hardness of various materials on Mohs' scale	332
Composition of the Earth's atmosphere	328		
The Greek alphabet	328		

Solubility products

Substance	Solubility product at temperature noted
aluminium hydroxide, $Al(OH)_3$	$3{,}7 \times 10^{-15}$ at 25°C
barium carbonate, $BaCO_3$	$8{,}1 \times 10^{-9}$ at 25°C
barium sulphate, $BaSO_4$	$1{,}08 \times 10^{-10}$ at 25°C
calcium carbonate, $CaCO_3$	$0{,}87 \times 10^{-8}$ at 25°C
calcium ethanedioate, $CaC_2O_4 \cdot H_2O$	$2{,}57 \times 10^{-9}$ at 25°C
calcium fluoride, CaF_2	$3{,}95 \times 10^{-11}$ at 26°C
copper(I) sulphide, Cu_2S	2×10^{-47} at 17°C
copper(II) sulphide, CuS	$8{,}5 \times 10^{-45}$ at 18°C
iron(II) hydroxide, $Fe(OH)_2$	$1{,}64 \times 10^{-14}$ at 18°C
iron(III) hydroxide, $Fe(OH)_3$	$1{,}1 \times 10^{-36}$ at 18°C
iron(II) sulphide, FeS	$3{,}7 \times 10^{-19}$ at 18°C
lead(II) carbonate, $PbCO_3$	$3{,}3 \times 10^{-14}$ at 18°C
lead(II) sulphate, $PbSO_4$	$1{,}06 \times 10^{-8}$ at 18°C
lead(II) sulphide, PbS	$3{,}4 \times 10^{-28}$ at 18°C
magnesium ammonium phosphate, $MgNH_4PO_4$	$2{,}5 \times 10^{-13}$ at 25°C
magnesium carbonate, $MgCO_3$	$2{,}6 \times 10^{-5}$ at 12°C
silver bromide, $AgBr$	$7{,}7 \times 10^{-13}$ at 25°C
silver chloride, $AgCl$	$1{,}56 \times 10^{-10}$ at 25°C
silver iodide, AgI	$1{,}5 \times 10^{-16}$ at 25°C
silver sulphide, Ag_2S	$1{,}6 \times 10^{-49}$ at 18°C
zinc hydroxide, $Zn(OH)_2$	$1{,}8 \times 10^{-14}$ at 19°C
zinc sulphide, ZnS	$1{,}2 \times 10^{-23}$ at 18°C

Density of water

Pure air-free water under 1 atm. Unit 1 g cm^{-3}

Temp./°C	0	2	4	6	8
0	0,999 84	0,999 94	0,999 97	0,999 94	0,999 85
10	0,999 70	0,999 50	0,999 24	0,998 94	0,998 59
20	0,998 20	0,997 77	0,997 30	0,996 78	0,996 23
30	0,995 65	0,995 03	0,994 37	0,993 69	0,992 97
40	0,992 22	0,991 44	0,990 63	0,989 79	0,988 93
50	0,988 04	0,987 12	0,986 18	0,985 21	0,984 22
60	0,983 20	0,982 16	0,981 09	0,980 01	0,978 90
70	0,977 77	0,976 61	0,975 44	0,974 24	0,973 03
80	0,971 79	0,970 53	0,969 26	0,967 96	0,966 65
90	0,965 31	0,963 96	0,962 59	0,961 20	0,959 79
100	0,958 36				

Molecular boiling-point elevations for some solvents

Solvent	Elevation, °C/mole solute/kg solvent
benzene, C_6H_6	2,53
carbon disulphide, CS_2	2,34
tetrachloromethane, CCl_4	5,03
ethanol, C_2H_5OH	1,22
methanol, CH_3OH	0,83
nitrobenzene, $C_6H_5NO_2$	5,24
water, H_2O	0,51

Molecular freezing-point depressions for some solvents

Solvent	Depression, °C/mole solute/kg solvent
benzene, C_6H_6	4,9
phenylbenzene, $(C_6H_5)_2$	8,0
1,2-dibromoethane, Br—CH_2—CH_2—Br	11,8
naphthalene, $C_{10}H_8$	6,8
water, H_2O	1,86

Values of some common standard electrode potentials at 25°C

Equation	Volts	Equation	Volts
$Li^+ + 1e \rightleftharpoons Li$	−3,04	$Fe^{2+} + 2e \rightleftharpoons Fe$	−0,44
$K^+ + 1e \rightleftharpoons K$	−2,92	$Co^{2+} + 2e \rightleftharpoons Co$	−0,28
$Ba^{2+} + 2e \rightleftharpoons Ba$	−2,90	$Ni^{2+} + 2e \rightleftharpoons Ni$	−0,25
$Ca^{2+} + 2e \rightleftharpoons Ca$	−2,76	$Sn^{2+} + 2e \rightleftharpoons Sn$	−0,14
$Na^+ + 1e \rightleftharpoons Na$	−2,71	$Pb^{2+} + 2e \rightleftharpoons Pb$	−0,13
$Mg^{2+} + 2e \rightleftharpoons Mg$	−2,37	$H^+ + 1e \rightleftharpoons \tfrac{1}{2}H_2$	0,00
$Al^{3+} + 3e \rightleftharpoons Al$	−1,70	$Cu^{2+} + 2e \rightleftharpoons Cu$	+0,34
$Zn^{2+} + 2e \rightleftharpoons Zn$	−0,76	$Ag^+ + 1e \rightleftharpoons Ag$	+0,80

Note that the more electropositive metals have the higher negative standard electrode potentials.

International system of units (SI)

Physical quantity	Unit	Symbol	Equivalent form
Length (l, x)	metre	m	base units
Mass (M, m)	kilogram	kg	
Time (t)	second	s	
Electric current (I)	ampere	A	
Thermodynamic temperature (T)	kelvin	K	
Luminous intensity (I)	candela	cd	
Amount of substance (n, v)	mole	mol	
Frequency, cycles per sec. (f, v)	hertz	Hz	s^{-1}
Force (F)	newton	N	$kg\,m\,s^{-2}$
Pressure (p, P)	pascal	Pa	$N\,m^{-2}$
Work, energy (W, E, U), quantity of heat (Q)	joule	J	$N\,m$
Power (P)	watt	W	$J\,s^{-1}$
Electric charge (Q)	coulomb	C	$A\,s$
Electric potential, p.d., e.m.f. (φ, V)	volt	V	$J\,C^{-1}, W\,A^{-1}$
Capacitance (C)	farad	F	$C\,V^{-1}$
Electric resistance (R)	ohm	Ω	$V\,A^{-1}$
Electric conductance (G)	siemens	S	$Ω^{-1}$
Magnetic flux (Φ)	weber	Wb	$V\,s$
Magnetic flux density (B)	tesla	T	$Wb\,m^{-2}$
Inductance (L)	henry	H	$Wb\,A^{-1}$

Multiples and submultiples: Prefixes are used to indicate multiples (and submultiples) of units (e.g. microfarad, µF). The following prefixes are in general use:

Factor	10^{12}	10^{9}	10^{6}	10^{3}	10^{-3}	10^{-6}	10^{-9}	10^{-12}
Prefix	tera	giga	mega	kilo	milli	micro	nano	pico
Symbol	T	G	M	k	m	µ	n	p

Other prefixes which are used less widely are hecto (h, 10^2), deca (da, 10^1), deci (d, 10^{-1}), centi (c, 10^{-2}), femto (f, 10^{-15}) and atto (a, 10^{-18}).

Compound units can be formed from multiples of named units (e.g. $km\,s^{-2}$), but prefixes are not used to denote multiples of compound units.

Other units

Unit	SI equivalent	Unit	SI equivalent
Length		*Mass and density*	
ångström (Å)	0,1 nm	gram (g)	10^{-3} kg
inch (in)	25,4 mm	tonne (t)	10^{3} kg
foot (ft)	0,3048 m	ounce (oz)	28,35 g
yard	0,9144 m	pound (lb)	0,4536 kg
mile	1,6093 km	ton	1,016 t
nautical mile (Int.)	1,852 km	lb ft^{-3}	16,02 kg m^{-3}
Area		*Force*	
hectare (ha)	10^4 m^2	poundal (pdl)	0,1383 N
square foot	$9,290 \times 10^{-2}$ m^2	lbf	4,448 N
acre	0,4047 ha	tonf	9,964 kN
Volume		*Energy and power*	
litre (l)	1 dm^3 or 10^{-3} m^3	ft pdl	0,04214 J
millilitre (ml)	1 cm^3 or 10^{-6} m^3	ft lbf	1,356 J
cubic foot	$2,832 \times 10^{-2}$ m^3	calorie 15 °C (cal$_{15}$)	4,1855 J
pint	0,568 l	calorie I.T. (cal$_{IT}$)	4,1868 J
gallon (UK)	4,546 l	Btu	1,055 kJ
gallon (US)	3,785 l	horsepower (hp)	745,7 W
Speed		*Pressure and stress*	
km/hour	0,2778 m s^{-1}	bar	10^5 Pa
mile/hour (m.p.h.)	0,4470 m s^{-1}	atmosphere (atm)	101,33 kPa
foot/second	0,3048 m s^{-1}	torr (mmHg)	133,3 Pa
knot (UK)	0,5148 m s^{-1}	lbf in^{-2} (p.s.i.)	6,895 kPa
rev/min (r.p.m.)	0,1047 rad s^{-1}	tonf ft^{-2}	107,3 kPa

The geological periods and time scale

Years ago (millions)	Era	Period
0		Present time
2		Quaternary
7		Pliocene
26		Miocene
38		Oligocene
54		Eocene
64		Palaeocene
136		Cretaceous
193		Jurassic
225		Triassic
280		Permian
345		Carboniferous
395		Devonian
430		Silurian
500		Ordovician
570		Cambrian
		Precambrian

The Greek alphabet

Letters		Name
A	α	alpha
B	β	beta
Γ	γ	gamma
Δ	δ	delta
E	ε	epsilon
Z	ζ	zeta
H	η	eta
Θ	θ	theta
I	ι	iota
K	κ	kappa
Λ	λ	lambda
M	μ	mu
N	ν	nu
Ξ	ξ	xi
O	o	omicron
Π	π	pi
P	ρ	rho
Σ	σ	sigma
T	τ	tau
Y	υ	upsilon
Φ	ϕ	phi
X	χ	chi
Ψ	ψ	psi
Ω	ω	omega

Composition of the Earth's atmosphere

Substance	Parts in 10^6 of dry air by volume
N_2	780 900
O_2	209 500
Ar	9400
CO_2	300
Ne	18
He	5,2
CH_4	1,5
Kr	1,14
N_2O	0,5
H_2	0,5
O_3	0,4
Xe	0,086
Ra	6×10^{-14}

The commonest elements in the Earth's crust

Element	Weight %
O	46,60
Si	27,72
Al	8,13
Fe	5,00
Mg	2,09
Ca	3,63
Na	2,83
K	2,59

Terms used in the classification of animals and plants

Kingdom
Division or phylum
Class
Order
Family
Genus
Species

The Periodic Table of the Elements

Groups	IA	IIA	IIIB	IVB	VB	VIB	VIIB		VIII		IB	IIB	IIIA	IVA	VA	VIA	VIIA	VIIIA
Period 1	1.01 H 1																	4.00 He 2
Period 2	6.94 Li 3	9.01 Be 4											10.81 B 5	12.01 C 6	14.01 N 7	16.00 O 8	19.00 F 9	20.18 Ne 10
Period 3	22.99 Na 11	24.31 Mg 12											26.98 Al 13	28.09 Si 14	30.97 P 15	32.06 S 16	35.45 Cl 17	39.95 Ar 18
Period 4	39.10 K 19	40.08 Ca 20	44.96 Sc 21	47.90 Ti 22	50.94 V 23	52.00 Cr 24	54.94 Mn 25	55.85 Fe 26	58.93 Co 27	58.71 Ni 28	63.55 Cu 29	65.37 Zn 30	69.72 Ga 31	72.59 Ge 32	74.92 As 33	78.96 Se 34	79.91 Br 35	83.80 Kr 36
Period 5	85.47 Rb 37	87.62 Sr 38	88.91 Y 39	91.22 Zr 40	92.91 Nb 41	95.94 Mo 42	97 Tc 43	101.07 Ru 44	102.91 Rh 45	106.4 Pd 46	107.87 Ag 47	112.40 Cd 48	114.82 In 49	118.69 Sn 50	121.75 Sb 51	127.6 Te 52	126.90 I 53	131.30 Xe 54
Period 6	132.91 Cs 55	137.34 Ba 56	138.91 La 57	178.49 Hf 72	180.95 Ta 73	183.85 W 74	186.2 Re 75	190.2 Os 76	192.2 Ir 77	195.09 Pt 78	196.97 Au 79	200.59 Hg 80	204.37 Tl 81	207.2 Pb 82	208.98 Bi 83	209 Po 84	210 At 85	222 Rn 86
Period 7	223 Fr 87	226 Ra 88	227 Ac 89	104	105													

— Transition elements —

Lanthanoids:
| 140.12 Ce 58 | 140.91 Pr 59 | 144.24 Nd 60 | 145 Pm 61 | 150.4 Sm 62 | 151.96 Eu 63 | 157.2 Gd 64 | 158.92 Tb 65 | 162.50 Dy 66 | 164.93 Ho 67 | 167.26 Er 68 | 168.94 Tm 69 | 173.04 Yb 70 | 174.97 Lu 71 |

Actinoids:
| 232.04 Th 90 | 231 Pa 91 | 238.03 U 92 | 237.05 Np 93 | 244 Pu 94 | 243 Am 95 | 247 Cm 96 | 247 Bk 97 | 251 Cf 98 | 254 Es 99 | 257 Fm 100 | 257 Md 101 | 259 No 102 | 260 Lr 103 |

From atomic number 104 the elements are called trans-actinoids

relative atomic mass → 1.01 H
atomic number → 1

Plant groups   Animal groups

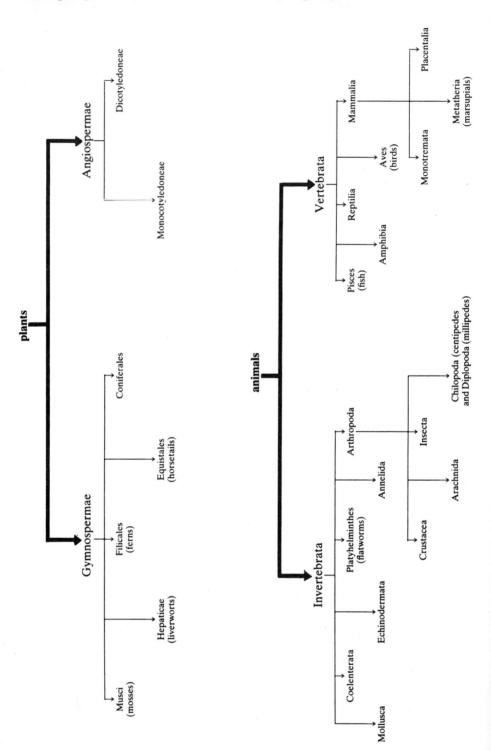

Melting-points and boiling-points of some elements

Element	Melting-point/°C	Boiling-point/°C
aluminium	660,37	2467
argon	−189,2	−185,7
barium	725	1640
bismuth	271,3	1560 ± 5
boron	2300	2550 (sub.)
bromine	−7,2	58,78
cadmium	320,9	765
calcium	839 ± 2	1484
carbon	3550	4827
cerium	799 ± 3	3426
chlorine	−100,98	−34,6
chromium	1857 ± 20	2672
cobalt	1495	2870
copper	1083,4 ± 0,2	2567
fluorine	−219,62	−188,14
gallium	29,78	2403
germanium	937,4	2830
gold	1064,43	2807
helium	−272,2 (26 atm.)	−268,934
hydrogen	−259,14	−252,87
iodine	113,5	184,35
iridium	2410	4130
iron	1535	2750
krypton	−156,6	−152,30 ± 0,10
lead	327,502	1740
lithium	180,54	1347
magnesium	648,8 ± 0,5	1090
manganese	1244 ± 3	1962
mercury	−38,87	356,58
neon	−248,67	−246,048
nickel	1453	2732
nitrogen	−209,86	−195,8
oxygen	−218,4	−182,962
phosphorus (white)	44,1	280
platinum	1772	3827 ± 100
plutonium	641	3232
potassium	63,65	774
radium	700	1140
radon	−71	−61,8
rubidium	38,89	688
scandium	1541	2831
selenium (grey)	217	684,9 ± 1,0
silicon	1410	2355
silver	961,93	2212
sodium	97,81 ± 0,03	882,9
strontium	769	1384
sulphur (rhombic)	112,8	444,674
sulphur (monoclinic)	119,0	444,674
tin	231,9681	2270
titanium	1660 ± 10	3287
tungsten	3410 ± 20	5660
uranium	1132,3 ± 0,8	3818
vanadium	1890 ± 10	3380
xenon	−111,9	−107,1 ± 3
zinc	419,58	907

Some common acid-base indicators

Name	Approximate pH range	Colour change	Preparation
Methyl violet	0,0–1,6	yellow–blue	0,01–0,05% in water
Crystal violet	0,0–1,8	yellow–blue	0,02% in water
Methyl orange	3,2–4,4	red–yellow	0,01% in water
Methyl red	4,8–6,0	red–yellow	0,02 g in 60 ml ethanol + 40 ml water
Alizarin	5,6–7,2	yellow–red	0,1% in methanol
Neutral red	6,8–8,0	red–amber	0,01 g in 50 ml ethanol + 50 ml water
Phenolphthalein	8,2–10,0	colourless–pink	0,05 g in 50 ml ethanol + 50 ml water
Thymolphthalein	9,4–10,6	colourless–blue	0,04 g in 50 ml ethanol + 50 ml water

Hardness of various materials on Mohs' scale

Substance	Formula	Mohs value
talc	$3MgO \cdot 4SiO_2 \cdot H_2O$	1
gypsum	$CaSO_4 \cdot 2H_2O$	2
cadmium	Cd	..
silver	Ag	..
zinc	Zn	..
calcite	$CaCO_3$	3
fluorite	CaF_2	4
copper	Cu	..
magnesia	MgO	..
apatite	$CaF_2 \cdot 3Ca_3(PO_4)_2$	5
nickel	Ni	..
glass (soda lime)	—	..
feldspar (orthoclase)	$K_2O \cdot Al_2O_3 \cdot 6SiO_2$	6
quartz	SiO_2	7
chromium	Cr	..
topaz	$(AlF)_2SiO_4(OH)_2$	8
tungsten carbide alloy	WC, Co	..
zirconium boride	ZrB_2	..
titanium nitride	TiN	9
tungsten carbide	WC	..
zirconium carbide	ZrC	..
alumina	Al_2O_3	..
beryllium carbide	Be_2C	..
titanium carbide	TiC	..
silicon carbide	SiC	..
aluminum boride	AlB	..
boron carbide	B_4C	..
diamond	C	10